Automorphic Forms and Related Topics

CONTEMPORARY MATHEMATICS

732

Automorphic Forms and Related Topics

ilding Bridges: 3rd EU/US Summer School and Workshop
Automorphic Forms and Related Topics
July 11–22, 2016
iversity of Sarajevo, Sarajevo, Bosnia and Herzegovina

Samuele Anni
Jay Jorgenson
Lejla Smajlović
Lynne Walling
Editors

AMS AMERICAN MATHEMATICAL SOCIETY

2010 *Mathematics Subject Classification.* Primary 11Fxx, 11Mxx, 11Gxx, 14Gxx, 14Kxx, 22E45, 22E50, 22E55, 22E57, 32Nxx.

Library of Congress Cataloging-in-Publication Data

Names: EU-US Summer School and Workshop on Automorphic Forms and Related Topics (3rd : 2016 : Sarajevo, Bosnia and Herzegovina) | Anni, Samuele, 1985- editor. | Jorgenson, Jay, editor. | Smajlovic, Lejla, 1974- | Walling, Lynne, 1958- editor.

Title: Automorphic forms and related topics : building bridges : 3rd EU-US Summer School and Workshop on Automorphic Forms and Related Topics, July 11-22, 2016, University of Sarajevo, Sarajevo, Bosnia and Herzegovina / Samuele Anni, Jay Jorgenson, Lejla Smajlovic, Lynne Walling, editors.

Description: Providence, Rhode Island : American Mathematical Society, [2019] | Series: Contemporary mathematics ; volume 732 | Includes bibliographical references.

Identifiers: LCCN 2019005230 | ISBN 9781470435257 (alk. paper)

Subjects: LCSH: Automorphic forms–Congresses. | Automorphic functions–Congresses. | AMS: Number theory – Discontinuous groups and automorphic forms – Discontinuous groups and automorphic forms. msc | Number theory – Zeta and L-functions: analytic theory – Zeta and L-functions: analytic theory. msc | Number theory – Arithmetic algebraic geometry (Diophantine geometry) – Arithmetic algebraic geometry (Diophantine geometry). msc | Algebraic geometry – Arithmetic problems. Diophantine geometry – Arithmetic problems. Diophantine geometry. msc | Algebraic geometry – Abelian varieties and schemes – Abelian varieties and schemes. msc | Topological groups, Lie groups – Lie groups – Representations of Lie and linear algebraic groups over real fields: analytic methods. msc | Topological groups, Lie groups – Lie groups – Representations of Lie and linear algebraic groups over local fields. msc | Topological groups, Lie groups – Lie groups – Representations of Lie and linear algebraic groups over global fields and adèle rings. msc | Topological groups, Lie groups – Lie groups – Geometric Langlands program: representation-theoretic aspects. msc | Several complex variables and analytic spaces – Automorphic functions – Automorphic functions. msc

Classification: LCC QA353.A9 E945 2016 | DDC 512.7–dc23

LC record available at https://lccn.loc.gov/2019005230

Contemporary Mathematics ISSN: 0271-4132 (print); ISSN: 1098-3627 (online)

DOI: https://doi.org/10.1090/conm/732

∞ The paper used in this book is acid-free and falls within the guidelines
established to ensure permanence and durability.
Visit the AMS home page at `https://www.ams.org/`

10 9 8 7 6 5 4 3 2 1 24 23 22 21 20 19

Contents

Preface

This volume of papers grew from the meeting entitled *Building Bridges: 3rd EU/US Summer School + Workshop on Automorphic Forms and Related Topics,* which was held from 11 July 2016 to 22 July 2016 at the University of Sarajevo in Bosnia and Herzegovina. For brevity, we refer to the conference as *BB3*. As the title indicates, *BB3* was one in a series of meetings which are organized biennially: *BB1* was held in Aachen, Germany in 2012, *BB2* was held in Bristol, England in 2014 and more recently, *BB4* was held in Budapest, Hungary in 2018. Tentative plans are in place to hold *BB5* in Sarajevo in 2020.

Each of the *Building Bridges* events has the same organizational theme. The first week is a summer school consisting of three different two day topics which are given at the advanced graduate student or early postdoctoral researcher level. The goal is to provide young mathematicians a means by which they can "build a mathematical bridge" from their current topic of study to other areas of research. The second week is a somewhat standard, while strikingly non-standard, workshop with numerous lectures of various length.

Beyond the mathematical content of each *BB* event, the organizers purposefully devote considerable attention and energy to create an environment of diversity and welcomeness for all participants. There are no invited lectures for the workshop, and everyone who wishes to speak is given an opportunity. Each talk is for at most 40 minutes, and for several days there are sessions of "speed talks", which are strictly timed to go for no more than two minutes. In fact, a panel of judges is formed to enforce the "rules" of speed talks and award prizes for certain dynamically created categories such as creativity, timing, and entertainment. The speed talks are a lot of fun.

Additionally, *BB3* scheduled a panel discussion on diversity which included Dr. Adriana Espinosa from The Department of Psychology at The City College of New York. The organizers wanted the topic of diversity, in its broadest sense, to have a prominent role in the conversation of the workshop. We were very happy to read that many of the (anonymous) reviews which were submitted after the workshop to praise the diversity panel. We have invited Dr. Espinosa to include in this volume an article which highlights the discussion from the diversity panel, and to provide additional ideas for future consideration.

The concept of *Building Bridges* begins with Lynne Walling, who remains the principal organizer no matter where *BB* is held. Lynne is the source of energy and inspiration for the *Building Bridges* events. The other editors of this volume (S.A., J.J., and L.S.) are following Lynne's lead, and are extremely grateful to be included in making her ideas and vision come to life.

We wish to thank the following financial sponsors of the *BB3*: European Mathematical Society, The Heilbronn Institute for Mathematical Research, The School of Economics and Business at the University of Sarajevo, the Number Theory Foundation, The National Science Foundation, Foundation Compositio Mathematica, and the American Institute of Mathematics. Their generosity is what made *BB3* happen.

As stated, the workshop solicited anonymous comments from all participants. We did so in order to learn ways in which the next *Building Bridges* could be improved. The broad consensus is that Sarajevo is wonderful. With that said, we wish to express our thanks to the School of Economics and Business at the University of Sarajevo for hosting *BB3*. For many of the participants, this event was their first trip to Sarajevo, and everyone is looking forward to their next visit.

This volume commemorates a great meeting held at a wonderful place, and each of us are happy to have been part of it.

Samuele Anni
Jay Jorgenson
Lejla Smajlović
Lynne Walling
November 2018

Contemporary Mathematics
Volume **732**, 2019
https://doi.org/10.1090/conm/732/14812

A note on the minimal level of realization for a mod ℓ eigenvalue system

Samuele Anni

Abstract. In this article we give a criterion to decide when the mod ℓ eigenvalue system attached to a mod ℓ Katz cuspform arises from lower level or weight. Namely, we prove the following. The eigenvalue system associated to a ring homomorphism $f : \mathbb{T} \to \overline{\mathbb{F}}_\ell$ from the Hecke algebra of level $\Gamma_1(n)$ and weight k to $\overline{\mathbb{F}}_\ell$, where ℓ is a prime not dividing n and $1 \leq k \leq \ell + 1$, arises from lower level or weight if there exists a prime r dividing $n\ell$ such that

$$\dim_{\overline{\mathbb{F}}_\ell} \bigcap_{p \neq r} \ker\left(T_p - f(T_p), S(n,k)_{\overline{\mathbb{F}}_\ell}\right) > 1,$$

where T_p denotes the p-th Hecke operator and $S(n,k)_{\overline{\mathbb{F}}_\ell}$ is the space of mod ℓ Katz cuspforms of level $\Gamma_1(n)$ and weight k.

1. Introduction and preliminaries

In this article we analyse degeneracy maps between modular curves in positive characteristic in order to deduce statements about residual modular Galois representations and their level of realization. In [**Edi06**], the study of the Hasse invariant and the Frobenius morphism led to similar results on the minimal weight at which the mod ℓ eigenvalue system attached to a Katz cuspform arises.

This article is based on the theory of Katz modular forms ([**Kat73**], [**Kat77**], [**Edi92**]) and modular curves ([**KM85**], [**Gro90**]). Analogous results in characteristic zero of the some presented in this article are well-known ([**DS05**, Section 5.7]). Anyway, in positive characteristic they can be obtained through the study of modular curves over finite fields and a geometric interpretation of degeneracy maps. The statements that we will prove do not rely on lifting to characteristic zero, and so, in particular, they hold also in weight 1.

For any pair of positive integers n and k, the level and the weight, we denote by $M(n,k)_\mathbb{C}$ the complex vector space of modular forms of weight k on $\Gamma_1(n)$ and by $S(n,k)_\mathbb{C}$ the subspace of cuspforms. Let $\mathbb{T}$ be the associated Hecke algebra, i.e. the $\mathbb{Z}$-subalgebra of $\mathrm{End}_\mathbb{C}(S(n,k)_\mathbb{C})$ generated by the Hecke operators T_p for every

---

2010 *Mathematics Subject Classification.* Primary 11F03, 11F80; Secondary 14J15, 14H25.
Key words and phrases. Katz modular forms, Galois representation, modular curves.
The author was supported by the EPSRC Programme Grant 'LMF: L-Functions and Modular Forms' EP/K034383/1 during his position at the University of Warwick, where most of this research was carried.

prime p and the diamond operators $\langle d\rangle$ for every $d \in (\mathbb{Z}/n\mathbb{Z})^*$. The Hecke algebra $\mathbb{T}$ is finitely generated as a $\mathbb{Z}$-module ([**EC11**, Theorem 2.5.11], [**DS05**, p.234]).

Let us fix an algebraic closure $\overline{\mathbb{Q}}$ of $\mathbb{Q}$ and denote by $G_{\mathbb{Q}}$ the absolute Galois group $\mathrm{Gal}(\overline{\mathbb{Q}}/\mathbb{Q})$. The following result, due to Shimura and Deligne ([**DS74**, Théorème 6.7], [**EC11**, Theorem 2.5.2]), associates a continuous semisimple Galois representation to any ring homomorphism from the Hecke algebra $\mathbb{T}$ to a finite field of characteristic not dividing the level.

THEOREM 1.1 (Deligne, Shimura). *Let n and k be positive integers. Let $\mathbb{F}$ be a finite field of characteristic ℓ and $f : \mathbb{T} \to \mathbb{F}$ a morphism of rings. Then there is a continuous semi-simple representation $\rho_f : G_{\mathbb{Q}} \to \mathrm{GL}_2(\mathbb{F})$, unramified outside $n\ell$, such that for all primes p not dividing $n\ell$ we have:*

$$\mathrm{Trace}(\rho_f(\mathrm{Frob}_p)) = f(T_p) \text{ and } \det(\rho_f(\mathrm{Frob}_p)) = f(\langle p\rangle)p^{k-1} \text{ in } \mathbb{F}.$$

Such a ρ_f is unique up to isomorphism.

A ring morphism as in Theorem 1.1 corresponds to an eigenform with coefficients in $\mathbb{F}$, for more details see [**EC11**, p.58]. We will denote by $\mathrm{N}(\rho_f)$ the Artin conductor of ρ_f away from ℓ and by $k(\rho_f)$ the weight of ρ_f, as in [**Edi92**].

The space $S(n,k)_{\mathbb{C}}$ is decomposed by characters, let $S(n,k,\tilde{\epsilon})_{\mathbb{C}}$ denote the space of cuspforms with character $\tilde{\epsilon}$. We associate a Hecke algebra to each $S(n,k,\tilde{\epsilon})_{\mathbb{C}}$, as before. This Hecke algebra, that we will denote as $\mathbb{T}_{\tilde{\epsilon}}$, is the $\mathcal{O}_{\tilde{\epsilon}}$-subalgebra in the complex endomorphism ring of $S(n,k,\tilde{\epsilon})_{\mathbb{C}}$ generated by the Hecke operators T_n for $n \geq 1$, where $\mathcal{O}_{\tilde{\epsilon}}$ is ring of integers of the number field containing the image of $\tilde{\epsilon}$. Note that the algebra $\mathbb{T}_{\tilde{\epsilon}}$ is free of finite rank as a $\mathbb{Z}$-module. From Theorem 1.1 it follows that:

COROLLARY 1.2. *Let n and k be positive integers, let $\tilde{\epsilon} : (\mathbb{Z}/n\mathbb{Z})^* \to \mathbb{C}^*$ be a character and let $f : \mathbb{T}_{\tilde{\epsilon}} \to \overline{\mathbb{F}}_\ell$ be a morphism of rings. Then there is a continuous semi-simple representation $\rho_f : G_{\mathbb{Q}} \to \mathrm{GL}_2(\overline{\mathbb{F}}_\ell)$, unramified outside $n\ell$, such that for all primes p not dividing $n\ell$ we have:*

$$\mathrm{Trace}(\rho_f(\mathrm{Frob}_p)) = f(T_p) \text{ and } \det(\rho_f(\mathrm{Frob}_p)) = f(\langle p\rangle)p^{k-1} \text{ in } \overline{\mathbb{F}}_\ell.$$

Such a ρ_f is unique up to isomorphism.

A ring morphism as in Corollary 1.2 corresponds to a mod ℓ eigenform with character $\epsilon : (\mathbb{Z}/n\mathbb{Z})^* \to \overline{\mathbb{F}}_\ell$ given by $a \mapsto \epsilon(a) := f(\langle a\rangle)$ for $a \in (\mathbb{Z}/n\mathbb{Z})^*$. Equivalently, $\det(\rho_f) = \epsilon \cdot \chi_\ell^{k-1}$ where χ_ℓ is the cyclotomic character mod ℓ. We will abuse the notation and denote by f the mod ℓ cuspidal eigenform corresponding to the ring homomorphism f as in the above corollary.

Let us recall that the Hecke algebras $\mathbb{T}$ and $\mathbb{T}_{\tilde{\epsilon}}$ for level n, weight $k \geq 2$ and character $\tilde{\epsilon} : (\mathbb{Z}/n\mathbb{Z})^* \to \mathcal{O}_{\tilde{\epsilon}}$, can be computed in polynomial time in n and k, using deterministic algorithms based on modular symbols, see [**Ste07**, Theorem 9.23 and Theorem 9.22].

Let ℓ be a prime not dividing n and $\overline{\mathbb{F}}_\ell$ an algebraic closure of $\mathbb{F}_\ell$. We will denote by $M(n,k)_{\overline{\mathbb{F}}_\ell}$ be the space of mod ℓ Katz modular forms of weight k for $\Gamma_1(n)$. The cuspidal subspace of $M(n,k)_{\overline{\mathbb{F}}_\ell}$, i.e. the subspace whose elements have q-expansions that are power series with constant term zero at all cusps, will be denoted by $S(n,k)_{\overline{\mathbb{F}}_\ell}$. For every character $\epsilon : (\mathbb{Z}/n\mathbb{Z})^* \to \overline{\mathbb{F}}_\ell^*$, let $S(n,k,\epsilon)_{\overline{\mathbb{F}}_\ell}$ be the space of Katz cuspforms of weight k for $\Gamma_1(n)$ and character ϵ.

If $n>4$ then $S(n,k)_{\overline{\mathbb{F}}_\ell}$ is isomorphic to $\mathrm{H}^0(X_1(n)_{\overline{\mathbb{F}}_\ell}, \omega^{\otimes k}(-\operatorname{Cusps}))$, the space of global sections of the line bundle $\omega^{\otimes k}(-\operatorname{Cusps})$, while the space of Katz modular forms $M(n,k)_{\overline{\mathbb{F}}_\ell}$ is isomorphic to $\mathrm{H}^0(X_1(n)_{\overline{\mathbb{F}}_\ell}, \omega^{\otimes k})$, see [**Gro90**, Proposition 2.2], where $X_1(n)_{\overline{\mathbb{F}}_\ell}$ is the base change of the modular curve $X_1(n)$ over $\overline{\mathbb{F}}_\ell$.

The main result of this article is the following:

THEOREM 1.3. *Let n and k be positive integers. Let ℓ be a prime not dividing n, such that $1 \leq k \leq \ell+1$. Let $f : \mathbb{T}_{\tilde{\epsilon}} \to \overline{\mathbb{F}}_\ell$ be a ring homomorphism from the Hecke algebra of level $\Gamma_1(n)$, weight k and character $\tilde{\epsilon} : (\mathbb{Z}/n\mathbb{Z})^* \to \mathbb{C}^*$ to $\overline{\mathbb{F}}_\ell$. Let $\epsilon : (\mathbb{Z}/n\mathbb{Z})^* \to \overline{\mathbb{F}}_\ell^*$ be the character defined by $a \mapsto f(\langle a \rangle)$ for all $a \in (\mathbb{Z}/n\mathbb{Z})^*$.*

Assume that there exists a prime r dividing $n\ell$ such that the $\overline{\mathbb{F}}_\ell$-vector space

$$V := \bigcap_{p \neq r} \ker\left(T_p - f(T_p), S(n,k,\epsilon)_{\overline{\mathbb{F}}_\ell}\right),$$

has dimension bigger than 1.

If r is different from ℓ then there exists $g \in S(n/r, k, \operatorname{Res}(\epsilon))_{\overline{\mathbb{F}}_\ell}$ such that f belongs to the span of $B_r^* g$ and $\alpha^* g$, where B_r^* and α^* are the maps induced by the degeneracy maps.

If $r = \ell$ then $k \in \{\ell, \ell+1\}$. In the first case there exists $g \in S(n,1,\epsilon)_{\overline{\mathbb{F}}_\ell}$ such that $F(g) = f$, where F is the Frobenius morphism. In the second case there exists $g \in S(n,2,\epsilon)_{\overline{\mathbb{F}}_\ell}$ such that $A_\ell g = f$, where A_ℓ is the Hasse invariant.

From the previous theorem we deduce the following results:

COROLLARY 1.4. *The Galois representation $\rho_f : \operatorname{Gal}(\overline{\mathbb{Q}}/\mathbb{Q}) \to \mathrm{GL}_2(\overline{\mathbb{F}}_\ell)$ attached to f, under the same hypotheses of Theorem 1.3, arises from a lower level or weight.*

We say that a form is minimal with respect to weight and level if the associate mod ℓ eigenvalue system does not arise from forms of lower level or weight.

COROLLARY 1.5. *Under the same notation of Theorem 1.3, if V has dimension 1 then f is of minimal weight and level and $\mathrm{N}(\rho_f) = n$ and $k(\rho_f) = k$.*

2. Notation and preliminaries on modular curves

Let n and k be positive integers, let ℓ be a prime not dividing n and let ϵ be a finite order character with values in $\overline{\mathbb{F}}_\ell$ and conductor dividing n.

Let us assume that $n = mp^r$ with $r \geq 1$ and p not dividing m. We split the level structure according to this factorization, and denote the space of mod ℓ cuspforms as $S(m, p^r, k, \overline{\epsilon})_{\overline{\mathbb{F}}_\ell}$. We will use analogous notation for modular curves.

There are two degeneracy maps at p on the modular curve $X_1(n)_{\overline{\mathbb{F}}_\ell}$. We will denote them by B_p and α:

$$\begin{array}{ccc} & X_1(m,p^r)_{\overline{\mathbb{F}}_\ell} & \\ \swarrow^{B_p} & & \searrow^{\alpha} \\ X_1(m,p^{r-1})_{\overline{\mathbb{F}}_\ell} & & X_1(m,p^{r-1})_{\overline{\mathbb{F}}_\ell} \end{array} \tag{1}$$

Let us recall the definition of the degeneracy maps via the moduli interpretation for $X_1(n)_{\overline{\mathbb{F}}_\ell}$. Let E/S be an elliptic curve over an $\overline{\mathbb{F}}_\ell$-scheme S, with P and Q respectively points of order m and p^r. The map α is the forgetful map

$\alpha\colon (E,P,Q) \longmapsto (E,P,pQ)$ and the map B_p is the p-th degeneracy map defined by $B_p\colon (E,P,Q) \mapsto (E/\langle p^{r-1}Q\rangle, \beta(P), \beta(Q))$, where

$$\langle p^{r-1}Q\rangle \rightarrowtail E \xrightarrow{\beta} E/\langle p^{r-1}Q\rangle$$

is the degree p isogeny whose kernel is generated by $p^{r-1}Q$.

Let us fix ζ_{p^r} a root of unity of order p^r in $\overline{\mathbb{F}}_\ell$. The Atkin-Lehner involution $w_{\zeta_{p^r}}$ on $X_1(n)_{\overline{\mathbb{F}}_\ell}$ is defined as follows:

$$\begin{array}{ccc} X_1(m,p^r)_{\overline{\mathbb{F}}_\ell} & \overset{w_{\zeta_{p^r}}}{\longrightarrow} & X_1(m,p^r)_{\overline{\mathbb{F}}_\ell} \\ (E,P,Q) & \longmapsto & (E/\langle Q\rangle, \beta_r(P), Q') \end{array} \tag{2}$$

where β_r is the degree p^r isogeny whose kernel is generated by Q, while Q' belongs to $\ker(\beta_r^t)(S)$:

$$\langle Q\rangle \rightarrowtail E \underset{\beta_r^t}{\overset{\beta_r}{\rightleftarrows}} E/\langle Q\rangle \leftarrowtail \langle Q'\rangle$$

and it is the unique element such that $e_\beta(Q,Q') = \zeta_{p^r}$, where e_β is the Weil pairing related to β i.e. the perfect μ_{p^r} valued pairing between $\ker(\beta)$ and $\ker(\beta^t)$, see [**KM85**, Section 2.8].

Let us recall some basic facts about Tate curves that will be relevant in the next sections, see [**DR73**, Chapter VII] and [**Sil94**, Chapter V] for details. For every positive integer d, the d-th Tate curve is a generalized elliptic curve $\mathrm{Tate}(q^d) \to \operatorname{Spec}\mathbb{Z}[[q]]$ that becomes a Néron d-gon after base change and that is a smooth elliptic curve over $\operatorname{Spec}\mathbb{Z}[[q]][q^{-1}]$. Let d, e and n be positive integers such that n is the least common multiple of d and e. Then the curve $\mathrm{Tate}(q^d)$ over $\operatorname{Spec}\mathbb{Z}[[q,\zeta_e]]$, where ζ_e is a e-th root of unity, admits a $\Gamma_1(n)$-structure. For each choice of d, of e and of the $\Gamma_1(n)$-structure, there is an injective map from $M(\Gamma_1(n),k)_{\overline{\mathbb{F}}_\ell}$ to $\overline{\mathbb{F}}_\ell \otimes_{\mathbb{Z}} \mathbb{Z}[[q,\zeta_e]]$, which is called the q-expansion map relative to $\mathrm{Tate}(q^d)$ with the given $\Gamma_1(n)$-structure. Evaluating any mod ℓ modular form in $M(\Gamma_1(n),k)_{\overline{\mathbb{F}}_\ell}$ on the Tate curve $\mathrm{Tate}(q^n)$ over $\operatorname{Spec}\mathbb{Z}[[q]]$ with respect to the $\Gamma_1(n)$-structure given by $\psi(i) = q^i$ for $i \in \mathbb{Z}/n\mathbb{Z}$ we obtain a particular q-expansion called the q-expansion at ∞ ([**KM85**, Sections 8.8 and 8.11]). For any $f \in M(\Gamma_1(n),k)_{\overline{\mathbb{F}}_\ell}$ we define $a_m(f)$ to be the m-th coefficient of this q-expansion.

3. Level lowering for Katz cuspforms

LEMMA 3.1. *Let n and k be positive integers and let p be a prime strictly dividing n with $n/p > 4$. Let ℓ be a prime not dividing n and let $f \in S(n,k)_{\overline{\mathbb{F}}_\ell}$ such that*

$$f(\mathrm{Tate}(q), \zeta_{n/p}, \zeta_p) = \sum_{j\geq 1} a_j(f) q^j \left(\frac{dt}{t}\right)^{\otimes k} \in \overline{\mathbb{F}}_\ell\,[[q^p]] \left(\frac{dt}{t}\right)^{\otimes k},$$

where $\zeta_{n/p}$ and ζ_p are respectively n/p-th and p-th roots of unity. Then there exists a unique cuspform $g \in S(n/p,k)_{\overline{\mathbb{F}}_\ell}$, such that $B_p^* g = f$, where B_p^* is the pullback of $B_p\colon X_1(n)_{\overline{\mathbb{F}}_\ell} \to X_1(n/p)_{\overline{\mathbb{F}}_\ell}$, the p-th degeneracy map.

PROOF. Since p strictly divides n, let us write $n = mp$ with $(m,p) = 1$. By hypothesis, the q-expansion of f at the cusp ∞ is such that $a_j(f) = 0$ for all positive integers j not divisible by p. As $m > 4$ by hypothesis, then f is an element of $\mathrm{H}^0(X_1(n)_{\overline{\mathbb{F}}_\ell}, \omega^{\otimes k}(-\mathrm{Cusps}))$. It is easy to show, using the description of the degeneracy maps in Section 2, that the following diagram commutes:

$$\begin{array}{ccc} X_1(n)_{\overline{\mathbb{F}}_\ell} & \xrightarrow{w_{\zeta_p}} & X_1(n)_{\overline{\mathbb{F}}_\ell} \\ {\scriptstyle B_p}\downarrow & & \downarrow{\scriptstyle \alpha} \\ X_1(m)_{\overline{\mathbb{F}}_\ell} & = & X_1(m)_{\overline{\mathbb{F}}_\ell} \end{array}$$

where B_p and α are the degeneracy maps, defined in (1), and w_{ζ_p} is the Atkin-Lehner map associated to the isogeny β, defined in (2). The isogeny β induces a morphism β^* from the line bundle $\omega^{\otimes k}(-\mathrm{Cusps})$ on $X_1(m)_{\overline{\mathbb{F}}_\ell}$ to the line bundle $\omega^{\otimes k}(-\mathrm{Cusps})$ on $X_1(n)_{\overline{\mathbb{F}}_\ell}$, since for each $(E/S, P, Q)$ we have an isomorphism of invertible O_S-modules β^* between $\omega_{E/\langle Q\rangle}$ and ω_E. The pullback $w^*_{\zeta_p} : S(n,k)_{\overline{\mathbb{F}}_\ell} \to S(n,k)_{\overline{\mathbb{F}}_\ell}$ is defined by

$$(w^*_{\zeta_p} h)(E,P,Q) = \beta^*(h(w_{\zeta_p}(E,P,Q))) = \beta^*(h(E/\langle Q\rangle, \beta(P), Q')),$$

where h belongs to $S(n,k)_{\overline{\mathbb{F}}_\ell}$, see [**Gro90**, Section 6]. There exists a unique form $f' \in S(n,k)_{\overline{\mathbb{F}}_\ell}$ satisfying $w^*_{\zeta_p} f' = f$. The q-expansion at the cusp ∞ is:

$$\begin{aligned} (w^*_{\zeta_p} f')(\infty) &= (w^*_{\zeta_p} f')(\mathrm{Tate}(q), \zeta_m, \zeta_p) = \beta^*(f'(\infty')) = \\ &= \beta^*(f'(\mathrm{Tate}(q^p), \zeta_m^p, q)) = \beta^*\left(\sum_{j\geq 1} a_j(f') q^j \left(\frac{dt}{t}\right)^{\otimes k}\right) = \\ &= \left(\sum_{j\geq 1} a_j(f') q^j\right) p^k \left(\frac{dt}{t}\right)^{\otimes k}, \end{aligned}$$

where $\infty' := w_{\zeta_p}(\infty)$ and the last equality follows from the exact sequence

$$\mu_p \rightarrowtail \mathrm{Tate}(q) \xrightarrow{\beta} \mathrm{Tate}(q^p),$$

which gives $\beta^*(dt/t) = p(dt/t)$. Since $f(\mathrm{Tate}(q), \zeta_m, \zeta_p) \in \overline{\mathbb{F}}_\ell\,[[q^p]]\,(dt/t)^{\otimes k}$, we deduce that $a_j(f')p^k = a_j(f)$ for all j and

$$f'(\mathrm{Tate}(q^p), \zeta_m^p, q) \in \overline{\mathbb{F}}_\ell\,[[q^p]] \left(\frac{dt}{t}\right)^{\otimes k}.$$

In order to prove that f comes from $X_1(m)_{\overline{\mathbb{F}}_\ell}$ via B_p^*, it is enough to show that f' does via α. To prove this claim, consider the following covering of modular curves:

$$\begin{array}{ccccc} X(\Gamma_1(m), \Gamma(p)^{\zeta_p-\mathrm{can}})_{\overline{\mathbb{F}}_\ell} & & (E,P,Q_1,Q_2) & & (\mathrm{Tate}(q^p), \zeta_m^p, q, \zeta_p) \\ {\scriptstyle\gamma}\downarrow & & \downarrow & & \downarrow \\ X(\Gamma_1(m), \Gamma_1(p))_{\overline{\mathbb{F}}_\ell} & & (E,P,Q_1) & & (\mathrm{Tate}(q^p), \zeta_m^p, q) \\ {\scriptstyle\alpha}\downarrow & & \downarrow & & \downarrow \\ X(\Gamma_1(m))_{\overline{\mathbb{F}}_\ell} & & (E,P) & & (\mathrm{Tate}(q^p), \zeta_m^p) \end{array}$$

where Q_1, Q_2 are such that $e_p(Q_1, Q_2) = \zeta_p$, and α, γ are both forgetful maps. On $X(\Gamma_1(m), \Gamma(p)^{\zeta_p-\mathrm{can}})_{\overline{\mathbb{F}}_\ell} \to X(\Gamma_1(m))_{\overline{\mathbb{F}}_\ell}$ there is the natural $\mathrm{SL}_2(\mathbb{F}_p)$ action, given by the action of $\mathrm{SL}_2(\mathbb{F}_p)$ on $\Gamma(p)^{\zeta_p-\mathrm{can}}$. This action is trivial on $X(\Gamma_1(m))_{\overline{\mathbb{F}}_\ell}$:

$$\begin{array}{ccc} & & X(\Gamma_1(m), \Gamma(p)^{\zeta_p-\mathrm{can}})_{\overline{\mathbb{F}}_\ell} \\ & & \downarrow \left(\begin{smallmatrix} 1 & * \\ 0 & 1 \end{smallmatrix}\right) \\ \mathrm{SL}_2(\mathbb{F}_p) & \curvearrowright & X(\Gamma_1(m), \Gamma_1(p))_{\overline{\mathbb{F}}_\ell} \\ & & \downarrow \\ & & X(\Gamma_1(m))_{\overline{\mathbb{F}}_\ell} \end{array}$$

Therefore, there exists $g \in S(m,k)_{\overline{\mathbb{F}}_\ell}$ such that $B_p^* g = f$ if and only if $\gamma^* f'$ is $\mathrm{SL}_2(\mathbb{F}_p)$-invariant.

The map $\gamma : X(\Gamma_1(m), \Gamma(p)^{\zeta_p-\mathrm{can}})_{\overline{\mathbb{F}}_\ell} \to X(\Gamma_1(m), \Gamma(p))_{\overline{\mathbb{F}}_\ell}$ corresponds to the action of elements of the form $\left(\begin{smallmatrix} 1 & * \\ 0 & 1 \end{smallmatrix}\right) \in \mathrm{SL}_2(\mathbb{F}_p)$ and $\gamma^* f'$ is invariant under the action of $\left(\begin{smallmatrix} 1 & * \\ 0 & 1 \end{smallmatrix}\right)$ by construction. Since $\{\left(\begin{smallmatrix} 1 & 0 \\ * & 1 \end{smallmatrix}\right), \left(\begin{smallmatrix} 1 & * \\ 0 & 1 \end{smallmatrix}\right)\}$ generate $\mathrm{SL}_2(\mathbb{F}_p)$, and $\left(\begin{smallmatrix} 1 & 0 \\ 1 & 1 \end{smallmatrix}\right)$ generates the subgroup $\left(\begin{smallmatrix} 1 & 0 \\ * & 1 \end{smallmatrix}\right)$ in $\mathrm{SL}_2(\mathbb{F}_p)$, in order to conclude it is enough to prove that $\gamma^* f'$ is invariant under the action of $\left(\begin{smallmatrix} 1 & 0 \\ 1 & 1 \end{smallmatrix}\right)$. Let us remark that $\left(\begin{smallmatrix} 1 & 0 \\ 1 & 1 \end{smallmatrix}\right)$ fixes $\infty' \in X(\Gamma_1(m), \Gamma(p)^{\zeta_p-\mathrm{can}})_{\overline{\mathbb{F}}_\ell}$ and it acts on $\overline{\mathbb{F}}_\ell[[q]]$ as

$$\begin{cases} \text{identity on } \overline{\mathbb{F}}_\ell \\ q \mapsto \zeta_p q \end{cases}$$

Therefore, $\left(\begin{smallmatrix} 1 & 0 \\ 1 & 1 \end{smallmatrix}\right)$ acts on $\overline{\mathbb{F}}_\ell[[q]](dt/t)^{\otimes k}$ by mapping $t \mapsto t$ and so $(dt/t)^{\otimes k} \mapsto (dt/t)^{\otimes k}$. Since $a_j(f') = 0$ for all positive integers j not divisible by p, then $\left(\begin{smallmatrix} 1 & 0 \\ 1 & 1 \end{smallmatrix}\right)^* f'$ has the same q-expansion as f'. Being $X(\Gamma_1(m), \Gamma(p)^{\zeta_p-\mathrm{can}})_{\overline{\mathbb{F}}_\ell}$ an irreducible and reduced curve, this means that $\left(\begin{smallmatrix} 1 & 0 \\ 1 & 1 \end{smallmatrix}\right)^* \gamma^* f' = \gamma^* f'$.

The cuspform $g \in S(m,k)_{\overline{\mathbb{F}}_\ell}$ such that $B_p^* g = f$ is unique: pullback does not change q-expansions and the q-expansion principle holds ([**Kat73**, Section 1.12]). □

Now we want to generalize the previous lemma for levels n and primes p such that n is highly divisible by p. In order to do so, we need some preliminary results which will help us understanding, on the one hand, the modular curves covering and, on the other hand, the interaction between the level structure and the Atkin-Lehner operators.

LEMMA 3.2. *Given $r \in \mathbb{Z}_{>0}$ and a prime p, the set*

$$\left\{ \begin{pmatrix} 1 & 0 \\ p^r & 1 \end{pmatrix}, \begin{pmatrix} 1 & 1 \\ 0 & 1 \end{pmatrix} \right\} \subset \mathrm{SL}_2(\mathbb{Z}/p^{r+1}\mathbb{Z})$$

generates the subgroup

$$\overline{\Gamma_1(p^r)} := \left\{ \begin{pmatrix} 1+p^r a & b \\ p^r c & 1+p^r d \end{pmatrix} \in \mathrm{SL}_2(\mathbb{Z}/p^{r+1}\mathbb{Z}) \right\}.$$

PROOF. This is a straightforward computation: every matrix in $\overline{\Gamma_1(p^r)}$ can be written as a finite product of $\left(\begin{smallmatrix} 1 & 0 \\ p^r & 1 \end{smallmatrix}\right)$ and $\left(\begin{smallmatrix} 1 & 1 \\ 0 & 1 \end{smallmatrix}\right)$. Indeed, let $\left(\begin{smallmatrix} 1+p^r a & b \\ p^r c & 1+p^r d \end{smallmatrix}\right)$ in

$SL_2(\mathbb{Z}/p^{r+1}\mathbb{Z})$, with $a, b, c, d \in \mathbb{F}_p$, then

$$\begin{pmatrix} 1+p^r a & b \\ p^r c & 1+p^r d \end{pmatrix} = \begin{pmatrix} 1 & 0 \\ p^r & 1 \end{pmatrix}^{c-1} \begin{pmatrix} 1 & 1 \\ 0 & 1 \end{pmatrix}^{a} \begin{pmatrix} 1 & 0 \\ p^r & 1 \end{pmatrix}^{-1} \begin{pmatrix} 1 & 1 \\ 0 & 1 \end{pmatrix}^{(p^r a-1)(a-b)}.$$

□

LEMMA 3.3. *Let E/S be an elliptic curve over an $\overline{\mathbb{F}}_\ell$-scheme S, let p be a prime different from ℓ. Let $r \geq 1$ be an integer and*

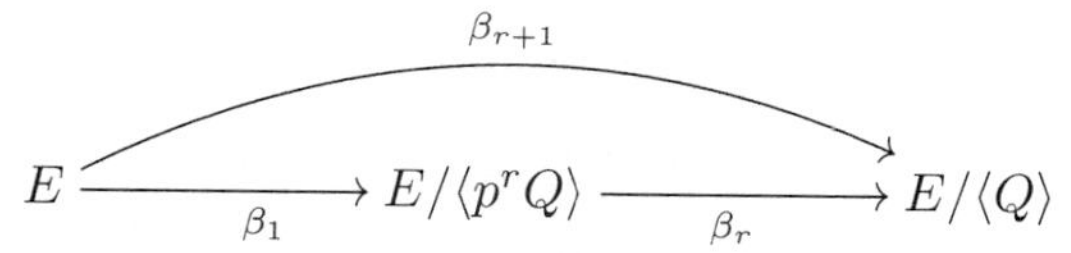

the standard factorization of a cyclic p^{r+1}-isogeny into a cyclic p-isogeny followed by a cyclic p^r-isogeny, where Q is a point of order p^{r+1}.

Let Q_{r+1} and Q_r be the unique points such that $e_{\beta_{r+1}}(Q, Q_{r+1}) = \zeta_{p^{r+1}}$ and $e_{\beta_r}(\beta_1(Q), Q_r) = \zeta_{p^r}$. Then $Q_r = pQ_{r+1}$.

PROOF. The Backing-up Theorem ([**KM85**, 6.7.11]) gives the following description of the kernels of the isogeny β_{r+1} and of its dual:

$$\langle \beta_1(Q)\rangle = \ker(\beta_r), \quad \langle p^r Q\rangle = \ker(\beta_1),$$

and

$$\langle \beta_r^t(Q_{r+1})\rangle = \ker(\beta_1^t), \quad \langle pQ_{r+1}\rangle = \ker(\beta_r^t).$$

For the sake of clarity, the following diagram resumes all the isogenies considered:

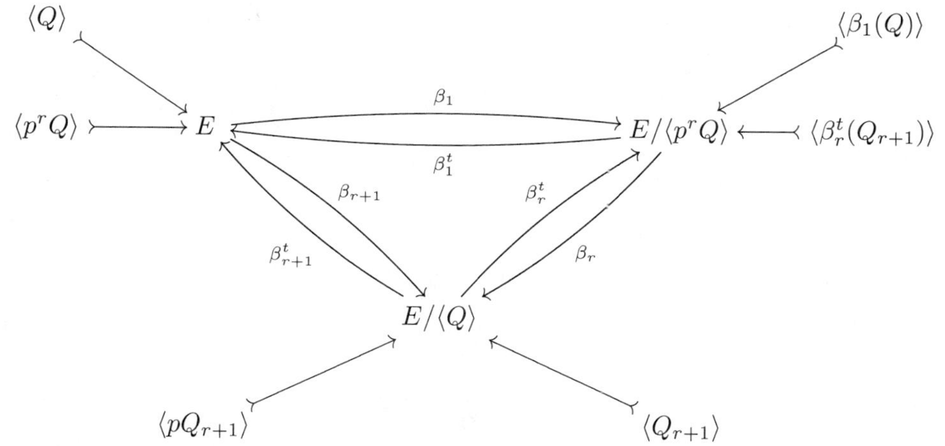

By definition of the Weil pairing, we have that $\langle Q_r\rangle = \ker(\beta_r^t) = \langle pQ_{r+1}\rangle$, so there exists $k \in (\mathbb{Z}/p^r\mathbb{Z})^*$ such that $kQ_r = pQ_{r+1}$, since they generate the same cyclic group of order p^r. Hence, by bilinearity of the pairing we have

$$e_{\beta_r}(\beta_1(Q), kQ_r) = \zeta_{p^r}^k = e_{\beta_r}(\beta_1(Q), pQ_{r+1}).$$

The compatibility of the pairing, see [**KM85**, 2.8] and [**Sil09**, III 8.1], implies that

$$e_{\beta_r}(\beta_1(Q), pQ_{r+1}) = e_{\beta_{r+1}}(Q, pQ_{r+1}) = \zeta_{p^{r+1}}^p = \zeta_{p^r}.$$

As $e_{\beta_r}(\beta_1(Q), Q_r) = \zeta_{p^r}$, then $k = 1$. Therefore, $Q_r = pQ_{r+1}$. □

LEMMA 3.4. *Let n and k be positive integers, and let p be a prime such that $n = mp^{r+1}$ with p and m coprime, $r \in \mathbb{Z}_{\geq 1}$ and $mp^r > 4$. Let ℓ be a prime not dividing n and let $f \in S(m, p^{r+1}, k)_{\overline{\mathbb{F}}_\ell}$ such that*

$$f(\mathrm{Tate}(q), \zeta_m, \zeta_{p^{r+1}}) = \sum_{j\geq 1} a_j(f) q^j \left(\frac{dt}{t}\right)^{\otimes k} \in \overline{\mathbb{F}}_\ell [[q^p]] \left(\frac{dt}{t}\right)^{\otimes k},$$

where ζ_m and $\zeta_{p^{r+1}}$ are fixed m-th and p^{r+1}-th roots of unity. Then there exists a unique cuspform $g \in S(m, p^r, k)_{\overline{\mathbb{F}}_\ell}$, such that $B_p^* g = f$, where B_p^* is the pullback of $B_p : X_1(m, p^{r+1})_{\overline{\mathbb{F}}_\ell} \to X_1(m, p^r)_{\overline{\mathbb{F}}_\ell}$ the p-th degeneracy map.

PROOF. Since $mp^r > 4$ we have the following diagram:

$$\begin{array}{ccc} X_1(m, p^{r+1})_{\overline{\mathbb{F}}_\ell} & \xrightarrow{\;w_{\zeta_{p^{r+1}}}\;} & X_1(m, p^{r+1})_{\overline{\mathbb{F}}_\ell} \\ {\scriptstyle B_p}\big\downarrow & & \big\downarrow{\scriptstyle \tilde{\alpha}} \\ X_1(m, p^r)_{\overline{\mathbb{F}}_\ell} & \xrightarrow{\;w_{\zeta_{p^r}}\;} & X_1(m, p^r)_{\overline{\mathbb{F}}_\ell} \end{array}$$

where B_p is the p-th degeneracy map, $w_{\zeta_{p^r}}$ and $w_{\zeta_{p^{r+1}}}$ are Atkin-Lehner involutions and $\tilde{\alpha} := w_{\zeta_{p^r}} \circ B_p \circ w^{-1}_{\zeta_{p^{r+1}}}$. We claim that $\tilde{\alpha}$ is the forgetful map α.

Through the moduli interpretation, we can write the previous diagram as follows. Let E be an elliptic curve over S, an $\overline{\mathbb{F}}_\ell$-scheme, with P and Q points of order m and p^{r+1} respectively, then we have:

$$\begin{array}{ccc} (E, P, Q) & \xmapsto{\;w_{\zeta_{p^{r+1}}}\;} & (E/\langle Q\rangle, \beta_{r+1}(P), Q_{r+1}) \\ {\scriptstyle B_p}\big\downarrow & & \big\downarrow{\scriptstyle \tilde{\alpha}} \\ (E/\langle p^r Q\rangle, \beta_1(P), \beta_1(Q)) & \xmapsto{\;w_{\zeta_{p^r}}\;} & (E/\langle Q\rangle, \beta_r(\beta_1(P)), Q_r) \end{array}$$

while the forgetful map gives

$$(E/\langle Q\rangle, \beta_{r+1}(P), Q_{r+1}) \xmapsto{\;\alpha\;} (E/\langle Q\rangle, \beta_{r+1}(P), pQ_{r+1}),$$

where the maps $\beta_1, \beta_r, \beta_{r+1}$ are isogenies defined by:

$$E \xrightarrow[\beta_1]{} E/\langle p^r Q\rangle \xrightarrow[\beta_r]{} E/\langle Q\rangle, \qquad E \xrightarrow{\;\beta_{r+1}\;} E/\langle Q\rangle,$$

so $\beta_r(\beta_1(P)) = \beta_{r+1}(P)$. By Lemma 3.3 we have that $Q_r = pQ_{r+1}$, so $\tilde{\alpha} = \alpha$.

We have that:

$$w_{\zeta_{p^{r+1}}}(\infty) = w_{\zeta_{p^{r+1}}}(\mathrm{Tate}(q), \zeta_m, \zeta_{p^{r+1}}) = (\mathrm{Tate}(q^{p^{r+1}}), \zeta_m^{p^{r+1}}, q) = \infty'.$$

There exists a unique Katz cuspform $f' \in S(m, p^{r+1}, k)_{\overline{\mathbb{F}}_\ell}$ such that $w_{\zeta_{p^{r+1}}}(f') = f$: by direct computation at the standard cusp, for all positive integers j we have

$$a_j(f') p^{(r+1)k} = a_j(f),$$

and hence $f'(\mathrm{Tate}(q), \zeta_m, \zeta_{p^{r+1}}) \in \overline{\mathbb{F}}_\ell [[q^p]] (dt/t)^{\otimes k}$.

Let us consider the following covering of modular curves:

$$\begin{array}{ccc}
X(\Gamma_1(m),\Gamma(p^{r+1})^{\zeta_{p^{r+1}}-\mathrm{can}})_{\overline{\mathbb{F}}_\ell} & & (E,P,Q_1,Q_2) \\
\downarrow \gamma \, \left(\begin{smallmatrix} 1 & * \\ 0 & 1 \end{smallmatrix}\right) & & \downarrow \\
\overline{\Gamma_1(p^r)} \quad X(\Gamma_1(m),\Gamma_1(p^{r+1}))_{\overline{\mathbb{F}}_\ell} & & (E,P,Q_1) \\
\downarrow \alpha & & \downarrow \\
X(\Gamma_1(m),\Gamma_1(p^r))_{\overline{\mathbb{F}}_\ell} & & (E,P,pQ_1)
\end{array}$$

where

$$\overline{\Gamma_1(p^r)}:=\left\{\begin{pmatrix} 1+p^r\cdot a & b \\ p^r\cdot c & 1+p^r\cdot d \end{pmatrix}\in \mathrm{SL}_2(\mathbb{Z}/p^{r+1}\mathbb{Z})\right\}.$$

There exists a Katz cuspform $g\in S(m,p^r,k)_{\overline{\mathbb{F}}_\ell}$ such that $B_p^*g=f$ if and only if γ^*f' is $\overline{\Gamma_1(p^r)}$-invariant.

The set $\{\left(\begin{smallmatrix} 1 & 0 \\ p^r & 1 \end{smallmatrix}\right),\left(\begin{smallmatrix} 1 & 1 \\ 0 & 1 \end{smallmatrix}\right)\}$ generates $\overline{\Gamma_1(p^r)}$ by Lemma 3.2. Therefore, it is enough to show that γ^*f' is invariant with respect to these generators. The covering map $\gamma: X(\Gamma_1(m),\Gamma(p^{r+1})^{\zeta_{p^{r+1}}-\mathrm{can}})_{\overline{\mathbb{F}}_\ell}\to X(\Gamma_1(m),\Gamma(p^{r+1}))_{\overline{\mathbb{F}}_\ell}$ corresponds to elements of the form $\left(\begin{smallmatrix} 1 & * \\ 0 & 1 \end{smallmatrix}\right)\in\mathrm{SL}_2(\mathbb{Z}/p^{r+1}\mathbb{Z})$. By construction, γ^*f' is invariant under the action of these elements. The action of $\left(\begin{smallmatrix} 1 & 0 \\ p^r & 1 \end{smallmatrix}\right)$ fixes $\infty'\in X(\Gamma_1(m),\Gamma(p^{r+1})^{\zeta_{p^{r+1}}-\mathrm{can}})_{\overline{\mathbb{F}}_\ell}$. Moreover, the corresponding action on $\overline{\mathbb{F}}_\ell[[q]]$ is given by

$$\begin{cases} \text{identity on } \overline{\mathbb{F}}_\ell \\ q\mapsto \zeta_{p^{r+1}}^{p^r}q=\zeta_p q. \end{cases},$$

Therefore, $\left(\begin{smallmatrix} 1 & 0 \\ p^r & 1 \end{smallmatrix}\right)$ acts on $\overline{\mathbb{F}}_\ell[[q]](dt/t)^{\otimes k}$ as $t\mapsto t$. Since $a_j(f')=0$ for all positive integers j not divisible by p, then $\left(\begin{smallmatrix} 1 & 0 \\ p^r & 1 \end{smallmatrix}\right)^*f'$ has the same q-expansion as f'. Now, since $X(\Gamma_1(m),\Gamma(p^{r+1})^{\zeta_{p^{r+1}}-\mathrm{can}})_{\overline{\mathbb{F}}_\ell}$ is an irreducible reduced curve, this means that $\left(\begin{smallmatrix} 1 & 0 \\ p^r & 1 \end{smallmatrix}\right)^*\gamma^*f'=\gamma^*f'$.

The uniqueness of $g\in S(m,p^r,k)_{\overline{\mathbb{F}}_\ell}$ such that $B_p^*g=f$ follows, as in the previous lemma, by the q-expansion principle. □

In the previous lemmas, we did not consider the action of diamond operators. First of all, let us recall briefly how this operators are defined. For n a positive integer, and for all $d\in(\mathbb{Z}/n\mathbb{Z})^*$, we define an automorphism

$$\begin{aligned} r_d: X(n)_{\overline{\mathbb{F}}_\ell} &\to X(n)_{\overline{\mathbb{F}}_\ell} \\ (E,P) &\mapsto r_d(E,P)=(E,dP), \end{aligned}$$

for all generalized elliptic curves E together with a $\Gamma_1(n)$-structure P. For every $d\in(\mathbb{Z}/n\mathbb{Z})^*$, the diamond operator $\langle d\rangle$ on $M(\Gamma_1(n),k)_{\overline{\mathbb{F}}_\ell}$ is defined as the automorphism of $M(\Gamma_1(n),k)_{\overline{\mathbb{F}}_\ell}$ induced by pullback via the automorphism r_d. If we are dealing with Katz cuspforms with a character $\epsilon:(\mathbb{Z}/n\mathbb{Z})^*\to\overline{\mathbb{F}}_\ell^*$, we are imposing that for all $d\in(\mathbb{Z}/n\mathbb{Z})^*$, the diamond operator $\langle d\rangle$ acts as $\epsilon(d)$.

LEMMA 3.5. *Let n and k be positive integers, let p be a prime dividing n such that $n/p>4$ and let ℓ be a prime not dividing n. Let $\epsilon:(\mathbb{Z}/n\mathbb{Z})^*\to\overline{\mathbb{F}}_\ell^*$ and*

$\chi : (\mathbb{Z}/(n/p)\mathbb{Z})^* \to \overline{\mathbb{F}}_\ell^*$ *be characters. Let* $f \in S(n,k,\epsilon)_{\overline{\mathbb{F}}_\ell}$ *and* $g \in S(n/p,k,\chi)_{\overline{\mathbb{F}}_\ell}$ *such that* $B_p^*(g) = f$. *Then*

$$\chi = \mathrm{Res}^{\mathbb{Z}/(n/p)\mathbb{Z}}_{\mathbb{Z}/n\mathbb{Z}}(\epsilon).$$

PROOF. Let E/S be an elliptic curve over an $\overline{\mathbb{F}}_\ell$-scheme S and let P be a point of order n. The degeneracy map $B_p : X_1(n)_{\overline{\mathbb{F}}_\ell} \to X_1(n/p)_{\overline{\mathbb{F}}_\ell}$ defined in (1) through the isogeny β is given by

$$(B_p^* g)(E,P) = \beta^*(g(E/\langle (n/p)P\rangle, \beta(P))).$$

The action of the diamond operator $\langle d\rangle$ for $d \in (\mathbb{Z}/(n/p)\mathbb{Z})^*$ is given by

$$\begin{aligned}\langle d\rangle (B_p^* g)(E,P) &= (B_p^* g)(E,dP) = \beta^*(g(E/\langle (n/p)P\rangle, \beta(dP))) = \\ &= \beta^*(g(E/\langle (n/p)P\rangle, d\beta(P))) = \\ &= \beta^*(\chi(d) g(E/\langle (n/p)P\rangle, \beta(P))) = \chi(d)(B_p^* g)(E,P);\end{aligned}$$

hence, for all $d \in (\mathbb{Z}/(n/p)\mathbb{Z})^*$ we have $\langle d\rangle (B_p^* g) = \chi(d)(B_p^* g)$. Since $B_p^*(g) = f$, then for all $d \in (\mathbb{Z}/(n/p)\mathbb{Z})^*$ it follows that $\langle d\rangle f = \chi(d) f = \epsilon(d) f$. Therefore, for all $d \in (\mathbb{Z}/(n/p)\mathbb{Z})^*$ the equality $\chi(d) = \epsilon(d)$ holds. □

Therefore, we can state Lemma 3.1 and Lemma 3.4 for forms in $S(n,k,\epsilon)_{\overline{\mathbb{F}}_\ell}$:

LEMMA 3.6. *Let* n *and* k *be positive integers. Let* p *be a prime such that* n *factors as* mp^{r+1} *with* p *and* m *coprime,* $r \in \mathbb{Z}_{\geq 0}$ *and* $mp^r > 4$. *Let* ℓ *be a prime not dividing* n *and let* $\epsilon : (\mathbb{Z}/n\mathbb{Z})^* \to \overline{\mathbb{F}}_\ell^*$ *be a finite order character. Let* $f \in S(m, p^{r+1}, k, \epsilon)_{\overline{\mathbb{F}}_\ell}$ *such that*

$$f(\mathrm{Tate}(q), \zeta_m, \zeta_{p^{r+1}}) = \sum_{j\geq 1} a_j(f) q^j \left(\frac{dt}{t}\right)^{\otimes k} \in \overline{\mathbb{F}}_\ell[[q^p]] \left(\frac{dt}{t}\right)^{\otimes k},$$

where ζ_m *and* $\zeta_{p^{r+1}}$ *are fixed roots of unity of order* m *and* p^{r+1}. *Then there exists a unique form* $g \in S(m, p^r, k, \epsilon')_{\overline{\mathbb{F}}_\ell}$, *such that* $B_p^* g = f$, *where* $\epsilon' = \mathrm{Res}^{\mathbb{Z}/(mp^r)\mathbb{Z}}_{\mathbb{Z}/n\mathbb{Z}}(\epsilon)$.

In the lemmas above we assumed that the prime p, dividing the level n, is such that the quotient $n/p > 4$. This hypothesis implies that all the objects used in the proofs are modular curves. Without such hypothesis, we should use the theory of algebraic stacks, see [**Con07**, Theorem 1.2.1]. Anyway, the embedding of the space of cuspforms of level n into the spaces of cuspforms of levels multiple of n give us a way to circumvent this problem:

COROLLARY 3.1. *Lemma* 3.6 *holds with no assumption on* n/p.

PROOF. If $mp^r < 4$, let q be any prime not dividing the level n (in particular, $q \neq p$) such that $qmp^r > 4$. Since $S(m, p^{r+1}, k, \epsilon)_{\overline{\mathbb{F}}_\ell}$ embeds in $S(mq, p^{r+1}, k, \epsilon)_{\overline{\mathbb{F}}_\ell}$ via the trivial embedding, we have that $f \in S(mq, p^{r+1}, k, \epsilon)_{\overline{\mathbb{F}}_\ell}$ with the same q-expansion, where ϵ is considered as a character of modulus qn. The hypotheses of Lemma 3.6 are satisfied, so there exists a unique form $g \in S(mq, p^r, k, \epsilon')_{\overline{\mathbb{F}}_\ell}$, such that $B_p^* g = f$, where $\epsilon' = \mathrm{Res}^{\mathbb{Z}/(qmp^r)\mathbb{Z}}_{\mathbb{Z}/qn\mathbb{Z}}(\epsilon)$. Since f comes from level n, it is possible to show that g comes from level mp^r as in the previous lemmas (the forms considered at levels nq and qmp^r are images of forms in levels n and mp^r via the forgetful map). The character ϵ' is a character of modulus mp^r since ϵ has modulus n. □

Moreover, we can generalize the previous results, iterating through divisors and equating q-expansions:

LEMMA 3.7. *Let n and k be positive integers. Let d be a positive integer such that n factors as md with d and m coprime. Let ℓ be a prime not dividing n and let $\epsilon : (\mathbb{Z}/n\mathbb{Z})^* \to \overline{\mathbb{F}}_\ell^*$ be a character. Let $f \in S(m, d, k, \epsilon)_{\overline{\mathbb{F}}_\ell}$ such that*

$$f(\mathrm{Tate}(q), \zeta_m, \zeta_d) = \sum_{j\geq 1} a_j(f) q^j \left(\frac{dt}{t}\right)^{\otimes k} \in \overline{\mathbb{F}}_\ell\left[[q^d]\right] \left(\frac{dt}{t}\right)^{\otimes k},$$

where $\mathrm{Tate}(q)$ *is the Tate curve over $\overline{\mathbb{F}}_\ell((q))$, ζ_m and ζ_d are fixed roots of unity of order m and d. Then there exists a unique form $g \in S(m, k, \epsilon')_{\overline{\mathbb{F}}_\ell}$, such that $B_d^* g = f$, where $\epsilon' = \mathrm{Res}(\epsilon)$.*

The proof of this statement consists in the iteration of applications of Lemma 3.6 and Corollary 3.1, after factorizing d.

4. Proof of Theorem 1.3

PROOF. The operator T_r commutes with all T_p, and it acts on V. Let v be the dimension of V. By hypothesis $v > 1$, hence V contains a non-zero eigenvector f for all T_p. In particular $a_1(f) \neq 0$.

The subspace $V_1 \subset V$ consisting of the g in V with $a_1(g) = 0$ is of dimension $v - 1 > 0$. Every element $g \in V_1$ is such that $0 = a_1(g) = a_1(T_j(g))$ for all integer j not divisible by m. This means that g is an eigenform with q-expansion in $\mathbb{F}\left[[q^r]\right]((dt)/t)^{\otimes k}$ at the standard cusp.

Let us first suppose that $r \neq \ell$. Then we are in the hypotheses of Lemma 3.6 and Corollary 3.1, so there exists a non zero mod ℓ cuspform $g' \in S(n/r, k, \epsilon')_{\overline{\mathbb{F}}_\ell}$ such that $B_r^* g' = g$, where $\epsilon' = \mathrm{Res}(\epsilon)$ and B_r^* is the pullback of the degeneracy map B_r. This means that $V_1 = B_r^* V_2$ where V_2 is the eigenspace of $S(n, k, \overline{\epsilon})_{\overline{\mathbb{F}}_\ell}$ associated to the system of eigenvalues $\{(f(T_p), f(\langle p \rangle))\}$ for p prime, p different from r. Hence the system of eigenvalues occurs in level n/r.

Let us now suppose that r is equal to ℓ. As proved in [**Edi06**, Proposition 6.2], the only two following cases occur. First case: $k = \ell$. The form g, constructed as before, is in the image of the Frobenius F, i.e. there exists $g' \in S(n, 1, \epsilon)_{\mathbb{F}}$ such that $g = (g')^\ell = F(g')$. This means that the system of eigenvalues occurs already in weight 1. Second case: $k = \ell + 1$. We have that $g = A_\ell g'$ where A_ℓ is the Hasse invariant of weight $\ell - 1$ and $g' \in S(n, 2, \epsilon)_{\mathbb{F}}$, hence the system of eigenvalues occurs in weight 2. □

From the previous theorem, recursively applying the previous argument to the subspace V_1 constructed in the proof above, we deduce the following:

COROLLARY 4.1. *Under the same hypotheses of Theorem 1.3, if $r \neq \ell$ then the dimension of V is at most $a + 1$ where a is such that $r^a \mathrm{N}(\rho_f) = n$.*

EXAMPLE. In [**DS18**], Dieulefait and Soto gave an example of a mod 5 eigenvalue system attached to a modular form f of level $\Gamma_0(1406)$ and weight 2 which satisfies the hypotheses of Theorem 1.3. There exist a form g at the same level and weight 2 whose q-expansion coincide with the one of f away from 19, we have indeed $1 = f(T_{19}) = -g(T_{19})$. It is easy to check that the mod 5 eigenvalue system arises from a modular form of level $\Gamma_0(74)$ and weight 2: there are two Galois orbits of eigenforms in level $\Gamma_0(74)$ and weight 2, and the eigenvalue system arises by reduction of a characteristic zero eigenvalue system in the orbit with Hecke eigenvalue field is $\mathbb{Q}(\sqrt{5})$.

Acknowledgments

I would like to thank Bas Edixhoven for the long discussions about the material in this article. I also am very grateful to Gabor Wiese for all the fruitful conversations about this topic. Finally, I would like to thank the referee for the helpful comments and suggestions on the manuscript.

References

[Con07] Brian Conrad, *Arithmetic moduli of generalized elliptic curves*, J. Inst. Math. Jussieu **6** (2007), no. 2, 209–278, DOI 10.1017/S1474748006000089. MR2311664

[DR73] P. Deligne and M. Rapoport, *Les schémas de modules de courbes elliptiques* (French), Modular functions of one variable, II (Proc. Internat. Summer School, Univ. Antwerp, Antwerp, 1972), Springer, Berlin, 1973, pp. 143–316. Lecture Notes in Math., Vol. 349. MR0337993

[DS74] Pierre Deligne and Jean-Pierre Serre, *Formes modulaires de poids* 1 (French), Ann. Sci. École Norm. Sup. (4) **7** (1974), 507–530 (1975). MR0379379

[DS05] Fred Diamond and Jerry Shurman, *A first course in modular forms*, Graduate Texts in Mathematics, vol. 228, Springer-Verlag, New York, 2005. MR2112196

[DS18] Luis Dieulefait and Eduardo Soto, *On congruences between normalized eigenforms with different sign at a Steinberg prime*, Rev. Mat. Iberoam. **34** (2018), no. 1, 413–421, DOI 10.4171/RMI/990. MR3763351

[EC11] Bas Edixhoven and Jean-Marc Couveignes (eds.), *Computational aspects of modular forms and Galois representations*, Annals of Mathematics Studies, vol. 176, Princeton University Press, Princeton, NJ, 2011. How one can compute in polynomial time the value of Ramanujan's tau at a prime. MR2849700

[Edi92] Bas Edixhoven, *The weight in Serre's conjectures on modular forms*, Invent. Math. **109** (1992), no. 3, 563–594, DOI 10.1007/BF01232041. MR1176206

[Edi06] Bas Edixhoven, *Comparison of integral structures on spaces of modular forms of weight two, and computation of spaces of forms mod 2 of weight one* (English, with English and French summaries), J. Inst. Math. Jussieu **5** (2006), no. 1, 1–34, DOI 10.1017/S1474748005000113. With appendix A (in French) by Jean-François Mestre and appendix B by Gabor Wiese. MR2195943

[Gro90] Benedict H. Gross, *A tameness criterion for Galois representations associated to modular forms (mod p)*, Duke Math. J. **61** (1990), no. 2, 445–517, DOI 10.1215/S0012-7094-90-06119-8. MR1074305

[Kat73] Nicholas M. Katz, *p-adic properties of modular schemes and modular forms*, Modular functions of one variable, III (Proc. Internat. Summer School, Univ. Antwerp, Antwerp, 1972), Springer, Berlin, 1973, pp. 69–190. Lecture Notes in Mathematics, Vol. 350. MR0447119

[Kat77] Nicholas M. Katz, *A result on modular forms in characteristic p*, Modular functions of one variable, V (Proc. Second Internat. Conf., Univ. Bonn, Bonn, 1976), Springer, Berlin, 1977, pp. 53–61. Lecture Notes in Math., Vol. 601. MR0463169

[KM85] Nicholas M. Katz and Barry Mazur, *Arithmetic moduli of elliptic curves*, Annals of Mathematics Studies, vol. 108, Princeton University Press, Princeton, NJ, 1985. MR772569

[Sil94] Joseph H. Silverman, *Advanced topics in the arithmetic of elliptic curves*, Graduate Texts in Mathematics, vol. 151, Springer-Verlag, New York, 1994. MR1312368

[Sil09] Joseph H. Silverman, *The arithmetic of elliptic curves*, 2nd ed., Graduate Texts in Mathematics, vol. 106, Springer, Dordrecht, 2009. MR2514094

[Ste07] William Stein, *Modular forms, a computational approach*, Graduate Studies in Mathematics, vol. 79, American Mathematical Society, Providence, RI, 2007. With an appendix by Paul E. Gunnells. MR2289048

Mathematics Research Unit, University of Luxembourg, Maison du nombre, 6, avenue de la Fonte, L-4364 Esch-sur-Alzette, Grand-Duchy of Luxembourg
Email address: samuele.anni@gmail.com

Contemporary Mathematics
Volume **732**, 2019
https://doi.org/10.1090/conm/732/14797

A discussion on the number eta-quotients of prime level

Allison Arnold-Roksandich

ABSTRACT. It is known that all modular forms on $SL_2(Z)$ can be expressed as a rational function in $\eta(z)$, $\eta(2z)$ and $\eta(4z)$. By using a theorem about eta-quotients being modular forms, and calculating the order of vanishing, we can compute the eta-quotients for a given level. Using this count, knowing how many eta-quotients are linearly independent and using the dimension formula, we can figure out a subspace spanned by the eta-quotients. In this paper, we primarily focus on the case where $N = p$ a prime.

Modular forms and cusp forms encode important arithmetic information, and are therefore important to study. An easy way to accomplish this is to study the Dedekind eta-function

$$\eta(z) := q^{1/24} \prod_{n\geq 1} (1 - q^n), \quad \text{where } q = e^{2\pi i z}. \tag{0.1}$$

In particular, we focus on functions of the form

$$f(z) = \prod_{d|N} \eta^{r_d}(dz), \quad r_d \in \mathbb{Z} \tag{0.2}$$

which we call eta-quotients, as they provide nice examples of modular forms.

The following theorem is the primary motivation behind the results mentioned in this paper.

THEOREM 1. [**Ono04**, Thm. 1.67] *Every modular form on* $SL_2(\mathbb{Z})$ *may be expressed as a rational function in* $\eta(z)$, $\eta(2z)$, *and* $\eta(4z)$.

While the recent work of Rouse and Webb ([**RW15**]) has shown that Theorem 1 does not generalize to all levels, the subspace of eta-quotients for fixed level at least 2 is still interesting. The goal of this paper is to discuss the vector spaces of modular forms with prime level, $M_k(\Gamma_1(p))$, and count the number of eta-quotients for fixed weight, k, and level p, and compare the span of these eta-quotients with $M_k(\Gamma_1(p))$. In other words, this discussion focuses on the explicit count of the eta-quotients that are modular forms for the congruence subgroups $\Gamma_0(p)$ and $\Gamma_1(p)$ where p is a prime.

To make this computation easier, the following two theorems are tools used to compute the eta-quotients that are in $\Gamma_0(p)$. Theorem 2 provides sufficient

---

2010 *Mathematics Subject Classification.* Primary 11F20, 11F11, 11F37.
Key words and phrases. Dedekind eta-function, eta-quotients, modular forms.

conditions for an eta-quotient being in $\Gamma_0(p)$ and Theorem 3 tells us the order of vanishing.

THEOREM 2. [**Ono04**, Thm. 1.64] *Define the an eta-quotient*

$$f(z) = \prod_{\delta|N} \eta^{r_\delta}(\delta z),$$

and set $k = \frac{1}{2}\sum_{\delta|N} r_\delta \in \mathbb{Z}$. *Suppose our exponents* $r_1, ..., r_N$ *satisfy*

$$\sum_{\delta|N} \delta r_\delta \equiv 0 \pmod{24}, \quad \textit{and} \quad \sum_{\delta|N} \frac{N}{\delta} r_\delta \equiv 0 \pmod{24}.$$

Then,

$$f|_k\gamma(z) = \chi(d)f(z)$$

for all $\gamma = \begin{pmatrix} a & b \\ c & d \end{pmatrix} \in \Gamma_0(N)$, *where* $\chi(n) = \left(\frac{(-1)^k s}{n}\right)$ *with* $s = \prod_{\delta|N} \delta^{r_\delta}$.

THEOREM 3. [**Ono04**, Thm. 1.65] *Let* $f(z)$ *be an eta-quotient satisfying the the conditions of Theorem 2. Let* $c, d \in \mathbb{N}$ *with* $d|N$ *and* $(c, d) = 1$. *Then, the order of vanishing of* $f(z)$ *at the cusp* $\frac{c}{d}$ *is*

$$v_d = \frac{N}{24}\sum_{\delta|N} \frac{(d,\delta)^2 r_\delta}{(d, \frac{N}{d})d\delta}.$$

We can then make conditions on k that become necessary and sufficient for showing that the space of weakly holomorphic weight k and level $\Gamma_0(p)$ contains eta-quotients.

THEOREM 4. [**ARJK18**, Thm. 1.2] *Let* $p > 3$ *be a prime and* k *be an integer. Then there exists* $f(z) = \eta^{r_1}(z)\eta^{r_p}(pz)$ *such that* $f(z)$ *is a weakly holomorphic modular form with weight* k *of level* $\Gamma_1(p)$ *if and only if* k *is divisible by* $h = \frac{1}{2}\gcd(p-1, 24)$.

With some effort, it is possible create similar conditions to guarantee when $f(z)$ is in $M_k(\Gamma_1(p))$; and the details for this can be found in [**ARJK18**]. The next theorem gives an explicit count of the number of eta-quotients that are cusp forms of weight k and level $\Gamma_1(p)$.

THEOREM 5. [**ARJK18**, Thm. 1.3] *Let* $p > 3$ *be a prime. Let* $k = hk'$ *where* h *is the needed divisor of* k *given by Theorem 4. Let* c *be the smallest positive integer representative of* $\frac{k'h}{12}$ *modulo* $\frac{p-1}{2h}$. *The number of eta-quotients in* $S_k(\Gamma_1(p))$ *is*

$$\left\lfloor \frac{k(p+1)}{12\frac{p-1}{2h}} \right\rfloor + \varepsilon,$$

where $\varepsilon \in \{-1, 0, 1\}$ *and depends on* p *and* k.

There are also eta-quotients in $M_k(\Gamma_1(p))$ that are not cusp forms that are given by the following theorem.

THEOREM 6. [**ARJK18**, Thm. 1.4] *Let* $p > 3$ *be a prime. Then,* $M_k(\Gamma_1(p)) \setminus S_k(\Gamma_1(p))$ *contains at least one eta-quotient if and only if* $\frac{p-1}{2}|k$. *Furthermore, for*

$k > 0$ and $\frac{p-1}{2} | k$, there are exactly two eta-quotients in $M_k(\Gamma_1(p)) \setminus S_k(\Gamma_1(p))$, which are of the form

$$\frac{\eta^{\frac{2pk}{p-1}}(pz)}{\eta^{\frac{2k}{p-1}}(z)}, \quad \text{and} \quad \frac{\eta^{\frac{2pk}{p-1}}(z)}{\eta^{\frac{2k}{p-1}}(pz)}.$$

Finally, the following theorem also tells us the size of the subspace spanned by eta-quotients.

THEOREM 7. [**ARJK18**, Thm. 1.5] *Let $p > 3$ be a prime. Then, the eta-quotients in $M_k(\Gamma_1(p))$ given by the previous theorems are linearly independent.*

Thus, we know the dimension of the subspace spanned by eta-quotients is the same as the count given in Theorems 5 and 6.

References

[ARJK18] Allison Arnold-Roksandich, Kevin James, and Rodney Keaton, *Counting eta-quotients of prime level*, Involve **11** (2018), no. 5, 827–844, DOI 10.2140/involve.2018.11.827. MR3784029

[Ono04] Ken Ono, *The web of modularity: arithmetic of the coefficients of modular forms and q-series*, CBMS Regional Conference Series in Mathematics, vol. 102, Published for the Conference Board of the Mathematical Sciences, Washington, DC; by the American Mathematical Society, Providence, RI, 2004. MR2020489

[RW15] Jeremy Rouse and John J. Webb, *On spaces of modular forms spanned by eta-quotients*, Adv. Math. **272** (2015), 200–224, DOI 10.1016/j.aim.2014.12.002. MR3303232

DEPARTMENT OF MATHEMATICS, OREGON STATE UNIVERSITY, CORVALLIS, OREGON 97331-4605

Email address: `arnoldra@math.oregonstate.edu`

Contemporary Mathematics
Volume **732**, 2019
https://doi.org/10.1090/conm/732/14796

Dedekind sums, reciprocity, and non-arithmetic groups

Claire Burrin

ABSTRACT. Dedekind sums, arithmetic correlation sums that arose in Dedekind's study of the modular transformation of the logarithm of the η-function [De1892], are surprisingly ubiquitous. Their arithmetic properties attracted the attention of number theorists, combinatorists, and theoretical computer scientists alike [RG1972,Me1957,Po1993,Kn1981,BR2004], and they appear more broadly in geometry, topology, and physics [At1987,KM1994,BG1992]. Accordingly, there is a similarly vast literature on variations and generalizations of Dedekind sums. It is the goal of this note to survey some of the aspects of Dedekind sums for (non-uniform) lattices in $\mathrm{SL}_2(\mathbb{R})$, otherwise referred to as Dedekind symbols. Intrinsically, this gives us a framework in which to investigate the rôle 'arithmeticity' plays in defining properties of Dedekind sums. In this note, we discuss the reciprocity law for Dedekind symbols associated to lattices that are not necessarily arithmetic. We will see that the reciprocity law holds given an algebraic structure similar to that of $\mathrm{SL}_2(\mathbb{Z})$, e.g. Hecke triangle groups.

1. Dedekind sums

Let $x \mapsto ((x))$ be the function that assigns to any real number x the value

$$((x)) = \begin{cases} 0 & x \in \mathbb{Z} \\ x - \lfloor x \rfloor - 1/2 & x \in \mathbb{R} \setminus \mathbb{Z}, \end{cases}$$

For any pair of coprime integers (a, c) with $c > 0$, the Dedekind sum is

$$s(a,c) = \sum_{k=1}^{c-1} \left(\left(\frac{k}{c}\right)\right)\left(\left(\frac{ak}{c}\right)\right). \tag{1.1}$$

An intuitive way of thinking of Dedekind sums is perhaps the following. Consider the sequence $(y_n) = \left(\frac{x_n}{c}\right)$ of primitive rational points given iteratively by the linear congruential rule

$$x_0 = 1$$
$$x_n \equiv a x_{n-1} \bmod c$$

defined by a pair of coprime integers (a, c). By Euler's theorem, (y_n) is a periodic sequence, whose period is completely determined by the modulus c. More precisely, the period is given by Euler's totient function $\varphi(c)$, and hence is maximal if the modulus c is a prime. Moreover, in this case, (y_n) runs over all primitive rational points with denominator c. Then the Dedekind sum $s(a, c)$ is the covariance of successive points in (y_n). In other words, a large Dedekind sum expresses high correlation between successive points, which suggests recurring patterns in the distribution of these points in the unit interval, while a small Dedekind sum indicates the possible asymptotic uniformly distribution of these points in the unit interval if we let $c \to \infty$. If such is the case, (y_n) is called pseudo-random, meaning that although it is the result of a deterministic process, the output values look random. In short, Dedekind sums measure the pseudorandomness of (y_n) depending on the input pair (a, c) of coprime integers.

2. ...and modular transformations

As a guiding introduction to the construction of Dedekind symbols, we will see where Dedekind sums appear in analytic number theory. From this viewpoint, where best to start than with the Riemann ζ-function! Recall that

$$\zeta(s) = \sum_{n \geq 1} \frac{1}{n^s}$$

is defined for $\mathfrak{Re}(s) > 1$, has a meromorphic continuation to the whole complex plane that is regular except for a simple pole at $s = 1$ with residue 1, and the constant term in the Laurent expansion of $\zeta(s)$ at 1 is the Euler–Mascheroni constant C, i.e.

$$\lim_{s \to 1} \left(\zeta(s) - \frac{1}{s-1} \right) = C.$$

More generally, any real positive-definite binary quadratic form $Q(x, y) = ax^2 + bxy + cy^2$, defines a ζ-function

$$\zeta_Q(s) = \sum_{m,n \in \mathbb{Z}} \frac{1}{Q(m,n)^s} = \sum_{\substack{m,n \in \mathbb{Z} \\ \text{not both } 0}} \frac{y^s}{|m + nz|^{2s}} \qquad \left(\begin{array}{ccc} z & = & x + iy \\ & = & \frac{1}{a}\left(b + \sqrt{-d}\right) \end{array} \right)$$

for $\mathfrak{Re}(s) > 1$. The first limit formula of Kronecker asserts that

$$\lim_{s \to 1} \left(\zeta_Q(s) - \frac{\pi}{s-1} \right) = 2\pi \left(C - \log 2 - \log \sqrt{y}\, |\eta(z)|^2 \right)$$

where $\eta(z)$ is Dedekind's η-function

$$\eta(z) = e^{\frac{\pi i z}{12}} \prod_{n \geq 1} (1 - e^{2\pi i n z}).$$

From Kronecker's first limit formula, one can show that for any $\gamma \in \mathrm{SL}_2(\mathbb{Z})$,

$$\eta(\gamma z) = \varepsilon (cz + d)^{1/2} \eta(z)$$

with $\varepsilon = \varepsilon(a, b, c, d)$ and $|\varepsilon| = 1$ (see for instance [**Si1961**]). Dedekind [**De1892**] determined ε explicitly, by showing that

$$\log \eta(\gamma z) = i\pi\phi(\gamma) + \frac{1}{2} \log \left(\frac{cz + d}{i \operatorname{sign}(c)} \right) + \log \eta(z) \tag{2.1}$$

where log denotes the principal branch of the logarithm, $\operatorname{sign}(c) = \frac{c}{|c|}$ and

$$\phi \begin{pmatrix} a & b \\ c & d \end{pmatrix} = \begin{cases} \dfrac{b}{d} & c = 0 \\[2ex] \dfrac{1}{12} \dfrac{a + d}{c} - \operatorname{sign}(c) s(a, |c|) & c \neq 0. \end{cases}$$

The modular transformation (2.1) of $\log \eta$ can be exploited to say things about Dedekind sums. A famous example is Dedekind's reciprocity law

$$s(a,c) + s(c,a) = \frac{1}{12}\left(\frac{a}{c} + \frac{1}{ac} + \frac{c}{a}\right) - \frac{1}{4} \tag{2.2}$$

for a, c positive and coprime.

3. Dedekind symbols

Let $\Gamma \subset \mathrm{SL}_2(\mathbb{R})$ be a lattice (i.e. a discrete subgroup of finite covolume). Then Γ acts on the (hyperbolic) upper half-plane $\mathbb{H} = \{x + iy : y > 0\}$ properly discontinuously by fractional linear transformation, $\left(\begin{smallmatrix} a & b \\ c & d \end{smallmatrix}\right).z = \frac{az+b}{cz+d}$, and this action extends to $\overline{\mathbb{H}} = \mathbb{H} \cup \partial\mathbb{H} = \mathbb{H} \cup \mathbb{R} \cup \{\infty\}$. For simplicity, we will first assume that Γ has only one cusp (i.e. the action of Γ has a single fixed point in $\partial\mathbb{H}$) and that this cusp is the point at ∞. Let $\Gamma_\infty \subset \Gamma$ be the stabilizer subgroup of the cusp. Up to normalization, we may assume that $\Gamma_\infty = \left(\begin{smallmatrix} 1 & \mathbb{Z} \\ & 1 \end{smallmatrix}\right)$. The Eisenstein series

$$E(z,s) = \sum_{\gamma \in \Gamma_\infty \backslash \Gamma} \mathfrak{Im}(\gamma z)^s = \sum_{\gamma \in \Gamma_\infty \backslash \Gamma} \frac{y^s}{|cz+d|^{2s}}$$

is defined for $\mathfrak{Re}(s) > 1$, and has a meromorphic continuation to the whole complex plane that is holomorphic in $\mathfrak{Re}(s) \geq 1$ except for a simple pole at $s = 1$ with residue V^{-1}, where V is the covolume of Γ. Goldstein [**Gol1973**] derived (formally) the Kronecker first limit formula

$$\lim_{s \to 1}\left(E(z,s) - \frac{V^{-1}}{s-1}\right) = K(z)$$

to obtain Dedekind sums for principal congruence subgroups. The existence of this limit can be deduced from bounds on the Fourier coefficients of the Eisenstein series, see [**JO2005**]. We present a softer approach, using only that the Eisenstein series are eigenfunctions of the Laplacian on $\mathbb{H}$, expressed in local coordinates $z = x + iy$ by $\Delta = -y^2(\partial_{xx} + \partial_{yy})$. In fact,

$$\Delta E(z,s) = s(1-s)E(z,s)$$

and note that Δ and the Γ-action commute, i.e. $\Delta f(\gamma z) = (\Delta f)(\gamma z)$, and $E(\gamma z, s) = E(z,s)$ for all $\gamma \in \Gamma$. We can observe that

(1) $K(\gamma z) = K(z)$ for all $\gamma \in \Gamma$,
(2) $K(z)$ is real-valued and real-analytic,
(3) $\Delta K(z) = -V^{-1}$.

The last two facts imply that

$$H(z) = VK(z) + \ln y + c,$$

for any fixed real constant c, is harmonic, i.e. $\Delta H(z) = 0$, and that it can be realized as the real part of a holomorphic function $F : \mathbb{H} \to \mathbb{C}$, i.e. $\mathfrak{Re}(F(z)) = H(z)$. Hence

$$\lim_{s \to 1}\left(E(z,s) - \frac{V^{-1}}{s-1}\right) = V^{-1}\left(c - \ln y + \mathfrak{Re}F(z)\right). \tag{3.1}$$

(The constant c can be made explicit by looking at the Fourier coefficients of the Eisenstein series.)

To obtain Dedekind symbols, we study the automorphic transformation of the holomorphic function F. The Kronecker limit formula (3.1) and Fact (1) imply

$$\mathfrak{Re}\,(F(z) - F(\gamma z)) = \ln |cz+d|^2 = \mathfrak{Re}\left(\log(-(cz+d)^2)\right),$$

where we fix log to be the principal branch of the logarithm. As a result, for any fixed group element $\gamma \in \Gamma$, the function

$$\phi_\gamma(z) = F(z) - F(\gamma z) - \log(-(cz+d)^2),$$

is holomorphic and takes values on the imaginary axis. Therefore, by the Open Mapping Theorem, it is constant, i.e. $\phi(\gamma) \equiv \phi_\gamma(z)$. Since what we are really interested in is the variation in the argument, we replace ϕ by the normalization $\frac{\phi}{2\pi i}$, so that

$$\phi(\gamma) = \frac{1}{2\pi i}\left(F(z) - F(\gamma z) - \log(-(cz+d)^2)\right). \tag{3.2}$$

By analogy with Dedekind's formula (2.1) for the transformation of $\log \eta$, we define for any $\left(\begin{smallmatrix} a & b \\ c & d \end{smallmatrix}\right) \in \Gamma$ such that $c \neq 0$,

$$\mathcal{S}(a,b,c,d) = \frac{V}{4\pi}\frac{a+d}{c} - \phi\begin{pmatrix} a & b \\ c & d \end{pmatrix}.$$

The right-hand side (RHS) of this equation factors through $\Gamma_\infty \backslash \Gamma / \Gamma_\infty$ [**Bu2017**, Theorem 2]. Observe that for $\gamma = \left(\begin{smallmatrix} a & b \\ c & d \end{smallmatrix}\right)$ and $\gamma' = \left(\begin{smallmatrix} a & b' \\ c & d' \end{smallmatrix}\right)$ in Γ,

$$\gamma^{-1}\gamma' = \begin{pmatrix} d & -b \\ -c & a \end{pmatrix}\begin{pmatrix} a & b' \\ c & d' \end{pmatrix} = \begin{pmatrix} 1 & * \\ 0 & 1 \end{pmatrix} \in \Gamma_\infty$$

meaning that any double coset representative $[\![\left(\begin{smallmatrix} a & b \\ c & d \end{smallmatrix}\right)]\!] = \Gamma_\infty \left(\begin{smallmatrix} a & b \\ c & d \end{smallmatrix}\right) \Gamma_\infty$ is completely determined by the column vector $(a,c)^T$. The same argument can be used to show that the double coset representative $[\![\left(\begin{smallmatrix} a & b \\ c & d \end{smallmatrix}\right)]\!]$ can as well be completely determined by the row vector (c,d). Hence, we have the Dedekind symbol

$$\mathcal{S}_c(a,c) = \mathcal{S}_r(c,d) = \mathcal{S}\left[\!\!\left[\begin{pmatrix} a & b \\ c & d \end{pmatrix}\right]\!\!\right] = \frac{V}{4\pi}\frac{a+d}{c} - \phi\begin{pmatrix} a & b \\ c & d \end{pmatrix}$$

for Γ. More generally, if $\Gamma \subset \mathrm{SL}_2(\mathbb{R})$ has cusps, then each cusp $\mathfrak{a}$ gives rise to a Dedekind symbol

$$\mathcal{S}_{\mathfrak{a}}([\![\gamma]\!]) = \frac{V}{4\pi}\frac{a_\gamma + d_\gamma}{c_\gamma} - \phi\begin{pmatrix} a_\gamma & b_\gamma \\ c_\gamma & d_\gamma \end{pmatrix} \tag{3.3}$$

for each non-trivial double coset $[\![\gamma]\!]$ in $\Gamma_{\mathfrak{a}} \backslash \Gamma / \Gamma_{\mathfrak{a}}$ and

$$\begin{pmatrix} a_\gamma & b_\gamma \\ c_\gamma & d_\gamma \end{pmatrix} = \sigma_{\mathfrak{a}}^{-1}\gamma\sigma_{\mathfrak{a}},$$

where $\sigma_{\mathfrak{a}} \in \mathrm{SL}_2(\mathbb{R})$ is a scaling, i.e. $\sigma_{\mathfrak{a}}(\infty) = \mathfrak{a}$ and $\sigma_{\mathfrak{a}}^{-1}\Gamma_{\mathfrak{a}}\sigma_{\mathfrak{a}} = \left(\begin{smallmatrix} 1 & \mathbb{Z} \\ & 1 \end{smallmatrix}\right)$. The definition (3.3) does not depend on the particular choice of $\sigma_{\mathfrak{a}}$.

4. Equidistribution mod 1

The perhaps *a priori* artificial definition (3.3) of Dedekind symbols above is justified by the following theorem (which is the effective version of [**Bu2017**, Theorem 4] and can be compared to [**Va1987**, **Va1993**].)

THEOREM 4.1. *Let $\Gamma \subset \mathrm{SL}_2(\mathbb{R})$ be a non-uniform lattice, with a cusp at $\mathfrak{a}$ and associated Dedekind symbol $\mathcal{S}_{\mathfrak{a}}$. The values of*

$$\mathcal{S}_{\mathfrak{a}}([\![\gamma]\!]) = \mathcal{S}_c(a_\gamma, c_\gamma),$$

running over ordered admissible pairs (a_γ, c_γ) as in (3.3), become equidistributed mod 1, as $c_\gamma \to \infty$. More precisely, let μ_X be the normalized counting measure defined on the subset

$$\mathcal{D}_X = \{\mathcal{S}_c(a_\gamma, c_\gamma) \text{ mod } 1 : 0 \le a_\gamma < c_\gamma \le X\}$$

of $\mathbb{T} = \mathbb{R}/\mathbb{Z}$, and let μ denote the Lebesgue measure on $\mathbb{T}$. Then for any $f \in C^\infty(\mathbb{T})$,

$$\mu_X(f) = \mu(f) + O\left(X^{-\frac{2}{3}+\varepsilon}\right)$$

as $X \to \infty$. The implied constant depends on a Sobolev norm of f and Γ.

PROOF. Any $f \in C^\infty(\mathbb{T})$ admits a Fourier series

$$f(x) = \sum_{n\in\mathbb{Z}} a_n e(nx) := \sum_{n\in\mathbb{Z}} a_n e^{2\pi i n x}$$

that converges uniformly. Then

$$\begin{aligned}
\mu_X(f) &= \frac{1}{|\mathcal{D}_X|} \sum_{x\in\mathcal{D}_X} f(x) \\
&= \frac{1}{|\mathcal{D}_X|} \sum_{n\in\mathbb{Z}} a_n \sum_{\substack{a_\gamma < c_\gamma \\ c_\gamma \le X}} e(n\mathcal{S}_c(a_\gamma, c_\gamma)) \\
&= \frac{1}{|\mathcal{D}_X|} \sum_{n\in\mathbb{Z}} a_n \sum_{\substack{a_\gamma < c_\gamma \\ c_\gamma \le X}} e\left(n\frac{V}{4\pi}\frac{a+d}{c}\right) e(-n\phi(\gamma)) \\
&= \frac{1}{|\mathcal{D}_X|} \sum_{n\in\mathbb{Z}} a_n \sum_{c_\gamma \le X} S(n, n, c_\gamma, \phi)
\end{aligned}$$

is expressed in function of sums of ϕ-twisted Selberg–Kloosterman sums [**Se1965**]. The constant term in this expansion is

$$\frac{a_0}{|\mathcal{D}_X|} \sum_{c_\gamma \le X} S(0, 0, c_\gamma, \phi) = a_0 = \int_{\mathbb{T}} f(x)dx = \mu(f).$$

Subconvex bounds on the sums of ϕ-twisted Selberg–Kloosterman sums sums not only yield $\mu_X(f) \to \mu(f)$ as $X \to \infty$, they also control the rate of equidistribution, as we will now see. Thanks to the spectral theory of automorphic forms, subconvex bounds for (general) sums of Kloostermans sums are available, and in the specific case of the ϕ-twisted Selberg–Kloosterman sums considered here, [**GS1983**, Theorem 2] together with [**Ro1966**, Satz 5.5], yield, for every pair $m, n \in \mathbb{Z}$ such that $mn \neq 0$,

$$\sum_{c_\gamma \le X} S(m, n, c_\gamma, \phi) \ll |mn|\, X^{\frac{4}{3}+\varepsilon},$$

$\varepsilon > 0$, with an implied constant depending on Γ.

In [**Bu2017**], we showed that

$$|\mathcal{D}_X| \sim \frac{V}{4\pi} X^2$$

as $X \to \infty$. Combining these results yields

$$|\mu_X(f) - \mu(f)| \ll \sum_{n\neq 0} |a_n|\,|n|^2\, X^{-\frac{2}{3}+\varepsilon} \ll \left(\sum_{n\neq 0} |a_n|^2\,|n|^8\right)^{\frac{1}{2}} X^{-\frac{2}{3}+\varepsilon} \le S(f) X^{-\frac{2}{3}+\varepsilon},$$

where we used Cauchy–Schwarz in the second step, $S(f)$ is the adequate L^2-Sobolev norm, and the implied constant in the first step depends on Γ. $\square$

5. Reciprocity of Dedekind symbols

We will now see that there is an analogue of the reciprocity law (2.2) for Dedekind symbols. We extract from the construction of Dedekind symbols the formula

$$F(z) - F(\gamma z) = \log(-(cz+d)^2) + 2\pi i \phi(\gamma) = 2\log(cz+d) + 2\pi i \psi(\gamma)$$

with

$$\psi(\gamma) = \phi(\gamma) - \frac{1}{4}\mathrm{sign}(c),$$

for which one can check directly that

$$\psi(\gamma\tau) - \psi(\gamma) - \psi(\tau) = \frac{1}{2\pi i}\left(\log j(\gamma, \tau z) + \log j(\tau, z) - \log j(\gamma\tau, z)\right), \tag{5.1}$$

where $j(g,z) = cz+d$, $g = \left(\begin{smallmatrix} * & * \\ c & d \end{smallmatrix}\right)$. The RHS is independent of z and extends naturally to a function on $\mathrm{SL}_2(\mathbb{R}) \times \mathrm{SL}_2(\mathbb{R})$. Moreover, it can be computed on explicit elements of $\mathrm{SL}_2(\mathbb{R})$. In fact, in [**Bu2018**, Theorem 1.2], we show that, for any $g = \left(\begin{smallmatrix} a_g & b_g \\ c_g & d_g \end{smallmatrix}\right)$, $h = \left(\begin{smallmatrix} a_h & b_h \\ c_h & d_h \end{smallmatrix}\right)$ in $\mathrm{SL}_2(\mathbb{R})$, and any $z \in \mathbb{H}$,

$$\begin{aligned}\omega(g,h) &:= \frac{1}{2\pi i}\left(\log j(g, hz) + \log j(h,z) - \log j(gh, z)\right)\\ &= \frac{1}{4}\{\mathrm{sign}(c_g(-d_g)) + \mathrm{sign}(c_h(-d_h)) - \mathrm{sign}(c_{gh}(-d_{gh}))\\ &\qquad\qquad -\mathrm{sign}(c_g(-d_g)c_h(-d_h)c_{gh}(-d_{gh}))\}\end{aligned}$$

where

$$c(-d) = \begin{cases} c & c \neq 0,\\ -d & c = 0.\end{cases}$$

Let $\gamma, \tau \in \Gamma$ such that $[\![\gamma]\!]$, $[\![\tau]\!]$, $[\![\gamma\tau]\!]$ are non-trivial double cosets in $\Gamma_{\mathfrak{a}}\backslash\Gamma/\Gamma_{\mathfrak{a}}$. As before, write $\left(\begin{smallmatrix} a_\gamma & b_\gamma \\ c_\gamma & d_\gamma \end{smallmatrix}\right) = \sigma_{\mathfrak{a}}^{-1}\gamma\sigma_{\mathfrak{a}}$. Then

$$\begin{aligned}\mathcal{S}_{\mathfrak{a}}([\![\gamma]\!]) + \mathcal{S}_{\mathfrak{a}}([\![\tau]\!]) - \mathcal{S}_{\mathfrak{a}}([\![\gamma\tau]\!]) &= \frac{V}{4\pi}\left(\frac{c_\gamma}{c_{\gamma\tau}c_\tau} + \frac{c_\tau}{c_\gamma c_{\gamma\tau}} + \frac{c_{\gamma\tau}}{c_\tau c_\gamma}\right)\\ &\qquad - \phi\begin{pmatrix} a_\gamma & b_\gamma \\ c_\gamma & d_\gamma\end{pmatrix} - \phi\begin{pmatrix} a_\tau & b_\tau \\ c_\tau & d_\tau\end{pmatrix} + \phi\begin{pmatrix} a_{\gamma\tau} & b_{\gamma\tau} \\ c_{\gamma\tau} & d_{\gamma\tau}\end{pmatrix}\\ &= \frac{V}{4\pi}\left(\frac{c_\gamma}{c_{\gamma\tau}c_\tau} + \frac{c_\tau}{c_\gamma c_{\gamma\tau}} + \frac{c_{\gamma\tau}}{c_\tau c_\gamma}\right)\\ &\qquad + \omega(\gamma,\tau) + \frac{1}{4}\left(\mathrm{sign}(c_{\gamma\tau}) - \mathrm{sign}(c_\gamma) - \mathrm{sign}(c_\tau)\right)\\ &= \frac{V}{4\pi}\left(\frac{c_\gamma}{c_{\gamma\tau}c_\tau} + \frac{c_\tau}{c_\gamma c_{\gamma\tau}} + \frac{c_{\gamma\tau}}{c_\tau c_\gamma}\right) - \frac{1}{4}\mathrm{sign}(c_\gamma c_\tau c_{\gamma\tau}),\end{aligned}$$

where the last step follows from [**Bu2018**, Theorem 1.2] as cited above.

THEOREM 5.1. *Let $\Gamma \subset \mathrm{SL}_2(\mathbb{R})$ be a non-uniform lattice with a cusp at ∞ and such that $\iota = \left(\begin{smallmatrix} & -1 \\ 1 & \end{smallmatrix}\right) \in \Gamma$. Then the reciprocity law*

$$\begin{aligned}\mathcal{S}_\infty(\llbracket\gamma\rrbracket) - \mathcal{S}_\infty(\llbracket\gamma\iota\rrbracket) &= \mathcal{S}_r(c_\gamma, d_\gamma) - \mathcal{S}_r(d_\gamma, -c_\gamma) \\ &= \frac{V}{4\pi}\left(\frac{c_\gamma}{d_\gamma} + \frac{1}{c_\gamma d_\gamma} + \frac{d_\gamma}{c_\gamma}\right) - \frac{1}{4}\mathrm{sign}\,(c_\gamma d_\gamma)\end{aligned}$$

holds for any non-trivial double coset $\llbracket\gamma\rrbracket$ in $\Gamma_\infty\backslash\Gamma/\Gamma_\infty$ such that $\llbracket\gamma\iota\rrbracket$ is also non-trivial (i.e. $c_\gamma d_\gamma \neq 0$).

PROOF. If we choose $\tau = \iota$ in the general formula above, then

$$\mathcal{S}_\infty(\llbracket\gamma\rrbracket) + \mathcal{S}_\infty(\llbracket\iota\rrbracket) - \mathcal{S}_\infty(\llbracket\gamma\iota\rrbracket) = \frac{V}{4\pi}\left(\frac{c_\gamma}{d_\gamma} + \frac{1}{c_\gamma d_\gamma} + \frac{d_\gamma}{c_\gamma}\right) - \frac{1}{4}\mathrm{sign}\,(c_\gamma d_\gamma)\,.$$

By definition, we have

$$\mathcal{S}_\infty(\llbracket\iota\rrbracket) = \mathcal{S}_r(1,0) = -\phi(\iota) = -\psi(\iota) - \frac{1}{4}$$

and, applying (5.1) repeatedly,

$$\begin{aligned}-2\psi(\iota) = \omega(\iota,\iota) - \psi(-I) &= \omega(\iota,\iota) + \frac{1}{2}\omega(-I,-I) - \frac{1}{2}\psi(I) \\ &= \omega(\iota,\iota) + \frac{1}{2}\omega(-I,-I) + \frac{1}{2}\omega(I,I).\end{aligned}$$

We can then read from [**Bu2018**, Theorem 2.1] that $\psi(\iota) = -\frac{1}{4}$. □

The reciprocity law, applied together with the periodicity relation $\mathcal{S}_r(c,d) = \mathcal{S}_r(c,d')$ for all $d \equiv d' \bmod c$ terminating when $\mathcal{S}_r(1,0) = 0$, goes to show that such Dedekind symbols are rational in the matrix coefficients of Γ. In the case of a Hecke triangle group H_q, which is generated by ι and $\tau_q = \left(\begin{smallmatrix} 1 & \lambda_q \\ & 1 \end{smallmatrix}\right)$, $\lambda_q = 2\cos\frac{\pi}{q}$, the reciprocity law implies that up to a constant scaling by λ_q, the Dedekind symbol is rational. Alternatively, a simple cohomological argument leads to the same conclusion. Observe that the RHS of (5.1) defines an integral 2-cocycle on Γ. The rational class obtained under the natural map $H^2(\Gamma,\mathbb{Z}) \to H^2(\Gamma,\mathbb{Q})$ is necessarily trivial, since $H^2(\Gamma,\mathbb{Q}) = 0$. This means that there exists a function $f : \Gamma \to \mathbb{Q}$ such that $f(\gamma\tau) - f(\gamma) - f(\tau) = \psi(\gamma) - \psi(\gamma) - \psi(\tau)$. In the case $\Gamma = H_q$, the fact that $H_q = C_2 * C_q$ implies that $H^1(H_q,\mathbb{Q}) = 0$, and hence that the map f is unique, i.e. $f = \psi$.

References

[At1987] M. Atiyah, *The logarithm of the Dedekind η-function*, Math. Ann. **278** (1987), no. 1-4, 335–380, DOI 10.1007/BF01458075. MR909232

[BG1992] J. Barge and É. Ghys, *Cocycles d'Euler et de Maslov* (French), Math. Ann. **294** (1992), no. 2, 235–265, DOI 10.1007/BF01934324. MR1183404

[BR2004] M. Beck and S. Robins, *Dedekind sums: a combinatorial-geometric viewpoint*, Unusual applications of number theory, DIMACS Ser. Discrete Math. Theoret. Comput. Sci., vol. 64, Amer. Math. Soc., Providence, RI, 2004, pp. 25–35. MR2063199

[Bu2017] C. Burrin, *Generalized Dedekind sums and equidistribution* mod 1, J. Number Theory **172** (2017), 270–286, DOI 10.1016/j.jnt.2016.08.003. MR3573154

[Bu2018] C. Burrin, *Reciprocity of Dedekind sums and the Euler class*, Proc. Amer. Math. Soc. **146** (2018), no. 4, 1367–1376, DOI 10.1090/proc/13880. MR3754325

[De1892] R. Dedekind, *Erläuterungen zu zwei Fragmenten von Riemann.* Riemann's Gesammelte Math. Werke (1892), 466–472.

[GS1983] D. Goldfeld and P. Sarnak, *Sums of Kloosterman sums*, Invent. Math. **71** (1983), no. 2, 243–250, DOI 10.1007/BF01389098. MR689644

[Gol1973] L. J. Goldstein, *Dedekind sums for a Fuchsian group. I*, Nagoya Math. J. **50** (1973), 21–47. MR0318065

[Gol1974] L. J. Goldstein, *Errata for "Dedekind sums for a Fuchsian group. I" (Nagoya Math. J.* **50** *(1973), 21–47)*, Nagoya Math. J. **53** (1974), 235–237. MR0347739

[JO2005] J. Jorgenson and C. O'Sullivan, *Convolution Dirichlet series and a Kronecker limit formula for second-order Eisenstein series*, Nagoya Math. J. **179** (2005), 47–102, DOI 10.1017/S0027763000025605. MR2164401

[KM1994] R. Kirby and P. Melvin, *Dedekind sums, μ-invariants and the signature cocycle*, Math. Ann. **299** (1994), no. 2, 231–267, DOI 10.1007/BF01459782. MR1275766

[Kn1981] D. E. Knuth, *The art of computer programming. Vol. 2*, 2nd ed., Addison-Wesley Publishing Co., Reading, Mass., 1981. Seminumerical algorithms; Addison-Wesley Series in Computer Science and Information Processing. MR633878

[Me1957] C. Meyer, *Über einige Anwendungen Dedekindscher Summen* (German), J. Reine Angew. Math. **198** (1957), 143–203, DOI 10.1515/crll.1957.198.143. MR0104643

[Po1993] J. E. Pommersheim, *Toric varieties, lattice points and Dedekind sums*, Math. Ann. **295** (1993), no. 1, 1–24, DOI 10.1007/BF01444874. MR1198839

[RG1972] H. Rademacher and E. Grosswald, *Dedekind sums*, The Mathematical Association of America, Washington, D.C., 1972. The Carus Mathematical Monographs, No. 16. MR0357299

[Ro1966] W. Roelcke, *Das Eigenwertproblem der automorphen Formen in der hyperbolischen Ebene, I* (German), Math. Ann. **167** (1966), no. 4, 292–337, DOI 10.1007/BF01364540. MR1513277

[Se1965] A. Selberg, *On the estimation of Fourier coefficients of modular forms*, Proc. Sympos. Pure Math., Vol. VIII, Amer. Math. Soc., Providence, R.I., 1965, pp. 1–15. MR0182610

[Si1961] C. L. Siegel, *Lectures on advanced analytic number theory*, Notes by S. Raghavan. Tata Institute of Fundamental Research Lectures on Mathematics, No. 23, Tata Institute of Fundamental Research, Bombay, 1965. MR0262150

[Va1987] I. Vardi, *A relation between Dedekind sums and Kloosterman sums*, Duke Math. J. **55** (1987), no. 1, 189–197, DOI 10.1215/S0012-7094-87-05510-4. MR883669

[Va1993] I. Vardi, *Dedekind sums have a limiting distribution*, Internat. Math. Res. Notices **1** (1993), 1–12, DOI 10.1155/S1073792893000017. MR1201746

Department of Mathematics, Rutgers University, 110 Frelinghuysen Road, Piscataway, New Jersey 08854

Email address: `claire.burrin@rutgers.edu`

Contemporary Mathematics
Volume **732**, 2019
https://doi.org/10.1090/conm/732/14784

Noncommutative modular symbols and Eisenstein series

Gautam Chinta, Ivan Horozov, and Cormac O'Sullivan

Abstract. We form real-analytic Eisenstein series twisted by Manin's noncommutative modular symbols. After developing their basic properties, these series are shown to have meromorphic continuations to the entire complex plane and satisfy functional equations in some cases. This theory neatly contains and generalizes earlier work in the literature on the properties of Eisenstein series twisted by classical modular symbols.

1. Introduction

1.1. Background. Let Γ be a congruence subgroup acting on the complex upper half plane $\mathbb{H}$ together with its cusps: $\overline{\mathbb{H}} = \mathbb{H} \cup \mathbb{P}^1(\mathbb{Q})$. The quotient is the modular curve $X = \Gamma\backslash\overline{\mathbb{H}}$. Manin [**Man72**] introduced *modular symbols*, which are elements of $H_1(X, \mathbb{Q})$. There is a natural pairing between modular symbols and holomorphic weight two cusp forms on Γ, obtained by integration. This pairing, combined with the action of Hecke operators, yields rich arithmetic information about periods of elliptic curves, special values of L-functions and Fourier coefficients of modular forms.

In [**Gol99**] Goldfeld began a program of studying the distribution of modular symbols. For $\gamma \in \Gamma$ we use the notation

$$\langle \gamma, f \rangle := 2\pi i \int_{z_0}^{\gamma z_0} f(u)\, du \tag{1.1}$$

for the pairing between the modular symbol $\{z_0, \gamma z_0\}$ and the weight two cusp form f for Γ. It may be seen that (1.1) is independent of $z_0 \in \overline{\mathbb{H}}$, and since f is a cusp form, $\langle \gamma, f \rangle$ depends only on the bottom row of the matrix γ. Following earlier work of O'Sullivan [**O'S00**] and Petridis [**Pet02**], it is shown in Petridis-Risager [**PR04**, Thms. A,G] that as $T \to \infty$ the values in

$$\left\{ \frac{\operatorname{Re}\langle \left(\begin{smallmatrix} * & * \\ c & d \end{smallmatrix}\right), f \rangle}{\sqrt{\log(c^2+d^2)}} \; : \; c^2 + d^2 \leqslant T \right\}$$

follow a normal distribution, and that

$$\sum_{c^2+d^2 \leqslant T} |\langle \left(\begin{smallmatrix} * & * \\ c & d \end{smallmatrix}\right), f \rangle|^2 = \frac{16\pi^2}{\operatorname{vol}(\Gamma\backslash\mathbb{H})} \|f\|^2 T \log T + O(T) \quad \text{as} \quad T \to \infty.$$

2010 *Mathematics Subject Classification.* Primary 11F11, 11F67, 30F35.
Key words and phrases. Modular symbols, iterated integrals, Eisenstein series.

Very recently, Petridis and Risager in [**PR18**] have also verified, exactly and on average, some of the conjectures of Mazur, Rubin and Stein on the statistics of modular symbols.

These results depend on properties of *Eisenstein series twisted by modular symbols*, that is, functions of the form

$$E(z,s;f) = \sum_{\gamma \in \Gamma_\infty \backslash \Gamma} \langle \gamma, f \rangle \operatorname{Im}(\gamma z)^s \tag{1.2}$$

and their generalizations, as in [**Gol99**]. Here z is in the upper half plane and s is a complex number, initially with real part bigger than 1 to ensure convergence. The function $E(z,s;f)$ is not automorphic in z, but it does satisfy the more complicated relation

$$E(\gamma_1\gamma_2 z, s; f) - E(\gamma_1 z, s; f) - E(\gamma_2 z, s; f) + E(z, s; f) = 0 \tag{1.3}$$

for all $\gamma_1, \gamma_2 \in \Gamma$.

Define $(f|_k\gamma)(z)$ as $j(\gamma,z)^{-k} f(\gamma z)$ for $j((\begin{smallmatrix} a & b \\ c & d \end{smallmatrix}), z) := cz + d$ and extend this action to the group ring $\mathbb{C}[\Gamma]$ by linearity. The relation (1.3) led Chinta-Diamantis-O'Sullivan [**CDO02**] to define a *second-order modular form* on Γ of weight k as a holomorphic function on the upper half plane which satisfies

$$f|_k(\gamma_1 - I)(\gamma_2 - I) = 0, \quad \text{for all } \gamma_1, \gamma_2 \in \Gamma. \tag{1.4}$$

An *nth-order modular form* of weight k on Γ satisfies

$$f|_k(\gamma_1 - I)(\gamma_2 - I)\cdots(\gamma_n - I) = 0, \quad \text{for all } \gamma_i \in \Gamma. \tag{1.5}$$

Higher-order modular forms were also independently defined by Kleban and Zagier [**KZ03**]. Their motivation comes from the study of modular properties in crossing probabilities on 2 dimensional lattices. One can similarly define higher-order (real-analytic) Eisenstein series generalizing the second-order series in (1.2) as in [**PR04**] and [**JO08**]. Very general series of this form are studied by Diamantis and Sim [**DS08**].

Manin [**Man06**] has initiated the development of a theory of *noncommutative modular symbols*. These arise as iterated integrals of cusp forms and Eisenstein series. The principal motivation for their study comes from *multiple zeta values*. See, for example, the paper of Choie and Ihara [**CI13**] which gives explicit formulas for iterated integrals in terms of multiple Hecke L-functions. Generalizing in another direction, Horozov [**Hor15**] defines *noncommutative Hilbert modular symbols*.

In the present paper we construct Eisenstein series twisted by Manin's noncommutative modular symbols. The result is a generating series whose coefficients contain the higher-order Eisenstein series studied in [**PR04, JO08, DS08**] as well as further new series not studied before. We find that the automorphic properties and functional equations of the earlier higher-order Eisenstein series are elegantly and succinctly encapsulated in our new formalism.

1.2. Main results. We introduce some basic notation in order to state our main results; see e.g. Iwaniec [**Iwa02**] for a fuller background. The group $\mathrm{SL}_2(\mathbb{R})$ acts by linear fractional transformations on $\mathbb{H} \cup \mathbb{P}^1(\mathbb{R})$. Let $\Gamma \subset \mathrm{SL}_2(\mathbb{R})$ be a Fuchsian group of the first kind, i.e. a discrete subgroup of $\mathrm{SL}_2(\mathbb{R})$ with quotient $\Gamma\backslash\mathbb{H}$ of finite hyperbolic volume. We may fix a finite set of inequivalent cusps for Γ, which we assume to be nonempty. Let $\Gamma_\mathfrak{a}$ be the subgroup of Γ fixing the cusp $\mathfrak{a}$. Then $\overline{\Gamma}_\mathfrak{a}$ is isomorphic to $\mathbb{Z}$, where the bar means the image under the map

$\mathrm{SL}_2(\mathbb{R}) \to \mathrm{SL}_2(\mathbb{R})/\pm I$. (We don't assume $-I \in \Gamma$.) This isomorphism can be seen explicitly as there exists a scaling matrix $\sigma_{\mathfrak{a}} \in \mathrm{SL}_2(\mathbb{R})$ such that $\sigma_{\mathfrak{a}}\infty = \mathfrak{a}$ and

$$\sigma_{\mathfrak{a}}^{-1}\overline{\Gamma}_{\mathfrak{a}}\sigma_{\mathfrak{a}} = \{\pm\left(\begin{smallmatrix}1 & \ell \\ 0 & 1\end{smallmatrix}\right) \mid \ell \in \mathbb{Z}\}.$$

A function f on $\mathbb{H}$ has weight k for Γ if f satisfies $(f|_k\gamma)(z) = f(z)$ for all $\gamma \in \Gamma$. If f is also holomorphic on $\mathbb{H}$ and has a Fourier expansion at each cusp $\mathfrak{a}$ of the form

$$(f|_k\sigma_{\mathfrak{a}})(z) = \sum_{m=0}^{\infty} c_{\mathfrak{a}}(m;f)e^{2\pi imz} \tag{1.6}$$

then f is a modular form for Γ. If $c_{\mathfrak{a}}(0;f) = 0$ for every cusp $\mathfrak{a}$ then f is called a cusp form.

For a and b in $\mathbb{H} \cup \{\texttt{cusps}\}$, set

$$C_a^b(f_1, f_2, \ldots, f_n) := \int_a^b f_n(z_n) \int_a^{z_n} f_{n-1}(z_{n-1}) \cdots \int_a^{z_2} f_1(z_1)\, dz_1 \cdots dz_n \tag{1.7}$$

where $f_1, f_2, \ldots, f_n$ are weight 2 holomorphic cusp forms for Γ. Near end-points that are cusps, we assume the path of integration follows a geodesic to ensure we remain in the region where the cusp forms have exponential decay. In this way (1.7) is always absolutely convergent.

Let $\mathbf{f} = (f_1, f_2, \ldots, f_r)$ be an r-tuple of not necessarily distinct, weight 2 holomorphic cusp forms for Γ. To each f_i we associate the variable X_i and define

$$I_a^b = I_a^b(\mathbf{f}) := 1 + \sum_i C_a^b(f_i)\cdot X_i + \sum_{i,j} C_a^b(f_i, f_j)\cdot X_iX_j + \sum_{i,j,k} C_a^b(f_i, f_j, f_k)\cdot X_iX_jX_k + \cdots. \tag{1.8}$$

Thus I_a^b is an element of the ring $\mathbb{C}\langle\langle X_1, \ldots, X_r\rangle\rangle$ of formal power series in the non-commuting variables X_i with coefficients in $\mathbb{C}$. We will also use the notation $I_a^b(f_1, f_2, \ldots, f_r)$ for (1.8). In the case when $r = 0$ it is convenient to set

$$I_a^b() := 1, \qquad C_a^b() := 1. \tag{1.9}$$

As we will see, these symbols satisfy the two key relations of *concatenation* and Γ-*invariance*:

$$I_a^b I_b^c = I_a^c, \tag{1.10}$$

$$I_{\gamma a}^{\gamma b} = I_a^b \tag{1.11}$$

for all $a, b, c \in \mathbb{H} \cup \{\texttt{cusps}\}$ and all $\gamma \in \Gamma$. Manin's *noncommutative modular symbol* is $I_{\gamma a}^a$ for $\gamma \in \Gamma$ and base point $a \in \mathbb{H} \cup \{\texttt{cusps}\}$. As described in [**Man06**, Prop. 2.5.1], $I_{\gamma a}^a$ is a representative of a nonabelian cohomology class in $H^1(\Gamma, \Pi)$ where Π is the group of invertible elements of $\mathbb{C}\langle\langle X_1, \ldots, X_n\rangle\rangle$ with constant term 1.

For each cusp $\mathfrak{b}$ we define the *Eisenstein series with noncommutative twists* as

$$\mathcal{E}_{\mathfrak{b}}(z,s) := \sum_{\gamma \in \Gamma_{\mathfrak{b}}\backslash\Gamma} I_{\gamma a}^a \cdot \mathrm{Im}(\sigma_{\mathfrak{b}}^{-1}\gamma z)^s. \tag{1.12}$$

Our first result proves the basic properties of this series. Recall that the hyperbolic Laplacian Δ is given by $y^2(d^2/dx^2 + d^2/dy^2)$ for $z = x + iy$.

THEOREM 1.1. *The series $\mathcal{E}_{\mathfrak{b}}(z,s)$ is well-defined and absolutely convergent for* $\mathrm{Re}(s) > 1$. *It is an eigenfunction of the Laplacian:*

$$\Delta\mathcal{E}_{\mathfrak{b}}(z,s) = s(s-1)\mathcal{E}_{\mathfrak{b}}(z,s) \tag{1.13}$$

and for all $\delta \in \Gamma$ *satisfies*

$$\mathcal{E}_{\mathfrak{b}}(\delta z, s) = I_a^{\delta a} \cdot \mathcal{E}_{\mathfrak{b}}(z, s). \tag{1.14}$$

In the above statement, the absolute convergence of $\mathcal{E}_{\mathfrak{b}}(z,s)$ means that the coefficient of each $X_{i_1} \cdots X_{i_n}$ in the formal power series defining $\mathcal{E}_{\mathfrak{b}}(z,s)$ is an absolutely convergent sum over $\gamma \in \Gamma_{\mathfrak{b}} \backslash \Gamma$. The meromorphic continuation of $\mathcal{E}_{\mathfrak{b}}(z,s)$ described below has the same meaning.

Note that changing the base point in (1.12) has a simple effect: if $\mathcal{E}'_{\mathfrak{b}}(z,s)$ is defined as in (1.12), but with base point c instead of a, then $\mathcal{E}'_{\mathfrak{b}}(z,s) = I_c^a \cdot \mathcal{E}_{\mathfrak{b}}(z,s) \cdot I_a^c$ since

$$I_{\gamma c}^c = I_{\gamma c}^{\gamma a} \cdot I_{\gamma a}^a \cdot I_a^c = I_c^a \cdot I_{\gamma a}^a \cdot I_a^c \tag{1.15}$$

by (1.10) and (1.11).

THEOREM 1.2. *The series* $\mathcal{E}_{\mathfrak{b}}(z,s)$ *admits a meromorphic continuation to all* $s \in \mathbb{C}$.

Theorem 1.2 is shown in Section 4.2. In fact much more is proved there, including bounds on the continued series $\mathcal{E}_{\mathfrak{b}}(z,s)$ as well as its Fourier coefficients.

When $r = 1$, so there is just one cusp form $f_1 = f$ and I_a^b is in the commutative power series ring $C[\![X]\!]$, the function $\mathcal{E}_{\mathfrak{b}}(z,s)$ makes sense not just as a formal power series but also as a convergent series in the real parameter X. In fact, this is the approach taken by Petridis [**Pet02**] and Petridis-Risager [**PR04**] who use spectral deformation theory to study generalizations of the series (1.2). Thus our work may be considered an extension of their ideas in using a generating series to simultaneously study the properties of a collection of modular symbols or iterated integrals. See also our remarks following Theorems 4.3 and 4.4.

Acknowledgments. We thank the organizers of *Building Bridges: 3rd EU/US Summer School and Workshop on Automorphic Forms and Related Topics* for encouraging us to prepare and submit this work. We also thank Nikolaos Diamantis and Yuri Manin for their comments on an earlier draft. All three authors were partially supported by PSC-CUNY Awards and the first named author acknowledges support from NSF-DMS 1601289.

2. Properties of iterated integrals

2.1. Basic properties. We collect in this section properties of iterated integrals which we will need in our study of noncommutative modular symbols and the associated Eisenstein series. Note that the definition (1.7) makes sense for general holomorphic functions $f_1, \dots, f_n$ on $\mathbb{C}$ and endpoints a, $b \in \mathbb{C}$. So, just in this subsection, we include these more general functions f_i.

PROPOSITION 2.1. *Let* $a, b, c \in \mathbb{C}$ *and let* $f_1, \dots, f_n$ *be holomorphic functions on* $\mathbb{C}$. *Then we have*

(i) $C_a^b(\overbrace{f_1, f_1, \dots, f_1}^{n}) = \dfrac{1}{n!}\left(\displaystyle\int_a^b f_1(u)\, du\right)^n$,

(ii) $C_a^b(f_1, f_2, \dots, f_n) = (-1)^n C_b^a(f_n, f_{n-1}, \dots, f_1)$,

(iii) $C_a^c(f_1, f_2, \dots, f_n) = \sum_{0 \leqslant i \leqslant n} C_a^b(f_1, \dots, f_i) \cdot C_b^c(f_{i+1}, \dots, f_n)$.

PROOF. Part (i) follows by induction and the change of variables $w = \int_a^z f(u)\,du$. Part (ii) is [**DH13**, Lemma 1.1, (v)]. We may also give an elementary proof by induction as follows. The cases $n = 0, 1$ are true, so assume $n \geqslant 2$. Set

$$F_j(z) := \int_{a_0}^{z} f_j(w)\,dw \tag{2.1}$$

for some fixed a_0. Then

$$C_a^b(f_1, f_2, \ldots, f_n) = C_a^b(F_1 \cdot f_2, f_3, \ldots, f_n) - F_1(a) C_a^b(f_2, \ldots, f_n). \tag{2.2}$$

The induction hypothesis yields

$$C_a^b(f_1, f_2, \ldots, f_n) = (-1)^{(n-1)} \left(C_b^a(f_n, \ldots, f_2 \cdot F_1) - F_1(a) C_b^a(f_n, \ldots, f_2) \right). \tag{2.3}$$

Applying (2.2) to each term on the right of (2.3) and using induction again shows

(2.4)
$$\begin{aligned} C_a^b(f_1, f_2, \ldots, f_n) = -C_a^b(F_1 \cdot f_2, f_3, \ldots, f_{n-2}, f_{n-1} \cdot F_n) + F_n(b) C_a^b(F_1 \cdot f_2, f_3, \ldots, f_n) \\ + F_1(a) C_a^b(f_1, \ldots, f_{n-2}, f_{n-1} \cdot F_n) - F_1(a) F_n(b) C_a^b(f_2, \ldots, f_{n-1}). \end{aligned}$$

Expressing $C_b^a(f_n, f_{n-1}, \ldots, f_1)$ with formula (2.4) and using the induction hypothesis completes the proof of (ii).

Part (iii) is [**DH13**, Lemma 1.1, (iii)]. We may easily prove it by taking $a_0 = a$ in the definition (2.1) of F_1 to get

$$C_a^c(f_1, f_2, \ldots, f_n) = C_a^c(F_1 \cdot f_2, f_3, \ldots, f_n). \tag{2.5}$$

Applying (2.5) to both sides of (iii) and using induction completes the proof. □

Proposition 2.1 (iii) implies the concatenation relation (1.10). With Proposition 2.1 (i) we have, at least formally,

$$I_a^b(f_1) = 1 + C_a^b(f_1) \cdot X + C_a^b(f_1, f_1) \cdot X^2 + C_a^b(f_1, f_1, f_1) \cdot X^3 + \cdots = \exp\left(\int_a^b f_1(u)\,du \cdot X \right).$$

We will see that the iterated integrals $C_a^b(f_1, f_2, \ldots, f_n)$ satisfy a further family of general identities in (5.8).

2.2. Modular properties. The iterated integrals will have certain modular invariance properties when $f_1, \ldots, f_n$ are weight 2 cusp forms, as we assume for the rest of the paper. The Γ-invariance relation (1.11) is equivalent to

$$C_{\gamma a}^{\gamma b}(f_1, f_2, \ldots, f_n) = C_a^b(f_1, f_2, \ldots, f_n)$$

which may be easily verified directly.

PROPOSITION 2.2. *For any $a \in \mathbb{H}$ and any parabolic $\gamma \in \Gamma$, when $n \geqslant 1$ we have* $C_a^{\gamma a}(f_1, f_2, \ldots, f_n) = 0$.

PROOF. Suppose γ fixes the cusp $\mathfrak{b}$. Then $\sigma_{\mathfrak{b}}^{-1} \gamma \sigma_{\mathfrak{b}} = \pm \left(\begin{smallmatrix} 1 & \ell \\ 0 & 1 \end{smallmatrix} \right)$ for some $\ell \in \mathbb{Z}$. We first consider C_a^b for $b \in \mathbb{H}$. Using the change of variables $z_i = \sigma_{\mathfrak{b}} w_i$ in (1.7) yields

$$\begin{aligned} &C_a^b(f_1, f_2, \ldots, f_n) \\ &= \int_{\sigma_{\mathfrak{b}}^{-1} a}^{\sigma_{\mathfrak{b}}^{-1} b} f_n(\sigma_{\mathfrak{b}} w_n) \int_{\sigma_{\mathfrak{b}}^{-1} a}^{w_n} f_{n-1}(\sigma_{\mathfrak{b}} w_{n-1}) \cdots \int_{\sigma_{\mathfrak{b}}^{-1} a}^{w_2} f_1(\sigma_{\mathfrak{b}} w_1)\, d\sigma_{\mathfrak{b}} w_1 \cdots d\sigma_{\mathfrak{b}} w_n \\ &= \int_{\sigma_{\mathfrak{b}}^{-1} a}^{\sigma_{\mathfrak{b}}^{-1} b} (f_n|_2 \sigma_{\mathfrak{b}})(w_n) \int_{\sigma_{\mathfrak{b}}^{-1} a}^{w_n} (f_{n-1}|_2 \sigma_{\mathfrak{b}})(w_{n-1}) \cdots \int_{\sigma_{\mathfrak{b}}^{-1} a}^{w_2} (f_1|_2 \sigma_{\mathfrak{b}})(w_1)\, dw_1 \cdots dw_n. \end{aligned}$$

Writing $u = \sigma_{\mathfrak{b}}^{-1}a$, $v = \sigma_{\mathfrak{b}}^{-1}b$, $\mathbf{e}(z) := e^{2\pi i z}$ and using (1.6) shows

$$C_a^b(f_1, f_2, \dots, f_n) = \sum_{m_1,\dots,m_n \in \mathbb{Z}_{\geqslant 1}} c_{\mathfrak{b}}(m_1; f_1) \cdots c_{\mathfrak{b}}(m_n; f_n) \times \int_u^v \mathbf{e}(m_n w_n) \int_u^{w_n} \mathbf{e}(m_{n-1} w_{n-1}) \cdots \int_u^{w_2} \mathbf{e}(m_1 w_1)\, dw_1 \cdots dw_n. \tag{2.6}$$

Note that (2.6) is absolutely convergent because the coefficients $c_{\mathfrak{b}}(m_i; f_i)$ have polynomial growth in m_i and the integrals have exponential decay when $u, v \in \mathbb{H}$. Taking $b = \gamma a$ we find

$$v = \sigma_{\mathfrak{b}}^{-1}\gamma a = \sigma_{\mathfrak{b}}^{-1}\gamma\sigma_{\mathfrak{b}} u = u + \ell.$$

Since the integrand in (2.6) is a polynomial in $\mathbf{e}(w_n)$ with constant term zero, it follows that (2.6) is zero for $b = \gamma a$ as desired. ☐

COROLLARY 2.3. *For any $a, b \in \mathbb{H} \cup \{\text{cusps}\}$ and any parabolic $\gamma \in \Gamma$ we have*

$$I_a^{\gamma b} = I_{\gamma a}^b = I_a^b, \tag{2.7}$$

$$C_a^{\gamma b}(f_1, f_2, \dots, f_n) = C_{\gamma a}^b(f_1, f_2, \dots, f_n) = C_a^b(f_1, f_2, \dots, f_n). \tag{2.8}$$

PROOF. Let $c \in \mathbb{H}$ and we may change to this base point using

$$I_a^{\gamma b} = I_a^c \cdot I_c^{\gamma c} \cdot I_c^b \tag{2.9}$$

which follows as in (1.15). Proposition 2.2 implies that $I_c^{\gamma c} = 1$ and hence $I_a^{\gamma b} = I_a^b$. The second identity in (2.7) is similar and then (2.8) follows from (2.7). ☐

EXAMPLE 2.4. To give an example of the relation (2.6), we compute

$$\int_u^v \mathbf{e}(m_2 w_2) \int_u^{w_2} \mathbf{e}(m_1 w_1)\, dw_1 dw_2$$
$$= \frac{1}{(2\pi i)^2} \left(\frac{\mathbf{e}((m_1 + m_2)v) - \mathbf{e}((m_1 + m_2)u)}{m_1(m_1 + m_2)} + \frac{\mathbf{e}((m_1 + m_2)u) - \mathbf{e}(m_1 u + m_2 v)}{m_1 m_2} \right).$$

With the identity

$$-\frac{1}{m_1(m_1 + m_2)} + \frac{1}{m_1 m_2} = \frac{1}{m_2(m_1 + m_2)}$$

we obtain, with $u = \sigma_{\mathfrak{b}}^{-1}a$, $v = \sigma_{\mathfrak{b}}^{-1}b$ as before,

$$C_a^b(f_1, f_2) = \sum_{m_1, m_2 \in \mathbb{Z}_{\geqslant 1}} \frac{c_{\mathfrak{b}}(m_1; f_1) c_{\mathfrak{b}}(m_2; f_2)}{(2\pi i)^2} \left(\frac{\mathbf{e}((m_1 + m_2)v)}{m_1(m_1 + m_2)} + \frac{\mathbf{e}((m_1 + m_2)u)}{m_2(m_1 + m_2)} - \frac{\mathbf{e}(m_1 u + m_2 v)}{m_1 m_2} \right).$$

See [**CI13**] for more examples of expressing iterated integrals as multiple L-functions.

2.3. Growth estimates for iterated integrals. Here we give some growth estimates which will be needed for the convergence of Eisenstein series formed with noncommutative modular symbols.

PROPOSITION 2.5. *Let $\mathfrak{b}$ be a cusp of Γ and let $a \in \mathbb{H} \cup \{\text{cusps}\}$. For an implied constant independent of $b \in \mathbb{H}$,*

$$C_a^b(f_1, f_2, \dots, f_n) \ll 1 + |\log \operatorname{Im}(\sigma_{\mathfrak{b}}^{-1}b)|^n. \tag{2.10}$$

PROOF. Recall that for any cusp form f of weight 2, we have $y|f(z)| \ll 1$ for all $z \in \mathbb{H}$. We use induction on n to prove the proposition, the case $n = 0$ being clear. First consider $C^b_{\sigma_\mathfrak{b} i}(f_1, \dots, f_n)$ with imaginary number i. There exists $\gamma \in \Gamma_\mathfrak{b}$ such that $\mathrm{Re}(\sigma_\mathfrak{b}^{-1}\gamma\sigma_\mathfrak{b} i) = \mathrm{Re}(\sigma_\mathfrak{b}^{-1} b) + t$ for $|t| \leqslant 1/2$. By Corollary 2.3, $C^b_{\gamma\sigma_\mathfrak{b} i}(f_1, \dots, f_n) = C^b_{\sigma_\mathfrak{b} i}(f_1, \dots, f_n)$. Hence

$$\begin{aligned} C^b_{\sigma_\mathfrak{b} i}(f_1, \dots, f_n) &= \int_{\gamma\sigma_\mathfrak{b} i}^{b} f_n(w) \cdot C^w_{\gamma\sigma_\mathfrak{b} i}(f_1, \dots, f_{n-1})\, dw \\ &= \int_{\sigma_\mathfrak{b}^{-1}\gamma\sigma_\mathfrak{b} i}^{\sigma_\mathfrak{b}^{-1} b} (f_n|_2\sigma_\mathfrak{b})(z) \cdot C^{\sigma_\mathfrak{b} z}_{\sigma_\mathfrak{b} i}(f_1, \dots, f_{n-1})\, dz. \end{aligned}$$

With our cusp form bound and induction, the integrand is $\ll (1 + |\log y|^{n-1})/y$. Hence

$$\begin{aligned} C^b_{\sigma_\mathfrak{b} i}(f_1, \dots, f_n) &\ll \int_{i-1/2}^{i+1/2} \frac{1 + |\log y|^{n-1}}{y}\, dx + \int_1^{\mathrm{Im}(\sigma_\mathfrak{b}^{-1} b)} \frac{1 + |\log y|^{n-1}}{y}\, dy \\ &\ll 1 + |\log \mathrm{Im}(\sigma_\mathfrak{b}^{-1} b)|^n. \end{aligned}$$

Finally, use the identity

$$C^b_a(f_1, f_2, \dots, f_n) = \sum_{0 \leqslant j \leqslant n} C^{\sigma_\mathfrak{b} i}_a(f_1, \dots, f_j) C^b_{\sigma_\mathfrak{b} i}(f_{j+1}, \dots, f_n) \tag{2.11}$$

to obtain (2.10) and complete the induction. □

Note that, with Proposition 2.1 (ii) it is clear that we may replace $C^b_a(\dots)$ by $C^a_b(\dots)$ in the statement of Proposition 2.5.

COROLLARY 2.6. *Let $\mathfrak{b}$ be a cusp of Γ. For an implied constant independent of a, $b \in \mathbb{H}$,*

$$C^b_a(f_1, f_2, \dots, f_n) \ll 1 + |\log \mathrm{Im}(\sigma_\mathfrak{b}^{-1} a)|^n + |\log \mathrm{Im}(\sigma_\mathfrak{b}^{-1} b)|^n.$$

PROOF. This follows from Proposition 2.5 and (2.11). □

COROLLARY 2.7. *Let $\mathfrak{b}$ be a cusp of Γ and let $a \in \mathbb{H} \cup \{\mathtt{cusps}\}$. For an implied constant independent of $\gamma \in \Gamma$ and $z \in \mathbb{H}$,*

$$C^a_{\gamma a}(f_1, f_2, \dots, f_n) \ll 1 + |\log \mathrm{Im}(\sigma_\mathfrak{b}^{-1} z)|^n + |\log \mathrm{Im}(\sigma_\mathfrak{b}^{-1} \gamma z)|^n.$$

PROOF. Since $I^a_{\gamma a} = I^{\gamma z}_{\gamma a} I^z_{\gamma z} I^a_z = I^z_a I^z_{\gamma z} I^a_z$, we have

$$C^a_{\gamma a}(f_1, f_2, \dots, f_n) = \sum_{0 \leqslant i \leqslant j \leqslant n} C^z_a(f_1, \dots, f_i) C^z_{\gamma z}(f_{i+1}, \dots, f_j) C^a_z(f_{j+1}, \dots, f_n).$$

With Proposition 2.5 and Corollary 2.6, this is

$$\begin{aligned} \ll \sum_{0 \leqslant i \leqslant j \leqslant n} \left(1 + |\log \mathrm{Im}(\sigma_\mathfrak{b}^{-1} z)|^i\right) &\left(1 + |\log \mathrm{Im}(\sigma_\mathfrak{b}^{-1} z)|^{j-i} + |\log \mathrm{Im}(\sigma_\mathfrak{b}^{-1} \gamma z)|^{j-i}\right) \\ &\times \left(1 + |\log \mathrm{Im}(\sigma_\mathfrak{b}^{-1} z)|^{n-j}\right) \end{aligned}$$

and the result follows. □

The implied constants in Proposition 2.5 and Corollaries 2.6, 2.7 may depend on the remaining parameters. For example, the implied constant in Corollary 2.7 depends on $\mathfrak{b}$, a, n and $f_1, \dots, f_n$.

3. Eisenstein series twisted by noncommutative modular symbols

In this section we introduce our primary object of study, the Eisenstein series formed with noncommutative modular symbols. The series we define here are slightly more general than those described in the introduction.

3.1. Definition and convergence. Let $\mathbf{f}=(f_1,f_2,\dots,f_r)$ and $\mathbf{g}=(g_1,g_2,\dots,g_t)$ be lists of weight 2 cusp forms. To each f_i we associate the variable X_i and to each g_i we associate Y_i. Let $I_a^b=I_a^b(\mathbf{f})$ be as before in (1.8) and put

$$J_a^b = J_a^b(\mathbf{g}) = 1+\sum_i C_a^b(g_i)\cdot Y_i + \sum_{i,j} C_a^b(g_i,g_j)\cdot Y_iY_j+\cdots. \tag{3.1}$$

Then $I_a^b\cdot\overline{J_a^b}$ is an element of the ring of formal power series in the variables X_i and $\overline{Y_i}$ with coefficients in $\mathbb{C}$. We take these variables to be non-commuting, except for the relations $X_i\overline{Y_j}=\overline{Y_j}X_i$ for $1\leqslant i\leqslant r$, $1\leqslant j\leqslant t$. We may write

$$\begin{aligned}I_a^b\cdot\overline{J_a^b} = 1&+\sum_i C_a^b(f_i)\cdot X_i+\sum_i \overline{C_a^b(g_i)}\cdot\overline{Y_i}\\ &+\sum_{i,j} C_a^b(f_i,f_j)\cdot X_iX_j+\sum_{i,j}C_a^b(f_i)\overline{C_a^b(g_j)}\cdot X_i\overline{Y_j}+\sum_{i,j}\overline{C_a^b(g_i,g_j)}\cdot\overline{Y_iY_j}+\cdots.\end{aligned}\tag{3.2}$$

Fix a base point $a\in\mathbb{H}\cup\{\mathsf{cusps}\}$. For $z\in\mathbb{H}$ and any cusp $\mathfrak{b}$ of Γ, define the *Eisenstein series twisted by noncommutative modular symbols* as

$$\mathcal{E}_{\mathfrak{b}}(z,s):=\sum_{\gamma\in\Gamma_{\mathfrak{b}}\backslash\Gamma} I_{\gamma a}^a\cdot\overline{J_{\gamma a}^a}\cdot\operatorname{Im}(\sigma_{\mathfrak{b}}^{-1}\gamma z)^s. \tag{3.3}$$

This agrees with our earlier definition (1.12) when $\mathbf{g}$ is empty. The summands are well-defined by Corollary 2.3. Expanding with (3.2) gives

$$\begin{aligned}\mathcal{E}_{\mathfrak{b}}(z,s) &= E_{\mathfrak{b}}(z,s)+\sum_i E_{\mathfrak{b}}(z,s;f_i)\cdot X_i+\sum_i E_{\mathfrak{b}}(z,s;\overline{g_i})\cdot\overline{Y_i}\\ &+\sum_{i,j}E_{\mathfrak{b}}(z,s;f_i,f_j)\cdot X_iX_j+\sum_{i,j}E_{\mathfrak{b}}(z,s;f_i,\overline{g_j})\cdot X_i\overline{Y_j}+\sum_{i,j}E_{\mathfrak{b}}(z,s;\overline{g_i},\overline{g_j})\cdot\overline{Y_iY_j}+\cdots\end{aligned}\tag{3.4}$$

where

$$\begin{aligned}&E_{\mathfrak{b}}(z,s;f_{i_1},\dots,f_{i_m},\overline{g_{j_1}},\dots,\overline{g_{j_n}})\\ &\qquad:=\sum_{\gamma\in\Gamma_{\mathfrak{b}}\backslash\Gamma}C_{\gamma a}^a(f_{i_1},\dots,f_{i_m})\overline{C_{\gamma a}^a(g_{j_1},\dots,g_{j_n})}\operatorname{Im}(\sigma_{\mathfrak{b}}^{-1}\gamma z)^s.\end{aligned}\tag{3.5}$$

THEOREM 3.1. *The series $E_{\mathfrak{b}}(z,s;f_1,\dots,f_m,\overline{g_1},\dots,\overline{g_n})$ from (3.5) is absolutely convergent for $\operatorname{Re}(s)>1$. This convergence is uniform for s in compact sets.*

PROOF. In the case $m=n=0$ the theorem is true for $E_{\mathfrak{b}}(z,s)$, which is just the ordinary real-analytic Eisenstein series for Γ. By Corollary 2.7, the summand of the series corresponding to γ is bounded by a constant times

$$\left(1+|\log\operatorname{Im}(\sigma_{\mathfrak{b}}^{-1}z)|^{m+n}+|\log\operatorname{Im}(\sigma_{\mathfrak{b}}^{-1}\gamma z)|^{m+n}\right)\operatorname{Im}(\sigma_{\mathfrak{b}}^{-1}\gamma z)^{\operatorname{Re}(s)}. \tag{3.6}$$

The elementary inequality

$$|\log y|^k<(2/\varepsilon)^k(y^{\varepsilon k}+y^{-\varepsilon k})$$

for all $y,\varepsilon>0$ and $k\geqslant 0$, implies the first factor in (3.6) is

$$\ll \operatorname{Im}(\sigma_{\mathfrak{b}}^{-1}z)^{\varepsilon(m+n)}+\operatorname{Im}(\sigma_{\mathfrak{b}}^{-1}z)^{-\varepsilon(m+n)}+\operatorname{Im}(\sigma_{\mathfrak{b}}^{-1}\gamma z)^{\varepsilon(m+n)}+\operatorname{Im}(\sigma_{\mathfrak{b}}^{-1}\gamma z)^{-\varepsilon(m+n)}.$$

Hence

$$\sum_{\gamma\in\Gamma_{\mathfrak{b}}\backslash\Gamma}\left|C^{a}_{\gamma a}(f_1,\dots,f_m)\overline{C^{a}_{\gamma a}(g_1,\dots,g_n)}\,\mathrm{Im}(\sigma_{\mathfrak{b}}^{-1}\gamma z)^s\right|$$
$$\ll E_{\mathfrak{b}}(z,\mathrm{Re}(s))+E_{\mathfrak{b}}(z,\mathrm{Re}(s)+\varepsilon(m+n))+E_{\mathfrak{b}}(z,\mathrm{Re}(s)-\varepsilon(m+n))$$

for any $\varepsilon>0$, with the implied constant depending on ε and z. The result follows. □

3.2. Transformation properties.

PROPOSITION 3.2. *For* $\mathrm{Re}(s)>1$ *and all* $\delta\in\Gamma$

$$\mathcal{E}_{\mathfrak{b}}(\delta z,s)=I^{\delta a}_{a}\cdot\overline{J^{\delta a}_{a}}\cdot\mathcal{E}_{\mathfrak{b}}(z,s). \tag{3.7}$$

PROOF. We have

$$\begin{aligned}\mathcal{E}_{\mathfrak{b}}(\delta z,s)&=\sum_{\gamma\in\Gamma_{\mathfrak{b}}\backslash\Gamma}I^{a}_{\gamma a}\cdot\overline{J^{a}_{\gamma a}}\cdot\mathrm{Im}(\sigma_{\mathfrak{b}}^{-1}\gamma\delta z)^s\\&=\sum_{\gamma\in\Gamma_{\mathfrak{b}}\backslash\Gamma}I^{a}_{\gamma\delta^{-1}a}\cdot\overline{J^{a}_{\gamma\delta^{-1}a}}\cdot\mathrm{Im}(\sigma_{\mathfrak{b}}^{-1}\gamma z)^s.\end{aligned}$$

Then $I^{a}_{\gamma\delta^{-1}a}=I^{\gamma a}_{\gamma\delta^{-1}a}I^{a}_{\gamma a}=I^{a}_{\delta^{-1}a}I^{a}_{\gamma a}=I^{\delta a}_{a}I^{a}_{\gamma a}$ and similarly $J^{a}_{\gamma\delta^{-1}a}=J^{\delta a}_{a}J^{a}_{\gamma a}$. Using these relations we obtain

$$\begin{aligned}\mathcal{E}_{\mathfrak{b}}(\delta z,s)&=\sum_{\gamma\in\Gamma_{\mathfrak{b}}\backslash\Gamma}I^{\delta a}_{a}I^{a}_{\gamma a}\cdot\overline{J^{\delta a}_{a}J^{a}_{\gamma a}}\cdot\mathrm{Im}(\sigma_{\mathfrak{b}}^{-1}\gamma z)^s\\&=I^{\delta a}_{a}\overline{J^{\delta a}_{a}}\sum_{\gamma\in\Gamma_{\mathfrak{b}}\backslash\Gamma}I^{a}_{\gamma a}\cdot\overline{J^{a}_{\gamma a}}\cdot\mathrm{Im}(\sigma_{\mathfrak{b}}^{-1}\gamma z)^s.\end{aligned}$$ □

The relation (3.7) elegantly encapsulates a lot of information. Comparing corresponding coefficients shows

$$\begin{aligned}&E_{\mathfrak{b}}(\delta z,s;f_1,\dots,f_m,\overline{g_1},\dots,\overline{g_n})\\&=\sum_{i=0}^{m}\sum_{j=0}^{n}C^{\delta a}_{a}(f_1,\dots,f_i)\overline{C^{\delta a}_{a}(g_1,\dots,g_j)}E_{\mathfrak{b}}(z,s;f_{i+1},\dots,f_m,\overline{g_{j+1}},\dots,\overline{g_n}).\end{aligned} \tag{3.8}$$

When $\mathbf{f}=()$, a similar formula is given by Diamantis and Sim [**DS08**, Eq. (3.30)].

EXAMPLE 3.3. Another special case of Proposition 3.2 appears in [**JO08**]. Suppose that $\mathcal{E}_{\mathfrak{b}}(z,s)$ is defined with modular symbols $I^{b}_{a}(f)$ and $J^{b}_{a}(g)$ with each depending on single cusp forms. Then the coefficients in the series (3.4) take the form

$$\begin{aligned}E_{\mathfrak{b}}(z,s;\underbrace{f,\dots,f}_{m},\underbrace{\overline{g},\dots,\overline{g}}_{n})&=\sum_{\gamma\in\Gamma_{\mathfrak{b}}\backslash\Gamma}C^{a}_{\gamma a}(f,\dots,f)\overline{C^{a}_{\gamma a}(g,\dots,g)}\,\mathrm{Im}(\sigma_{\mathfrak{b}}^{-1}\gamma z)^s\\&=\sum_{\gamma\in\Gamma_{\mathfrak{b}}\backslash\Gamma}\frac{1}{m!}\left(\int_{\gamma a}^{a}f(u)\,du\right)^m\frac{1}{n!}\left(\overline{\int_{\gamma a}^{a}g(u)\,du}\right)^n\mathrm{Im}(\sigma_{\mathfrak{b}}^{-1}\gamma z)^s\\&=\frac{(-1)^m}{(2\pi i)^{m+n}}\frac{1}{m!n!}\sum_{\gamma\in\Gamma_{\mathfrak{b}}\backslash\Gamma}\langle\gamma,f\rangle^m\overline{\langle\gamma,g\rangle}^n\,\mathrm{Im}(\sigma_{\mathfrak{b}}^{-1}\gamma z)^s\end{aligned} \tag{3.9}$$

where the modular symbol pairing was defined in (1.1). The series in (3.9) is studied in detail in [**PR04**], [**JO08**] and written as

$$E^{m,n}_{\mathfrak{b}}(z,s;f,g):=\sum_{\gamma\in\Gamma_{\mathfrak{b}}\backslash\Gamma}\langle\gamma,f\rangle^m\overline{\langle\gamma,g\rangle}^n\,\mathrm{Im}(\sigma_{\mathfrak{b}}^{-1}\gamma z)^s. \tag{3.10}$$

It follows from relation (3.7), by translating the terms of (3.8) with (3.9), that

$$E_{\mathfrak{b}}^{m,n}(\delta z,s;f,g)=\sum_{i=0}^{m}\sum_{j=0}^{n}\binom{m}{i}\binom{n}{j}(-\langle\gamma,f\rangle)^i(-\overline{\langle\gamma,g\rangle})^j E_{\mathfrak{b}}^{m-i,n-j}(z,s;f,g)$$

for all $\delta\in\Gamma$. This is Lemma 4.1 of [**JO08**].

The results in Sections 3.1 and 3.2 complete the proof of Theorem 1.1, except for the verification of (1.13). But this follows from $\Delta y^s = s(s-1)y^s$ as in the classical case for $E_{\mathfrak{b}}(z,s)$.

3.3. Fourier expansions. Write $z = x + iy$ for $z\in\mathbb{H}$. Let $\mathfrak{a}$ and $\mathfrak{b}$ be cusps for Γ. By [**Iwa02**, Thm. 3.4], the Fourier expansion of the classical real-analytic Eisenstein series is

$$E_{\mathfrak{a}}(\sigma_{\mathfrak{b}}z,s)=\delta_{\mathfrak{a}\mathfrak{b}}y^s+\phi_{\mathfrak{a}\mathfrak{b}}(s)y^{1-s}+\sum_{k\in\mathbb{Z}_{\neq 0}}\phi_{\mathfrak{a}\mathfrak{b}}(k,s)W_s(kz). \tag{3.11}$$

The function $W_s(z)$ is a Whittaker function and $\phi_{\mathfrak{a}\mathfrak{b}}(s)$, $\phi_{\mathfrak{a}\mathfrak{b}}(k,s)$ may be expressed in terms of Kloosterman sums. Also $\delta_{\mathfrak{a}\mathfrak{b}}$ takes the value 1 if $\mathfrak{a}$ and $\mathfrak{b}$ are Γ-equivalent and is 0 otherwise.

The same proof gives the expansion

$$\begin{aligned}(3.12)\quad & E_{\mathfrak{a}}(\sigma_{\mathfrak{b}}z,s;f_1,\dots,f_m,\overline{g_1},\dots,\overline{g_n}))\\ &=\phi_{\mathfrak{a}\mathfrak{b}}(s;f_1,\dots,\overline{g_n}))y^{1-s}+\sum_{k\in\mathbb{Z}_{\neq 0}}\phi_{\mathfrak{a}\mathfrak{b}}(k,s;f_1,\dots,\overline{g_n}))W_s(kz)\end{aligned}$$

for $m+n\geqslant 1$. The coefficients $\phi_{\mathfrak{a}\mathfrak{b}}(s;f_1,\dots,\overline{g_n}))$ and $\phi_{\mathfrak{a}\mathfrak{b}}(k,s;f_1,\dots,\overline{g_n}))$ may be written in terms of Kloosterman sums twisted by iterated integrals, similarly to [**O'S00**, Eqs. (1.2),(1.3)]. It is an interesting question to find explicit forms for these coefficients. See [**GO03**, Remark 5.5] and [**Ris03**, Sect. 4.1] for the determination of $\phi_{\mathfrak{a}\mathfrak{b}}(s;f_1)$. In Bruggeman-Diamantis [**BD16**], $\phi_{\mathfrak{a}\mathfrak{b}}(s;f_1)$ and $\phi_{\mathfrak{a}\mathfrak{b}}(k,s;f_1)$ are expressed in terms of L-functions and shifted convolution sums.

Combining the expansions (3.12) gives the Fourier expansion of $\mathcal{E}_{\mathfrak{a}}(z,s)$ as

$$\mathcal{E}_{\mathfrak{a}}(\sigma_{\mathfrak{b}}z,s)=\delta_{\mathfrak{a}\mathfrak{b}}y^s+\Phi_{\mathfrak{a}\mathfrak{b}}(s)y^{1-s}+\sum_{k\in\mathbb{Z}_{\neq 0}}\Phi_{\mathfrak{a}\mathfrak{b}}(k,s)W_s(kz) \tag{3.13}$$

where

$$\begin{aligned}(3.14)\quad \Phi_{\mathfrak{a}\mathfrak{b}}(s)&=\phi_{\mathfrak{a}\mathfrak{b}}(s)+\sum_i\phi_{\mathfrak{a}\mathfrak{b}}(s;f_i)\cdot X_i+\sum_i\phi_{\mathfrak{a}\mathfrak{b}}(s;\overline{g_i})\cdot\overline{Y_i}\\ &+\sum_{i,j}\phi_{\mathfrak{a}\mathfrak{b}}(s;f_i,f_j)\cdot X_iX_j+\sum_{i,j}\phi_{\mathfrak{a}\mathfrak{b}}(s;f_i,\overline{g_j})\cdot X_i\overline{Y_j}+\sum_{i,j}\phi_{\mathfrak{a}\mathfrak{b}}(s;\overline{g_i},\overline{g_j})\cdot\overline{Y_iY_j}+\cdots\end{aligned}$$

and $\Phi_{\mathfrak{a}\mathfrak{b}}(k,s)$ is a similar series.

4. Meromorphic continuation

The meromorphic continuation of the classical Eisenstein series $E_{\mathfrak{a}}(z,s)$ to all $s\in\mathbb{C}$ may be shown using its Fourier expansion in simple cases, such as when $\Gamma=\mathrm{SL}_2(\mathbb{Z})$. Selberg proved the continuation for general groups Γ and one of Selberg's methods, as described in [**Iwa02**, Chap. 6], was extended in [**JO08**] to prove the continuation of $E_{\mathfrak{a}}^{m,n}(z,s;f,g)$. We recall from (3.9) that this is a constant times

$$E_{\mathfrak{a}}(z,s;f_1,\dots,f_m,\overline{g_1},\dots,\overline{g_n}) \tag{4.1}$$

where $f_1 = f_2 = \cdots = f_m = f$ and $g_1 = g_2 = \cdots = g_n = g$. The proof in [**JO08**] goes through almost without change in the general case of distinct functions f_i and g_j. The two main properties that the proof needs are that (4.1) is an eigenfunction of the Laplacian and that it transforms into itself as z is replaced by γz for $\gamma \in \Gamma$ except for the addition of a lower-order term. The proof also requires some growth estimates that we describe next.

4.1. Growth estimates for twisted Eisenstein series. Following [**Iwa02**], we may measure the growth of Γ-invariant functions in terms of the *invariant height* function defined by

$$y_\Gamma(z) := \max_{\mathfrak{a}} \max_{\gamma\in\Gamma}(\operatorname{Im}({\sigma_{\mathfrak{a}}}^{-1}\gamma z))$$

for $z \in \mathbb{H}$ where the outer maximum is taken over our fixed set of inequivalent cusps. Thus $y_\Gamma(z)$ approaches ∞ as z approaches any cusp.

The twisted Eisenstein series are not Γ-invariant and it is more convenient to fix a fundamental domain $\mathbb{F}$ and examine their growth there. Let $\mathcal{P}_Y \subset \mathbb{H}$ denote the strip with $|x| \leqslant 1/2$ and $y \geqslant Y$. We choose $\mathbb{F}$ so that its closure contains the cuspidal zones $\sigma_{\mathfrak{a}}\mathcal{P}_Y$ for all $\mathfrak{a}$ and Y large enough; see [**Iwa02**, Section 2.2]. For $z \in \mathbb{F}$ we define the *domain height* function

$$y_{\mathbb{F}}(z) := \max_{\mathfrak{a}}(\operatorname{Im}({\sigma_{\mathfrak{a}}}^{-1}z)).$$

Clearly, $y_{\mathbb{F}}(z)$ is bounded below by a positive constant for $z \in \mathbb{F}$ and bounded above by $y_\Gamma(z)$. In fact $y_{\mathbb{F}}(z) = y_\Gamma(z)$ when z is in the cuspidal zones $\sigma_{\mathfrak{a}}\mathcal{P}_Y$ for all $\mathfrak{a}$ and Y large enough. Hence $y_\Gamma(z) \ll y_{\mathbb{F}}(z)$ for $z \in \mathbb{F}$. It is also shown in [**JO08**, Lemma A.1] that, for any cusp $\mathfrak{b}$ and all $z \in \mathbb{H}$

$$y_\Gamma(\sigma_{\mathfrak{b}}z) \leqslant (c_\Gamma + 1/c_\Gamma)(y + 1/y) \tag{4.2}$$

for c_Γ depending only on Γ.

With Proposition 2.5 we have shown that, for any cusp $\mathfrak{b}$,

$$C_a^z(f_1, f_2, \dots, f_n) \ll 1 + |\log \operatorname{Im}(\sigma_{\mathfrak{b}}^{-1}z)|^n \tag{4.3}$$

for all $z \in \mathbb{H}$. It follows from (4.3) that

$$C_a^z(f_1, f_2, \dots, f_n) \ll \log^n(y_{\mathbb{F}}(z) + e) \tag{4.4}$$

for all $z \in \mathbb{F}$. We have, writing $\sigma = \operatorname{Re}(s)$ as usual,

$$E_{\mathfrak{a}}(z,s) \ll y_\Gamma(z)^\sigma \tag{4.5}$$

as in [**Iwa02**, Corollary 3.5] for example. Using (4.4) in the proof of [**JO08**, Prop. 3.3] shows

$$E_{\mathfrak{a}}(z,s; f_1, \dots, f_m, \overline{g_1}, \dots, \overline{g_n}) \ll y_{\mathbb{F}}(z)^{1-\sigma+\varepsilon} \tag{4.6}$$

for all $\varepsilon > 0$ and all $z \in \mathbb{F}$ when $m + n > 0$.

Our next goal is to extend (4.6) to a bound on all of $\mathbb{H}$. Similarly to [**JO08**, Sect. 5.1] we may do this neatly using the series

$$Q_{\mathfrak{b}}(z,s; f_1, \dots, f_m, \overline{g_1}, \dots, \overline{g_n}) := \sum_{\gamma\in\Gamma_{\mathfrak{b}}\backslash\Gamma} C_{\gamma z}^a(f_1, \dots, f_m)\cdot\overline{C_{\gamma z}^a(g_1, \dots, g_n)}\cdot\operatorname{Im}(\sigma_{\mathfrak{b}}^{-1}\gamma z)^s. \tag{4.7}$$

PROPOSITION 4.1. *The series* $Q_{\mathfrak{b}}(z,s;f_1,\dots,f_m,\overline{g_1},\dots,\overline{g_n})$ *converges absolutely for* $\mathrm{Re}(s)>1$ *to a* Γ*-invariant function of* z*. It satisfies the bound*

$$Q_{\mathfrak{b}}(z,s;f_1,\dots,f_m,\overline{g_1},\dots,\overline{g_n}) \ll \log^{m+n}(y_\Gamma(z)+e)\cdot y_\Gamma(z)^\sigma \tag{4.8}$$

for all $z\in\mathbb{H}$ *and is related to the twisted Eisenstein series through the identities*

$$\begin{aligned} &Q_{\mathfrak{b}}(z,s;f_1,\dots,f_m,\ \overline{g_1},\dots,\overline{g_n})\\ &=\sum_{i=0}^{m}\sum_{j=0}^{n} C_z^{\mathfrak{a}}(f_1,\dots,f_i)\cdot\overline{C_z^{\mathfrak{a}}(g_1,\dots,g_j)}\cdot E_{\mathfrak{b}}(z,s;f_{i+1},\dots,f_m,\overline{g_{j+1}},\dots,\overline{g_n}),\end{aligned} \tag{4.9}$$

$$\begin{aligned} &E_{\mathfrak{b}}(z,s;f_1,\dots,f_m,\ \overline{g_1},\dots,\overline{g_n})\\ &=\sum_{i=0}^{m}\sum_{j=0}^{n} C_{\mathfrak{a}}^{z}(f_1,\dots,f_i)\cdot\overline{C_{\mathfrak{a}}^{z}(g_1,\dots,g_j)}\cdot Q_{\mathfrak{b}}(z,s;f_{i+1},\dots,f_m,\overline{g_{j+1}},\dots,\overline{g_n}).\end{aligned} \tag{4.10}$$

PROOF. We first see that (4.9) follows from the identities $I_{\gamma z}^{\mathfrak{a}} = I_{\gamma z}^{\gamma\mathfrak{a}} I_{\gamma\mathfrak{a}}^{\mathfrak{a}} = I_z^{\mathfrak{a}} I_{\gamma\mathfrak{a}}^{\mathfrak{a}}$. Hence (4.7) converges absolutely for $\mathrm{Re}(s)>1$. It is Γ-invariant because replacing z by δz for $\delta\in\Gamma$ just reorders the series.

Use (4.4), (4.5) and (4.6) to bound the right side of (4.9) and obtain

$$Q_{\mathfrak{b}}(z,s;f_1,\dots,f_m,\overline{g_1},\dots,\overline{g_n}) \ll \log^{m+n}(y_{\mathbb{F}}(z)+e)\cdot y_{\mathbb{F}}(z)^\sigma \tag{4.11}$$

for all $z\in\mathbb{F}$. Then (4.8) is a consequence of (4.11) since $Q_{\mathfrak{b}}$ is Γ-invariant.

The relation (4.10) follows similarly to (4.9). □

COROLLARY 4.2. *For every cusp* $\mathfrak{b}$ *and all* $z\in\mathbb{H}$ *we have*

$$E_{\mathfrak{a}}(\sigma_{\mathfrak{b}}z,s;f_1,\dots,f_m,\overline{g_1},\dots,\overline{g_n}) \ll \log^{m+n}(y+1/y)\cdot(y^\sigma+y^{-\sigma}). \tag{4.12}$$

PROOF. Use (4.2) in (4.8) to show that

$$Q_{\mathfrak{a}}(\sigma_{\mathfrak{b}}z,s;f_1,\dots,f_m,\overline{g_1},\dots,\overline{g_n}) \ll \log^{m+n}(y+1/y)\cdot(y^\sigma+y^{-\sigma}) \tag{4.13}$$

for all $z\in\mathbb{H}$. Then (4.13) and (4.3) along with the inequality $|\log y|+1<3\log(y+1/y)$ bound the right side of (4.10). □

The estimates we have developed in this subsection are the same as those found in Sections 3.2 and 5.1 of [**JO08**] for the case of equal cusp forms: $f_1=\dots=f_m$ and $g_1=\dots=g_n$.

4.2. Continuation using the Fredholm theory of integral equations. We are following Sections 5.2 and 5.3 of [**JO08**] in the next discussion. Set

$$u(z,w):=\frac{|z-w|^2}{4\,\mathrm{Im}(z)\,\mathrm{Im}(w)},\qquad G_\alpha(u):=\frac{1}{4\pi}\int_0^1 (t(1-t))^{\alpha-1}(t+u)^{-\alpha}\,dt.$$

It is convenient to let $\mathbf{f}=(f_1,\dots,f_m)$ and $\mathbf{g}=(g_1,\dots,g_n)$. The resolvent of the Laplacian may be written as an integral operator with kernel given by the above Green function $G_\alpha(u)$. Since $E_{\mathfrak{a}}(z,s;\mathbf{f},\overline{\mathbf{g}})$ is an eigenfunction of Δ and Corollary 4.2 holds, we obtain

$$\frac{-E_{\mathfrak{a}}(z,s;\mathbf{f},\overline{\mathbf{g}})}{(\alpha(1-\alpha)-s(1-s))}=\int_{\mathbb{H}} G_\alpha(u(z,z'))E_{\mathfrak{a}}(z',s;\mathbf{f},\overline{\mathbf{g}})\,d\mu(z') \tag{4.14}$$

for $1<\mathrm{Re}(s)<\alpha-2$. To remove a logarithmic singularity at $u=0$ we set $G_{\alpha\beta}(u):=G_\alpha(u)-G_\beta(u)$ for $\beta<\alpha$. Also put

$$\nu_{\alpha\beta}(s):=(\alpha(1-\alpha)-s(1-s))^{-1}-(\beta(1-\beta)-s(1-s))^{-1}$$

and, on breaking up $\mathbb{H}$ into the images of $\mathbb{F}$ under Γ, (4.14) becomes

$$-\nu_{\alpha\beta}(s)E_{\mathfrak{a}}(z,s;\mathbf{f},\overline{\mathbf{g}}) = \int_{\mathbb{F}} \sum_{\gamma\in\Gamma} G_{\alpha\beta}(u(z,\gamma z'))E_{\mathfrak{a}}(\gamma z',s;\mathbf{f},\overline{\mathbf{g}})\,d\mu(z') \tag{4.15}$$

for $1 < \mathrm{Re}(s) < \beta-2 < \alpha-2$. Use the transformation property (3.8) of $E_{\mathfrak{a}}(\gamma z',s;\mathbf{f},\overline{\mathbf{g}})$ to expand this. With

$$G^{i,j}_{\alpha\beta}(z,z') := \sum_{\gamma\in\Gamma} C^{\gamma a}_{a}(f_1,\dots,f_i)\overline{C^{\gamma a}_{a}(g_1,\dots,g_j)}G_{\alpha\beta}(u(z,\gamma z'))$$

and $G_{\alpha\beta}(z,z') := G^{0,0}_{\alpha\beta}(z,z')$ we find

$$E_{\mathfrak{a}}(z,s;\mathbf{f},\overline{\mathbf{g}}) = q^{m,n}(z,s) + \lambda\int_{\mathbb{F}} G_{\alpha\beta}(z,z')E_{\mathfrak{a}}(\gamma z',s;\mathbf{f},\overline{\mathbf{g}})\,d\mu(z') \tag{4.16}$$

for $\lambda = \lambda(s) := -1/\nu_{\alpha\beta}(s)$, a polynomial in s of degree 4, and

$$q^{m,n}(z,s) := \lambda\sum_{i,j}\int_{\mathbb{F}} G^{i,j}_{\alpha\beta}(z,z')E_{\mathfrak{a}}(z',s;f_{i+1},\dots,f_m,\overline{g_{j+1}},\dots,\overline{g_n})\,d\mu(z') \tag{4.17}$$

where the sum in (4.17) is over all i, j that satisfy $0\leqslant i\leqslant m$, $0\leqslant j\leqslant n$ and are not both zero.

The integral equation (4.16) is valid for $1<\mathrm{Re}(s)<\beta-2$. However, it determines $E_{\mathfrak{a}}(z,s;\mathbf{f},\overline{\mathbf{g}})$ uniquely and by the Fredholm theory there exists a kernel $D_\lambda(z,z')$ and a function $D(\lambda)\not\equiv 0$, which are both built from $G_{\alpha\beta}(z,z')$ and analytic in λ, so that

$$E_{\mathfrak{a}}(z,s;\mathbf{f},\overline{\mathbf{g}}) = q^{m,n}(z,s) + \frac{\lambda}{D(\lambda)}\int_{\mathbb{F}} D_\lambda(z,z')q^{m,n}(z',s)\,d\mu(z'). \tag{4.18}$$

This provides the desired analytic continuation of $E_{\mathfrak{a}}(z,s;\mathbf{f},\overline{\mathbf{g}})$ with an induction argument when we know that the lower-order series $q^{m,n}(z,s)$ already has a continuation.

We simplified the presentation above by omitting a step. The kernel $G_{\alpha\beta}(z,z')$ in (4.16) is not bounded, as required by the Fredholm theory we are using, and must be replaced by a truncated version $\eta(z)\eta(z')H_s(z,z')$ as described in [**Iwa02**, p. 84] and [**JO08**, Sect. 5.3].

Theorem 1.2 follows immediately from the next result.

THEOREM 4.3. *Let* $\mathbf{f}=(f_1,\dots,f_m)$ *and* $\mathbf{g}=(g_1,\dots,g_n)$ *for* $m,n\in\mathbb{Z}_{\geqslant 0}$. *For every ball* $\mathbb{B}_r\subset\mathbb{C}$ *of radius* r *about the origin there exist functions* $A_{\mathfrak{a}}(s)$, $\phi_{\mathfrak{ab}}(s;\mathbf{f},\overline{\mathbf{g}})$ *and* $\phi_{\mathfrak{ab}}(k,s;\mathbf{f},\overline{\mathbf{g}})$ *for all* $k\in\mathbb{Z}_{\neq 0}$ *so that the following assertions hold:*

(i) $A_{\mathfrak{a}}(s)$ *is analytic on* $\mathbb{B}_r$ *and not identically* 0.
(ii) $\phi_{\mathfrak{ab}}(s;\mathbf{f},\overline{\mathbf{g}})$ *and* $\phi_{\mathfrak{ab}}(k,s;\mathbf{f},\overline{\mathbf{g}})$ *are meromorphic functions of* s *on* $\mathbb{B}_r$.
(iii) *For all* $s\in\mathbb{B}_r$

$$A^{m+n+1}_{\mathfrak{a}}(s)\cdot\phi_{\mathfrak{ab}}(s;\mathbf{f},\overline{\mathbf{g}}) \ll 1,$$
$$A^{m+n+1}_{\mathfrak{a}}(s)\cdot\phi_{\mathfrak{ab}}(k,s;\mathbf{f},\overline{\mathbf{g}}) \ll (\log^{m+n}|k|+1)(|k|^{\sigma}+|k|^{1-\sigma}).$$

(iv) *The Fourier expansion*

$$E_{\mathfrak{a}}(\sigma_{\mathfrak{b}}z,s;\mathbf{f},\overline{\mathbf{g}}) = \delta^{m,n}_{0,0}\cdot\delta_{\mathfrak{ab}}y^s + \phi_{\mathfrak{ab}}(s;\mathbf{f},\overline{\mathbf{g}})y^{1-s} + \sum_{k\neq 0}\phi_{\mathfrak{ab}}(k,s;\mathbf{f},\overline{\mathbf{g}})W_s(kz) \tag{4.19}$$

agrees with (3.11), (3.12) *for* $\mathrm{Re}(s)>1$ *and, for all* $z\in\mathbb{H}$, *converges to a meromorphic function of* $s\in\mathbb{B}_r$.

(v) *For all $s \in \mathbb{B}_r$ and $z \in \mathbb{F}$ we have*

$$A_{\mathfrak{a}}^{m+n+1}(s) \cdot E_{\mathfrak{a}}(z,s;\mathbf{f},\overline{\mathbf{g}}) \ll y_{\mathbb{F}}(z)^{|\sigma-1/2|+1/2}. \tag{4.20}$$

PROOF. The case $m+n=0$ is given in [**JO08**, Thm. 6.1]. Note that it should have been stated there that $A_{\mathfrak{a}}(s)$ is not identically 0. This follows from its construction from $D(\lambda)$ in [**JO08**, Eq. (6.6)] and the fact that $D(0)=1$ from [**Iwa02**, p. 193].

The rest of the proof uses induction on $m+n$. This is the same as the proof of [**JO08**, Thm. 6.5], but based on the integral equation (4.16) and using the bounds from Section 4.1. □

The techniques of Diamantis-Sim [**DS08**] should also give the meromorphic continuation of $E_{\mathfrak{a}}(z,s;\mathbf{f},\overline{\mathbf{g}})$ to all $s \in \mathbb{C}$. There, they essentially find the continuation of $Q_{\mathfrak{a}}(z,s;\mathbf{f},\overline{\mathbf{g}})$, as defined in (4.7), by means of its spectral expansion. Their proof is for $\mathbf{f}$ empty but should carry over to our setting without difficulty; see [**DS08**, Theorem 3.4].

The first proof of the continuation of $E_{\mathfrak{a}}^{m,n}(z,s;f,g)$, corresponding to the case of equal cusp forms $f_1=\cdots=f_m$ and $g_1=\cdots=g_n$, was given in Petridis-Risager [**PR04**], following earlier work in Petridis [**Pet02**]. This method exploits the fact that

$$\chi_\varepsilon^f(\gamma) = \chi_\varepsilon(\gamma) := \exp\left(\varepsilon \int_{z_0}^{\gamma z_0} f(u)\,du\right) \tag{4.21}$$

is a character with $\chi_\varepsilon(\gamma_1\gamma_2) = \chi_\varepsilon(\gamma_1)\chi_\varepsilon(\gamma_2)$. Setting

$$E_{\mathfrak{b}}(z,s;\chi_\varepsilon) := \sum_{\gamma\in\Gamma_{\mathfrak{b}}\backslash\Gamma} \chi_\varepsilon(\gamma)\cdot \operatorname{Im}(\sigma_{\mathfrak{b}}^{-1}\gamma z)^s, \tag{4.22}$$

we find that $E_{\mathfrak{b}}(z,s;f)$ can be recovered as $\frac{d}{d\varepsilon}E_{\mathfrak{b}}(z,s;\chi_\varepsilon)|_{\varepsilon=0}$. To obtain $E_{\mathfrak{a}}^{m,n}(z,s;f,g)$, the character χ_ε is replaced with a product $\chi_{\varepsilon_1}\cdot\chi_{\varepsilon_2}\cdots$ with different parameters; see [**PR04**, Eq. (1.10)]. The meromorphic continuation of (4.22) and this many parameter generalization is achieved in [**Pet02**, **PR04**] by employing spectral deformation theory. Taking derivatives then gives the continuation of $E_{\mathfrak{a}}^{m,n}(z,s;f,g)$.

It is not clear if the methods of [**Pet02**, **PR04**] extend to proving the continuation of the general Eisenstein series $E_{\mathfrak{a}}(z,s;\mathbf{f},\overline{\mathbf{g}})$ studied in this paper. For example, if the integral in (4.21) is replaced by $C_{z_0}^{\gamma z_0}(f_1,f_2)$ then $\chi_\varepsilon(\gamma)$ will not be a character in general. By Example 5.2 (iv) and (5.3), it will satisfy the more complicated relation

$$\chi_\varepsilon(\gamma_1\gamma_2\gamma_3) = \frac{\chi_\varepsilon(\gamma_1\gamma_2)\chi_\varepsilon(\gamma_1\gamma_3)\chi_\varepsilon(\gamma_2\gamma_3)}{\chi_\varepsilon(\gamma_1)\chi_\varepsilon(\gamma_2)\chi_\varepsilon(\gamma_3)}.$$

4.3. Functional equations. By analogy with the classical Eisenstein series, we expect a relationship between the values of $E_{\mathfrak{a}}(z,s;\mathbf{f},\overline{\mathbf{g}})$ at s and $1-s$. In some cases we do have such a functional equation and to express it we set up the following notation.

Suppose Γ has h inequivalent cusps. Let $E(z,s;\mathbf{f},\overline{\mathbf{g}})$ be the $h\times 1$ column vector with entries $E_{\mathfrak{a}}(z,s;\mathbf{f},\overline{\mathbf{g}})$ as $\mathfrak{a}$ lists the inequivalent cusps. With the same ordering, let $\phi(s;\mathbf{f},\overline{\mathbf{g}})$ be the $h\times h$ matrix with entries $\phi_{\mathfrak{a}\mathfrak{b}}(s;\mathbf{f},\overline{\mathbf{g}})$.

The formal series version of E above is $\mathcal{E}(z,s)$, the $h\times 1$ column vector with entries $\mathcal{E}_{\mathfrak{a}}(z,s)$. It satisfies the vector version of (3.4):

$$\mathcal{E}(z,s)=\sum_{c,d\geqslant 0}\sum_{i_1,\dots,i_c,j_1,\dots,j_d} E(z,s;f_{i_1},\dots,f_{i_c},\overline{g_{j_1}},\dots,\overline{g_{j_d}})\cdot X_{i_1}\cdots X_{i_c}\overline{Y_{j_1}}\cdots\overline{Y_{j_d}}.$$

The formal series version of ϕ is $\Phi(s)$, the $h\times h$ matrix with entries $\Phi_{\mathfrak{a}\mathfrak{b}}(s)$. It satisfies the matrix version of (3.14):

$$\Phi(s)=\sum_{c,d\geqslant 0}\sum_{i_1,\dots,i_c,j_1,\dots,j_d} \phi(s;f_{i_1},\dots,f_{i_c},\overline{g_{j_1}},\dots,\overline{g_{j_d}})\cdot X_{i_1}\cdots X_{i_c}\overline{Y_{j_1}}\cdots\overline{Y_{j_d}}.$$

We may give a simple reformulation of [**JO08**, Thm. 7.1] as follows.

THEOREM 4.4. *Define $\mathcal{E}(z,s)$ with modular symbols $I^{a}_{\gamma a}(f)$ and $J^{a}_{\gamma a}(g)$ depending on single cusp forms. It satisfies the functional equation*

$$\Phi(1-s)\mathcal{E}(z,s)=\mathcal{E}(z,1-s) \tag{4.23}$$

with

$$\Phi(1-s)\Phi(s)=I \tag{4.24}$$

for I the $h\times h$ identity matrix.

The formal series equations (4.23) and (4.24) are equivalent to showing the matrix equations

(4.25)

$$\sum_{i=0}^{m}\sum_{j=0}^{n}\phi(1-s;\overbrace{f,\dots,f}^{i},\overbrace{\overline{g},\dots,\overline{g}}^{j})E(z,s;\overbrace{f,\dots,f}^{m-i},\overbrace{\overline{g},\dots,\overline{g}}^{n-j})=E(z,1-s;\overbrace{f,\dots,f}^{m},\overbrace{\overline{g},\dots,\overline{g}}^{n}),$$

(4.26)

$$\sum_{i=0}^{m}\sum_{j=0}^{n}\phi(1-s;\overbrace{f,\dots,f}^{i},\overbrace{\overline{g},\dots,\overline{g}}^{j})\phi(s;\overbrace{f,\dots,f}^{m-i},\overbrace{\overline{g},\dots,\overline{g}}^{n-j})=\begin{cases} I, & \text{if } m=n=0;\\ 0, & \text{otherwise}\end{cases}$$

for all $m,\,n\in\mathbb{Z}_{\geqslant 0}$. The $m=n=0$ cases of (4.25) and (4.26) are the classical functional equations

$$\phi(1-s)E(z,s)=E(z,1-s),\qquad \phi(1-s)\phi(s)=I$$

as shown in [**Iwa02**, Sect. 6.3].

In fact the functional equation of Theorem 4.4 appears already in Petridis [**Pet02**]. We may write the functional equation [**Pet02**, Eq. 1.9] of the series (4.22) as the matrix equation

$$\phi(1-s;\chi^{f}_{\varepsilon}\cdot\chi^{\overline{g}}_{\varepsilon'})E(z,s;\chi^{f}_{\varepsilon}\cdot\chi^{\overline{g}}_{\varepsilon'})=E(z,1-s;\chi^{f}_{\varepsilon}\cdot\chi^{\overline{g}}_{\varepsilon'}). \tag{4.27}$$

Now we see that (4.27), treated as a relation of formal series in ε and ε', agrees with (4.23).

See also [**Ris03**, Thm. 43] where a functional equation equivalent to (4.23) is proved, but with scattering matrix $\Phi(1-s)$ defined differently.

It is natural to ask whether $\mathcal{E}(z,s)$ satisfies the functional equation (4.23), or some other one, when it contains modular symbols $I^{a}_{\gamma a}(\mathbf{f})$ and $J^{a}_{\gamma a}(\mathbf{g})$ depending on more than one cusp form.

5. Higher-order automorphic forms and maps

We see next how the series $E_{\mathfrak{b}}(z,s;\mathbf{f},\overline{\mathbf{g}})$ and the iterated integrals $C_a^z(\mathbf{f})$ and $C_a^{\gamma a}(\mathbf{f})$ fit into a larger framework.

5.1. Higher-order forms. Following the description in [**JO08**, Sect. 3], we may define a sequence $\mathcal{A}^n(\Gamma)$ of sets of smooth functions from $\mathbb{H} \to \mathbb{C}$ recursively as follows. Let $\mathcal{A}^0(\Gamma) := \{\mathbb{H} \to 0\}$ and for $n \in \mathbb{Z}_{\geqslant 1}$ set

$$\mathcal{A}^n(\Gamma) := \left\{\psi \,\middle|\, \psi(\gamma z) - \psi(z) \in \mathcal{A}^{n-1}(\Gamma) \quad \text{for all} \quad \gamma \in \Gamma\right\}.$$

Elements of $\mathcal{A}^n(\Gamma)$ are called *nth-order automorphic forms*, naturally forming a vector space over $\mathbb{C}$. The classical Γ-invariant functions, such as $E_{\mathfrak{a}}(z,s)$, are in $\mathcal{A}^1(\Gamma)$ and so are first-order. If we let $\gamma \in \Gamma$ act on ψ by $(\psi|\gamma)(z) := \psi(\gamma z)$, and extend this action to all $\mathbb{C}[\Gamma]$ by linearity, then we see that $\psi \in \mathcal{A}^n(\Gamma)$ if and only if

$$\psi\big|(\gamma_1 - I)(\gamma_2 - I)\cdots(\gamma_n - I) = 0 \quad \text{for all} \quad \gamma_1, \gamma_2, \ldots, \gamma_n \in \Gamma.$$

Inductive arguments show that

$$\mathcal{A}^m(\Gamma) \subseteq \mathcal{A}^n(\Gamma) \quad \text{for all} \quad 0 \leqslant m \leqslant n \tag{5.1}$$

and if $\phi(z) \in \mathcal{A}^m(\Gamma)$ and $\psi(z) \in \mathcal{A}^n(\Gamma)$ then

$$\phi(z)\cdot\psi(z) \in \mathcal{A}^{m+n-1}(\Gamma)$$

for $m+n \geqslant 1$. With (3.8), $E_{\mathfrak{a}}(\gamma z,s;f) - E_{\mathfrak{a}}(z,s;f) \in \mathcal{A}^1(\Gamma)$ implying that $E_{\mathfrak{a}}(z,s;f) \in \mathcal{A}^2(\Gamma)$. So $E_{\mathfrak{a}}(z,s;f)$ is a second-order form satisfying (1.3).

PROPOSITION 5.1. *For all m, $n \in \mathbb{Z}_{\geqslant 0}$ we have*

(i) $C_a^z(f_1,\ldots,f_m) \in \mathcal{A}^{m+1}(\Gamma)$,
(ii) $E_{\mathfrak{a}}(z,s;f_1,\ldots,f_m,\overline{g_1},\ldots,\overline{g_n}) \in \mathcal{A}^{m+n+1}(\Gamma)$.

PROOF. Part (i) is proved by induction on m. The $m = 0$ case is true since $C_a^z() = 1$, so assume $m \geqslant 1$. With

$$I_a^{\gamma z} = I_a^{\gamma a} I_{\gamma a}^{\gamma z} = I_a^{\gamma a} I_a^z$$

we find

$$C_a^{\gamma z}(f_1,f_2,\ldots,f_m) - C_a^z(f_1,f_2,\ldots,f_m) = \sum_{j=1}^{m} C_a^{\gamma a}(f_1,\ldots,f_j) C_a^z(f_{j+1},\ldots,f_m). \tag{5.2}$$

By induction $C_a^z(f_{j+1},\ldots,f_m) \in \mathcal{A}^{m-j+1}(\Gamma)$. With (5.1), the right side of (5.2) is in $\mathcal{A}^m(\Gamma)$ and hence $C_a^z(f_1,\ldots,f_m) \in \mathcal{A}^{m+1}(\Gamma)$, completing the induction.

Part (ii) has a similar proof using (3.8). □

Note that results equivalent to Proposition 5.1 are proved in [**DS08**, Sect. 3.2].

5.2. Higher-order maps. As in [**IO09**, Sect. 10], one can define a related sequence $\mathrm{Hom}^{[n]}(\Gamma,\mathbb{C})$ of sets of functions from Γ to $\mathbb{C}$ as follows. Let $\mathrm{Hom}^{[0]}(\Gamma,\mathbb{C}) := \{\Gamma \to 0\}$. With the notation $L_\delta(\gamma) := L(\gamma\delta) - L(\gamma)$ and $n \in \mathbb{Z}_{\geqslant 1}$ define

$$\mathrm{Hom}^{[n]}(\Gamma,\mathbb{C}) := \left\{L : \Gamma \to \mathbb{C} \,\middle|\, L_\delta \in \mathrm{Hom}^{[n-1]}(\Gamma,\mathbb{C}) \quad \text{for all} \quad \delta \in \Gamma\right\}.$$

For $L:\Gamma\to\mathbb{C}$ and $\gamma\in\Gamma$, set $L|\gamma := L(\gamma)$ and extend this linearly to all $\mathbb{C}[\Gamma]$. Then $L\in\mathrm{Hom}^{[n]}(\Gamma,\mathbb{C})$ if and only if

$$L\big|(\gamma_1-I)(\gamma_2-I)\cdots(\gamma_n-I)=0 \quad \text{for all} \quad \gamma_1,\gamma_2,\dots,\gamma_n\in\Gamma.$$

We see that $\mathrm{Hom}^{[1]}(\Gamma,\mathbb{C})$ is the space of constant functions. Elements L of $\mathrm{Hom}^{[2]}(\Gamma,\mathbb{C})$ satisfy

$$L(\gamma_1\gamma_2)-L(\gamma_1)-L(\gamma_2)+L(I)=0 \quad \text{for all} \quad \gamma_1,\gamma_2\in\Gamma,$$

making $\gamma\mapsto L(\gamma)-L(I)$ a homomorphism into the additive part of $\mathbb{C}$. Similarly, elements L of $\mathrm{Hom}^{[3]}(\Gamma,\mathbb{C})$ satisfy

$$L(\gamma_1\gamma_2\gamma_3)-L(\gamma_1\gamma_2)-L(\gamma_1\gamma_3)-L(\gamma_2\gamma_3)+L(\gamma_1)+L(\gamma_2)+L(\gamma_3)-L(I)=0 \tag{5.3}$$

for all $\gamma_1,\gamma_2,\gamma_3\in\Gamma$. We may call elements of $\mathrm{Hom}^{[n]}(\Gamma,\mathbb{C})$ *nth-order maps* from Γ to $\mathbb{C}$ and they form a complex vector space. Inductive arguments demonstrate that

$$\mathrm{Hom}^{[m]}(\Gamma,\mathbb{C})\subseteq\mathrm{Hom}^{[n]}(\Gamma,\mathbb{C}) \quad \text{for all} \quad 0\leqslant m\leqslant n$$

and if $L\in\mathrm{Hom}^{[m]}(\Gamma,\mathbb{C})$ and $L'\in\mathrm{Hom}^{[n]}(\Gamma,\mathbb{C})$ then

$$L\cdot L'\in\mathrm{Hom}^{[m+n-1]}(\Gamma,\mathbb{C}) \tag{5.4}$$

for $m+n\geqslant 1$. If γ is an elliptic element of Γ, with $\gamma^N=I$ for some $N>0$, then a similar proof to [**IO09**, Lemma 17] shows $L(\gamma)=L(I)$ for all $L\in\mathrm{Hom}^{[n]}(\Gamma,\mathbb{C})$ with $n\geqslant 0$.

EXAMPLE 5.2. We have the following examples of higher-order maps:

(i) The modular symbol map $\gamma\mapsto\langle\gamma,f\rangle$ is in $\mathrm{Hom}^{[2]}(\Gamma,\mathbb{C})$. In fact, if we define

$$\mathrm{Hom}_0^{[n]}(\Gamma,\mathbb{C}):=\left\{L\in\mathrm{Hom}^{[n]}(\Gamma,\mathbb{C})\,\middle|\,L(\gamma)=0 \quad \text{for all parabolic} \quad \gamma\in\Gamma\right\}$$

then $\langle\gamma,f\rangle$ is in $\mathrm{Hom}_0^{[2]}(\Gamma,\mathbb{C})$. Note that $L\in\mathrm{Hom}_0^{[n]}(\Gamma,\mathbb{C})$ implies $L(I)=0$.

(ii) It now follows from (5.4) that, for example,

$$\langle\gamma,f\rangle^m\overline{\langle\gamma,f\rangle}^n\in\mathrm{Hom}_0^{[m+n+1]}(\Gamma,\mathbb{C}). \tag{5.5}$$

(iii) If $\psi\in\mathcal{A}^n(\Gamma)$ then, for fixed $z_0\in\mathbb{H}$, $L(\gamma)$ defined as $\psi(\gamma z_0)$ is in $\mathrm{Hom}^{[n]}(\Gamma,\mathbb{C})$.

(iv) As functions from Γ to $\mathbb{C}$, the iterated integral $C_a^{\gamma b}(f_1,f_2,\dots,f_n)$ and its complex conjugate are in $\mathrm{Hom}^{[n+1]}(\Gamma,\mathbb{C})$. This follows from the previous example and Proposition 5.1 (i). With the identity (2.9), we may allow a and b to be in $\mathbb{H}\cup\{\mathtt{cusps}\}$. By Corollary 2.3 these functions are in the subspace $\mathrm{Hom}_0^{[n+1]}(\Gamma,\mathbb{C})$ for $n\geqslant 1$.

(v) An interesting third-order map $\theta_{\mathfrak{a}}$ is obtained in the paper [**JOS**]. The Kronecker limit formula gives the first two terms in the Laurent expansion of $E_{\mathfrak{a}}(z,s)$ at $s=1$. From the second term we may derive the *modular Dedekind symbol* $S_{\mathfrak{a}}$ which is a map from Γ to $\mathbb{R}$. Doing the same with the Eisenstein series $E_{\mathfrak{a}}^{m,m}(z,s;f,f)$ from (3.10) for $m\geqslant 1$ produces the *higher-order modular Dedekind symbol* $S_{\mathfrak{a}}^*$ which is independent of m and is also a map from Γ to $\mathbb{R}$. Then, as shown in [**JOS**, Sect. 5.5]

$$\theta_{\mathfrak{a}}:=S_{\mathfrak{a}}^*-S_{\mathfrak{a}}\in\mathrm{Hom}^{[3]}(\Gamma,\mathbb{C}).$$

However $\theta_\mathfrak{a}$ is not zero on all parabolic elements and so cannot be expressed in terms of products such as (5.5) or the iterated integral $C_a^{\gamma b}(f_1, f_2)$. It would be interesting to understand how these higher-order modular Dedekind symbols fit into the context of the noncommutative Dedekind symbols Manin introduces in [**Man14**].

5.3. The subspace $\mathrm{H}^{[n]}(\Gamma, \mathbb{C})$. For fixed $a \in \mathbb{H} \cup \{\mathtt{cusps}\}$, we see by Example 5.2 (iv) and (5.4) that

$$C_a^{\gamma a}(f_1, \dots, f_j)\overline{C_a^{\gamma a}(f_{j+1}, \dots, f_m)} \tag{5.6}$$

is in $\mathrm{Hom}_0^{[m+1]}(\Gamma, \mathbb{C})$ for $m \geqslant 1$. For $n \geqslant 1$, let $\mathrm{H}^{[n]}(\Gamma, \mathbb{C})$ be the subspace of $\mathrm{Hom}_0^{[n]}(\Gamma, \mathbb{C})$ spanned by maps of the form (5.6) for $0 \leqslant m \leqslant n-1$. It follows from the identity (2.9) for changing the base point that $\mathrm{H}^{[n]}(\Gamma, \mathbb{C})$ is independent of a.

The next result shows that including more iterated integrals in the product and changing endpoints does not take you out of the space $\mathrm{H}^{[n]}(\Gamma, \mathbb{C})$, provided that the number of cusp forms used is $\leqslant n-1$.

PROPOSITION 5.3. *We have*

$$\prod_{i=1}^{u} C_{a_i}^{\gamma b_i}(f_{i_1}, \dots, f_{i_{r(i)}}) \prod_{j=1}^{v} \overline{C_{a'_j}^{\gamma b'_j}(f_{j_1}, \dots, f_{j_{t(j)}})} \in \mathrm{H}^{[n]}(\Gamma, \mathbb{C}) \tag{5.7}$$

for a_i, b_i, a'_j, b'_j in $\mathbb{H} \cup \{\mathtt{cusps}\}$ and $r(1) + \dots + r(u) + t(1) + \dots + t(v) \leqslant n-1$.

PROOF. With (2.9) we may rewrite (5.7) as a linear combination of terms of the same form as (5.7) but with every a_i, b_i, a'_j, b'_j replaced by a fixed a.

We next introduce the *shuffle permutations*; see for example [**DH13**, Lemma 1.1, (iv)]. Let $sh(j,k)$ be the set of all permutations ρ of $\{1, 2, \dots, j+k\}$ such that

$$\rho(1) < \dots < \rho(j) \quad \text{and} \quad \rho(j+1) < \dots < \rho(j+k).$$

Then the *shuffle relation* implies

$$C_a^{\gamma a}(f_1, \dots, f_j)C_a^{\gamma a}(f_{j+1}, \dots, f_k) = \sum_{\rho \in sh(j,k)} C_a^{\gamma a}(f_{\rho(1)}, f_{\rho(2)}, \dots, f_{\rho(j+k)}). \tag{5.8}$$

Therefore products of two iterated integrals may be expressed as a sum of single iterated integrals. Applying this repeatedly then gives (5.7) as a linear combination of terms of the form (5.6) as desired. □

For the inclusion

$$\mathrm{H}^{[n]}(\Gamma, \mathbb{C}) \subseteq \mathrm{Hom}_0^{[n]}(\Gamma, \mathbb{C}), \tag{5.9}$$

we clearly have equality when $n = 1$ and also when $n = 2$ by [**GO03**, Prop. 2.1], for example. Do we have equality in (5.9) for higher values of n? If not, how are the extra nth-order maps on the right described?

References

[BD16] Roelof Bruggeman and Nikolaos Diamantis, *Fourier coefficients of Eisenstein series formed with modular symbols and their spectral decomposition*, J. Number Theory **167** (2016), 317–335, DOI 10.1016/j.jnt.2016.03.009. MR3504050

[CDO02] G. Chinta, N. Diamantis, and C. O'Sullivan, *Second order modular forms*, Acta Arith. **103** (2002), no. 3, 209–223, DOI 10.4064/aa103-3-2. MR1905087

[CI13] YoungJu Choie and Kentaro Ihara, *Iterated period integrals and multiple Hecke L-functions*, Manuscripta Math. **142** (2013), no. 1-2, 245–255, DOI 10.1007/s00229-013-0605-2. MR3081007

[DH13] Anton Deitmar and Ivan Horozov, *Iterated integrals and higher order invariants*, Canad. J. Math. **65** (2013), no. 3, 544–552, DOI 10.4153/CJM-2012-020-8. MR3043039

[DS08] Nikolaos Diamantis and David Sim, *The classification of higher-order cusp forms*, J. Reine Angew. Math. **622** (2008), 121–153, DOI 10.1515/CRELLE.2008.067. MR2433614

[GO03] Dorian Goldfeld and Cormac O'Sullivan, *Estimating additive character sums for Fuchsian groups*, Ramanujan J. **7** (2003), no. 1-3, 241–267, DOI 10.1023/A:1026255414488. Rankin memorial issues. MR2035805

[Gol99] Dorian Goldfeld, *The distribution of modular symbols*, Number theory in progress, Vol. 2 (Zakopane-Kościelisko, 1997), de Gruyter, Berlin, 1999, pp. 849–865. MR1689548

[Hor15] Ivan Horozov, *Noncommutative Hilbert modular symbols*, Algebra Number Theory **9** (2015), no. 2, 317–370, DOI 10.2140/ant.2015.9.317. MR3320846

[IO09] Özlem Imamoḡlu and Cormac O'Sullivan, *Parabolic, hyperbolic and elliptic Poincaré series*, Acta Arith. **139** (2009), no. 3, 199–228, DOI 10.4064/aa139-3-1. MR2545926

[Iwa02] Henryk Iwaniec, *Spectral methods of automorphic forms*, 2nd ed., Graduate Studies in Mathematics, vol. 53, American Mathematical Society, Providence, RI; Revista Matemática Iberoamericana, Madrid, 2002. MR1942691

[JO08] Jay Jorgenson and Cormac O'Sullivan, *Unipotent vector bundles and higher-order non-holomorphic Eisenstein series* (English, with English and French summaries), J. Théor. Nombres Bordeaux **20** (2008), no. 1, 131–163. MR2434161

[JOS] Jay Jorgenson, Cormac O'Sullivan, and Lejla Smajlovic. Modular Dedekind symbols associated to Fuchsian groups and higher-order Eisenstein series. Submitted. Available on the arXiv.

[KZ03] Peter Kleban and Don Zagier, *Crossing probabilities and modular forms*, J. Statist. Phys. **113** (2003), no. 3-4, 431–454, DOI 10.1023/A:1026012600583. MR2013692

[Man72] Ju. I. Manin, *Parabolic points and zeta functions of modular curves* (Russian), Izv. Akad. Nauk SSSR Ser. Mat. **36** (1972), 19–66. MR0314846

[Man06] Yuri I. Manin, *Iterated integrals of modular forms and noncommutative modular symbols*, Algebraic geometry and number theory, Progr. Math., vol. 253, Birkhäuser Boston, Boston, MA, 2006, pp. 565–597, DOI 10.1007/978-0-8176-4532-8_10. MR2263200

[Man14] Yuri I. Manin, *Non-commutative generalized Dedekind symbols*, Pure Appl. Math. Q. **10** (2014), no. 2, 245–258, DOI 10.4310/PAMQ.2014.v10.n2.a3. MR3267346

[O'S00] Cormac O'Sullivan, *Properties of Eisenstein series formed with modular symbols*, J. Reine Angew. Math. **518** (2000), 163–186, DOI 10.1515/crll.2000.003. MR1739405

[Pet02] Yiannis N. Petridis, *Spectral deformations and Eisenstein series associated with modular symbols*, Int. Math. Res. Not. **19** (2002), 991–1006, DOI 10.1155/S1073792802111159. MR1903327

[PR04] Y. N. Petridis and M. S. Risager, *Modular symbols have a normal distribution*, Geom. Funct. Anal. **14** (2004), no. 5, 1013–1043, DOI 10.1007/s00039-004-0481-8. MR2105951

[PR18] Yiannis N. Petridis and Morten S. Risager, *Arithmetic statistics of modular symbols*, Invent. Math. **212** (2018), no. 3, 997–1053, DOI 10.1007/s00222-017-0784-7. MR3802302

[Ris03] M. S. Risager. *Automorphic forms and modular symbols.* PhD thesis, University of Aarhus, 2003.

Department of Mathematics, The City College of New York, New York, New York 10031

Email address: `gchinta@ccny.cuny.edu`

Department of Mathematics, City University of New York, Bronx Community College, New York, New York 10453

Email address: `ivan.horozov@bcc.cuny.edu`

Department of Mathematics, The CUNY Graduate Center, New York, New York 10016

Email address: `cosullivan@gc.cuny.edu`

Contemporary Mathematics
Volume **732**, 2019
https://doi.org/10.1090/conm/732/14795

An annotated discussion of a panel presentation on improving diversity in mathematics

Adriana Espinosa

ABSTRACT. On July 19, 2016 an inter-disciplinary panel met during the 3rd Building Bridges conference in Sarajevo, BiH to discuss issues related to the encouragement and retention of a diverse community in mathematics. This article summarizes the points discussed by the panel, highlighting some empirical evidence related to these points. The article concludes with a call to continue these types of discussions, and to raise awareness about the need to broaden the definition of diversity to include aspects other than gender and/or ethnicity. We hope this article encourages the participation of other mathematicians, and scientists alike to join the discussion and help generate policies and/or programs in efforts to increase diversity in mathematics and all other STEM fields.

1. Introduction

Over the past decades, jobs in science, technology, engineering and mathematics (STEM) have increased faster than jobs in non-STEM occupations in the US ([**22**], [**64**]; [**73**]), representing a large increase in demand for STEM-educated individuals. Women are crucial in satisfying this increased demand. Yet, while women make up approximately half of the general work force in the US, they represent less than a quarter of the STEM workforce ([**39**], [**74**]). According to the National Academy of Sciences, a nation's chance of remaining globally competitive rests on its success to address the underrepresentation of women and other minority groups in STEM ([**52**]). Such declaration underscores an instrumental role for the scientific community toward increasing diversity in mathematics; the core of all STEM disciplines. Along these lines, the organizers of the *3rd EU/US Summer School and Workshop on Automorphic forms and Related Topics* arranged a cross-disciplinary panel on the topic of encouraging and retaining a diverse community in mathematics. The panel, which addressed all conference attendees on July 19th, 2016, was moderated by L. Walling from the University of Bristol. Members of the panel included L. Smajlović, Department of Economics, University of Sarajevo; L. Thompson, Department of Mathematics, Oberlin College; G. Davidoff, Department of Mathematics, Mount Holyoke College; J. Jorgenson, Department of Mathematics, The City College of New York, and myself. As the only non-mathematician in the panel, I am honored to have been included in the discussion, and remain extremely appreciative to know that my contribution was well-received. Through conversations

2010 *Mathematics Subject Classification.* Primary 00-01.

with individuals who thoroughly know the field, I have gained a new outlook on how to visualize and attempt to address the problem of increasing diversity in mathematics; one of my research interests. I will certainly use this newfound knowledge to further shape aspects of my research going forward. As the discourse regarding the diversification of STEM is generally divided across two areas, recruitment and retention, the panel discussion was structured accordingly. Below I summarize the key points discussed by the panel, while highlighting relevant empirical evidence from various fields.

2. Issues of retention

Multiple studies have found that females in STEM occupations, including academia, have higher exit rates than males as well as other females in non-STEM related occupations ([**24**], [**33**]). While several interpersonal factors can in part explain the disproportionate attrition of women in STEM occupations, researchers have identified several salient environmental indicators resulting from implicit biases (i.e., unconscious attitudes and stereotypes that affect our judgement). These are described below.

2.1. Invalidation of women's work. A well-recognized hindrance to the retention of women in STEM careers is due to a systemic lack of recognition or the invalidation of their work. Before Andrew Wiles solved Fermat's Last Theorem, Sophie Germain provided a general approach. Before Anthony Hewish and Martin Ryle received the Nobel Prize in physics, Jocelyn Bell Burnell discovered pulsars. Before Joshua Lederberg, George Beadle and Edward Tatum received the Nobel Prize for physiology, Esther Lederberg's work on replica plating had laid the foundation for their discoveries. Nettie Stevens's work on mealworms led to groundbreaking results in genetics accredited to Thomas Hunt Morgan; and the list goes on ([**41**], [**63**]).

The invalidation of women's scientific contributions relative to that of men's is referred to in the Social Sciences as The Matilda Effect ([**63**]); a corollary of a much broader concept known as The Matthew Effect ([**59**], [**50**]). One of the best examples known to date of the Matilda Effect is encapsulated in the narrative of Ben Barres ([**2**]). Ben Barres was a renowned neurobiologist at Stanford University's School of Medicine, and the first openly transgendered member of the US National Academy of Sciences. Prior to transitioning to male, Ben went by the name of Barbara who after the transition was known to most colleagues as Ben's sister. In his narrative Barres explains how, despite the transition having no impact on his scientific output, his colleagues often described his work to be stronger than that of his sister's. In fact, as he explains, once Barbara became Ben, the scientific value of his work was validated by his academic peers. In many cases, the Matilda Effect is a result of implicit bias, which often manifests as disproportionate awards, citations, funding, promotions, conference invitations, and ultimately lower retention of women in STEM ([**47**], [**38**], [**42**], [**2**]).

Racial ethnic minorities (REMs) are also considerably underrepresented, with 6 percent of the STEM workforce being Black and 7 percent Hispanic relative to their share in the general workforce of 11 and 15 percent, respectively ([**39**]). The extent to which ethnicity interplays with gender in observed manifestations of the Matilda Effect is unclear, yet worthy of further investigation. For example, a well-documented racial stereotype suggests that Asian individuals (in particular

Chinese) work harder and are better at math than others ([**12**]). The extent to which this stereotype translates into Asian women not being as susceptible to the Matilda Effect as non-Asian women is under-investigated. Nevertheless, one should not take comfort in the realization that some implicit biases cancel each other out. Instead, we should acknowledge that nobody is free from biases, understand our own, and recognize how they may affect our behaviors, at least in the context of increasing diversity in mathematics and in STEM more generally. As a key example, implicit bias can lead to differences in the evaluation and writing of letters of recommendation for junior male and female scientists, and these differences have been shown to obstruct retention efforts for women ([**43**]).

2.2. Inequitable letters of recommendation. Letters of recommendation are crucial for decisions related to the retention and promotion of women and other underrepresented minorities in STEM ([**63**]). Empirical evidence has demonstrated that unconscious gender biases can impact the way in which letters of recommendation are written and interpreted ([**5**], [**29**]), which in turn affects how applicants are rated ([**35**]). More specifically, empirical findings show that within male-dominated fields, letters using comparable language to describe equally-qualified male and female applicants may translate into lower ratings of female qualifications relative to that of males.

Moreover, content analyses of letters of recommendation for academic positions in various fields show differences in how males and females are described in said letters ([**43**]). Specifically, women are habitually described with communal language (i.e., nurturing, good colleague, nice) and men with agency language (i.e., leader, smart, etc.) ([**65**]). Research also indicates that letters of recommendation for females tend to be shorter on average (i.e., less than 10 lines) than letters for males which are 50 or more lines on average ([**72**]).

2.3. Identifying our biases. While a variety of scientific approaches to gauge implicit bias exist, including physiological responses to external stimuli, a widely used measure is the Implicit Association Test (IAT) ([**53**], [**25**]). The IAT is currently made available by a non-profit organization called Project Implicit, which was founded by three scientists with extensive research on implicit bias, and their manifestations within society. A call to action was made from the panel to all attendees to go to the Project Implicit website [1] and take one of the many IAT tests available. It is only through the collective effort of all members of the STEM domain that we will begin to rectify the retention problem from within.

Yet, the diversification of any field can not only be centered on issues pertaining to retention. Efforts must also be placed to recruit a new diverse population of individuals. As recruitment begins in school, the panel discussion morphed into what we can do as faculty to engage students, particularly those who traditionally avoid mathematics and other STEM areas.

3. Issues of recruitment

A large longitudinal study of over 6,000 high school students in Australia revealed that only 13.8% students expressed interest in pursuing a STEM career, with only 0.12% (8 students) expressing interest in mathematics ([**32**]). The likelihood of expressing interest in a STEM career was higher for males than it was for females,

[1] https://implicit.harvard.edu/implicit/aboutus.html

with males up to 3 times more likely than females to express such interest. The literature has identified several key factors explaining the disparate recruitment of women and other minorities in STEM.

3.1. The role of introductory mathematics. Introductory mathematics courses have been identified as a "critical filter" that thwarts the entrance of women and other minorities into quantitative disciplines ([**26**], [**46**], [**34**], [**60**], [**71**]). Among all introductory mathematics courses, first year calculus has been substantially recognized as a catalyst ([**69**], [**11**], [**17**]). This issue was raised within the panel discussion, when one panelist recounted that in this panelist's college, introductory mathematics was a consistent deterrent for females to pursue a STEM major. In accordance with such observation, a recent article using a sample of 2,266 students, reports that females are 1.5 times more likely than males to be deterred from a STEM major after taking introductory calculus ([**18**]). Yet, the same article indicates that among students with higher than average performance in mathematics, females reported lower self-confidence in their ability than males, and they did so before and after the course ended. In contrast to common beliefs, this latter finding suggests that first semester calculus doesn't weed out those without the ability, but rather it weeds out those without the self-confidence, many of whom are female. The extent to which the same reason applies for individuals from other underrepresented groups is yet to be understood, but worthy of investigation. According to the findings stated above, if females were to continue in their intended STEM majors beyond calculus I at the same rate as males, women could represent as high as 37% of the STEM workforce rather than the 25% they represent today ([**18**], [**39**]).

Similarly, studies also suggest improving student perceptions regarding the usefulness of mathematics may increase the pool of women who pursue mathematics and STEM-related disciplines altogether ([**60**]). These studies highlight the role of attitudinal factors in early mathematics education; a segment of the literature that began in the mid-1970's as an attempt to explain the gender gap in mathematics performance, and consequently in the pursuit of mathematics-related careers.

3.2. Attitudes about mathematics. Unlike other disciplines, mathematics is socially viewed as the source of an identity that is either adopted or rejected, and may even be treated with disgust. Whereas it is unprecedented to self-declare oneself as not a words person, many individuals accept the identity of not being a numbers person ([**6**]); an identification that affects academic achievement and ultimately preparation for the workforce ([**36**], [**58**], [**54**]). This segment of the discussion began with a description of a meme depicting a librarian who complains about the library's mathematics books being placed in the incorrect section (i.e., the horror books section). Such image was circulated in social media at the time of the conference, and albeit humorous is an example of a socially acceptable association of mathematics with fear-inducing and avoidable objects.

Similar negative expressions about mathematics are captured in a compelling article highlighting reasons given by fifteen thousand adolescents in England for not continuing their study of mathematics ([**8**]). Declarations such as "because it SUCKS and I wouldn't want to spend any more of my time looking at algebra and other crap", as well as "I hate mathematics and I would rather die" were among them ([**8**] page 8).

Yet, such emotions are not innate, but rather learned as eloquently stated by S. Lang's Math Talks to Undergraduates ([**40**]) "... there is considerable evidence that we are programmed naturally to like mathematics... Every 5 year old kid I have ever met likes to add numbers, or subtract numbers. What happens afterwards is something else." ([**40**], p. vii). According to a panelist experienced in teaching mathematics to younger children, a child can develop a dislike of mathematics as early as 7 years of age. Empirical studies confirm the panelist's observation, and show that although attitudes about mathematics worsen with age, for many students they begin to deteriorate during the primary school years ([**37**], [**14**], [**75**]). Psychologists have identified the main reason for such decline in the appreciation of mathematics to be math anxiety, feelings of fear, apprehension and tension experienced when performing mathematical calculations ([**4**], [**1**]). Evidence suggests that students begin to experience math anxiety in first grade and such anxiety leads to a decline in their appreciation of the subject, and consequently its avoidance ([**61**]).

Not surprisingly, many adults share negative views about mathematics, and overwhelming evidence shows that these attitudes map onto that of children exposed to them (e.g., as parents, teachers, and caretakers) ([**45**], [**4**], [**16**]). A survey conducted by Change the Equation, an organization seeking to improve STEM literacy in the US, shows that approximately 30% of American adults report they would rather clean a bathroom than engage in mathematics problem-solving. As the literature on math anxiety would predict, feelings of anxiety, frustration, inadequacy and worry were commonly reported as the result of doing mathematics, and females were more likely than males to report such feelings[2].

Although in the classroom setting educators may not be able to prevent the negative influence of social attitudes about mathematics on that of their students', they can promote change by highlighting the positive side of learning and doing mathematics. In fact, one panelist shared that they had a dislike of mathematics upon taking calculus for the first time. However, a professor introduced the panelist to more provocative calculus applications that helped this panelist develop an appreciation for mathematics, and pursue a career in mathematics.

3.3. Mentors and role models. In the case described above, the panelist's professor became a mentor, and mentors are integral in influencing students' career interests and choices, including the pursuit of STEM careers ([**70**], [**10**], [**7**], [**62**], [**21**]). In addition to capturing the interest of students via thought-provoking applications of basic concepts, as was the case for the panelist mentioned above, mentors have two other key roles within the context of STEM.

First, a mentor can help a student develop self-confidence in their own mathematics ability ([**10**], [**13**], [**31**]), commonly lacking among underrepresented groups, especially among women. Empirical evidence has demonstrated that self-confidence in mathematics ability is associated with improvements in student performance and learning of mathematics ([**23**], [**20**], [**57**], [**76**], [**56**]), and negatively related to math anxiety and math avoiding behaviors ([**1**]). In the adolescent study mentioned in the prior section, reasons given by female students for not further pursuing mathematics studies were also indicative of a lack of self-confidence in the subject: "I don't feel confident enough to do so even if I'm at an A grade" and "because I have no confidence I will get stressed..." ([**8**], page 10).

[2]http://changetheequation.org/press/new-survey-americans-say-%E2%80%9Cwe%E2%80%99re-not-good-math%E2%80%9D

Female's deficiencies in self-confidence in mathematics relative to their male counterparts have also been found to exist among high achieving students interested in pursuing a STEM major ([**18**], [**39**]). Therefore, by asserting their confidence in the student's ability, a mentor can efficiently prevent the attrition of those who lack the confidence, but are well-suited to succeed in a STEM career. However, mentors should also be cognizant that the relation between self-confidence, performance, and math avoidance is not monotonic. Overconfidence may be as bad as or worse than low self-confidence ([**19**]). A mentor must therefore help the student develop a growth mindset, which will equip the student with the right amount of self-confidence to properly address failure and other barriers ([**15**]).

Second, a mentor can serve as a role model that mitigates the effect of stereotypes. Within the academic context, studies have shown that stereotypes hinder performance, and elicit disengagement and avoidant behaviors among stereotyped individuals ([**55**], [**68**]). The social psychology literature labels the path leading to such outcomes as stereotype threat; an individual's awareness that their behavior will be judged in accordance with common negative stereotypes about their group, which in turn puts the individual in a situational dilemma that increases the risk of confirming such negative stereotypes ([**66**], [**67**]). Experimental studies have demonstrated that role models, by promoting the success of members of the stereotyped group within the context in question (e.g., STEM), can reduce and/or eliminate the negative effects of stereotype threat, and increase the recruitment of individuals in careers where they are underrepresented ([**9**], [**48**], [**49**], [**44**]). Yet, instructors, and other faculty need not formally be mentors to serve as role models. By definition, women and other racial minorities in STEM serve the role described here, and an effective way to promote diversity is to frequently make students aware of their achievements.

Nevertheless, the benefits of a mentor-protégé relation far exceed that which has been stated herein, and due to space restrictions cannot be further detailed. I therefore end this section by echoing an audience member's plea for faculty to mentor as many students as possible; a statement that everyone in the panel agreed with and which marked the conclusion of the panel discussion.

4. Conclusion

This article summarizes the key points discussed during a diversity panel that met at the 3rd Building Bridges conference in Sarajevo, BiH, and presents some empirical evidence around the points discussed. The focus of the panel was on the retention and recruitment of diverse individuals in mathematics, and issues such as validation of contributions, attitudes about mathematics, self-confidence among students, and the role of mentors were highlighted. Although panel discussions such as the one described herein evidence significant progress in addressing the lack of diversity within mathematics and STEM more generally, we are far from done. It is imperative that the scientific community spread awareness through the continuation of these discussions, along with the creation of policies and interventions that encourage increasing diversity in STEM.

More importantly, and as stated by a member of the panel, the scientific community must make a concerted effort to broaden the definition of diversity in the discussions. Thus far, the scope of the research, discussions and programs aiming to increase diversity in STEM are based on sex (i.e., male, female) and ethnicity.

However, diversity implies the recognition of individual differences more broadly. As individuals are not only characterized by their biological sex and/or ethnicity, such narrow definition misses the point. If global competitiveness relies on the diversification of STEM, then the conversations need to be based on a global definition of diversity, including religion, culture, gender identity, age, physical ability and others. I sincerely hope to, in conjunction with other scientists, continue to be part of such discussions, the generation of ideas, their implementation, evaluation and improvement.

References

[1] Ashcraft, M H. (2002). Math anxiety: personal, educational, and cognitive consequences. Current Directions in Psychological Science, 11(5): 181–185.

[2] Barres, B. (2006). Does Gender Matter? Nature, 442, 133–136. doi:10.1038/442133a

[3] Beilock, L.S. and Maloney, E.A. (2012). Math Anxiety: A Factor in Math Achievement Not to Be Ignored. Policy Insights from the Behavioral and Brain Sciences, 2(1) 4–12.

[4] Beilock, L. S., Gunderson, E. A., Ramirez, G. and Levine, S. C. (2009). Female teachers' math anxiety affects girls' math achievement. Proceedings of the National Academy of Sciences, 107(5): 1860–1863.

[5] Biernat, M. and Eidelman, S. (2007). Translating subjective language in letters of recommendation: The case of the sexist professor. European Journal of Social Psychology, 37, 1149–1175.

[6] Boaler, J. Mathematical Mindsets - summer 2015, Wiley Press. Foreward by Carol Dweck.

[7] Bonetta, L. (2010, February). Reaching Gender Equity in Science: The Importance of Role Models and Mentors. Science. Retrieved from https://www.sciencemag.org/careers/features/2010/02/reaching-gender-equity-science-importance-role-models-and-mentors

[8] Brown, M., Brown, P., and Bibby, T. (2008). "I would rather die": Reasons given by 16 year olds for not continuing their study of mathematics. Research in Mathematics Education, 10(1): 3–18.

[9] Campbell, A. and Skoog, G. (2004). Preparing undergraduate women for science careers. Journal of College Science Teaching, 33: 24–26

[10] Charleson, L. and Leon, R. (2016). Constructing self-efficacy in STEM graduate education. Journal for Multicultural Education, 10(2): 152–166.

[11] Chen, X. (2013). STEM Attrition: College students' paths into and out of STEM fields. Statistical Analysis Report. NCES 2014–001. National Center for Education Statistics. 2013 Nov.

[12] Cvencek, D., Nasir, N., O'Connor, K., Wischnia, S. and Meltzoff, A. N. (2014). The Development of Math-Race Stereotypes: "The Say Chinese People Are the Best at Math". Journal of Research on Adolescence, 1–8. DOI: 10.1111/jora.12151

[13] DiRenzo, M. S., Linnehan, F., Shao, P., Rosenberg, W. L. (2010). A moderated mediation model of e-mentoring. Journal of Vocational Behavior, 76(2): 292–305.

[14] Dowker, A. Individual Differences in Arithmetic: Implications for Psychology, Neuroscience and Education, Psychology Press, New York, NY, USA, 2005.

[15] Dweck, C. (2006). Mindset: The New Psychology of Success, Ballantine Books: New York.

[16] Eccles, J. and Jacobs, J. (1986). Social forces shape math attitudes and performance, Signs, 11(2): 367–380.

[17] Ehrenberg, R. G. (2010). Analyzing the factors that influence persistence rates in STEM field, majors: Introduction to the symposium. Economics of Education Review, 29(6): 888–891.

[18] Ellist, J., Foskick, B. K., Rasmussen, C. (2016). Women 1.5 Times More Likely to Leave STEM Pipeline after Calculus Compared to Men Lack of Mathematical Confidence a Potential Culprit. PLOS ONE, 11(7): e0157447. Doi: 10.1371/journal.pone.0157447.

[19] Erikson, S. and Heit, E. (2015). Metacognition and confidence: comparing math to other academic subjects. Frontiers in Psychology, 6: 742. doi: 10.3389/fpsyg.2015.00742

[20] Espinosa, A., Tikhonov, A. and Jorgenson, J. (2016). Increasing Retention in Mathematics Courses: The role of self-confidence in mathematics on academic performance. Proceedings

of the 21st International Academic Conference of the International Institute of Social and Economic Sciences (February), pp. 50–62 (March 13, 2016). doi: 10.20472/IAC.2016.021.012

[21] Etzkowitz, H., Kemelgor, C., Neuschatz, M., and Uzzi, B. (1994). Barriers to women's participation in academic science and enginnering. In E. W. Fechter, Who Will Do In Science? Educating the Next Generation (pp. 43–67). Baltimore, MS: Johns Hopkins University.

[22] Fayer, S., Lacey, A., and Watson, A. (2017). STEM occupations: Past, present, and future. Spotlight on statistics. Washington, DC: U.S. Department of Labor, Bureau of Labor Statistics.

[23] Ganley, C. M. and Lubienski, S. T. (2016). Mathematics confidence, interest, and performance: Examining gender patterns and reciprocal relations. Learning and Individual Differences, 47: 182–193.

[24] Glass, J. L., Sassler, S., Levitte, Y., and Michelmore, K. M. (2013). What's so special about STEM? A comparison of women's retention in STEM and professional occupations. Social Forces, 92(2), 723–756.

[25] Greenwald, A. G., McGhee, D. E. and Schwartz, J. L. K. (1998). Measuring individual differences in implicit cognition: The Implicit Association Test. Journal of Personality and Social Psychology, 74(6): 1465–1480.

[26] Griffith , A. L., (2010). Persistence of women and minorities in STEM field majors: Is it the school that matters? Economics of Education Review, 29(6): 911–922.

[27] Heilman, M.E. (1983). Sex bias in work setting: The Lack of fit model. Research in Organizational Behavior, 5, 269–298.

[28] Heilman, M.E. (1985). Sex stereotypes and their effects in the workplace: What we know and what we don't know. Journal of Social Behavior and Personality, 10(6), 3–26.

[29] Heilman, M.E., Martell, R.F. and Simon, M.C. (1988). The vagaries of sex bias: Conditions regulating the undervaluation, equivalence, and overvaluation of female job applicants. Organizational Behavior and Human Decision Processes, 41, 98–110.

[30] Heilman, M.E. (2001). Description and prescription: How gender stereotypes prevent women's ascent up the organizational ladder. Journal of Social Issues, 57, 657–674.

[31] Hill, C., Corbett, C. and St. Rose, A. (2010). Why So Few? Women in Science, Technology, Engineering and Mathematics. Washington, DC: AAUW

[32] Holmes, K., Gore, J., Simth, M., and Lloyd, A. (2018). An integrated analysis of school students' aspirations for STEM careers: Which student and school factors are most predictive? International Journal of Science and Mathematics Education, 16(4), 655–675.

[33] Hunt, J. (2010). Why Do Women Leave Science and Engineering? NBER Working Paper No. 15853. March 2010.

[34] Hyde, J. S., Fennema E., Ryan, M., Frost, L. A. and Hopp, C. (1990). Gender Comparisons of Mathematics Attitudes and Affect: A Meta Analysis, Psychology of Women Quarterly, 14: 299–324.

[35] Isaac, C. and Carnes, M. (2015). Interventions That Affect Gender Bias in Hiring: A Systematic Review. Acad Med, 84(10), 1440–1446.

[36] Job, V., Walton, G.M., Bernecker, K., and Dweck, C.S. (2015). Implicit theories about willpower predict self -regulation and grades in everyday life. Journal of Personality and Social Psychology, 108(4): 637–647.

[37] Krinzinger, H. Kaufmann, L. and Willmes, K. (2009). Math anxiety and math ability in early primary school years, Journal of Psychoeducational Assessment, 27(3): 206–225.

[38] Knobloch-Westerwich, S. and Glynn, C. J. (2013). The Matilda Effect-Role Congruity Effects on Scholarly Communication: A Citation Analysis of Communication Research and Journal of Communication Articles. Communications Research, 40(1), 3–26.

[39] Landivar, L. C. (2013). Disparities in STEM employment by sex, race, and Hispanic origin. Educ Rev, 26(6): 911–922.

[40] Lang, S. (1999). Math Talks for Undergraduates. Springer: New York.

[41] Lee, J. L. (2013). 6 Women Scientists Who Were Snubbed Due to Sexism. National Geographic - http://news.nationalgeographic.com/news/2013/13/130519-women-scientists-overlooked-dna-history-science/

[42] Lincoln, A. E., Pincus, S., Koster, J. B., and Leboy, P. S. (2012). The Matilda Effect in science: Awards and prizes in the US, 1990s and 2000s. Social Studies of Science, 42(2), 307–320.

[43] Madera, J.M., Hebl, M.R. and Martin, R.C. (2009). Gender and Letters of Recommendation for Academia: Agentic and Communal Differences. Journal of Applied Psychology, 94(6), 1591-1599.

[44] Marx, D.M. and Goff, P.A. (2005). Clearing the air: The effect of experimenter race on target's test performance and subjective experience. British Journal of Social Psychology, 44: 645–657.

[45] Maloney, E. A., Ramirez, G., Gunderson, E. A., Levine, S.C. and Beilock, S. L. (2015). Intergenerational Effects of Parent's Math Anxiety on Children's Math Achievement and Anxiety, Psychological Science, 1–9.

[46] Maltese, A. V., Tai, R. H., Pipeline persistence: Examining the association of educational experiences with earned degrees in STEM among US students. Science Education, 95(5): 877–907.

[47] Martin, G. (2015). Addressing the underrepresentation of women in mathematics conferences, https://arxiv.org/abs/1502.06326

[48] Marx, D.M., Roman, J.S (2002). Female role models: Protecting women's math test performance. Personality and Social Psychology Bulletin, 28: 1183–1193.

[49] McIntyre, R.B., Lord, C.G., Gresky, D.M., Ten Eyck, L.L. and Bond, C.F. (2005). A social impact trend in the effects of role models on alleviating women's mathematics stereotype threat. Current Research in Social Psychology, 10: 116–136.

[50] Merton, R. K. (1968). The Matthew Effect in Science: The reward and communications systems of science are considered. Science, 159(3810), 56–63.

[51] National Academy of Sciences. (2007). Rising above the gathering storm: Energizing and employing America for a brighter economic future. National Academy of Engineering and Institute of Medicine of the National Academies. Committee on Prospering in the Global Economy of the 21st Century: An Agenda for American Science and Technology. Committee on Science, Engineering, and Public Policy. Washington D.C.: National Academies Press.

[52] National Science Foundation. (2006). Women, minorities, and persons with disabilities in science and engineering. Arlington, VA: Division of Science Resources Statistics. Retrieved November, 16, 2016, from http://www.nsf.gov/statistics/wmpd/

[53] Nosek, B.A., Greenwald, A.G. and Banaji, M. R. (2006). The Implicit Association Test at age 7: A methodological and conceptual review. In J.A. Bargh (Ed.), Social psychology and the unconscious: The automaticity of higher mental processes, pp. 265–292, Psychology Press.

[54] OECD Skills Outlook 2015: First Results from the Survey of Adult Skills.

[55] Osborne, J. and Walker, C. (2006). Stereotype Threat, Identification with Academics, and Withdrawal from School: Why the most successful students of colour might be the most likely to withdraw. Educational Psychology, 26(4): 563–577.

[56] Pajares, F. and Miller, M. D. (1995). Mathematics Self-Efficacy and Mathematics Performance - The Need for Specificity of Assessment. Journal of Counseling Psychology, 42(2): 190–198.

[57] Parsons, S., Croft, T., and Harrison, M. (2009). Does students' confidence in their ability in mathematics matter? Teaching Mathematics and its Applications, 28: 53–58.

[58] Paunesku, D., Walton, G., Romero, R., Smith, E.,Yeager, D., and Dweck, C.S. (2015). Mindset interventions are a scalable treatment for academic underachievement. Psychological Science: 1–10.

[59] Perc, M. (2014). The Matthew effect in empirical data. J. R. Soc. Interface, 11: 20140378. http://dx.doi.org/10.1098/rsif.2014.0378

[60] Perl, T. H. (1982). Discriminating factors and sex differences in electing mathematics. Journal for Research in Mathematics Education, 12: 66–74.

[61] Ramirez, G., Gunderson, E.A., Levine, S.C. and Beilock, S. L. (2013). Math Anxiety, Working Memory, and Math Achievement in Early Elementary School. Journal of Cognition and Development, 14(2): 187–202.

[62] Robst, J., Keil, J., and Russo, D. (1998). The effect of gender composition of faculty on student retention. Economics of Education Review, 29(4), 429–439.

[63] Rossiter, M. W. (1993). The Matthew Matilda Effect in Science. Social Studies of Science, 23(2), 325–341.

[64] Rothwell, J. (2013). The Hidden STEM Economy. Metropolitan Policy Program at Brookings, 1–37.

[65] Schmader, T., Whitehead, J. and Wysocki, V. (2007). A Linguistic Comparison of Letters of Recommendation for Male and Female Chemistry and Biochemistry Job Applicants. Sex Roles, 57(7-8), 509–514.

[66] Steele, C.M. (1997). A Threat in the Air: How Stereotypes Shape Intellectual Identity and Performance. American Psychologist, 52(6): 613–629.

[67] Steele, C.M. and Aronson, J. (1995). Stereotype threat and the intellectual test performance of African Americans, Journal of Personal Psychology, 69(5): 797–811.

[68] Steele, C.M., James, J.B. and Barnett, R.C. (2002). Learning in a Man'S World: Examining the Perceptions of Undergraduate Women in Male-Dominated Academic Areas. Psychology of Women, 26(1): 46–50.

[69] Steen, L. A. (1988). Calculus for a New Century: A Pump Not a Filter. Mathematical Association of America, Washington, D. C.

[70] Swanson, L. H. and Coddington, L. R. (2016). Creating partnerships between teachers and undergraduates interested in secondary math and science education. Teaching and Teacher Education, 59: 285–294.

[71] Tinto, V. (1975). Dropout from higher education: A theoretical synthesis of recent research. Review of Educational Research, 45(1): 89–125.

[72] Trix, F., and Psenka, C. (2003). Exploring the color of glass: Letters of recommendation for female and male medical faculty. Discourse and Society, 14, 191–220.

[73] U.S. Department of Commerce / Economics and Statistics Administration. (2017). STEM Jobs: 2017 Update. Retrieved on July 9, 2018 via http://www.esa.doc.gov/sites/default/files/stem-jobs-2017-update.pdf

[74] U.S. Department of Commerce / Economics and Statistics Administration. (2011). Women in STEM: A Gender Gap to Innovation (ESA Issue Brief #04-11). Retrieved on November 13, 2016 via http://www.esa.doc.gov/sites/default/files/womeninstemagaptoinnovation8311.pdf

[75] Wigfield, A. and Meece, J. L. (1998). Math anxiety in elementary and secondary school students, Journal of Educational Psychology, 80(2): 210–216.

[76] Wilkins, L. (2000). Preparing for the 21st century: The status of quantitative literacy in the United States. School Science and Mathematics, 100(8): 405–418.

Department of Psychology, Scholar in Residence, Master's Program in Public Administration, The City College of New York, Convent Avenue at 138th Street, New York, New York 10031

Email address: aespinosa@ccny.cuny.edu

Contemporary Mathematics
Volume **732**, 2019
https://doi.org/10.1090/conm/732/14785

Superzeta functions, regularized products, and the Selberg zeta function on hyperbolic manifolds with cusps

Joshua S. Friedman, Jay Jorgenson, and Lejla Smajlović

ABSTRACT. Let $\Lambda = \{\lambda_k\}$ denote a sequence of complex numbers and assume that that the counting function $\#\{\lambda_k \in \Lambda \mid |\lambda_k| < T\} = O(T^n)$ for some integer n. From Hadamard's theorem, we can construct an entire function f of order at most n such that Λ is the divisor f. In this article we prove, under reasonably general conditions, that the superzeta function $\mathcal{Z}_f(s, z)$ associated to Λ admits a meromorphic continuation. Furthermore, we describe the relation between the regularized product of the sequence $z - \Lambda$ and the function f as constructed as a Weierstrass product. In the case f admits a Dirichlet series expansion in some right half-plane, we derive the meromorphic continuation in s of $\mathcal{Z}_f(s, z)$ as an integral transform of f'/f. We apply these results to obtain superzeta product evaluations of Selberg zeta function associated to finite volume hyperbolic manifolds with cusps.

1. Introduction

1.1. Background discussion. The zeta-regularized product approach to define the product of an infinite sequence of complex numbers is, at this point, a well-accepted mathematical concept. Let $\Lambda = \{\lambda_k\}$ denote a sequence of complex numbers and assume that that the counting function $\#\{\lambda_k \in \Lambda \mid |\lambda_k| < T\} = O(T^n)$ for some integer n. For any complex number s with $\text{Re}(s) > n$, the function

$$\zeta_\Lambda(s) = \sum_{\lambda_k \in \Lambda, \lambda_k \neq 0} \lambda_k^{-s}$$

is convergent. Under additional conditions, $\zeta_\Lambda(s)$ admits a meromorphic continuation to a region which includes $s = 0$ and is holomorphic at $s = 0$. In such a case, one defines the zeta-regularized product of Λ, which is denoted by $\det^*(\Lambda)$, to be $\det^*(\Lambda) = \exp(-\zeta'_\Lambda(0))$ where the prime denotes a derivative with respect to the variable s. The use of the term *determinant* stems from example when Λ is a finite sequence of non-zero numbers associated to a finite dimensional linear operator T in which case $\exp(-\zeta'_\Lambda(0))$ is, indeed, the determinant of T. In general, the use of the asterisk signifies that one considers only the non-zero elements of Λ.

2010 *Mathematics Subject Classification.* Primary 11M36.

First-named author: The views expressed in this article are the author's own and not those of the U.S. Merchant Marine Academy, the Maritime Administration, the Department of Transportation, or the United States government.

Second-named author's research is supported by NSF and PSC-CUNY grants.

As stated, the zeta-regularized product of Λ requires that Λ satisfies additional conditions which are listed as general axioms in [**11**]. One of the conditions is that the associated theta function $\theta_\Lambda(t) = \sum_{\lambda_k \in \Lambda} e^{-\lambda_k t}$ is defined for real and positive values of t and admits an asymptotic expansion as t approaches zero. This condition allows one to compute $\zeta_\Lambda(s)$ as the Mellin transform of $\theta_\Lambda(t)$. If Λ is the sequence of eigenvalues of some Laplacian operator $\mathbf{\Delta}$ acting of a Hilbert space associated to compact Riemannian manifolds, then one views $\det^*(\Lambda)$ as the determinant of $\mathbf{\Delta}$ and one writes $\det^* \mathbf{\Delta} = \det^*(\Lambda)$.

In [**23**] the authors consider, among other ideas, the zeta-regularized product of the sequence of eigenvalues associated to the Laplacians acting on smooth sections of flat line bundles over genus one curves, when equipped with a unit volume flat metric, and higher genus Riemann surfaces, when equipped with hyperbolic metrics of constant negative curvature. In the genus one case, the evaluation of the determinant of the Laplacian amounts to Kronecker's second limit formula, resulting in the evaluation of determinant of the Laplacian in terms of Riemann's theta function. For higher genus surfaces, the determinant of the Laplacian is expressed in terms of Selberg's zeta function. Far-reaching applications of determinants of the Laplacian, and various sums known as analytic torsion, provide the underpinnings of the analytic aspect of Arakelov theory.

1.2. Our main results. If we continue to consider the setting of Laplacians on finite volume Riemann surfaces, we note that the zeta-regularized approach to defining a determinant of the Laplacian does not extend to the setting when the surface admits cusps. Articles exist where the authors propose means to define determinants of the Laplacian in the non-compact setting by regularizing the trace of the heat kernel in some sense; see, for example, [**12**], [**13**], [**14**] and [**21**]. The purpose of the present article is to provide another approach to the problem. The point of view we consider can be explained through the following example. Let M denote a finite volume hyperbolic Riemann surface, and let $\mathbf{\Delta}$ denote the associated Laplacian which acts on the space of smooth functions on M. If M is compact, the spectrum of $\mathbf{\Delta}$ consists of discrete eigenvalues. If M is not compact, then the corresponding spectrum has discrete eigenvalues, whose nature is the subject of the Phillips-Sarnak philosophy [**22**], as well as a continuous spectrum with associated measure $d\mu$ which is absolutely continuous with respect to Lebesgue measure on $\mathbb{R}$. Furthermore, one can write $d\mu(r) = \phi'/\phi(1/2 + ir)dr$ where dr denotes Lebesgue measure and $\phi(s)$ is a meromorphic function of $s \in \mathbb{C}$ but whose restriction to the line $\text{Re}(s) = 1/2$ appears in the evaluation of the spectral measure. The set Λ to be constructed consists of the poles of the function $\phi(s)$ for all $s \in \mathbb{C}$ together with additional points in $\mathbb{C}$ associated to the L^2 spectrum, appropriately computed in the parameter r.

Once the set Λ has been determined, then the regularized product we define follows the superzeta function construction which was systematically developed by Voros; see [**24**], [**25**] and [**26**]. In vague terms, which will be clarified below, one defines $\mathcal{Z}_f(s, z) = \sum_{\lambda_k \in \Lambda} (z - \lambda_k)^{-s}$ for $s \in \mathbb{C}$ with $\text{Re}(s)$ sufficiently large and $z \in \mathbb{C}$ suitably restricted. The superzeta regularized product of $z - \Lambda$ is obtained by proving the meromorphic continuation in s and employing a special value at $s = 0$.

The purpose of this paper is to establish general circumstances under which the proposed construction is valid. In section 2 we prove that for a general entire function f of finite order, one can define the superzeta regularized product of its zeros. Furthermore, we establish a precise relation between the Hadamard product representation of f and $\exp(-\mathcal{Z}_f'(0,z))$ where the prime denotes a derivative in s. In section 3 we present an example of the results from section 2 taking f to be the Selberg zeta function associated to any co-finite Kleinian group. In section 4 we generalize results of section 2, by assuming that f is a finite order meromorphic function which admits a general Dirichlet series representation. We then prove, in Theorem 4.1, that one has the meromorphic continuation of the corresponding superzeta function. Section 5 yields an example of the setting of section 4 obtained by considering the Selberg zeta function associated to any finite volume hyperbolic manifold of arbitrary dimension.

To summarize, the main results of this article combine to prove the analytic continuation of the superzeta regularized product for any meromorphic function which admits a general Dirichlet series representation. The result is more general than the so-called "ladder theorem" from [**10**] which requires a functional equation.

Finally, let us note that circumstances may occur when a sequence Λ may arise from an operator. In [**5**] we took Λ to be the divisor of the Selberg zeta function for a finite volume, non-compact hyperbolic Riemann surface. In this case, it was discussed in [**5**] how Λ can be viewed in terms of the Lax-Phillips scattering operator in [**19**], so then we obtain a construction of what could be viewed as the determinant of this operator.

2. Zeta regularization of entire functions

Let $\mathbb{R}^- = (-\infty, 0]$ be the non-positive real numbers. Let $\{y_k\}_{k\in\mathbb{N}}$ be the sequence of zeros of an entire function f of order $\kappa \geq 1$, repeated with their multiplicities. Let

$$X_f = \{z \in \mathbb{C} \mid (z - y_k) \notin \mathbb{R}^- \text{ for all } y_k\}.$$

For $z \in X_f$, and $s \in \mathbb{C}$ (where convergent) consider the series

$$\mathcal{Z}_f(s,z) = \sum_{k=1}^{\infty} (z - y_k)^{-s}, \tag{2.1}$$

where the complex exponent is defined using the principal branch of the logarithm with $\arg z \in (-\pi, \pi)$ in the cut plane $\mathbb{C} \setminus \mathbb{R}^-$.

Since f is of order κ, $\mathcal{Z}_f(s,z)$ converges absolutely for $\text{Re}(s) > \kappa$. The series $\mathcal{Z}_f(s,z)$ is called the zeta function associated to the zeros of f, or simply the *superzeta* function of f.

If $\mathcal{Z}_f(s,z)$ has a meromorphic continuation which is regular at $s = 0$, we define, for $z \in X_f$ the *zeta regularized product* associated to f as

$$D_f(z) = \exp\left(-\frac{d}{ds}\, \mathcal{Z}_f(s,z)\big|_{s=0}\right).$$

Let $m = \lfloor \kappa \rfloor$, where $\lfloor \kappa \rfloor$ denotes the largest integer less than or equal to κ. Hadamard's product formula allows us to write

$$f(z) = \Delta_f(z) = e^{g(z)} z^r \prod_{k=1}^{\infty} \left(\left(1 - \frac{z}{y_k}\right) \exp\left[\frac{z}{y_k} + \ldots + \frac{z^m}{m {y_k}^m}\right]\right), \tag{2.2}$$

where $g(z)$ is a polynomial of degree m or less, $r \geq 0$ is the order of the eventual zero of f at $z = 0$, and the other zeros y_k are listed with multiplicity. A simple calculation shows that when $z \in X_f$,

$$\mathcal{Z}_f(m+1, z) = \frac{(-1)^m}{m!} \left(\log \Delta_f(z)\right)^{(m+1)}. \tag{2.3}$$

The following proposition is due to Voros ([**24**], [**26**], [**27**]). We also give a different proof.

PROPOSITION 1. *Let f be an entire function of order $\kappa \geq 1$, and for $k \in \mathbb{N}$, let y_k be the sequence of zeros of f. Let $\Delta_f(z)$ denote the Hadamard product representation of f. Assume that for $n > m = \lfloor \kappa \rfloor$ we have the following asymptotic expansion:*

$$\log \Delta_f(z) = \sum_{j=0}^{m} \widetilde{a}_j z^j (\log z - H_j) + \sum_{j=0}^{m} b_j z^j + \sum_{k=1}^{n-1} a_k z^{\mu_k} + h_n(z), \tag{2.4}$$

where $H_0 = 0$, $H_j = \sum_{l=1}^{j}(1/l)$, for $j \geq 1$, $1 > \mu_1 > \ldots > \mu_n \to -\infty$, and $h_n(z)$ is a sequence of holomorphic functions in the sector $|\arg z| < \theta < \pi$, $(\theta > 0)$ such that $h_n^{(j)}(z) = O(|z|^{\mu_n - j})$, as $|z| \to \infty$ in the above sector, for all integers $j \geq 0$.

Then, for all $z \in X_f$, the superzeta function $\mathcal{Z}_f(s,z)$ has a meromorphic continuation to the half-plane $\mathrm{Re}(s) < \kappa$ which is regular at $s = 0$. Furthermore, the zeta regularized product $D_f(z)$ associated to $\mathcal{Z}_f(s,z)$ is related to $\Delta_f(z)$ through the formula

$$D_f(z) = e^{-\left(\sum_{j=0}^{m} b_j z^j\right)} \Delta_f(z) \tag{2.5}$$

which also provides analytic continuation of $D_f(z)$ from X_f to the whole complex z-plane.

PROOF. For any $z \in X_f$, the series

$$\mathcal{Z}_f(m+1, z+y) = \sum_{k=1}^{\infty} (z + y - y_k)^{-(m+1)} \tag{2.6}$$

converges absolutely and uniformly for $y \in (0, \infty)$. Furthermore, application of [**8**, Formula 3.194.3], with $\mu = m+1-s$, $\nu = m+1$ and $\beta = (z - y_k)^{-1}$ yields, for all y_k,

$$\int_0^{\infty} \frac{y^{m-s}\, dy}{(z+y-y_k)^{m+1}} = \frac{1}{m!}(z-y_k)^{-s}\Gamma(m+1-s)\Gamma(s).$$

Absolute and uniform convergence of the series (2.6) for $\mathrm{Re}(s) > \kappa$ implies that

$$\begin{aligned} \mathcal{Z}_f(s,z) &= \frac{m!}{\Gamma(m+1-s)\Gamma(s)} \int_0^{\infty} \mathcal{Z}_f(m+1, z+y) y^{m-s} dy \\ &= \frac{(-1)^m}{\Gamma(m+1-s)\Gamma(s)} \int_0^{\infty} \left(\log \Delta_f(z+y)\right)^{(m+1)} y^{m-s} dy, \end{aligned} \tag{2.7}$$

for $\kappa < \mathrm{Re}(s) < m+1$.

Next, we use (2.7) together with (2.4) in order to get the meromorphic continuation of $Z_f(s,z)$ to the half plane $\mathrm{Re}(s) < m+1$. We start with (2.3) and differentiate Equation (2.4) $(m+1)$ times to get

$$(\log \Delta_f(z+y))^{(m+1)} = \sum_{j=0}^{m} \frac{(-1)^{m-j} j!(m-j)!\widetilde{a}_j}{(z+y)^{(m+1-j)}} + \sum_{k=1}^{n-1} \frac{a_k \mu_k(\mu_k-1)\cdot\ldots\cdot(\mu_k-m)}{(z+y)^{m+1-\mu_k}} + h_n^{(m+1)}(z+y),$$

for any $n > m$.

Since $\mu_k \searrow -\infty$, for an arbitrary $\mu < 0$ there exists k_0 such that $\mu_k \leq \mu$ for all $k \geq k_0$, hence we may write

$$(\log \Delta_f(z+y))^{(m+1)} y^{m+1} = y^{m+1} \left(\sum_{j=0}^{m} \frac{(-1)^{m-j} j!(m-j)!\widetilde{a}_j}{(z+y)^{m+1-j}} + \sum_{k=1}^{n-1} \frac{a_k \mu_k(\mu_k-1)\cdot\ldots\cdot(\mu_k-m)}{(z+y)^{m+1-\mu_k}} \right) + g_\mu(z+y),$$

where $g_\mu(z+y) = y^{m+1} h_{k_0}^{(m+1)}(z+y)$.

Note that

$$g_\mu(z+y) = O(y^\mu) \quad \text{as } y \to \infty, \quad \text{and} \quad g_\mu(z+y) = O(y^{m+1}) \text{ as } y \searrow 0. \tag{2.8}$$

Application of [**8**, Formula 3.194.3] yields

$$\frac{(-1)^m}{\Gamma(m+1-s)\Gamma(s)} \int_0^\infty (\log \Delta_f(z+y))^{(m+1)} y^{m-s} dy = \sum_{j=0}^{m} (-1)^j j! \widetilde{a}_j \frac{\Gamma(s-j)}{\Gamma(s)} z^{j-s} - \sum_{k=1}^{k_0-1} a_k \frac{\Gamma(s-\mu_k)}{\Gamma(s)\Gamma(-\mu_k)} z^{\mu_k - s} + \frac{(-1)^m}{\Gamma(m+1-s)\Gamma(s)} \int_0^\infty g_\mu(z+y) y^{-s-1} dy. \tag{2.9}$$

The integral on the right hand side of (2.9) is the Mellin transform of the function g_μ. By (2.8) this integral represents a holomorphic function in s for all s in the half strip $\mu < \mathrm{Re}(s) < m+1$. The other terms on the right hand side of (2.9) are meromorphic in s, hence, by (2.7), the right-hand side of (2.9) provides meromorphic continuation of integral $\int_0^\infty \mathcal{Z}_f(m+1, z+y) y^{m-s} dy$ from the strip $\kappa < \mathrm{Re}(s) < m+1$ to the strip $\mu < \mathrm{Re}(s) < m+1$. Since $\mu < 0$ was chosen arbitrarily, we can let $\mu \to -\infty$ and obtain the meromorphic continuation of this integral to the half plane $\mathrm{Re}(s) < m+1$.

Formula (2.9), together with (2.7), now yields the following representation of $\mathcal{Z}_f(s,z)$, for an arbitrary, fixed $z \in X_f$, valid in the half plane $\mathrm{Re}(s) < m+1$:

$$\mathcal{Z}_f(s,z) = \sum_{j=0}^{m} (-1)^j j! \widetilde{a}_j \frac{\Gamma(s-j)}{\Gamma(s)} z^{j-s} - \sum_{k=1}^{k_0-1} a_k \frac{\Gamma(s-\mu_k)}{\Gamma(s)\Gamma(-\mu_k)} z^{\mu_k - s} + \frac{(-1)^m}{\Gamma(m+1-s)\Gamma(s)} \int_0^\infty h_{k_0}^{(m+1)}(z+y) y^{m-s} dy. \tag{2.10}$$

From the decay properties of $h_{k_0}^{(m+1)}(z+y)$, it follows that $\mathcal{Z}_f(s,z)$ is holomorphic at $s=0$. Furthermore since $\frac{1}{\Gamma(s)}$ has a zero at $s=0$, the derivative of the last term in (2.10) is equal to

$$\left(\frac{d}{ds}\frac{1}{\Gamma(s)}\bigg|_{s=0}\right)\frac{(-1)^m}{\Gamma(m+1)}\int_0^\infty h_{k_0}^{(m+1)}(z+y)y^m dy = \frac{(-1)^m}{m!}\int_0^\infty h_{k_0}^{(m+1)}(z+y)y^m dy$$
$$= -h_{k_0}(z),$$

where the last equality is obtained from integration by parts m times, and using the decay of $h_{k_0}(z+y)$ and its derivatives as $y \to +\infty$, for $\mu_{k_0} < 0$. Moreover, since

$$\frac{d}{ds}\frac{\Gamma(s-\mu_k)}{\Gamma(s)}\bigg|_{s=0} = \Gamma(-\mu_k)\cdot\frac{d}{ds}\frac{1}{\Gamma(s)}\bigg|_{s=0} = \Gamma(-\mu_k),$$

elementary computations yield that

$$\frac{d}{ds}\left(\sum_{k=1}^{k_0-1} a_k \frac{\Gamma(s-\mu_k)}{\Gamma(s)\Gamma(-\mu_k)} z^{\mu_k - s}\right)\bigg|_{s=0} = \sum_{k=1}^{k_0-1} a_k z^{\mu_k}.$$

This shows that
(2.11)

$$-\frac{d}{ds}\mathcal{Z}_f(s,z)\bigg|_{s=0} = -\frac{d}{ds}\left(\sum_{j=0}^{m}(-1)^j j!\widetilde{a}_j\frac{\Gamma(s-j)}{\Gamma(s)}z^{j-s}\right)\bigg|_{s=0} + \sum_{k=1}^{k_0-1} a_k z^{\mu_k} + h_{k_0}(z),$$

for z in the sector $|\arg z| < \theta < \pi$, $(\theta > 0)$.

Now, for $j \in \{1, ..., m\}$ one has

$$\frac{d}{ds}\left(\frac{\Gamma(s-j)}{\Gamma(s)}z^{j-s}\right)\bigg|_{s=0} = (-\log z)z^j \prod_{k=1}^{j}(0-k)^{-1} + z^j \frac{d}{ds}\frac{\Gamma(s-j)}{\Gamma(s)}\bigg|_{s=0}.$$

A straightforward computation shows that

$$\frac{d}{ds}\frac{\Gamma(s-j)}{\Gamma(s)}\bigg|_{s=0} = \frac{d}{ds}\prod_{k=1}^{j}(s-k)^{-1}\bigg|_{s=0} = \frac{(-1)^j}{j!}\sum_{k=1}^{j}\frac{1}{k} = H_j\cdot\frac{(-1)^j}{j!}.$$

Therefore,

$$\frac{d}{ds}\left((-1)^j j!\widetilde{a}_j\frac{\Gamma(s-j)}{\Gamma(s)}z^{j-s}\right)\bigg|_{s=0} = \widetilde{a}_j(-\log z)z^j + \widetilde{a}_j z^j H_j,$$

for $j \in \{1, ..., m\}$ and hence

$$-\frac{d}{ds}\left(\sum_{j=0}^{m}(-1)^j j!\widetilde{a}_j\frac{\Gamma(s-j)}{\Gamma(s)}z^{j-s}\right)\bigg|_{s=0} = \sum_{j=0}^{m}\widetilde{a}_j z^j(\log z - H_j).$$

Now, (2.5) follows from equation (2.11) and uniqueness of analytic continuation.

□

3. Superzeta functions constructed from confinite Kleinian groups

Let Γ be a cofinite Kleinian group. Suppose $T \in \Gamma$ is loxodromic. Then T is conjugate in $\mathrm{PSL}(2,\mathbb{C})$ to a unique element of the form $\begin{pmatrix} a(T) & 0 \\ 0 & a(T)^{-1}\end{pmatrix}$ such that $a(T) \in \mathbb{C}$ has $|a(T)| > 1$. Let $N(T) = |a(T)|^2$, and let $\mathcal{C}(T)$ denote the centralizer of T in Γ. There exists a (primitive) loxodromic element T_0, and a finite cyclic elliptic subgroup $\mathcal{E}(T)$ of order $m(T)$, generated by an element E_T such that $\mathcal{C}(T) = \langle T_0\rangle \times \mathcal{E}(T)$, where $\langle T_0\rangle = \{\, T_0^n \mid n \in \mathbb{Z}\,\}$.

The elliptic element E_T is conjugate in $\mathrm{PSL}(2,\mathbb{C})$ to an element of the form $\begin{pmatrix} \zeta(T_0) & 0 \\ 0 & \zeta(T_0)^{-1} \end{pmatrix}$ where here $\zeta(T_0)$ is a primitive $2m(T)$-th root of unity.

DEFINITION 2. *For* $\mathrm{Re}(s) > 1$ *the Selberg zeta-function* $Z(s)$ *is defined by*

$$Z(s) = \prod_{\{T_0\}\in\mathcal{R}} \prod_{\substack{l,k\geq 0 \\ c(T,l,k)=1}} \left(1 - a(T_0)^{-2k}\overline{a(T_0)^{-2l}}N(T_0)^{-s-1}\right).$$

Here the product with respect to T_0 *extends over a maximal reduced system* $\mathcal{R}$ *of* Γ*-conjugacy classes of primitive loxodromic elements of* Γ. *The system* $\mathcal{R}$ *is called reduced if no two of its elements have representatives with the same centralizer*[1]. *The function* $c(T,l,k)$ *is defined by* $c(T,l,k) = \zeta(T_0)^{2l}\zeta(T_0)^{-2k}$.

Let $\zeta \in \mathbb{P}^1$ be cusp of Γ, $\Gamma_\zeta = \{\ \gamma \in \Gamma \mid \gamma\zeta = \zeta\ \}$, and let Γ'_ζ be the maximal torsion-free parabolic subgroup of Γ_ζ. The possible values for the index of $[\Gamma_\zeta : \Gamma'_\zeta]$ are 1,2,3,4, and 6.

PROPOSITION 3 ([**3**]). *Let* Γ *be cofinite with one class of cusp at* ∞. *If* $[\Gamma_\infty : \Gamma'_\infty] = 1$ *or* $[\Gamma_\infty : \Gamma'_\infty] = 2$, *then* $Z(s)$ *is a meromorphic function. Furthermore, the zeros of the Selberg zeta-function are:*

(1) *zeros at the points* $\pm s_j$ *on the line* $\mathrm{Re}(s) = 0$ *and on the interval* $[-1,1]$. *Each point* s_j *is related to an eigenvalue* λ_j *of the discrete spectrum of (the self-adjoint extension) of Laplacian* $-\Delta$ *by* $1 - s_j^2 = \lambda_j$. *The multiplicity of each* $s_j > 0$, $m(s_j)$, *is equal to the multiplicity of the corresponding eigenvalue* $m(\lambda_j)$. *If* $-s_j < 0$ *happens to also be a zero of* $\phi(s)$ *of multiplicity* $q(-s_j)$ *then* $m(-s_j) = m(\lambda_j) - q(-s_j)$. *Here* $\phi(s)$ *is determinant of the automorphic scattering matrix (in this case, it is a scalar since we assume the number of cusps is one).*
(2) *zeros at the points* ρ_j, *that are poles of* $\phi(s)$, *which lie in the half-plane* $\mathrm{Re}(s) < 0$. *The multiplicity of each* ρ_j *is equal its multiplicity, as a pole, of* $\phi(s)$.
(3) *if* $[\Gamma_\infty : \Gamma'_\infty] = 1$ *then zeros at the points* $s = \mathbb{Z}_{<0}$ *with multiplicity 1; at the point* $s = 0$ *it could be a pole or a zero, with multiplicity* $\frac{1}{2}(\phi(0) - 1) + 2m(1)$, *where* $m(1)$ *denotes the multiplicity of the possible eigenvalue* $\lambda = 1$.
(4) *if* $[\Gamma_\infty : \Gamma'_\infty] = 2$ *then zeros at the points* $s = -1, -3, -5, \ldots$, *with multiplicity 1; at the point* $s = 0$ *it could be a pole or a zero with multiplicity* $\frac{1}{2}(\phi(0) - 1) + 2m(1)$.

Let $\begin{pmatrix} * & * \\ c_n & d \end{pmatrix} \in \mathcal{R}$ denote the representatives of $\Gamma'_\infty \setminus \Gamma / \Gamma'_\infty$. For $\mathrm{Re}(s) > 1$, the scattering matrix $\phi(s)$ can be written as a Dirichlet series

$$\phi(s) = \frac{\pi}{[\Gamma_\infty : \Gamma'_\infty]} \frac{m(c_0)}{|\mathcal{P}|^s} |c_0|^{-2-2s} (1 + \sum_{n\geq 1} q_n^{-2-2s}),$$

where $q_n = |c_n/c_0| > 1$, $|\mathcal{P}|$ is the co-area of the lattice associated to Γ_∞, c_0 is the minimal (non-zero) modulus lower left entry for Γ, and $m(c_0)$ counts the number of times c_0 is present as a representative in $\mathcal{R}$. See [**2**, p. 111 and p.234 Eq. 1.10]) for more details.

---

[1]See [**2**] section 5.4 for more details

3.1. The case $[\Gamma_\infty : \Gamma'_\infty] = 1$. When $[\Gamma_\infty : \Gamma'_\infty] = 1$, we define $Z_1^+(s) = Z(s)/\Gamma(s)$, and let $\mathcal{Z}_1^+(s,z)$ be its superzeta function. From the asymptotic version of Stirling's formula and from the product expansion of $Z(s)$, it follows that as $\mathrm{Re}(s) \to +\infty$, we have that

$$\log Z_1^+(s) \sim \frac{1}{2}(\log s - 0) - 1s(\log s - 1) - \frac{1}{2}\log 2\pi.$$

Hence applying Proposition 1 to the finite-order entire function $Z_1^+(s)$, we obtain the regularized product of the zeros of $Z_1^+(s)$ as $D_{Z_1^+}(s) = \sqrt{2\pi} Z_1^+(s)$.

Next let $Z_1^-(s) = \phi(s) Z_1^+(s)$. Hence it follows that as $\mathrm{Re}(s) \to +\infty$,

$$\log Z_1^-(s) \sim \frac{1}{2}(\log s - 0) - 1s(\log s - 1) - \frac{1}{2}\log 2\pi + \log\left(\frac{\pi m(c_0)}{|c_0|^2}\right) - s\log(|c_0|^2|\mathcal{P}|).$$

Once again, applying Proposition 1 we obtain that

$$D_{Z_1^-}(s) = \frac{|c_0|^{2s}|\mathcal{P}|^s|c_0|^2}{\pi m(c_0)}\sqrt{2\pi} Z_1^-(s).$$

Recalling that $Z_1^-(s) = \phi(s) Z_1^+(s)$ we obtain a formula for $\phi(s)$ as a quotient of regularized determinants,

$$\phi(s) = \frac{D_{Z_1^-}(s)}{D_{Z_1^+}(s)}\frac{\pi m(c_0)}{|c_0|^{2s+2}|\mathcal{P}|^s}$$

from which one can obtain a formula for the central value of $\phi(s)$; see [**6**].

3.2. The case $[\Gamma_\infty : \Gamma'_\infty] = 2$. When $[\Gamma_\infty : \Gamma'_\infty] = 2$, (this is the case e.g. for the Picard group) define

$$Z_2^+(s) = \frac{sZ(s)}{\Gamma\left(\frac{s-1}{2}\right)},$$

and let $\mathcal{Z}_2^+(s,z)$ be its superzeta function.

The asymptotic expansion of $\log\Gamma(\frac{s-1}{2})$ can be computed as

$$\log\Gamma(\frac{s-1}{2}) \sim \frac{1}{2}s\log s - \frac{s}{2} - \frac{1}{2}s\log 2 - \log s + \frac{1}{2}\log 2\pi + \log 2.$$

Hence from the product expansion of $Z(s)$, it follows that as $\mathrm{Re}(s) \to +\infty$,

$$\log Z_2^+(s) \sim 2\log s - \frac{1}{2}s\log s + \frac{s}{2} + \frac{1}{2}s\log 2 - \frac{1}{2}\log 2\pi - \log 2$$

and can be rewritten in the form required by Proposition 1

$$\log Z_2^+(s) \sim 2(\log s - 0) - \frac{1}{2}s(\log s - 1) + \frac{\log 2}{2}s - \frac{1}{2}\log 2\pi - \log 2.$$

Hence

$$D_{Z_2^+}(s) = \sqrt{\pi}\, 2^{\left(\frac{3-s}{2}\right)} Z_2^+(s).$$

If we define $Z_2^-(s) = \phi(s) Z_2^+(s)$ then we obtain

$$D_{Z_2^-}(s) = \frac{|c_0|^{2s}|\mathcal{P}|^s|c_0|^2}{\pi m(c_0)}\sqrt{\pi}\, 2^{\left(\frac{3-s}{2}\right)} Z_2^-(s).$$

Finally we obtain

$$\phi(s) = \frac{D_{Z_2^-}(s)}{D_{Z_2^+}(s)}\frac{\pi m(c_0)}{|c_0|^{2s+2}|\mathcal{P}|^s}.$$

4. Zeta regularization of zeta-type functions through integral representation

In this section we assume that f is a meromorphic function of finite order κ such that $\log f$ possesses a representation as a generalized Dirichlet series

$$\log f(z) = \sum_{n=1}^{\infty} \frac{c_n}{q_n^z} \tag{4.1}$$

converging absolutely and uniformly in any half plane $\mathrm{Re}(s) \geq \sigma + \epsilon$, for any $\epsilon > 0$. Here $\{q_n\}$ denotes an increasing sequence of real numbers with $q_1 > 1$ and c_n are complex numbers. For such a function f we say it is a *zeta-type function.*

We denote by N_f the set of zeros of f and by P_f the set of poles of f and define

$$X_f = \{z \in \mathbb{C} \mid (z - z_k) \notin \mathbb{R}^- \text{ for all } z_k \in N_f \cup P_f\}.$$

For $z \in X_f$, and $s \in \mathbb{C}$ with $\mathrm{Re}(s) > \kappa$ consider the function

$$\mathcal{Z}_f(s, z) = \sum_{\rho \in N_f} \mathrm{ord}(\rho)(z-\rho)^{-s} - \sum_{\rho \in P_f} \mathrm{ord}(\rho)(z-\rho)^{-s} = \mathcal{Z}_f^{\mathrm{N}}(s, z) - \mathcal{Z}_f^{\mathrm{P}}(s, z), \tag{4.2}$$

where $\mathrm{ord}(\rho)$ denotes the order of a zero or pole. We call $\mathcal{Z}_f(s, z)$ the *superzeta function* associated to the meromorphic function f. Note that both series in (4.2) are absolutely convergent for $\mathrm{Re}(s) > \kappa$.

Denote by W_f^{N} and W_f^{P} Weierstrass products of order $m = \lfloor \kappa \rfloor$ associated to set of zeros and set of poles of f, respectively. Then, a straightforward computation shows that for $z \in X_f$

$$\mathcal{Z}_f(m+1, z) = \frac{(-1)^m}{m!} \left((\log W_f^{\mathrm{N}}(z))^{(m+1)} - (\log W_f^{\mathrm{P}}(z))^{(m+1)} \right),$$

and, moreover

$$(\log f(z))^{(m+1)} = (\log W_f^{\mathrm{N}}(z))^{(m+1)} - (\log W_f^{\mathrm{P}}(z))^{(m+1)}.$$

We have the following theorem.

THEOREM 4. *Let f be a meromorphic function of zeta-type, of finite order κ. Fix $z \in X_f$. The superzeta function $\mathcal{Z}_f(s, z)$ defined by (4.2) has a holomorphic continuation to all $s \in \mathbb{C}$, through the equation*

$$\mathcal{Z}_f(s, z) = \frac{\sin \pi s}{\pi} \int_0^{\infty} \left(\frac{f'}{f}(z+y) \right) y^{-s} dy. \tag{4.3}$$

PROOF. For $z \in X_f$ and $\kappa < \mathrm{Re}(s) < m+1$, we apply equation (2.3) to $\mathcal{Z}_f^{\mathrm{N}}(s, z)$ and $\mathcal{Z}_f^{\mathrm{P}}(s, z)$ and proceed analogously as in the proof of (2.7) to get

$$\mathcal{Z}_f(s, z) = \mathcal{Z}_f^{\mathrm{N}}(s, z) - \mathcal{Z}_f^{\mathrm{P}}(s, z)$$

$$= \frac{(-1)^m}{\Gamma(m+1-s)\Gamma(s)} \int_0^{\infty} \left((\log W_f^{\mathrm{N}}(z+y))^{(m+1)} - (\log W_f^{\mathrm{P}}(z+y))^{(m+1)} \right) y^{m-s} dy$$

$$= \frac{(-1)^m}{\Gamma(m+1-s)\Gamma(s)} \int_0^{\infty} (\log f(z+y))^{(m+1)} y^{m-s} dy. \tag{4.4}$$

For a fixed $z \in X_f$, generalized Dirichlet series representation (4.1) yields the asymptotic behavior

$$(\log F(z+y))^{(l)} = O\left(\frac{1}{y^n}\right), \quad \text{as } y \to \infty, \text{ for all positive integers } n$$

and

$$(\log F(z+y))^{(l)} = O(1), \quad \text{as } y \searrow 0$$

for any $1 \le l \le m$. Therefore, for $m-1 < \mathrm{Re}(s) < m$ we may integrate by parts in (4.4) and obtain

$$\mathcal{Z}_f(s,z) = \frac{(-1)^m}{\Gamma(m-s)(m-s)\Gamma(s)} \int_0^\infty y^{m-s} d\left((\log f(z+y))^{(m)} \right) = \frac{(-1)^{m-1}}{\Gamma(m-s)\Gamma(s)} \int_0^\infty (\log f(z+y))^{(m)} y^{m-1-s} dy \tag{4.5}$$

Integrating by parts in (4.5) for $m-2 < \mathrm{Re}(s) < m-1$ (in case when $m \ge 2$) we obtain

$$\mathcal{Z}_f(s,z) = \frac{(-1)^{m-2}}{\Gamma(m-1-s)\Gamma(s)} \int_0^\infty (\log f(z+y))^{(m-1)} y^{m-2-s} dy.$$

Proceeding inductively in m, we deduce that

(4.6)

$$\mathcal{Z}_f(s,z) = \frac{1}{\Gamma(1-s)\Gamma(s)} \int_0^\infty (\log f(z+y))' y^{-s} dy = \frac{\sin \pi s}{\pi} \int_0^\infty \left(\frac{f'}{f}(z+y) \right) y^{-s} dy,$$

for $0 < \mathrm{Re}(s) < 1$.

First, we claim that for a fixed $z \in X_f$, the integral

$$I(s,z) = \int_0^\infty \left(\frac{f'}{f}(z+y) \right) y^{-s} dy$$

which appears on the right hand side of (4.6) is actually holomorphic function in the half-plane $\mathrm{Re}(s) < 1$. To see this, let $\mu \le 0$ be arbitrary. Decay properties of $(\log f(z+y))'$, as $y \to +\infty$ with $n > -\mu+2$, yield that $(\log f(z+y))' y^{-s} = O(y^{-2})$, as $y \to +\infty$, for all s such that $\mu < \mathrm{Re}(s) \le 0$. Moreover, the bound $\frac{f'(z+y)}{f(z+y)} = O(1)$, for fixed $z \in X_f$ implies that $(\log f(z+y))' y^{-s} = O(1)$, as $y \to 0$, for all s in the half plane $\mathrm{Re}(s) \le 0$. This shows that for $z \in X_f$ the integral $I(s,z)$ is absolutely convergent in the strip $\mu < \mathrm{Re}(s) \le 0$, hence represents a holomorphic function for all s in that strip. Since $\mu \le 0$ was arbitrarily chosen, we have proved that $I(s,z)$, for $z \in X_f$, is holomorphic function in the half plane $\mathrm{Re}(s) \le 0$.

Next, we claim that $I(s,z)$, for $z \in X_f$, can be meromorphically continued to the half-plane $\mathrm{Re}(s) > 0$ with simple poles at the points $s = 1, 2, \ldots$ and corresponding residues

$$\mathrm{Res}_{s=n} I(s,z) = -\frac{1}{(n-1)!} (\log f(z))^{(n)}. \tag{4.7}$$

Since the function $\sin(\pi s)$ has simple zeros at points $s = 1, 2, \ldots$ this would prove that $\mathcal{Z}_f(s,z)$, for $z \in X_f$ is actually an entire function of s and the proof would be complete.

Let $\mu > 0$ be arbitrary, put $n = \lfloor \mu \rfloor$ to be the integer part of μ and let $\delta > 0$ (depending upon $z \in X_f$ and μ) be such that for $y \in (0, \delta)$ we have the Taylor series expansion

$$(\log f(z+y))' = \sum_{j=1}^{n} \frac{(\log f(z))^{(j)}}{(j-1)!} y^{j-1} + R_1(z,y),$$

where $R_1(z,y) = O(y^n)$, as $y \to 0$. Then, for $0 < \mathrm{Re}(s) < 1$ we may write

$$I(s,z) = \sum_{j=1}^{n} \frac{(\log f(z))^{(j)}}{(j-1)!} \frac{\delta^{j-s}}{j-s} + \int_0^\delta R_1(z,y) y^{-s} dy + \int_\delta^\infty \left(\frac{f'}{f}(z+y)\right) y^{-s} dy.$$

The bound on $R_1(z,y)$ and the growth of $(\log f(z+y))'$ as $y \to \infty$ imply that the last two integrals are holomorphic functions of s for $\mathrm{Re}(s) \in (0,\mu)$. The first sum is meromorphic in s, for $\mathrm{Re}(s) \in (0,\mu)$, with simple poles at $s = j$, $j \in \{1,...,n\}$ and residues equal to $-(\log f(z))^{(j)}/(j-1)!$. Since $\mu > 0$ is arbitrary, this proves the claim and completes the proof.

□

REMARK 5. We often split a superzeta function $\mathcal{Z}_f(s,z)$ up as

$$\mathcal{Z}_f(s,z) = \mathcal{Z}_f^{\mathrm{NT}}(s,z) + \mathcal{Z}_f^{\mathrm{T}}(s,z) - \mathcal{Z}_f^{\mathrm{P}}(s,z),$$

where $\mathcal{Z}_f^{\mathrm{NT}}(s,z)$, $\mathcal{Z}_f^{\mathrm{T}}(s,z)$ and $\mathcal{Z}_f^{\mathrm{P}}(s,z)$ denote the superzeta functions associated to the non-trivial zeros, the trivial zeros, and the poles respectively. Often $\mathcal{Z}_f^{\mathrm{T}}(s,z)$ and $\mathcal{Z}_f^{\mathrm{P}}(s,z)$ can be expressed in terms of some special functions, namely, the Hurwitz zeta functions, as in [**5**]. Upon applying Theorem 4 one could deduce a relation giving a meromorphic continuation of the superzeta function associated to the non-trivial zeros of a zeta-type function. Moreover, by differentiating the equation (4.3) with respect to the $s-$variable one may directly compute the regularized determinant $D_f(z)$ associated to the superzeta function constructed over the set of non-trivial zeros.

The advantage of the full asymptotic expansion (2.4) is that it yields the coefficients in the regularized determinant expression, which can not immediately be derived from Theorem 4.

5. Superzeta functions constructed over non-trivial zeros of the Selberg zeta function on hyperbolic manifolds with cusps

In this section we show how to apply Theorem 4 to derive the meromorphic continuation of superzeta function constructed over the set of non-trivial zeros of the Selberg zeta function on a finite volume hyperbolic manifold with cusps. We also derive an expression for the regularized determinant associated to this superzeta function.

We will follow the notation from [**7**], where the exact divisor of the Selberg zeta function is given.

Consider real hyperbolic space of dimension d, which can be realized as the symmetric space G/K, where $G = SO_0(d,1)$ and $K = SO(d)$. By Γ, we denote a discrete, torsion-free subgroup of G such that $\mathrm{vol}(\Gamma\backslash G)$ is finite, where the volume is computed with respect to the standard normalized Haar measure on G. We assume that the metric induced from the Haar measure is such that the space G/K has constant curvature equal to -1 and that the group Γ satisfies the conditions $\Gamma_P = \Gamma \cap P = \Gamma \cap N(P)$ for $P \in \mathfrak{P}_\Gamma$, which ensure that the resulting manifold $X_\Gamma = \Gamma\backslash G/K$ is a non-compact hyperbolic manifold with cusps. Here, $\mathfrak{P}_\Gamma$ is the set of all $\Gamma-$conjugacy classes of $\Gamma-$cuspidal parabolic subgroups in G and $N(P)$ denotes the unipotent radical of P.

Let $\mathfrak{g} = \mathfrak{so}(d,1)$ and $\mathfrak{k} \cong \mathfrak{so}(d)$ be the Lie algebras of G and K respectively and let $\mathfrak{g} = \mathfrak{k} \oplus \mathfrak{p}$ be the decomposition of $\mathfrak{g}$ induced by the Cartan involution

θ. By $\mathfrak{a}$ we denote the (one-dimensional) maximal abelian subspace of $\mathfrak{p}$ and let $M \cong SO(d-1)$ be the centralizer of $A = \exp(\mathfrak{a})$. The positive restricted root of $(\mathfrak{g}, \mathfrak{a})$ with the normalized Cartan-Killing norm equal to one is denoted by β, whereas $\mathfrak{n}$ denotes the positive root space of β.

Every hyperbolic element $\gamma \in \Gamma$ has the form $m_\gamma a_\gamma \in MA^+$, where $A^+ = \{a \in A \mid a = \exp(tH),\ H \in \mathfrak{a},\ t > 0\}$. The length of a closed geodesic C_γ determined by γ is denoted by $l(C_\gamma)$, whereas $j(\gamma)$ is a positive integer such that $\gamma = \gamma_0^{j(\gamma)}$, for a primitive γ_0.

In the case when $d = 2n$ or $d = 2n+1$ and $l \neq n$, we denote by σ_l the irreducible fundamental representation of $M = SO(d-1)$ acting on $\bigwedge^l(\mathbb{C}^{d-1})$ and by $d(\sigma_l)$ the dimension of the representation space of σ_l. In the case when $d = 2n+1$ and $l = n$, we denote by $\sigma_n^{\pm}$ the two irreducible half spin representations acting on $\bigwedge^n(\mathbb{C}^{2n})$ and by $d(\sigma_n)$ the dimension of the representation space of $\sigma_n^{\pm}$.

Going further, let χ be a unitary representation of Γ on a Hermitian vector space V_χ of a finite dimension $\dim V_\chi$. For $P \in \mathfrak{P}_\Gamma$, let $V_P \subseteq V_\chi$ denote the maximal subspace of V_χ on which $\chi|_{\Gamma_P}$ acts trivially, and put $d_P(\chi) = \dim V_P$.

For $0 \leq k \leq \lfloor (d-1)/2 \rfloor$, the Selberg zeta function $Z_\chi(\sigma_k, s)$, for $\mathrm{Re}(s) > d-1$ is defined by the absolutely convergent product

$$Z_\chi(\sigma_k, s) = \prod_{\gamma_\Gamma \in \mathrm{P}\Gamma_{\mathrm{hyp}}} \prod_{k=0}^{\infty} \det\left(\mathrm{Id} - \overline{\sigma_k(m_\gamma)} \otimes \chi(\gamma) \otimes S^k(\mathrm{Ad}\,(m_\gamma a_\gamma)|_{\overline{\mathfrak{n}}}) e^{-sl(C_\gamma)}\right)$$

$$= \exp\left(-\sum_{\gamma \in \Gamma_{\mathrm{hyp}}} \mathrm{Tr}\chi(\gamma) j(\gamma)^{-1} D(\gamma)^{-1} \overline{\mathrm{Tr}\sigma_k(m_\gamma)} \exp(-(s-(d-1/2))l(C_\gamma))\right).$$

Here, $\mathrm{P}\Gamma_{\mathrm{hyp}}$ denotes the set of $\Gamma-$conjugacy classes of primitive hyperbolic elements of Γ, $\overline{\mathfrak{n}} = \theta\mathfrak{n}$, S^k is the $k-$th symmetric power of $\mathrm{Ad}\,(m_\gamma a_\gamma)|_{\overline{\mathfrak{n}}}$ and

$$D(\gamma) = D(m_\gamma a_\gamma) = \exp\left(\frac{d-1}{2} l(C_\gamma)\right) \left|\det(\mathrm{Ad}(m_\gamma a_\gamma)^{-1} - \mathrm{Id}|_{\mathfrak{n}})\right|.$$

Finally, we denote by $C_\chi^k(\sigma_k, s)$ the hyperbolic scattering operator (i.e. hyperbolic scattering matrix) defined in section 2.6. of [**7**], see also [**15**] where the divisor of $C_\chi^k(\sigma_k, s)$ is derived.

With the above notation, we state the divisor of $Z_\chi(\sigma_k, s)$, for $0 \leq k \leq \lfloor (d-1)/2 \rfloor$. The Selberg zeta function $Z_\chi(\sigma_k, s)$, defined for $\mathrm{Re}(s) > d-1$, has a meromorphic continuation to $\mathbb{C}$ with non-trivial zeros at

- $s = (d-1)/2 \pm i\lambda_j(k)$ of order $m_j(k)$, where $\lambda_j(k)^2 + ((d-1)/2 - k)^2$ is an eigenvalue of the Hodge Laplacian Δ_k acting on the L^2-space of $k-$forms twisted by χ, with multiplicity $m_j(k)$
- $s = (d-1)/2 + q$ of order $d(\sigma_k)b$, where $s = q$ is a pole of $\det C_\chi^k(\sigma_k, s)$ with $\mathrm{Re}(q) < 0$, of order b.

The location of poles and trivial zeros depends on the parity of d. When $d = 2n+1$ is odd, $Z_\chi(\sigma_k, s)$ has no trivial zeros, and has poles at

- $s = k$ of order $d_c(\chi)e(d,k)$, where $d_c(\chi) = \sum_{P \in \mathfrak{P}_\Gamma} d_P(\chi)$ and $e(d,k) = (-1)^k \left(\sum_{j=0}^k (-1)^j d(\sigma_j)\right) \geq 0$
- $s = n$ of order $\frac{1}{2} d(\sigma_k) \mathrm{Tr}\left(n(\sigma_k)\mathrm{Id} - C_\chi^k(\sigma_k, 0)\right) = \frac{1}{2} d(\sigma_k) a_k$

- $s = n - q_j$ of order $d(\sigma_k)b_j$, where $s = q_j$ is a pole of $\det C_\chi^k(\sigma_k, s)$ with $0 < q_j < n$, of order b_j, for finitely many j, say $j = 1, ..., K$
- $s = n - l$, for $l \in \mathbb{N} \setminus \{n-k\}$ of order $d_c(\chi)d(\sigma_k)$, when $k \neq n$ and in case $k = n$, order is $2d_c(\chi)d(\sigma_k)$.

When $d = 2n$ is even, $Z_\chi(\sigma_k, s)$ has poles at

- $s = n - \frac{1}{2} - q_j$ of order $d(\sigma_k)b_j$, where $s = q_j$ is a pole of $\det C_\chi^k(\sigma_k, s)$ with $0 < q_j < n$, of order b_j, for finitely many j
- $s = n - \frac{1}{2} - l$, for $l \in \mathbb{N}$ of order $d_c(\chi)d(\sigma_k)$

and possesses zeros or poles (according to the sign of their orders) at

- $s = k$ of order $d_c(\chi)d(d,k) + (-1)^{k+1} \dim V_\chi E(X_\Gamma)$, where $d(d,k) = d(\sigma_k) - e(d,k) \geq 0$
- $s=-l$, $l\in\mathbb{N}$ of order $-\dim V_\chi E(X_\Gamma)\left[\binom{2n+l-1}{l+k}\binom{l+k-1}{k}+\binom{2n+l-1}{k}\binom{2n+l-k-2}{l-1}\right]$, where $E(X_\Gamma)$ is the Euler characteristic of X_Γ.

We adopt the convention that in the case when the locations of zeros or poles coincide, their orders are added.

Obviously, the function $Z_\chi(\sigma_k, s)$, for $0 \leq k \leq \lfloor (d-1)/2 \rfloor$ is a zeta-type function. Assume that $0 \leq k \leq \lfloor (d-1)/2 \rfloor$ is fixed and denote by $\mathcal{Z}_f^{\mathrm{NT}}(s,z)$ the superzeta function associated to the set of its non-trivial zeros, i.e. for $\mathrm{Re}(s) > d-1$ and $z \in \mathbb{C}$ such that $z - s \notin \mathbb{R}^-$, for all non-trivial (NT) zeros s of $Z_\chi(\sigma_k, s)$ we have

$$\mathcal{Z}_{\chi,\sigma_k}^{\mathrm{NT}}(s,z) = \sum_j \left(\frac{m_j(k)}{(z-(d-1)/2 - i\lambda_j(k))^s} + \frac{m_j(k)}{(z-(d-1)/2 + i\lambda_j(k))^s} \right) + \sum_{\mathrm{Re}(q)<0} \frac{d(\sigma_k)b}{(z-(d-1)/2-q)^s}.$$

To define the superzeta function associated to the set of trivial zeros and poles we need to consider two cases, depending on the parity of d.

In case when d is odd, the superzeta function associated to the set of poles (P) of $Z_\chi(\sigma_k, s)$ is defined for $\mathrm{Re}(s) > d-1$ and $z \in \mathbb{C}$ such that $z - s \notin \mathbb{R}^-$, for all poles s of $Z_\chi(\sigma_k, s)$ by

$$\begin{aligned} \mathcal{Z}_{\chi,\sigma_k}^{\mathrm{P}}(s,z) = {} & \frac{d_c(\chi)e(d,k)}{(z-k)^s} + \frac{1}{2}d(\sigma_k)\mathrm{Tr}\left(n(\sigma_k)\mathrm{Id} - C_\chi^k(\sigma_k,0)\right)(z-n)^{-s} \\ & + \sum_{j=1}^{K} \frac{d(\sigma_k)b_j}{(z-(n-q_j))^s} + (1+\delta_{kn})d_c(\chi)d(\sigma_k)\zeta_H(s, z-n+1) \\ & - (1-\delta_{kn})d_c(\chi)d(\sigma_k)(z-k)^{-s}. \end{aligned}$$

Note that ζ_H is the Hurwitz zeta function $\zeta_H(s,z) = \sum_{l=0}^{\infty}(z+l)^{-s}$; and δ_{kn} denotes the Kronecker delta function.

In the case when $d = 2n+1$ is odd, we have the following corollary of Theorem 4.

COROLLARY 6. *For $d = 2n+1$, $0 \leq k \leq n$ and $z \in \mathbb{C}$ such that $z - s \notin \mathbb{R}^-$, for any non-trivial zero s of $Z_\chi(\sigma_k, s)$ the superzeta function $\mathcal{Z}_{\chi,\sigma_k}^{\mathrm{NT}}(s,z)$ associated to the set of non-trivial zeros of $Z_\chi(\sigma_k, s)$, initially defined for $\mathrm{Re}(s) > 2n+1$, possesses a meromorphic continuation to the whole complex s-plane, with only one simple* *pole at $s = 1$ with corresponding residue equal to $\beta(k,n,\chi) = (1+\delta_{kn})d_c(\chi)d(\sigma_k)$. Moreover, the zeta regularized product associated to this zeta function is given by*

$$\exp\left(-\frac{d}{ds}\mathcal{Z}^{\mathrm{NT}}_{\chi,\sigma_k}(s,z)\Big|_{s=0}\right) = (z-k)^{\alpha(k,n,\chi)}(z-n)^{\frac{1}{2}d(\sigma_k)a_k}\prod_{j=1}^{K}(z-(n-q_j))^{d(\sigma_k)b_j}.$$

$$\cdot\left(\frac{\Gamma(z-n+1)}{\sqrt{2\pi}}\right)^{-\beta(k,n,\chi)} Z_\chi(\sigma_k,z), \tag{5.1}$$

where $\alpha(k,n,\chi) = d_c(\chi)(e(d,k) - (1-\delta_{kn})d(\sigma_k))$.

PROOF. Remark 5, Theorem 4 together with definition of the superzeta function associated to the set of poles of $Z_\chi(\sigma_k,s)$ yield that

$$\begin{aligned}\mathcal{Z}^{\mathrm{NT}}_{\chi,\sigma_k}(s,z) &= d_c(\chi)(e(d,k) - (1-\delta_{kn})d(\sigma_k))(z-k)^{-s}\\ &+ d(\sigma_k)(a_k/2 - (1+\delta_{kn})d_c(\chi))(z-n)^{-s}\\ &+\sum_{j=1}^{K}\frac{d(\sigma_k)b_j}{(z-(n-q_j))^s} + (1+\delta_{kn})d_c(\chi)d(\sigma_k)\zeta_H(s,z-n)\\ &+\frac{\sin\pi s}{\pi}\int_0^\infty\left(\frac{Z_\chi(\sigma_k,z+y)'}{Z_\chi(\sigma_k,z+y)}\right)y^{-s}dy,\end{aligned}\tag{5.2}$$

and the above equation is valid for all admissible z, i.e. $z\in\mathbb{C}$ such that $z-s\notin\mathbb{R}^-$, for all non-trivial zeros s of $Z_\chi(\sigma_k,s)$.

For a fixed, admissible z, all terms on the right-hand side of (5.2), except the Hurwitz zeta function are holomorphic in s. Therefore, for a fixed, admissible z, the function $\mathcal{Z}^{\mathrm{P}}_{\chi,\sigma_k}(s,z)$ possesses a meromorphic continuation to the whole complex $s-$plane, with only one pole at $s=1$ and corresponding residue $(1+\delta_{kn})d_c(\chi)d(\sigma_k)$ stemming from the factor of the Hurwitz zeta function.

Differentiating equation (5.2) with respect to the s variable, inserting $s=0$, and employing equation $\frac{d}{ds}\zeta_H(s,z)|_{s=0} = \log\Gamma(z) - \log\sqrt{2\pi}$, we get (5.1). □

In the case when $d = 2n$ is even, the superzeta function constructed over poles/trivial zeros is somewhat more complicated. First, we introduce multiple Hurwitz zeta function and multiple gamma functions.

For a positive integer m, the multiple Hurwitz zeta function is defined for $\mathrm{Re}(s) > m$ and $z\in\mathbb{C}\setminus\{0,-1,-2,...\}$ by the absolutely convergent series

$$\zeta_m(s,z) = \sum_{l=0}^{\infty}\binom{m+l-1}{l}(z+l)^{-s},$$

which possesses a meromorphic continuation to the whole complex $s-$plane with simple poles at $s = 1,2,...,m$ (see [**1**], p. 505). In [**1**] it is also proved that the multiple Hurwitz zeta function is related to Hurwitz zeta function via the formula

$$\zeta_m(s,z) = \sum_{j=0}^{m-1}p_{m,j}(z)\zeta_H(s-j,z),$$

where $p_{m,j}(z)$ are polynomials in z defined by

$$p_{m,j}(z) = \frac{1}{(m-1)!}\sum_{l=j}^{m-1}(-1)^{m+1-j}\binom{l}{j}s(n,l+1)z^{l-j}$$

and $s(n,l)$ is the Stirling number of the first kind, i.e. the coefficient in the expansion

$$(z)_m = z(z+1)\cdots(z+m-1) = \sum_{l=0}^{m}(-1)^{m+l}s(m,l)z^l$$

of the Pochhammer symbol $(z)_m$.

Since the Hurwitz zeta function $\zeta_H(s,z)$ possesses only one simple pole at $s=1$, with residue one, it is immediate that residues of $\zeta_m(s,z)$ at $s=k\in\{1,...,m\}$ are

$$\mathrm{Res}_{s=k}\zeta_m(s,z) = p_{m,k-1}(z).$$

The multiple gamma function Γ_m is defined by

$$\Gamma_m(z) = \exp\left(\frac{d}{ds}\zeta_m(s,z)\bigg|_{s=0}\right).$$

COROLLARY 7. *For $d=2n$, $0\le k\le n-1$ and $z\in\mathbb{C}$ such that $z-s\notin\mathbb{R}^-$, for any non-trivial zero s of $Z_\chi(\sigma_k,s)$ the superzeta function $\mathcal{Z}^{\mathrm{NT}}_{\chi,\sigma_k}(s,z)$ associated to the set of non-trivial zeros of $Z_\chi(\sigma_k,s)$, initially defined for* $\mathrm{Re}(s)>2n$, *possesses a meromorphic continuation to the whole complex s-plane, with simple poles at $s_r=r$, for $r\in\{1,...,2n\}$ with corresponding residues equal to*

$$\text{(5.3)}\quad \mathrm{Res}_{s=r}\mathcal{Z}^{\mathrm{NT}}_{\chi,\sigma_k}(z,s) = \delta_{1r}d_c(\chi)d(\sigma_k) + \dim V_\chi E(X_\Gamma)\sum_{j=0}^{k}(-1)^j\binom{2n}{k-j}\left(p_{2n,r-1}(z-j)+p_{2n,r-1}(z+j+1)\right).$$

Moreover, the zeta regularized product associated to this superzeta function is given by

(5.4)

$$\exp\left(-\frac{d}{ds}\mathcal{Z}^{\mathrm{NT}}_{\chi,\sigma_k}(s,z)\bigg|_{s=0}\right) = Z_\chi(\sigma_k,z)\cdot(z-k)^{-\gamma(k,n,\chi)}\left(\prod_{j=1}^{K}(z-n-\frac{1}{2}-q_j)^{d(\sigma_k)b_j}\right)$$
$$\cdot\left(\frac{\Gamma(z-n+\frac{3}{2})}{\sqrt{2\pi}}\right)^{-d_c(\chi)d(\sigma_k)}\cdot\left(\prod_{j=0}^{K}\left(\Gamma_{2n}(z-j)\Gamma_{2n}(z+j+1)\right)^{(-1)^j\binom{2n}{k-j}}\right)^{\dim V_\Gamma E(X_\Gamma)},$$

where $\gamma(k,n,\chi) = d_c(\chi)d(d,k)+(-1)^{k+1}(1-\dim V_\chi E(X_\Gamma))$.

PROOF. From definition of divisor of $Z_\chi(\sigma_k,s)$, for all $z\in\mathbb{C}$ such that $z-s\notin\mathbb{R}^-$ for all w which belong to the set of non-trivial zeros or poles of $Z_\chi(\sigma_k,s)$ we define for $\mathrm{Re}(s)>2n$ (following notation set in Remark 5)

$$\mathcal{Z}^{\mathrm{P}}_{\chi,\sigma_k}(z,s) - \mathcal{Z}^{\mathrm{T}}_{\chi,\sigma_k}(s,z) = -\frac{d_c(\chi)d(d,k)+(-1)^{k+1}\dim V_\Gamma E(X_\Gamma)}{(z-k)^s}$$
$$+\sum_{j=1}^{K}\frac{d(\sigma_k)b_j}{(z-(n-1/2-q_j))^s} + d_c(\chi)d(\sigma_k)\zeta_H(s,z-n+3/2)$$
$$+\dim V_\chi E(X_\Gamma)\sum_{l\in\mathbb{N}}\left[\binom{2n+l-1}{l+k}\binom{l+k-1}{k} + \binom{2n+l-1}{k}\binom{2n+l-k-2}{l-1}\right](z+l)^{-s}$$

Next, we employ the following transformation formulas for the binomial coefficients:

$$\sum_{j=0}^{k-m}(-1)^j\binom{2n}{k-m-j}\binom{2n+j-1}{j}=0, \text{ for } m\in\{0,...,k-1\};$$

$$\sum_{j=0}^{k}(-1)^j\binom{2n}{k-j}\binom{2n+m+j-1}{m+j}=\binom{2n+m-1}{m+k}\binom{m+k-1}{k}$$

and

$$\sum_{j=0}^{\min\{k,m-1\}}(-1)^j\binom{2n}{k-j}\binom{2n+m-j-2}{m-j-1}=\binom{2n+m-1}{k}\binom{2n+m-k-2}{m-1},$$

for positive integers m to deduce, by equating coefficients of $(z+m)^{-s}$, for $m-1\geq k$, that

$$\sum_{l\in\mathbb{N}}\left[\binom{2n+l-1}{l+k}\binom{l+k-1}{k}+\binom{2n+l-1}{k}\binom{2n+l-k-2}{l-1}\right](z+l)^{-s}=$$

$$=\sum_{j=0}^{k}(-1)^j\binom{2n}{k-j}\left(\zeta_{2n}(s,z-j)+\zeta_{2n}(s,z+j+1)\right)+(-1)^{k+1}(z-k)^{-s}.$$

Therefore, application of Theorem 4 yields the equation

$$\mathcal{Z}^{\mathrm{NT}}_{\chi,\sigma_k}(s,z)=\sum_{j=1}^{K}\frac{d(\sigma_k)b_j}{(z-(n-1/2-q_j))^s}+d_c(\chi)d(\sigma_k)\zeta_H(s,z-n+3/2)+$$

$$+\dim V_\chi E(X_\Gamma)\sum_{j=0}^{k}(-1)^j\binom{2n}{k-j}\left(\zeta_{2n}(s,z-j)+\zeta_{2n}(s,z+j+1)\right)+$$

$$+\frac{(-1)^{k+1}(\dim V_\chi E(X_\Gamma)-1)-d_c(\chi)d(d,k)}{(z-k)^s}+\frac{\sin\pi s}{\pi}\int_0^\infty\left(\frac{Z_\chi(\sigma_k,z+y)'}{Z_\chi(\sigma_k,z+y)}\right)y^{-s}dy.$$

For admissible values of z, function on the right-hand side of the above equation is meromorphic in s with simple poles at $s\in\{1,...,2n\}$ and corresponding residues at $s=r\in\{1,...,2n\}$ given by equation (5.3), which proves the first part.

By taking derivatives with respect to the $s-$variable, inserting $s=0$ and using the definition of multiple gamma function, we immediately deduce (5.4). □

REMARK 8. The zeta regularized product $D_{\chi,\sigma_k}(z)=\exp\left(-\frac{d}{ds}\mathcal{Z}^{\mathrm{NT}}_{\chi,\sigma_k}(s,z)\big|_{s=0}\right)$ associated to the Selberg zeta function is evaluated in Corollaries 6 and 7 for admissible values of z, i.e. $z\in\mathbb{C}$ such that $z-s\notin\mathbb{R}^-$, for any non-trivial zero s of $Z_\chi(\sigma_k,s)$. The right-hand side of (5.1) and (5.4) provides meromorphic continuation of $D_{\chi,\sigma_k}(z)$ to the cut plane $\mathbb{C}\setminus(-\infty,n]$.

References

[1] J. Choi and H. M. Srivastava, *The multiple Hurwitz zeta function and the multiple Hurwitz-Euler eta function*, Taiwanese J. Math. **15** (2011), no. 2, 501–522, DOI 10.11650/twjm/1500406218. MR2810165

[2] J. Elstrodt, F. Grunewald, and J. Mennicke, *Groups acting on hyperbolic space*, Springer Monographs in Mathematics, Springer-Verlag, Berlin, 1998. Harmonic analysis and number theory. MR1483315

[3] J. S. Friedman, *The Selberg trace formula and Selberg zeta-function for cofinite Kleinian groups with finite-dimensional unitary representations*, Math. Z. **250** (2005), no. 4, 939–965, DOI 10.1007/s00209-005-0806-9. MR2180383

[4] J. S. Friedman, *Regularized determinants of the Laplacian for cofinite Kleinian groups with finite-dimensional unitary representations*, Comm. Math. Phys. **275** (2007), no. 3, 659–684, DOI 10.1007/s00220-007-0330-3. MR2336359

[5] J. S. Friedman, J. Jorgenson, L. Smajlović, *The determinant of the Lax-Phillips scattering operator*, preprint (arxiv.org/abs/1603.07613).

[6] J. S. Friedman, J. Jorgenson, L. Smajlović, *An evaluation of the central value of the automorphic scattering determinant*, preprint (arxiv.org/abs/1607.08053).

[7] Y. Gon and J. Park, *The zeta functions of Ruelle and Selberg for hyperbolic manifolds with cusps*, Math. Ann. **346** (2010), no. 3, 719–767, DOI 10.1007/s00208-009-0408-7. MR2578567

[8] I. S. Gradshteyn and I. M. Ryzhik, *Table of integrals, series, and products*, 7th ed., Elsevier/Academic Press, Amsterdam, 2007. Translated from the Russian; Translation edited and with a preface by Alan Jeffrey and Daniel Zwillinger; With one CD-ROM (Windows, Macintosh and UNIX). MR2360010

[9] G. Illies, *Regularized products and determinants*, Comm. Math. Phys. **220** (2001), no. 1, 69–94, DOI 10.1007/s002200100448. MR1882400

[10] J. Jorgenson and S. Lang, *On Cramér's theorem for general Euler products with functional equation*, Math. Ann. **297** (1993), no. 3, 383–416, DOI 10.1007/BF01459509. MR1245398

[11] J. Jorgenson and S. Lang, *Basic analysis of regularized series and products*, Lecture Notes in Mathematics, vol. 1564, Springer-Verlag, Berlin, 1993. MR1284924

[12] J. Jorgenson and R. Lundelius, *Convergence theorems for relative spectral functions on hyperbolic Riemann surfaces of finite volume*, Duke Math. J. **80** (1995), no. 3, 785–819, DOI 10.1215/S0012-7094-95-08027-2. MR1370116

[13] J. Jorgenson and R. Lundelius, *Continuity of relative hyperbolic spectral theory through metric degeneration*, Duke Math. J. **84** (1996), no. 1, 47–81, DOI 10.1215/S0012-7094-96-08402-1. MR1394748

[14] J. Jorgenson and R. Lundelius, *Convergence of the normalized spectral counting function on degenerating hyperbolic Riemann surfaces of finite volume*, J. Funct. Anal. **149** (1997), no. 1, 25–57, DOI 10.1006/jfan.1997.3098. MR1471098

[15] D. Kelmer, *On distribution of poles of Eisenstein series and the length spectrum of hyperbolic manifolds*, Int. Math. Res. Not. IMRN **23** (2015), 12319–12344, DOI 10.1093/imrn/rnv051. MR3431622

[16] K. Kimoto and M. Wakayama, *Remarks on zeta regularized products*, Int. Math. Res. Not. **17** (2004), 855–875, DOI 10.1155/S1073792804131565. MR2040075

[17] N. Kurokawa and M. Wakayama, *Zeta regularizations*, Acta Appl. Math. **81** (2004), no. 1-3, 147–166, DOI 10.1023/B:ACAP.0000024207.37694.3b. MR2069336

[18] N. Kurokawa, M. Wakayama, and Y. Yamasaki, *Milnor-Selberg zeta functions and zeta regularizations*, J. Geom. Phys. **64** (2013), 120–145, DOI 10.1016/j.geomphys.2012.10.015. MR3004020

[19] P. D. Lax and R. S. Phillips, *Scattering theory for automorphic functions*, Princeton Univ. Press, Princeton, N.J., 1976. Annals of Mathematics Studies, No. 87. MR0562288

[20] P. D. Lax and R. S. Phillips, *Scattering theory for automorphic functions*, Bull. Amer. Math. Soc. (N.S.) **2** (1980), no. 2, 261–295, DOI 10.1090/S0273-0979-1980-14735-7. MR555264

[21] W. Müller, *Relative zeta functions, relative determinants and scattering theory*, Comm. Math. Phys. **192** (1998), no. 2, 309–347, DOI 10.1007/s002200050301. MR1617554

[22] R. Phillips and P. Sarnak, *Perturbation theory for the Laplacian on automorphic functions*, J. Amer. Math. Soc. **5** (1992), no. 1, 1–32, DOI 10.2307/2152749. MR1127079

[23] D. B. Ray and I. M. Singer, *Analytic torsion for complex manifolds*, Ann. of Math. (2) **98** (1973), 154–177, DOI 10.2307/1970909. MR0383463

[24] A. Voros, *Spectral functions, special functions and the Selberg zeta function*, Comm. Math. Phys. **110** (1987), no. 3, 439–465. MR891947

[25] A. Voros, *Zeta functions for the Riemann zeros* (English, with English and French summaries), Ann. Inst. Fourier (Grenoble) **53** (2003), no. 3, 665–699. MR2008436

[26] A. Voros, *More zeta functions for the Riemann zeros*, Frontiers in number theory, physics, and geometry. I, Springer, Berlin, 2006, pp. 349–363, DOI 10.1007/978-3-540-31347-2_9. MR2261100

[27] A. Voros, *Zeta functions over zeros of zeta functions*, Lecture Notes of the Unione Matematica Italiana, vol. 8, Springer-Verlag, Berlin; UMI, Bologna, 2010. Appendix D: Hj. Mellin's "On the zeros of the zeta function" translated from the German original [Ann. Acad. Sci. Fenn. A **10** (1917), no. 11] and annotated by Voros. MR2572085

Department of Mathematics and Science, United States Merchant Marine Academy, 300 Steamboat Road, Kings Point, New York 11024
Email address: friedmanj@usmma.edu, crowneagle@gmail.com

Department of Mathematics, The City College of New York, Convent Avenue at 138th Street, New York, New York 10031
Email address: jjorgenson@mindspring.com

Department of Mathematics, University of Sarajevo, Zmaja od Bosne 35, 71 000 Sarajevo, Bosnia and Herzegovina
Email address: lejlas@pmf.unsa.ba

Contemporary Mathematics
Volume **732**, 2019
https://doi.org/10.1090/conm/732/14787

Computing p-adic periods of abelian varieties from automorphic forms

Xavier Guitart and Marc Masdeu

ABSTRACT. We give an overview of the techniques introduced in a previous paper of the authors, in which we exploit Darmon's p-adic $\mathcal{L}$-invariants to compute p-adic uniformizations of certain motives attached to modular forms. We illustrate our methods with new examples.

1. Introduction

Let E be an elliptic curve of conductor pN, with p a prime not dividing N, and let f be the newform for $\Gamma_0(pN)$ corresponding to (the isogeny class of) E under the Modularity Theorem. Let also K be a real quadratic field in which p is inert and all the primes dividing N split. In this setting, Darmon introduced in his seminal paper [**Dar01**] a construction of local points on $E(K_p)$, which he called Stark–Heegner points, and which are conjecturally global and defined over ring class fields of K.

Taking advantage of the isomorphism $E(K_p) \simeq K_p^\times/q_E^{\mathbb{Z}}$, where q_E is the Tate period of E, Darmon defined Stark–Heegner points as suitable values of certain $K_p^\times$-valued integrals of f. By construction, however, such values are only well defined (i.e., independent of any choices) modulo a certain p-adic lattice $q_f^{\mathbb{Z}}$, where q_f is a p-adic number whose definition depends on f and is a priori unrelated to q_E. The quantity $\mathcal{L}_p(f) = \log_p(q_f)/\operatorname{ord}_p(q_f)$ can be interpreted as a p-adic $\mathcal{L}$-invariant. Indeed, one of the main results of [**Dar01**] is the following relation between special values of classical and p-adic L-functions:

$$L_p'(f,1) = \mathcal{L}_p(f)L(f,1).$$

Therefore, as a consequence of the exceptional zero conjecture of Mazur–Tate–Teitelbaum proven by Greenberg-Stevens [**GS93**], Darmon's p-adic $\mathcal{L}$-invariant coincides with $\log_p(q_E)/\operatorname{ord}_p(q_E)$ and, in particular, $K_p^\times/q_E^{\mathbb{Z}}$ and $K_p^\times/q_f^{\mathbb{Z}}$ are isogenous elliptic curves. Thus the construction of Stark–Heegner points does produce (perhaps after composing with an isogeny) well defined points in $E(K_p)$.

Over the years there have been a number of generalizations and variants of Stark–Heegner points attached to more general motives (e.g., elliptic curves over number fields [**Gre09, Tri06, Gär12, GM17, GMS15**], modular abelian varieties [**Das05, LRV12**], motives attached to higher weight modular forms [**RS12**]), that have come to be known as Darmon points or, even more generally, Darmon cycles.

2010 *Mathematics Subject Classification.* Primary 11G40; Secondary 11F41, 11Y99.

A common feature of these constructions is that the points (or cycles) are defined as p-adic integrals which are only well defined modulo a Darmon-like $\mathcal{L}$-invariant given by certain period integrals of automorphic forms. That these integrals yield well defined cycles on the motive depends on the (in some cases still conjectural) relationship between the p-adic $\mathcal{L}$-invariant and the periods of the p-adic uniformization of the motive.

In this note we give an overview of [**GM17**], in which we explore the possibility of using Darmon's p-adic $\mathcal{L}$-invariants, in the form of suitable automorphic periods, and their conjectured properties to compute p-adic uniformizations of certain motives attached to modular forms for which no unconditional construction is yet known. More precisely, we consider the case of modular forms of weight two over number fields of mixed signature. The associated motives are expected to be abelian varieties of GL_2-type, although the Eichler–Shimura-type construction that associates to any such modular form an abelian variety is only known for totally real number fields. In some cases for which no Eichler–Shimura construction is known, such as number fields with one complex place, we have computed approximations to the p-adic automorphic periods conforming Darmon's $\mathcal{L}$-invariants. Granting the conjecture that these periods are isogenous to the p-adic periods of the abelian variety, we have been able to recover the global algebraic equations.

In Section 2 we recall the conjecture that associates to any modular form of weight two for GL_2 an abelian variety, and we collect a few facts on their p-adic uniformization. In Section 3 we describe the p-adic integrals attached to the modular forms and the construction of the periods that, conjecturally, coincide with the p-adic periods of the associated abelian variety. Finally, in Section 4, we illustrate the method by reporting on two new examples of the computation of $\mathcal{L}$-invariants and abelian varieties.

The authors would like to thank Victor Rotger for a careful reading of the introduction. Masdeu wishes to thank the organizers and participants of the *Building Bridges: Workshop on Automorphic Forms and Related Topics* for providing such a relaxed and comfortable atmosphere in which to discuss ideas. Guitart was supported by MTM2015-66716-P and MTM2015-63829, and Masdeu was supported by MSC–IF–H2020–ExplicitDarmonProg.

2. Automorphic forms and abelian varieties

In this section we follow the approach of [**Tay95**]. Let F be a number field of signature (r, s), and consider an ideal $\mathfrak{N} \subset \mathcal{O}_F$, which we will call a *level*. In order to avoid technical difficulties, we will assume that F has narrow class number one, and that $\mathfrak{N}$ is squarefree. After fixing $r+s$ embeddings of F into $\mathbb{C}$ corresponding to the places of F, the group $\Gamma_0(\mathfrak{N}) = \{\left(\begin{smallmatrix} a & b \\ c & d \end{smallmatrix}\right) \in \mathrm{SL}_2(\mathcal{O}_F) \colon \mathfrak{N} \mid c\}$ acts discretely on $\mathbb{H}^r \times \mathbb{H}_3^s$, where $\mathbb{H}$ (respectively $\mathbb{H}_3$) denotes the hyperbolic upper half plane (respectively upper half space). The cohomology of the quotient orbifold $Y_0(\mathfrak{N}) = \Gamma_0(\mathfrak{N})\backslash(\mathbb{H}^n \times \mathbb{H}_3^s)$ can be computed via group cohomology. This cohomology comes equipped with an action of the commutative Hecke algebra $\mathbb{T}$, generated by the Hecke operators $T_\mathfrak{l}$ for primes $\mathfrak{l} \nmid \mathfrak{N}$. The following conjecture is a particular case of [**Tay95**, Conjecture 3].

CONJECTURE 2.1. *Let $f \in H^{n+s}(\Gamma_0(\mathfrak{N}), \mathbb{C})$ be an eigenvector for all the Hecke operators, say $T_\mathfrak{l} f = a_\mathfrak{l} f$ for all $\mathfrak{l} \nmid \mathfrak{N}$, which is cuspidal, new, and without complex multiplication. Let d be the degree of the number field $K_f = \mathbb{Q}(\{a_\mathfrak{l}\}_\mathfrak{l})$. There is a* *simple abelian variety A_f/F of dimension d, of conductor $\mathfrak{N}^d$, and with* $\mathrm{End}(A_f)\otimes\mathbb{Q}\supseteq K_f$, *such that*

$$L(A_f,s)=\prod_{\sigma\colon K_f\hookrightarrow\mathbb{C}} L({}^{\sigma}f,s).$$

One should remark that in the original conjecture there is another possibility for the variety associated to f; namely, it could happen that it is of dimension $2d$ and has quaternionic multiplication. This possibility is excluded in our setting, thanks to the assumptions of $\mathfrak{N}$ being square-free and of f not having complex multiplication. In fact, this is the reason why we restrict to forms without complex multiplication (see [**GM17**, Proposition 2.2] and [**GM17**, Proposition 2.3] for more details).

Taylor's conjecture is true when $F=\mathbb{Q}$, thanks to the Eichler–Shimura construction. It is known to be true in other cases, namely when F is totally real and there exists a Jacquet–Langlands lift of f to a Shimura curve (e.g., this is always the case if $[F:\mathbb{Q}]$ is odd).

The aim of this short note is to give an overview of the explicit construction that we propose in [**GM17**] which gives a conjectural description of the variety A_f as a p-adic torus. At this point the reader might get unsettled by the possibility that A_f would not admit a p-adic uniformization. After all, here is one example of this situation: the elliptic curve E_1 given by the equation $y^2=x^3+1$ has conductor $N_1=36$. The elliptic curve E_2 given by the equation $y^2+xy=x^3-x^2-2x-1$ has conductor 49. Therefore the abelian surface $A=E_1\times E_2$ has conductor $1764=(2\cdot 3\cdot 7)^2$. However, since neither E_1 nor E_2 is p-adically uniformizable (because neither has multiplicative reduction at any p), the abelian surface A cannot be p-adically uniformizable. In our setting, however, the endomorphism structure of A will make the uniformization possible.

THEOREM 2.2 ([**GM17**], Prop. 2.4). *Let A/F be an abelian variety of dimension d and let $\mathfrak{p}$ be a prime of F. Denote by $\mathbb{Q}_{\mathfrak{p}}$ the completion of F at $\mathfrak{p}$. Suppose that*

- $\mathfrak{p}^d\parallel\mathrm{conductor}(A)$, *and*
- $\mathrm{End}(A)\otimes\mathbb{Q}$ *contains a totally real field K of degree d.*

Then there exists a discrete lattice $\Lambda\subset(\mathbb{Q}_{\mathfrak{p}}^{\times})^d$ such that

$$A(\bar{\mathbb{Q}}_{\mathfrak{p}})\cong(\bar{\mathbb{Q}}_{\mathfrak{p}}^{\times})^d/\Lambda.$$

3. Periods of automorphic forms

As in the previous section let $f\in H^{n+s}(\Gamma_0(\mathfrak{N}),\mathbb{C})$ be a cuspidal newform without complex multiplication, and suppose that $\mathfrak{p}$ is a prime that divides $\mathfrak{N}$ exactly. The periods are constructed by means of a $\mathfrak{p}$-adic integration pairing between certain homology and cohomology groups of a $\{\mathfrak{p}\}$-arithmetic subgroup $\Gamma\subset\mathrm{PGL}_2(\mathbb{Q}_{\mathfrak{p}})$ related to $\Gamma_0(\mathfrak{N})$.

3.1. The $\{\mathfrak{p}\}$-arithmetic subgroup. We now proceed to define a $\{\mathfrak{p}\}$-arithmetic group Γ. We start by choosing a factorization of the form $\mathfrak{N}=\mathfrak{p}\mathfrak{D}\mathfrak{m}$ and a choice of $n\le r$ real places, say $v_1,v_2,\ldots,v_r$, in such a way that the set of places $\{\mathfrak{q}\mid\mathfrak{D}\}\cup\{v_{n+1},\ldots,v_r\}$ has even cardinal. Let $B_{/F}$ be the quaternion algebra whose ramification locus is precisely this set. Fix Eichler orders $R_0^{\mathfrak{D}}(\mathfrak{p}\mathfrak{m})\subset R_0^{\mathfrak{D}}(\mathfrak{m})\subset B$, and an embedding $\iota_{\mathfrak{p}}\colon R_0^{\mathfrak{D}}(\mathfrak{m}) \to M_2(\mathbb{Z}_{\mathfrak{p}})$ (here $\mathbb{Z}_{\mathfrak{p}}$ stands for the ring of integers of $\mathbb{Q}_{\mathfrak{p}}$). Define $\Gamma_0^{\mathfrak{D}}(\mathfrak{pm}) = R_0^{\mathfrak{D}}(\mathfrak{pm})^\times/\mathcal{O}_F^\times$ and $\Gamma_0^{\mathfrak{D}}(\mathfrak{m}) = R_0^{\mathfrak{D}}(\mathfrak{m})^\times/\mathcal{O}_F^\times$. Finally, define

$$\Gamma = \iota_{\mathfrak{p}}(R_0^{\mathfrak{D}}(\mathfrak{m})[1/\mathfrak{p}]^\times/\mathcal{O}_F[1/\mathfrak{p}]^\times) \subset \mathrm{PGL}_2(\mathbb{Q}_{\mathfrak{p}}).$$

In the following subsections we will see how an integration pairing in the (co)homology of Γ gives rise to naturally defined periods of automorphic forms.

3.2. Integration on $\mathcal{H}_{\mathfrak{p}}$. Let $\mathbb{Q}_{\mathfrak{p}^2}$ be the quadratic unramified extension of $\mathbb{Q}_{\mathfrak{p}}$, and consider $\mathcal{H}_{\mathfrak{p}} = \mathbb{P}^1(\mathbb{Q}_{\mathfrak{p}^2}) \smallsetminus \mathbb{P}^1(\mathbb{Q}_{\mathfrak{p}})$. It is a $\mathfrak{p}$-adic analogue to $\mathbb{H}$ in many ways: it has a rigid-analytic structure, and an action of $\mathrm{PGL}_2(\mathbb{Q}_{\mathfrak{p}})$ by fractional linear transformations. There is a theory of rigid-analytic 1-forms $\omega \in \Omega^1_{\mathcal{H}_{\mathfrak{p}}}$, which replaces the complex-analytic theory. We denote by $\Omega^1_{\mathcal{H}_{\mathfrak{p}},\mathbb{Z}}$ the forms having $\mathbb{Z}$-valued residues. Coleman integration allows us to make sense of $\int_{\tau_1}^{\tau_2} \omega \in \mathbb{C}_{\mathfrak{p}}$. Moreover, Darmon [**Dar01**] constructed a $\mathrm{PGL}_2(\mathbb{Q}_{\mathfrak{p}})$-equivariant pairing

$$⨍ : \Omega^1_{\mathcal{H}_{\mathfrak{p}},\mathbb{Z}} \times \mathrm{Div}^0 \mathcal{H}_{\mathfrak{p}} \to \mathbb{Q}_{\mathfrak{p}^2}^\times \subset \mathbb{C}_{\mathfrak{p}}^\times,$$

which refines Coleman integration into a multiplicative variant. Cap product induces a pairing

$$\begin{array}{ccc} \mathrm{H}^i(\Gamma, \Omega^1_{\mathcal{H}_{\mathfrak{p}},\mathbb{Z}}) \times \mathrm{H}_i(\Gamma, \mathrm{Div}^0 \mathcal{H}_{\mathfrak{p}}) & \xrightarrow{\langle \cdot,\cdot \rangle} & \mathbb{C}_{\mathfrak{p}}^\times \\ \left(\phi, \sum_{\underline{\gamma}} \underline{\gamma} \otimes D_{\underline{\gamma}}\right) & \longmapsto & \sum_{\underline{\gamma}} ⨍_{D_{\underline{\gamma}}} \phi(\underline{\gamma}). \end{array}$$

3.3. The conjecture. Denote with a subindex f the f-isotypical part of a Hecke-module: the submodule on which the Hecke operator T_ℓ acts as multiplication by $a_\ell(f)$. The following result is [**GM17**, Proposition 3.1].

THEOREM 3.1. *There is a natural isomorphism of Hecke modules*

$$\mathbb{Q} \otimes_{\mathbb{Z}} \mathrm{H}^{n+s}(\Gamma_0^{\mathfrak{D}}(\mathfrak{pm}), \mathbb{Z})_f \cong \mathbb{Q} \otimes_{\mathbb{Z}} \mathrm{H}^{n+s}(\Gamma, \Omega^1_{\mathcal{H}_{\mathfrak{p}},\mathbb{Z}})_f.$$

Next, consider the Γ-equivariant short exact sequence that defines the degree 0 divisors on $\mathcal{H}_{\mathfrak{p}}$

$$0 \to \mathrm{Div}^0 \mathcal{H}_{\mathfrak{p}} \to \mathrm{Div}\, \mathcal{H}_{\mathfrak{p}} \xrightarrow{\deg} \mathbb{Z} \to 0.$$

Taking Γ-coinvariants yields a long exact sequence in homology, from which we isolate the following piece:

$$\cdots \to \mathrm{H}_{n+s+1}(\Gamma, \mathbb{Z}) \xrightarrow{\delta} \mathrm{H}_{n+s}(\Gamma, \mathrm{Div}^0 \mathcal{H}_{\mathfrak{p}}) \to \mathrm{H}_{n+s}(\Gamma, \mathrm{Div}\, \mathcal{H}_{\mathfrak{p}}) \to \mathrm{H}_{n+s}(\Gamma, \mathbb{Z}) \to \cdots$$

Set $\underline{\omega}_f$ to be a fixed basis of the non-torsion part of $\mathrm{H}^{n+s}(\Gamma, \Omega^1_{\mathcal{H}_{\mathfrak{p}},\mathbb{Z}})_f$.

CONJECTURE 3.2. *Set*

$$\Lambda_f = \left\{\langle \underline{\omega}_f, \delta(c)\rangle : c \in \mathrm{H}_{n+s+1}(\Gamma, \mathbb{Z})\right\} \subset (\bar{\mathbb{Q}}_{\mathfrak{p}}^\times)^d.$$

Then Λ_f is a lattice in $(\mathbb{Q}_{\mathfrak{p}}^\times)^d$ and $A_f(\bar{\mathbb{Q}}_{\mathfrak{p}})$ is isogenous to $(\bar{\mathbb{Q}}_{\mathfrak{p}}^\times)^d/\Lambda_f$.

This conjecture was proven by Darmon [**Dar01**] in the case $F = \mathbb{Q}$, $B = \mathrm{M}_2(\mathbb{Q})$ and $d = 1$; by Dasgupta [**Das05**] when $F = \mathbb{Q}$, $B = \mathrm{M}_2(\mathbb{Q})$, and $d > 1$; by Dasgupta–Greenberg [**DG12**] and Longo–Rotger–Vigni ([**LRV12**]) in the case $F = \mathbb{Q}$ and B a division algebra; by Spiess [**Spi14**] when F is totally real, $B = M_2(F)$, $\mathbb{Q}_{\mathfrak{p}} = \mathbb{Q}_p$ and $d = 1$. To the best of our knowledge, it is open in all other cases.

We have developed and implemented algorithms to compute approximations to the periods defining Λ_f. In the next section we illustrate how these can be used to find equations for the putative A_f, at least in favorable situations.

4. Examples

4.1. Elliptic curves. We explain first how to compute the equations of elliptic curves, from their p-adic periods. First, the knowledge of $\Lambda_f = \langle q_f \rangle$ gives us the p-adic period q_f, which we conjecture should be the same (up to taking a rational power whose denominator can be bounded by the p-adic valuation of q_f) as the period q_E attached to $E = E_f$. By possibly replacing q_f with q_f^{-1} we can assume that $v_{\mathfrak{p}}(q_f) > 0$. By using the q-expansion of the j-function we can recover

$$j(q_f) = q_f^{-1} + 744 + 196884 q_f + \cdots \in \mathbb{Q}_{\mathfrak{p}}^{\times}.$$

Instead of trying to recognize $j(q_f)$ algebraically, it is often better to proceed in a more indirect way, since $j(q_f)$ has much larger height than the minimal equation for the sought E. Instead, from the knowledge of $\mathfrak{N}$ (and therefore of the primes of bad reduction of E) we can guess the discriminant Δ_E of a minimal model. This just amounts to trying finitely many possibilities. One can then use the equation $j_E = c_4^3/\Delta_E$ to recover a p-adic approximation to c_4 (after taking a p-adic cube root). One can then recognize c_4 algebraically. Similarly, from the equation $1728\Delta_E = c_4^3 - c_6^2$ one may recover c_6. The candidate equation is then

$$E\colon Y^2 = X^3 - \frac{c_4}{48}X - \frac{c_6}{864}.$$

If E has the right conductor $\mathfrak{N}$ (which is easily computed using Tate's algorithm) one can then proceed to compare some terms of the L-series, until one is convinced that E is really attached to f. If one requires a proof, then one could use the Faltings'–Serre method, although we have not done this in practice.

Here is a new example of such a calculation. Let $F = \mathbb{Q}(\alpha)$, where α has minimal polynomial $f_\alpha(x) = x^6 - x^5 - 4x^4 + x^3 + 4x^2 + x - 1$. The discriminant of F is $\operatorname{disc}(F) = -367792$, and it has signature $(4,1)$. Consider the level $\mathfrak{N} = (\alpha^3 - \alpha^2 - 2\alpha)$, which has norm 7. We take $\mathfrak{p} = \mathfrak{N}$, and consider the quaternion algebra B/F whose locus of ramification is the set of real places of F. There is a rational eigenclass $f \in S_2(\Gamma_0^{\mathcal{O}_F}(\mathfrak{N}))$. From f we compute $\omega_f \in \mathrm{H}^1(\Gamma, \Omega^1_{\mathcal{H}_p,\mathbb{Z}})$ and an approximation to $\Lambda_f = \langle q_f \rangle$. This yields the quantity

$$q_E \stackrel{?}{=} q_f = 7^{-2} \cdot 68530475965426443260903897030400405725776366704466935859 44 + O(7^{67}).$$

Using the method described above we find

$$c_4 = 16\alpha^5 + 16\alpha^4 - 48\alpha^3 - 32\alpha^2 + 32\alpha + 32$$
$$c_6 = 160\alpha^5 + 264\alpha^4 - 32\alpha^3 - 336\alpha^2 - 256\alpha - 56$$

yielding the elliptic curve

$$E_0\colon y^2 = x^3 + \frac{1}{3}\left(-\alpha^5 - \alpha^4 + 3\alpha^3 + 2\alpha^2 - 2\alpha - 2\right)x + \frac{1}{108}\left(-20\alpha^5 - 33\alpha^4 + 4\alpha^3 + 42\alpha^2 + 32\alpha + 7\right)$$

with conductor $\mathfrak{N}$. We have verified that the Euler factors of the respective L-series match for the primes of norm up to 20.

Note that in this case the working precision allowed us to also recognize $j(q_f^{-1})$ directly, as the algebraic number

$$j_E = \frac{1}{49}\left(163840\alpha^5 - 180224\alpha^4 - 557056\alpha^3 + 139264\alpha^2 + 360448\alpha + 368640\right).$$

One could have then immediately written down the curve

$$E_1\colon y^2 = x^3 - 3j_E(j_E - 1728)x - 2j_E(j_E - 1728)^2,$$

which has the right j-invariant, but its conductor is[1]

$$\mathfrak{N} \cdot (\alpha^2 - \alpha - 1)_4^8 \cdot (3)^2 \cdot (\alpha^4 - 3\alpha^2 - 3\alpha)_{43}^2 \cdot (\alpha^5 - 5\alpha^3 - \alpha^2 + 3\alpha + 1)_{53}^2 \cdot (\alpha^5 - 4\alpha^3 - 2\alpha^2 + \alpha + 2)_{63}^2.$$

After performing a quadratic twist by the element

$$-\frac{3931}{78178816}\alpha^5 - \frac{12315}{156357632}\alpha^4 + \frac{77849}{234536448}\alpha^3 + \frac{182557}{469072896}\alpha^2 - \frac{3}{9772352}\alpha - \frac{64003}{469072896}$$

we would still recover the curve E_0.

Finally, the global minimal model (recall that F has class number one) of E_0 is

$$E_{/F}\colon y^2 + \left(\alpha^2 - 1\right) y = x^3 - \left(\alpha^5 - \alpha^4 - 3\alpha^3 + \alpha^2 - \alpha - 1\right) x^2 - \left(\alpha^5 + \alpha^4 - 3\alpha^3 - 6\alpha^2 - 2\alpha + 2\right) x - \alpha^5 - \alpha^4 + 3\alpha^3 + 4\alpha^2 - 1.$$

4.2. Abelian surfaces. Suppose that A_f is principally polarizable, so that $A_f = \mathrm{Jac}(X_f)$ for a genus 2 (hyperelliptic) curve X_f. In this section we explain a method for computing an equation of X_f.

We expect $A_f(\bar{\mathbb{Q}}_\mathfrak{p})$ to be isogenous to $(\bar{\mathbb{Q}}_\mathfrak{p}^\times)^2/\Lambda_f$. Let $\left(\frac{A}{B}\right), \left(\frac{B}{D}\right) \in (\mathbb{Q}_\mathfrak{p}^\times)^2$ be generators of Λ_f. From these generators, we may recover the so-called "half-periods" $p_1 = (BD)^{-1/2}$, $p_2 = (AB)^{-1/2}$, $p_3 = B^{1/2}$. Write

$$X_f\colon y^2 = x(x-1)(x-x_1)(x-x_2)(x-x_3),$$

and define

$$\lambda_1 = 1 - x_1^{-1}, \quad \lambda_2 = (1-x_2)^{-1}, \quad \lambda_3 = x_3.$$

Teitelbaum provides in [**Tei88**] three power series in the variables p_1, p_2, p_3 such that

$$\lambda_k = \sum_{(i,j)\in\mathbb{Z}^2} a_{i,j}^{(k)} p_1^i p_2^j p_3^{(i-j)}.$$

From this, one can compute the absolute invariants i_1, i_2, i_3 of X_f, defined as

$$i_1 = I_2^5/I_{10}, i_2 = I_2^3 I_4/I_{10}, \text{ and } i_3 = I_2^2 I_6/I_{10},$$

where I_2, I_4, I_6, I_{10} are the Igusa invariants. Next, from $\mathfrak{N} = \mathfrak{p}\mathfrak{D}$ one can again guess the discriminant I_{10}, which will be of the form $u \cdot 2^a \cdot \mathfrak{N}^2$, for some $u \in \mathcal{O}_F^\times$ and some $a \in \mathbb{Z}_{\geq 0}$. An approximation to $i_1 = I_2^5/I_{10}$ then allows us (after taking a fifth root) to recover I_2. The approximation to $i_2 = I_2^3 I_4/I_{10}$ gives us I_4, and finally $i_3 = I_2^2 I_6/I_{10}$ gives I_6.

The final step is to apply Mestre's algorithm to find a genus-2 hyperelliptic curve X_f with invariants $(I_2 : I_4 : I_6 : I_{10})$.

---

[1]The notation $(a)_n$ with $a \in \mathcal{O}_F$ denotes that the ideal $(a) \subset \mathcal{O}_F$ has norm n.

The problem we are faced with is that A_f is determined up to isogeny, and we are looking for a surface in the isogeny class which is the Jacobian of some hyperelliptic curve –being principally polarized is not preserved under isogeny–. Therefore we should allow for "isogenous" Λ_f, and this will allow us to find Igusa invariants of smaller height, which we have a chance to identify. Recall that in the elliptic curve case $\mathbb{C}_p^\times/q_1^{\mathbb{Z}} \sim \mathbb{C}_p^\times/q_2^{\mathbb{Z}}$ if and only if there exist $y, z \in \mathbb{Z} \setminus \{0\}$ with $q_1^y = q_2^z$. The right analogue in higher dimension is provided by the following result.

THEOREM 4.1 (Kadziela, [**Kad07**]). *Let $V_1, V_2 \in M_d(\mathbb{Q}_p^\times)$ whose columns generate lattices Λ_1 and Λ_2. Then $(\bar{\mathbb{Q}}_p^\times)^d/\Lambda_1$ is isogenous to $(\bar{\mathbb{Q}}_p^\times)^d/\Lambda_2$ if and only if*

$$V_1^Y = {}^Z V_2, \qquad \text{for some } Y, Z \in M_d(\mathbb{Z}).$$

REMARK 4.2. The notation used in the theorem requires some explanation. Given a matrix $V \in M_d(\mathbb{Q}_p^\times)$ and a matrix $X \in M_d(\mathbb{Z})$, we define the matrix V^X to be the unique matrix $W \in M_d(\mathbb{Q}_p^\times)$ satisfying

$$\ell(W) = X\ell(V)$$

for all characters $\ell\colon \mathbb{Q}_p^\times \to \mathbb{Q}_p$. Here ℓ applied to a matrix is used to mean the matrix obtained by applying ℓ to its coefficients. Similarly, by ${}^X V$ we mean the unique matrix $W \in M_d(\mathbb{Q}_p^\times)$ satisfying

$$\ell(W) = \ell(V)X,$$

for all characters ℓ as above.

In [**GM17**] we report on two numerical calculations that illustrate how the above method can be used to compute equations of genus two curves C whose Jacobian is attached to a modular form over a number field. We end this note with an example of a surface defined over $\mathbb{Q}$. Recall that in this case Conjecture 3.2 is proven. Therefore, the above method can in fact be regarded as a means of computing the p-adic $\mathcal{L}$-invariant of Mazur–Tate–Teitelbaum [**MTT86**]. Explicit examples of such $\mathcal{L}$-invariants where computed in [**Tei88**] when C is a modular curve; we remark that the method presented in this note can be used more generally when $\mathrm{Jac}(C)$ is a simple factor of a modular Jacobian.

Consider the hyperelliptic curve

$$C\colon y^2 = x^6 + 6x^5 + 11x^4 + 14x^3 + 5x^2 - 12x,$$

whose Jacobian has conductor $165 = 3\cdot 5\cdot 11$ and is a simple factor of $\mathrm{Jac}(X_0(165))$. This curve can be found in [**GJG03**, Page 412].

Let $B/\mathbb{Q}$ be the (indefinite) quaternion algebra of discriminant 15, which can be described as $\mathbb{Q}\langle i, j\rangle$, with $i^2 = -3$ and $j^2 = 5$.

At level 11 we find a two-dimensional Hecke-eigenspace for which the Hecke operator T_2 has characteristic polynomial $x^2 + 2x - 1$, of discriminant 8. The automorphic form f gives thus rise to the field $K_f \cong \mathbb{Q}(\sqrt{2})$. The integration pairing gives $\Lambda_f = \left(\begin{smallmatrix} A_0 & B_0 \\ C_0 & D_0 \end{smallmatrix}\right) = \mathbb{Z}[T_2]\cdot\left(\begin{smallmatrix} A_0 & B_0 \end{smallmatrix}\right)$, with

$$A_0 = 11^{-16}\cdot 1073553889889971382127802037769611157724713 + O(11^{25})$$

$$B_0 = 373933394789407595092699936123731095474 50 + O(11^{41}).$$

Note that the relative precision of the two quantities is the same, however since A_0 has negative valuation the absolute precision appears to be less. By guessing appropriate Kadziela matrices we obtain a new set of periods:

$$A = A_0 B_0^2 \qquad\qquad B = B_0^{-2}.$$

They give rise to the absolute invariants

$$(i_1, i_2, i_3) = \left(\frac{I_2^5}{I_{10}}, \frac{I_2^3 I_4}{I_{10}}, \frac{I_2^2 I_6}{I_{10}}\right) = \left(\frac{28125651982744}{13476375}, -\frac{634841652013}{26952750}, \frac{163196533921}{35937000}\right)$$

which match with those of C.

Finally, the p-adic $\mathcal{L}$-invariant of $A_f = \mathrm{Jac}(C)$ is the element in $\mathbb{Z}[T_2] \otimes_{\mathbb{Z}} \mathbb{Q}_{11}$ written $a + bT_2$, where:

$$a = 3798008844904804573589510615996706666264894 + O(11^{41})$$
$$b = 4491262769664482304376991401975428348750721 + O(11^{41}).$$

References

[Dar01] Henri Darmon, *Integration on $\mathcal{H}_p \times \mathcal{H}$ and arithmetic applications*, Ann. of Math. (2) **154** (2001), no. 3, 589–639, DOI 10.2307/3062142. MR1884617

[Das05] Samit Dasgupta, *Stark-Heegner points on modular Jacobians* (English, with English and French summaries), Ann. Sci. École Norm. Sup. (4) **38** (2005), no. 3, 427–469, DOI 10.1016/j.ansens.2005.03.002. MR2166341

[DG12] Samit Dasgupta and Matthew Greenberg, *$\mathcal{L}$-invariants and Shimura curves*, Algebra Number Theory **6** (2012), no. 3, 455–485, DOI 10.2140/ant.2012.6.455. MR2966706

[Gär12] Jérôme Gärtner, *Darmon's points and quaternionic Shimura varieties*, Canad. J. Math. **64** (2012), no. 6, 1248–1288, DOI 10.4153/CJM-2011-086-5. MR2994664

[GJG03] Enrique González-Jiménez and Josep González, *Modular curves of genus 2*, Math. Comp. **72** (2003), no. 241, 397–418, DOI 10.1090/S0025-5718-02-01458-8. MR1933828

[GM17] Xavier Guitart and Marc Masdeu, *Periods of modular* GL_2*-type abelian varieties and p-adic integration*, Exp. Math. **27** (2018), no. 3, 344–361, DOI 10.1080/10586458.2017.1284624. MR3857669

[GMS15] Xavier Guitart, Marc Masdeu, and Mehmet Haluk Şengün, *Darmon points on elliptic curves over number fields of arbitrary signature*, Proc. Lond. Math. Soc. (3) **111** (2015), no. 2, 484–518, DOI 10.1112/plms/pdv033. MR3384519

[Gre09] Matthew Greenberg, *Stark-Heegner points and the cohomology of quaternionic Shimura varieties*, Duke Math. J. **147** (2009), no. 3, 541–575, DOI 10.1215/00127094-2009-017. MR2510743

[GS93] Ralph Greenberg and Glenn Stevens, *p-adic L-functions and p-adic periods of modular forms*, Invent. Math. **111** (1993), no. 2, 407–447, DOI 10.1007/BF01231294. MR1198816

[Kad07] Samuel Kadziela, *Rigid analytic uniformization of curves and the study of isogenies*, Acta Appl. Math. **99** (2007), no. 2, 185–204, DOI 10.1007/s10440-007-9162-6. MR2350208

[LRV12] Matteo Longo, Victor Rotger, and Stefano Vigni, *On rigid analytic uniformizations of Jacobians of Shimura curves*, Amer. J. Math. **134** (2012), no. 5, 1197–1246, DOI 10.1353/ajm.2012.0033. MR2975234

[MTT86] B. Mazur, J. Tate, and J. Teitelbaum, *On p-adic analogues of the conjectures of Birch and Swinnerton-Dyer*, Invent. Math. **84** (1986), no. 1, 1–48, DOI 10.1007/BF01388731. MR830037

[RS12] Victor Rotger and Marco Adamo Seveso, *$\mathcal{L}$-invariants and Darmon cycles attached to modular forms*, J. Eur. Math. Soc. (JEMS) **14** (2012), no. 6, 1955–1999, DOI 10.4171/JEMS/352. MR2984593

[Spi14] Michael Spieß, *On special zeros of p-adic L-functions of Hilbert modular forms*, Invent. Math. **196** (2014), no. 1, 69–138, DOI 10.1007/s00222-013-0465-0. MR3179573

[Tay95] Richard Taylor, *Representations of Galois groups associated to modular forms*, Proceedings of the International Congress of Mathematicians, Vol. 1, 2 (Zürich, 1994), Birkhäuser, Basel, 1995, pp. 435–442. MR1403943

[Tei88] Jeremy Teitelbaum, *p-adic periods of genus two Mumford-Schottky curves*, J. Reine Angew. Math. **385** (1988), 117–151, DOI 10.1515/crll.1988.385.117. MR931217

[Tri06] Mak Trifković, *Stark-Heegner points on elliptic curves defined over imaginary quadratic fields*, Duke Math. J. **135** (2006), no. 3, 415–453, DOI 10.1215/S0012-7094-06-13531-7. MR2272972

Departament de Matemàtiques i Informàtica, Universitat de Barcelona, 08007 Barcelona, Catalonia, Spain

Email address: `xevi.guitart@gmail.com`

Mathematics Institute, University of Warwick, Coventry CV4 7AL, United Kingdom

Email address: `masdeu@mat.uab.cat`

Contemporary Mathematics
Volume **732**, 2019
https://doi.org/10.1090/conm/732/14794

An algebraic and analytic approach to spinor exceptional behavior in translated lattices

Anna Haensch and Ben Kane

ABSTRACT. In this announcement we discuss the representation problem for translations of positive-definite lattices via a discussion of representation by inhomogeneous quadratic polynomials. In particular, we give a survey of the extent to which algebraic and analytic methods are useful in determining how the behavior of the spinor genus contributes to failure of the local-global principle.

For a polynomial, f, with rational coefficients we say that f represents an integer, n, if $f(\vec{x}) = n$ has an integral solution. From the negative answer to Hilbert's 10th problem, we know that in general, there is no finite algorithm to determine the set of integers represented by f. However, in certain cases a solution can be determined. In this note, we will consider this so-called representation problem for inhomogeneous quadratic polynomials, that is, polynomials of the form

$$f(\vec{x}) = q(\vec{x}) + \ell(\vec{x}) + c,$$

where q is quadratic, ℓ is linear, and c is constant. Since it will not change the arithmetic of the problem, there is no harm in letting $c = 0$. The goal behind much of the work on this problem has been to find an integral analogue of Hasse's famous local-global principle.

We begin with a brief survey of results in the case when the linear part of f is identically 0. That is, $f(\vec{x})$ is just the homogenous quadratic polynomial $q(\vec{x})$. Then we let L be the quadratic lattice with associated quadratic map q, underlying rational quadratic space V and rank k. When L is indefinite, the representation problem is well understood; for details on this case, we direct the reader to a survey by Hsia [**12**]. Therefore, for the remainder of the note, we will assume that every quadratic map, q, is positive-definite. It is clear that n is represented locally by L at every prime p (including the infinite prime) when n is represented by L. Stated in the language of the arithmetic theory of quadratic forms, if n is represented by the global lattice L, then n is represented by $\mathrm{gen}(L)$, where $\mathrm{gen}(L)$

2010 *Mathematics Subject Classification.* Primary 11E20, 11E25, 11E45, 11E81, 11H55, 05A30.

Key words and phrases. Sums of polygonal numbers, ternary quadratic polynomials, almost universal forms, theta series, lattice theory and quadratic spaces, modular forms, spinor genus theory.

The research of the second author was supported by grant project numbers 27300314, 17302515, and 17316416 of the Research Grants Council.

is the set of all lattices on V which are locally isometric to L at every prime p. However the converse of this statement does not necessarily hold. In the case $k \geq 4$, Tartakowsky [**24**] proved an asymptotic local-global principle (with some added primitivity conditions when $k = 4$). Later, in [**14**], Jöchner and Kitaoka showed that for lattices with $k \geq 5$, there is an asymptotic local-global principle for representations which approximate a given (finite) set of local representations. For lattices of rank 4, a similar result is proved in [**13**] by Hsia and Jöchner, but in this case there are some additional primitivity conditions imposed on the local representations.

In the ternary case, representation by the genus of L is not sufficient to guarantee a global representation, not even if the representations are primitive. The adelic spin group of V acts on gen(L), and under this action, the genus of L is decomposed into finitely many spinor genera. The spinor genus containing L, denoted spn(L), plays a vital role in determining local representation behavior for ternary quadratic lattices. It it well known that an integer which is primitively represented by the genus of L will be represented primitively by either every spinor genus in the genus of L, or precisely half of the spinor genera. Integers satisfying the latter condition are called primitive spinor exceptions, which can be effectively determined by the work of Earnest and Hsia [**7**], Earnest, Hsia and Hung [**8**], and Kneser [**16**]. A theorem of Duke and Schulze-Pillot [**5**] says that there exists a constant C^*, depending only on L, such that an integer a will be represented by L if a is primitively represented by spn(L) and $a > C^*$. Unfortunately, the constant C^* is ineffective (it relies on Siegel's ineffective bound for the class numbers of imaginary quadratic fields [**21**]), so this does not lead to a full determination of the integers represented in the ternary case. However, it does lead to a determination of whether the local-global principle fails infinitely often [**19**], and hence leads to a measure of the extent of the failure of a local-global principle. For a comprehensive survey of the results mentioned above, the reader is directed to [**20**].

Armed with these results, we return to the question at hand, namely the question of representation of integers by inhomogeneous quadratic polynomials. In the positive-definite case, representation by an inhomogeneous quadratic polynomial is the same as representation of another integer (transformed due to completing a square) by a translated lattice, or lattice coset. Consequently, classical results for representation by lattices can be helpful in determining representation by inhomogeneous quadratic polynomials, and similarly, representations by translated lattices. When f is inhomogeneous with 4 or more variables then a solution to the representation problem follows immediately from the local-global representation with approximation from [**13**] and [**14**]. This is shown explicitly by Chan and Oh in [**3**]. Therefore only the ternary case remains of interest. For certain ternary inhomogeneous polynomials, specifically those that arise as sums of squares and triangular numbers, partial solutions are given in [**1**], [**2**], and [**15**]. In particular, these papers determine when such a polynomial represents all but finitely many positive integers; we call such a polynomial almost universal. Similar results in [**9**] and [**10**] determine when a polynomial is almost universal given that it satisfies a set of mild arithmetic conditions.

Our present work endeavors to solve the question of almost universality for ternary inhomogeneous f whose corresponding quadratic part has class number one. As a test case for this, we consider representations by $f = P_m$ where

$$P_m(x,y,z) := \frac{(m-2)x^2-(m-4)x}{2} + \frac{(m-2)y^2-(m-4)y}{2} + \frac{(m-2)z^2-(m-4)z}{2},$$

the sum of three generalized (i.e., $x, y, z \in \mathbb{Z}$) m-gonal numbers where m is even. An easy calculation shows that an integer n is represented by P_m if and only if

$$t_n := 3\left(\frac{(m-4)}{2}\right)^2 + 2(m-2)n$$

is represented by the translated lattice $L+\nu$, where $L \cong \langle (m-2)^2, (m-2)^2, (m-2)^2 \rangle$ in a basis $\{e_1, e_2, e_3\}$ and

$$\nu := \frac{(m-4)(e_1+e_2+e_3)}{2(m-2)}.$$

Here the underlying lattice L always has class number one, meaning that the genus, spinor genus, and class of L correspond, so any failure of local global principle by $L+\nu$ can't simply be reduced to a failure at the lattice level. As in the case with lattices, it is reasonable to expect that any infinite family of t_n for which the local-global principle fails results as a consequence of exceptional integers which are not represented by the spinor genus of $L+\nu$. However, the spinor theory in this case does not yet appear to be developed enough to conclude this claim. Motivated by the assumption that the local-global principle is governed by the spinor genus of $L+\nu$, we investigate the local-global principle via the analytic theory of modular forms. From an analytic perspective, we can consider representations of t_n by $L+\nu$ by considering the theta function $\Theta_{L+\nu}$, which is a weight $\frac{3}{2}$ modular form; the function $\Theta_{L+\nu}$ is the generating function for elements of $L+\nu$ of a given norm (introduced in [**22**] and investigated further in [**6**]). We can decompose the theta function as

$$\Theta_{L+\nu} = \mathcal{E}_{L+\nu} + \mathcal{U}_{L+\nu} + f_{L+\nu}, \tag{1}$$

where $\mathcal{E}_{L+\nu}$ is in the space spanned by Eisenstein series, $\mathcal{U}_{L+\nu}$ is in the space spanned by unary theta functions, and $f_{L+\nu}$ is a cusp form orthogonal to unary theta functions. If our motivation holds true, these components of the theta function can be interpreted as contributions to the representation by the genus, spinor genus and class of the translated lattice $L+\nu$, in the sense that the first component coincides whenever two lattice cosets are in the same genus, the second component coincides when they are in the same spinor genus, and the third component agrees when they are in the same class.

When $m \equiv 0 \mod 4$, it is always possible to find an entire congruence class of integers which fail to be represented locally by P_m and consequently almost universality is out of the questions. The case when $m \not\equiv 0 \mod 4$ then requires a careful melding of algebraic and analytic techniques; we are able to determine when local representations correspond to representations by the spinor genus and ultimately representations by the global lattice. To explain why a combination of these methods is necessary, we briefly contrast the strengths and weaknesses of the algebraic and analytic techniques. A primary contribution from the analytic side comes from the fact that Duke [**4**] has shown that the coefficients of $f_{L+\nu}$ grow slowly; this was a necessary ingredient needed by Duke and Schulze-Pillot [**5**] to overcome deficiencies on the algebraic side in the case of lattices. Using purely analytic techniques, one may be able to show that $\mathcal{U}_{L+\nu}$ vanishes identically for certain L and ν, in which case the slow-growing coefficients of $f_{L+\nu}$ imply that the local-global principle holds for t_n sufficiently large (with some primitivity condition). On the other hand, when $\mathcal{U}_{L+\nu}$ does not vanish identically, the analytic techniques fall short of providing an answer; namely, although the coefficients of $f_{L+\nu}$ are small, it is a historically difficult question to determine when these coefficients vanish. In cases where $\mathcal{U}_{L+\nu}$ does not vanish, we fall back on the motivation from spinor genera and combine with algebraic techniques to determine that certain coefficients of $f_{L+\nu}$ vanish.

Algebraic methods suffice in determining local representations by $L+\nu$. This can be done by considering representation of t_n by the so-called enveloping lattice $M := L+\mathbb{Z}\nu$ and comparing the the spinor genus and genus of M to that of $L+\nu$. In the case when m is even and $m \not\equiv 0 \mod 12$ the algebraic approach is even sufficient to determine that every t_n is represented primitively by $\mathrm{spn}(L+\nu)$ but the lack of any lattice coset analogue of Duke and Schulze-Pillot's result [**5**, Corollary] prevents further progress in this direction. In essence, the algebraic methods fails to gain any handle on the behavior of the spinor genus and the presence of something analogous to spinor exceptions, although heuristic evidence suggests that the locally represented integers which fail to be globally represented almost always fall inside one of finitely many square classes.

To understand the contribution of the spinor genus to the representation problem we turn to analytic method, specifically we look at the Siegel-Weil mass formula. Let $\mathbb{G}(L)$ be the set of representative of the classes in the genus of L. One formulation of the Siegel-Weil mass formula states that

$$\mathcal{E}_{\mathrm{gen}(L)} := \frac{1}{\sum_{M\in\mathbb{G}(L)}\omega_M^{-1}} \sum_{M\in\mathbb{G}(L)} \frac{\theta_M}{\omega_M} \tag{2}$$

is an Eisenstein series, where θ_L is the generating function for elements of L with a given norm, and ω_L is the number of automorphs of L. Due to work of Shimura [**23**], an analogue to equation (2) exists in the cases of translated lattices,

$$\mathcal{E}_{\mathrm{gen}(L+\nu)} := \frac{1}{\sum_{M+\nu'\in\mathbb{G}(L+\nu')}\omega_{M+\nu'}^{-1}} \sum_{M+\nu'\in\mathbb{G}(L+\nu)} \frac{\theta_{M+\nu'}}{\omega_{M+\nu'}}. \tag{3}$$

On the right-hand side of equation (3), the $t_n{}^{th}$ term in each summand counts (up to some scaling by the number of automorphs) the number of representations of n by a given lattice in the genus of $L+\nu$, and whether or not a coefficient of this average is positive can be completely determined using algebraic methods. So we already see a connection here between the local algebraic theory and its utility in building up the analytic theory. Summing instead over representations by the spinor genus, where $\mathbb{S}(L)$ is the set of representatives of the classes in the spinor genus of L, we obtain

$$\frac{1}{\sum_{M\in\mathbb{S}(L)}\omega_M^{-1}} \sum_{M\in\mathbb{S}(L)} \frac{\theta_M}{\omega_M} = \mathcal{E}_{\mathrm{gen}(L)} + \mathcal{U}_{\mathrm{spn}(L)} \tag{4}$$

where $\mathcal{U}_{\mathrm{spn}(L)}$ is a linear combination of unary theta functions, [**17, 18**]. The functions on the right-hand side are simply modular forms whose Fourier coefficients can be explicitly computed, but we point out that the role of $\mathcal{U}_{\mathrm{spn}(L)}$ is to count the excesses and deficiencies of representations by the spinor genus, as compared to representations by the weighted average of the genus. On the other hand, the left-hand side of the equation is a weighted average of sums of theta series, all of whose coefficients are positive. Therefore a zero ${t_n}^{th}$ coefficient in the left-hand side can only occur if the ${t_n}^{th}$ coefficient of each θ_M is zero. In this way, equation (4) illuminates the connection between the analytic theory and the algebraic theory of spinor exceptions and hints at how one would hope to make explicit the decomposition in (1).

This leads the authors to make the following conjecture regarding the representation of integers by the spinor genus of a lattice coset.

CONJECTURE 1. *We have*

$$\theta_{\mathrm{spn}(L+\nu)} := \frac{1}{\sum_{M+\nu' \in \mathbb{S}(L+\nu)} \omega_{M+\nu'}^{-1}} \sum_{M+\nu' \in \mathbb{S}(L+\nu)} \frac{\theta_{M+\nu'}}{\omega_{M+\nu'}} = \mathcal{E}_{\mathrm{gen}(L+\nu)} + \mathcal{U}_{\mathrm{spn}(L+\nu)},$$

where $\mathcal{U}_{\mathrm{spn}(L+\nu)}$ is a linear combination of unary theta functions.

A proof of a special cases of this conjecture for an infinite family of lattice cosets appears in the authors' recent publication [**11**].

References

[1] W. K. Chan and A. Haensch, *Almost universal ternary sums of squares and triangular numbers*, Quadratic and higher degree forms, Dev. Math., vol. 31, Springer, New York, 2013, pp. 51–62, DOI 10.1007/978-1-4614-7488-3_3. MR3156554

[2] W. K. Chan and B.-K. Oh, *Almost universal ternary sums of triangular numbers*, Proc. Amer. Math. Soc. **137** (2009), no. 11, 3553–3562, DOI 10.1090/S0002-9939-09-09990-0. MR2529860

[3] W. K. Chan and B.-K. Oh, *Representations of integral quadratic polynomials*, Diophantine methods, lattices, and arithmetic theory of quadratic forms, Contemp. Math., vol. 587, Amer. Math. Soc., Providence, RI, 2013, pp. 31–46, DOI 10.1090/conm/587/11684. MR3074801

[4] W. Duke, *Hyperbolic distribution problems and half-integral weight Maass forms*, Invent. Math. **92** (1988), no. 1, 73–90, DOI 10.1007/BF01393993. MR931205

[5] W. Duke and R. Schulze-Pillot, *Representation of integers by positive ternary quadratic forms and equidistribution of lattice points on ellipsoids*, Invent. Math. **99** (1990), no. 1, 49–57, DOI 10.1007/BF01234411. MR1029390

[6] M. Eichler, *Theta series of inhomogeneous quadratic forms*, Invent. Math. **66** (1982), no. 1, 99–113, DOI 10.1007/BF01404759. MR652649

[7] A. G. Earnest and J. S. Hsia, *Spinor norms of local integral rotations. II*, Pacific J. Math. **61** (1975), no. 1, 71–86. MR0404142

[8] A. G. Earnest, J. S. Hsia, and D. C. Hung, *Primitive representations by spinor genera of ternary quadratic forms*, J. London Math. Soc. (2) **50** (1994), no. 2, 222–230, DOI 10.1112/jlms/50.2.222. MR1291733

[9] A. Haensch, *A characterization of almost universal ternary inhomogeneous quadratic polynomials with conductor 2*, J. Number Theory **156** (2015), 247–262, DOI 10.1016/j.jnt.2015.04.016. MR3360338

[10] A. Haensch, *A characterization of almost universal ternary quadratic polynomials with odd prime power conductor*, J. Number Theory **141** (2014), 202–213, DOI 10.1016/j.jnt.2014.02.013. MR3195396

[11] A. Haensch and B. Kane, *Almost universal ternary sums of polygonal numbers*, Res. Number Theory **4** (2018), no. 1, Art. 4, 22, DOI 10.1007/s40993-018-0098-x. MR3757168

[12] J. S. Hsia, *Arithmetic of indefinite quadratic forms*, Integral quadratic forms and lattices (Seoul, 1998), Contemp. Math., vol. 249, Amer. Math. Soc., Providence, RI, 1999, pp. 1–15, DOI 10.1090/conm/249/03743. MR1732345

[13] J. S. Hsia and M. Jöchner, *Almost strong approximations for definite quadratic spaces*, Invent. Math. **129** (1997), no. 3, 471–487, DOI 10.1007/s002220050169. MR1465331

[14] M. Jöchner and Y. Kitaoka, *Representations of positive definite quadratic forms with congruence and primitive conditions*, J. Number Theory **48** (1994), no. 1, 88–101, DOI 10.1006/jnth.1994.1055. MR1284877

[15] B. Kane and Z.-W. Sun, *On almost universal mixed sums of squares and triangular numbers*, Trans. Amer. Math. Soc. **362** (2010), no. 12, 6425–6455, DOI 10.1090/S0002-9947-2010-05290-0. MR2678981

[16] M. Kneser, *Klassenzahlen indefiniter quadratischer Formen in drei oder mehr Veränderlichen* (German), Arch. Math. (Basel) **7** (1956), 323–332, DOI 10.1007/BF01900681. MR0082514

[17] R. Schulze-Pillot, *Darstellungsmaße von Spinorgeschlechtern ternärer quadratischer Formen* (German), J. Reine Angew. Math. **352** (1984), 114–132, DOI 10.1515/crll.1984.352.114. MR758697

[18] R. Schulze-Pillot, *Thetareihen positiv definiter quadratischer Formen* (German), Invent. Math. **75** (1984), no. 2, 283–299, DOI 10.1007/BF01388566. MR732548

[19] R. Schulze-Pillot, *Exceptional integers for genera of integral ternary positive definite quadratic forms*, Duke Math. J. **102** (2000), no. 2, 351–357, DOI 10.1215/S0012-7094-00-10227-X. MR1749442

[20] R. Schulze-Pillot, *Representation by integral quadratic forms—a survey*, Algebraic and arithmetic theory of quadratic forms, Contemp. Math., vol. 344, Amer. Math. Soc., Providence, RI, 2004, pp. 303–321, DOI 10.1090/conm/344/06226. MR2060206

[21] C. Siegel, *Über die Klassenzahl quadratischer Zahlenkörper*, Acta. Arith. **1** (1936), 83–86.

[22] G. Shimura, *On modular forms of half integral weight*, Ann. of Math. (2) **97** (1973), 440–481, DOI 10.2307/1970831. MR0332663

[23] G. Shimura, *Inhomogeneous quadratic forms and triangular numbers*, Amer. J. Math. **126** (2004), no. 1, 191–214. MR2033567

[24] W. Tartakowsky, *Die Gesamtheit der Zahlen, die durch eine positive quadratische Form* $F(x_1, ..., x_s)$ $(s \geq 4)$ *darstellbar sind*, Izv. Akad. Nauk SSSR **7**, (1929), 111-122, 165-195

Department of Mathematics and Computer Science, Duquesne University, Pittsburgh, Pennsylvania 15282

Email address: haenscha@duq.edu

Department of Mathematics, University of Hong Kong, Pokfulam, Hong Kong

Email address: bkane@hku.hk

Contemporary Mathematics
Volume **732**, 2019
https://doi.org/10.1090/conm/732/14793

Differential operators on Jacobi forms and special values of certain Dirichlet series

Abhash Kumar Jha and Brundaban Sahu

ABSTRACT. We construct Jacobi cusp forms by computing the adjoint of a certain linear map constructed using bilinear holomorphic differential operators with respect to the Petersson scalar product. The Fourier coefficients of the Jacobi cusp forms constructed involve special values of the shifted convolution of Dirichlet series of Rankin-Selberg type.

1. Introduction

The derivative of a modular form is not in general a modular form, but one can construct modular forms by using certain combinations of derivatives of modular forms. Rankin [**13, 14**] gave a general description of differential operators which map modular forms to modular forms. For every non-negative integer ν, Cohen [**7**] explicitly constructed certain bilinear operators from $M_k \times M_l$ to $M_{k+l+2\nu}$, where M_i denotes the space of holomorphic modular forms of weight i for the group $SL_2(\mathbb{Z})$. Zagier [**16**] studied algebraic properties of these bilinear operators and called them Rankin-Cohen brackets. For example, the first bracket $[\cdot,\cdot]_1$ satisfies the Jacobi identity

$$[[f,g]_1,h]_1 + [[g,h]_1,f]_1 + [[h,f]_1,g]_1 = 0, \quad (f \in M_k, g \in M_l, h \in M_m)$$

giving M_* the structure of a graded Lie algebra. Kohnen [**12**] constructed certain elliptic cusp forms whose Fourier coefficients involve special values of certain Dirichlet series of Rankin-Selberg type by computing the adjoint of the product map (i.e. the map $f \mapsto fg$, for a fixed modular form g) with respect to the Petersson scalar product. Recently, Herrero [**10**] generalized the work of Kohnen by computing the adjoint of certain linear maps constructed using Rankin-Cohen brackets with respect to the Petersson scalar product. Herrero constructed cusp forms whose Fourier coefficients involve special values of Dirichlet series similar to those which appeared in the work of Kohnen, with additional factors arising due to the binomial coefficients appearing in the Rankin-Cohen brackets.

The work of Kohnen [**12**] has been generalized by Choie, Kim and Knopp [**5**] and Sakata [**15**] to the case of Jacobi forms. Choie [**2, 3**] studied the Rankin-Cohen brackets for Jacobi forms by using the heat operator acting on Jacobi forms. Recently in [**11**], we generalized the work of Herrero to the case of Jacobi forms. We

2010 *Mathematics Subject Classification.* Primary 11F50; Secondary 11F25, 11F66.
Key words and phrases. Jacobi forms, differential operators, adjoint map.

explicitly computed the adjoint of certain linear maps constructed using Rankin-Cohen brackets with respect to the Petersson scalar product. Böecherer [1] studied the general bilinear holomorphic differential operators on the Jacobi group by using the Maass operator and proved that the space of bilinear holomorphic differential operators raising the weight by ν is in general of dimension equal to $\lfloor \frac{\nu}{2} \rfloor + 1$. Choie and Eholzer [4] explicitly constructed a family of dimension $\lfloor \frac{\nu}{2} \rfloor + 1$ of Rankin-Cohen type operators defined on the space of Jacobi forms raising the weight by $\nu \in \mathbb{N}$.

In this article, we consider certain linear maps defined on the space of Jacobi forms which are constructed using Rankin-Cohen type operators (studied in [4]). We explicitly compute their adjoints with respect to the Petersson scalar product. The Fourier coefficients of the image of a Jacobi cusp forms under the adjoint maps involve special values of shifted convolutions of Dirichlet series of Rankin-Selberg type.

2. Preliminaries on Jacobi forms of scalar index

Let $\mathbb{C}$ and $\mathcal{H}$ be the complex plane and the complex upper half-plane, respectively. The Jacobi group $\Gamma^J := SL_2(\mathbb{Z}) \ltimes (\mathbb{Z} \times \mathbb{Z})$ acts on $\mathcal{H} \times \mathbb{C}$ in the usual way by

$$\left(\begin{pmatrix} a & b \\ c & d \end{pmatrix}, (\lambda, \mu) \right) \cdot (\tau, z) = \left(\frac{a\tau + b}{c\tau + d}, \frac{z + \lambda\tau + \mu}{c\tau + d} \right).$$

Let k and m be fixed positive integers. If $\gamma = \left(\begin{pmatrix} a & b \\ c & d \end{pmatrix}, (\lambda, \mu) \right) \in \Gamma^J$ and ϕ is a complex-valued function defined on $\mathcal{H} \times \mathbb{C}$, then define

$$\phi|_{k,m}\gamma := (c\tau + d)^{-k} e^{2\pi i m (-\frac{c(z+\lambda\tau+\mu)^2}{c\tau+d} + \lambda^2\tau + 2\lambda z)} \phi(\gamma \cdot (\tau, z)).$$

Let $J_{k,m}$ be the space of Jacobi forms of weight k and index m on Γ^J, i.e. the space of holomorphic functions $\phi : \mathcal{H} \times \mathbb{C} \to \mathbb{C}$ satisfying $\phi|_{k,m}\gamma = \phi$, for all $\gamma \in \Gamma^J$ and having a Fourier expansion of the form

$$\phi(\tau, z) = \sum_{\substack{n,r \in \mathbb{Z}, \\ 4nm - r^2 \geq 0}} c(n, r) q^n \zeta^r \qquad (q = e^{2\pi i \tau}, \zeta = e^{2\pi i z}).$$

Furthermore, we say ϕ is a cusp form if and only if $c(n, r) \neq 0 \implies n > r^2/4m$. We denote the space of all Jacobi cusp forms by $J_{k,m}^{cusp}$. We define the Petersson scalar product on $J_{k,m}^{cusp}$ as

$$\langle \phi, \psi \rangle = \int_{\Gamma^J \backslash \mathcal{H} \times C} \phi(\tau, z) \overline{\psi(\tau, z)} v^k e^{\frac{-4\pi m y^2}{v}} dV_J,$$

where $\tau = u + iv, z = x + iy$ and $dV_J = \dfrac{dudvdxdy}{v^3}$ is an invariant measure under the action of Γ^J on $\mathcal{H} \times \mathbb{C}$. The space $(J_{k,m}^{cusp}, \langle . \, , \, . \rangle)$ is a finite dimensional Hilbert space. For more details on the theory of Jacobi forms, we refer the reader to [8]. The following lemma describes the growth of the Fourier coefficients of a Jacobi form.

LEMMA 2.1. *If $k > 3$ and $\phi \in J_{k,m}$ has Fourier coefficients $c(n, r)$, then*

$$c(n, r) \ll |r^2 - 4nm|^{k - \frac{3}{2}},$$

and, moreover if ϕ is a cusp form, then

$$c(n,r) \ll |r^2 - 4nm|^{\frac{k}{2}-\frac{1}{2}}.$$

For a proof, we refer to [**6**].

2.1. Poincaré series. Let m, n and r be fixed integers such that $r^2 < 4mn$. Let

$$\Gamma^J_\infty := \left\{ \left(\begin{pmatrix} 1 & t \\ 0 & 1 \end{pmatrix}, (0,\mu) \right) : t, \mu \in \mathbb{Z} \right\}$$

be the stabilizer of $q^n \zeta^r$ in Γ^J. Let

$$P_{k,m;(n,r)}(\tau, z) := \sum_{\gamma \in \Gamma^J_\infty \backslash \Gamma^J} e^{2\pi i(n\tau + rz)}|_{k,m}\gamma \tag{2.1}$$

be the (n,r)-th Poincaŕe series of weight k and index m. It is well-known that $P_{k,m;(n,r)} \in J^{cusp}_{k,m}$ for $k > 2$ [**9**].
This series has the following property:

LEMMA 2.2. *Let $\phi \in J^{cusp}_{k,m}$ has the Fourier expansion*

$$\phi(\tau, z) = \sum_{\substack{n,r \in \mathbb{Z}, \\ 4nm - r^2 > 0}} c(n,r) q^n \zeta^r.$$

Then

$$\langle \phi, P_{k,m;(n,r)} \rangle = \frac{m^{k-2}\Gamma(k-\frac{3}{2})(4mn - r^2)^{\frac{3}{2}-k}}{2\pi^{k-\frac{3}{2}}} c(n,r), \tag{2.2}$$

where $\Gamma(\cdot)$ denotes the usual gamma function.

One can get explicit Fourier expansion of $P_{k,m;(n,r)}$, for details and a proof of the Lemma 2.2, we refer to [**9**].

2.2. Differential operators on Jacobi forms. For an integer m, we define the heat operator

$$L_m := \frac{1}{(2\pi i)^2} (8\pi i m\, \partial_\tau - \partial_z^2),$$

where ∂_τ and ∂_z are the derivative with respect to τ and z, respectively. Note that

$$L_m(e^{(2\pi i(n\tau + rz))}) = (4mn - r^2) e^{(2\pi i(n\tau + rz))}. \tag{2.3}$$

Let k_1, k_2, m_1 and m_2 be positive integers, and let ϕ and ψ be two complex-valued holomorphic functions defined on $\mathcal{H} \times \mathbb{C}$. Then, for any complex number X and non-negative integer ν, define the 2ν-th and $(2\nu+1)$-th Rankin-Cohen type brackets of ϕ and ψ as

$$[\phi, \psi]^{k_1,k_2,m_1,m_2}_{X,\ 2\nu}$$
$$:= \sum_{\substack{\alpha,\beta,\gamma \in \mathbb{N}\cup\{0\} \\ \alpha+\beta+\gamma=\nu}} C_{\alpha,\beta,\gamma}(k_1,k_2)(1+m_1X)^\beta (2-m_2X)^\alpha L^\gamma_{m_1+m_2}\left(L^\alpha_{m_1}(\phi) L^\beta_{m_2}(\psi)\right),$$

$$[\phi,\ \psi]^{k_1,k_2,m_1,m_2}_{X,\ 2\nu+1} := m_1[\phi, \partial_z\psi]^{k_1,k_2,m_1,m_2}_{X,\ 2\nu} - m_2[\partial_z\phi,\ \psi]^{k_1,k_2,m_1,m_2}_{X,\ 2\nu},$$

where the coefficients $C_{\alpha,\beta,\gamma}(k_1,k_2)$ are given by

$$C_{\alpha,\beta,\gamma}(k_1,k_2) = \frac{(k_1+\nu-3/2)_{\beta+\gamma}}{\alpha!} \frac{(k_2+\nu-3/2)_{\alpha+\gamma}}{\beta!} \frac{(-(k_1+k_2+\nu-3/2))_{\alpha+\beta}}{\gamma!},$$

and

$$(x)_m = \prod_{0\leqslant i\leqslant m-1} (x-i),$$

with $x! = \Gamma(x+1)$. Using the action of the heat operator, one can verify that

$$(2.4)\quad [\phi|_{k_1,m_1}\gamma, \psi|_{k_2,m_2}\gamma]_{X,\nu}^{k_1,k_2,m_1,m_2} = [\phi,\psi]_{X,\nu}^{k_1,k_2,m_1,m_2}|_{k_1+k_2+\nu,m_1+m_2}\gamma,\ \forall\gamma\in\Gamma^J.$$

This implies the following result, which was proved in [**4**].

THEOREM 2.3. *Let $\phi \in J_{k_1,\ m_1}$ and $\psi \in J_{k_2,\ m_2}$, where k_1, k_2, m_1 and m_2 are positive integers. Then, for any $X \in \mathbb{C}$ and any non-negative integer ν, the function $[\phi,\psi]_{X,\nu}^{k_1,k_2,m_1,m_2}$ is a Jacobi form of weight $k_1+k_2+\nu$ and index m_1+m_2. Moreover, $[\phi,\psi]_{X,\nu}^{k_1,k_2,m_1,m_2}$ is a Jacobi cusp form for $\nu > 1$. In fact, $[\ ,\ ]_{X,\nu}^{k_1,k_2,m_1,m_2}$ is a bilinear map from $J_{k_1,\ m_1} \times J_{k_2,\ m_2}$ to $J_{k_1+k_2+\nu,\ m_1+m_2}$.*

REMARK 2.1. For $\nu \in 2\mathbb{N}$, the operator $\left(\frac{d}{dX}\right)^{\nu/2} [\phi,\psi]_{X,\nu}^{k_1,k_2,m_1,m_2}$ is, up to a scalar multiple of the Rankin-Cohen bracket, studied in [**2**]. If we take $\nu = 1$ in Theorem 2.3, we obtain Theorem 9.5 [**8**]. These are some applications of the Rankin-Cohen brackets.

3. Statement of the theorems

Let $\psi \in J_{k_2,\ m_2}^{cusp}$ and ν be a non-negative integer. For a complex number X, we define the map

$$T_{\psi,X,\nu} : J_{k_1,\ m_1}^{cusp} \to J_{k_1+k_2+\nu,\ m_1+m_2}^{cusp}$$

by

$$T_{\psi,X,\nu}(\phi) = [\phi,\psi]_{X,\nu}^{k_1,k_2,m_1,m_2}.$$

This is a $\mathbb{C}$-linear map of finite dimensional Hilbert spaces and therefore there exists an adjoint map

$$T_{\psi,X,\nu}^{*} : J_{k_1+k_2+\nu,\ m_1+m_2}^{cusp} \to J_{k_1,\ m_1}^{cusp},$$

such that

$$\langle\phi,\ T_{\psi,X,\nu}(\omega)\rangle = \langle T_{\psi,\ X,\ \nu}^{*}(\phi),\ \omega\rangle,$$

for all $\phi \in J_{k_1+k_2+\nu,\ m_1+m_2}^{cusp}$ and $\omega \in J_{k_1,\ m_1}^{cusp}$.

As an application one can obtain certain arithmetic information of Fourier coefficients of Jacobi forms using $T_{\psi,X,\nu}^{*}$ (refer to [**11**]) and analogous results in the case modular forms (refer to [**10, 12**]). In the following theorems, we exhibit the Fourier coefficients of $T_{\psi,X,\nu}^{*}(\phi)$ for $\phi \in J_{k_1+k_2+\nu,\ m_1+m_2}^{cusp}$ for ν even and odd. These Fourier coefficients involve special values of certain shifted convolution of Dirichlet series of Rankin-Selberg type associated to ϕ and ψ.

THEOREM 3.1. *Let k_1, k_2, m_1, m_2 be positive integers, such that $k_1 > 4, k_2 > 3$ and ν be a non-negative integer. Let $\psi \in J_{k_2,\ m_2}^{cusp}$ has Fourier expansion*

$$\psi(\tau,z) = \sum_{\substack{n_1,r_1\in\mathbb{Z},\\ 4m_2n_1-r_1^2>0}} a(n_1,r_1)q^{n_1}\zeta^{r_1}.$$

Then the image of any cusp form $\phi \in J^{cusp}_{k_1+k_2+2\nu,\ m_1+m_2}$ *with Fourier expansion*

$$\phi(\tau,z) = \sum_{\substack{n_2,r_2\in\mathbb{Z},\\ 4(m_1+m_2)n_2-r_2^2>0}} b(n_2,r_2)q^{n_2}\zeta^{r_2}$$

under $T^*_{\psi,X,2\nu}$ *is given by*

$$T^*_{\psi,X,2\nu}(\phi)(\tau,z) = \sum_{\substack{n,r\in\mathbb{Z},\\ 4m_1n-r^2>0}} c_{\nu,X}(n,r)q^n\zeta^r,$$

where

$$\begin{aligned} c_{\nu,X}(n,r) &= \frac{(4m_1n-r^2)^{k_1-3/2}\,(m_1+m_2)^{k_1+k_2+2\nu-2}\,\Gamma(k_1+k_2+2\nu-\frac{3}{2})}{\pi^{k_2+2\nu}\,m_1^{k_1-2}\,\Gamma(k_1-\frac{3}{2})} \\ &\times \sum_{\alpha+\beta+\gamma=\nu} C_{\alpha,\beta,\gamma}(k_1,k_2)(1+m_1X)^\beta(2-m_2X)^\alpha\,(4m_1n-r^2)^\alpha \\ &\times \sum_{\substack{n_1,r_1\in\mathbb{Z}\\ 4m_2n_1-r_1^2>0\\ 4(m_1+m_2)(n+n_1)-(r+r_1)^2>0}} \frac{(4m_2n_1-r_1^2)^\beta\,\overline{a(n_1,r_1)}\,b(n+n_1,r+r_1)}{(4(m_1+m_2)(n+n_1)-(r+r_1)^2)^{k_1+k_2+2\nu-(\gamma+\frac{3}{2})}}. \end{aligned}$$

THEOREM 3.2. *Let* k_1,k_2,m_1 *and* m_2 *be positive integers such that* $k_1>4, k_2>3$ *and* ν *be a non-negative integer. Let* $\psi\in J^{cusp}_{k_2,\ m_2}$ *has Fourier expansion*

$$\psi(\tau,z) = \sum_{\substack{n_1,r_1\in\mathbb{Z},\\ 4m_2n_1-r_1^2>0}} a(n_1,r_1)q^{n_1}\zeta^{r_1}.$$

Then the image of any cusp form $\phi \in J^{cusp}_{k_1+k_2+2\nu+1,\ m_1+m_2}$ *with Fourier expansion*

$$\phi(\tau,z) = \sum_{\substack{n_2,r_2\in\mathbb{Z},\\ 4(m_1+m_2)n_2-r_2^2>0}} b(n_2,r_2)q^{n_2}\zeta^{r_2}$$

under $T^*_{\psi,X,2\nu+1}$ *is given by*

$$T^*_{\psi,X,2\nu+1}(\phi)(\tau,z) = \sum_{\substack{n,r\in\mathbb{Z},\\ 4m_1n-r^2>0}} c_{\nu,X}(n,r)q^n\zeta^r,$$

where

$$\begin{aligned} c_{\nu,X}(n,r) &= \frac{2i(4m_1n-r^2)^{k_1-3/2}\,(m_1+m_2)^{k_1+k_2+2\nu-1}\,\Gamma(k_1+k_2+2\nu-\frac{1}{2})}{\pi^{k_2+2\nu}\,m_1^{k_1-2}\,\Gamma(k_1-\frac{3}{2})} \\ &\times \sum_{\alpha+\beta+\gamma=\nu} C_{\alpha,\beta,\gamma}(k_1,k_2)(1+m_1X)^\beta(2-m_2X)^\alpha\,(4m_1n-r^2)^\alpha \\ &\times \sum_{\substack{n_1,r_1\in\mathbb{Z}\\ 4m_2n_1-r_1^2>0\\ 4(m_1+m_2)(n+n_1)-(r+r_1)^2>0}} \frac{(4m_2n_1-r_1^2)^\beta\,(m_1r_1-m_2r)\overline{a(n_1,r_1)}\,b(n+n_1,r+r_1)}{(4(m_1+m_2)(n+n_1)-(r+r_1)^2)^{k_1+k_2+2\nu-(\gamma+\frac{1}{2})}}. \end{aligned}$$

4. Proofs

We follow the same exposition as in the proof of Theorem 3.1 in [**11**]. The following lemma is used to prove the above theorems.

LEMMA 4.1. *Using the same notation as in Theorem 3.1, we have*

$$\sum_{\gamma\in\Gamma^J_\infty\backslash\Gamma^J}\int_{\Gamma^J\backslash\mathcal{H}\times\mathbb{C}} \Big| \phi(\tau,z)\overline{\left[e^{2\pi i(n\tau+rz)}\,|_{k_1,m_1}\,\gamma,\ \psi(\tau,z)\right]^{k_1,k_2,m_1,m_2}_{X,2\nu}} \times \Im(\tau)^{k_1+k_2+2\nu}e^{\frac{-4\pi(m_1+m_2)(\Im(z))^2}{\Im(\tau)}}\Big|\, dV_J$$

converges.

PROOF. The proof is analogous to that of Lemma 4.1 in [**11**]. □

We now give a proof of Theorem 3.1. Write

$$T^*_{\psi,X,2\nu}(\phi)(\tau,z)=\sum_{\substack{n,r\in\mathbb{Z},\\ 4m_1n-r^2>0}} c_{\nu,X}(n,r)q^n\zeta^r.$$

Consider the (n,r)-th Poincaŕe series of weight k_1 and index m_1, as given in (2.1). Using Lemma 2.2, we have

$$\langle T^*_{\psi,\ X,\ 2\nu}\phi, P_{k_1,m_1;(n,r)}\rangle=\frac{m_1^{k_1-2}\Gamma(k_1-\frac{3}{2})(4m_1n-r^2)^{\frac{3}{2}-k_1}}{2\pi^{k_1-\frac{3}{2}}}\,c_{\nu,X}(n,r).$$

On the other hand, by the definition of the adjoint map, we have

$$\begin{aligned}\langle T^*_{\psi,\ X,\ 2\nu}\ \phi, P_{k_1,m_1;(n,r)}\rangle &= \langle \phi, T_{\psi,\ X,\ 2\nu}\ (P_{k_1,m_1;(n,r)})\rangle\\ &= \langle \phi, [P_{k_1,m_1;(n,r)},\psi]^{k_1,k_2,m_1,m_2}_{X,2\nu}\rangle.\end{aligned}$$

Hence, we obtain

$$c_{\nu,X}(n,r)=\frac{2\pi^{k_1-\frac{3}{2}}(4m_1n-r^2)^{k_1-\frac{3}{2}}}{m_1^{k_1-2}\Gamma(k_1-\frac{3}{2})}\langle\phi,[P_{k_1,m_1;(n,r)},\psi]^{k_1,k_2,m_1,m_2}_{X,2\nu}\rangle. \tag{4.1}$$

By definition,

$$\begin{aligned}\langle\phi,[P_{k_1,m_1;(n,r)},&\psi]^{k_1,k_2,m_1,m_2}_{X,\ 2\nu}\rangle\\ &=\int_{\Gamma^J\backslash\mathcal{H}\times\mathbb{C}}\phi(\tau,z)\overline{\left[P_{k_1,m_1;(n,r)}(\tau,z),\psi(\tau,z)\right]^{k_1,k_2,m_1,m_2}_{X,\ 2\nu}}\\ &\qquad\times\Im(\tau)^{k_1+k_2+2\nu}e^{\frac{-4\pi(m_1+m_2)(\Im(z))^2}{\Im(\tau)}}\,dV_J\\ &=\int_{\Gamma^J\backslash\mathcal{H}\times\mathbb{C}}\sum_{\gamma\in\Gamma^J_\infty\backslash\Gamma^J}\phi(\tau,z)\overline{\left[e^{2\pi i(n\tau+rz)}\,|_{k_1,m_1}\,\gamma,\ \psi(\tau,z)\right]^{k_1,k_2,m_1,m_2}_{X,\ 2\nu}}\\ &\qquad\times\Im(\tau)^{k_1+k_2+2\nu}e^{\frac{-4\pi(m_1+m_2)(\Im(z))^2}{\Im(\tau)}}\,dV_J.\end{aligned}$$

By Lemma 4.1, we can interchange the order of summation and integration in the above equation. Hence, $\langle\phi, [P_{k_1,m_1;(n,r)}, \psi]_{X,\ 2\nu}^{k_1,k_2,m_1,m_2}\rangle$ equals

$$\sum_{\gamma\in\Gamma_\infty^J\backslash\Gamma^J}\int_{\Gamma^J\backslash\mathcal{H}\times\mathbb{C}} \phi(\tau,z)\overline{\left[e^{2\pi i(n\tau+rz)}\,|_{k_1,m_1}\,\gamma, \psi(\tau,z)\right]_{X,\ 2\nu}^{k_1,k_2,m_1,m_2}} \times \Im(\tau)^{k_1+k_2+2\nu}\, e^{\frac{-4\pi(m_1+m_2)(\Im(z))^2}{\Im(\tau)}}\, dV_J.$$

Using equation (2.4), the fact that

$$\phi(\tau,z)\overline{\psi\,|_{k,m}\,\gamma(\tau,z)}\,\Im(\tau)^k\, e^{\frac{-4\pi m\Im(z)^2}{\Im(\tau)}} = \phi(\gamma(\tau,z))\overline{\psi\,|_{k,m}\,\gamma(\tau,z)}\,\Im(\gamma\tau)^k\, e^{\frac{-4\pi m\Im(z)^2}{\Im(\gamma\tau)}}$$

and the change of variable $(\tau,z)\mapsto\gamma^{-1}\cdot(\tau,z)$ and equation (2.4), the above equals

$$\sum_{\gamma\in\Gamma_\infty^J\backslash\Gamma^J}\int_{\gamma\cdot\Gamma^J\backslash\mathcal{H}\times\mathbb{C}} \phi(\tau,z)\overline{\left[e^{2\pi i(n\tau+rz)},\ \psi(\tau,z)\right]_{X,\ 2\nu}^{k_1,k_2,m_1,m_2}} \times \Im(\tau)^{k_1+k_2+2\nu}\, e^{\frac{-4\pi(m_1+m_2)(\Im(z))^2}{\Im(\tau)}}\, dV_J.$$

Using Rankin's unfolding argument, $\langle\phi, [P_{k_1,m_1;(n,r)}, \psi]_{X,2\nu}^{k_1,k_2,m_1,m_2}\rangle$ equals

$$\begin{aligned}
&\int_{\Gamma_\infty^J\backslash\mathcal{H}\times\mathbb{C}} \phi(\tau,z)\overline{\left[e^{2\pi i(n\tau+rz)}, \psi(\tau,z)\right]_{X,\ 2\nu}^{k_1,k_2,m_1,m_2}}\,\Im(\tau)^{k_1+k_2+2\nu}\, e^{\frac{-4\pi(m_1+m_2)(\Im(z))^2}{\Im(\tau)}}\, dV_J \\
&= \sum_{\substack{\alpha,\beta,\gamma\in\mathbb{N}\cup\{0\}\\ \alpha+\beta+\gamma=\nu}} C_{\alpha,\beta,\gamma}(k_1,k_2)\,(1+m_1X)^\beta\,(2-m_2X)^\alpha \\
&\times\int_{\Gamma_\infty^J\backslash\mathcal{H}\times\mathbb{C}} \phi(\tau,z)\overline{L_{m_1+m_2}^\gamma\left(L_{m_1}^\alpha(e^{2\pi i(n\tau+rz)})L_{m_2}^\beta(\psi(\tau,z))\right)} \\
&\qquad\times\Im(\tau)^{k_1+k_2+2\nu}\, e^{\frac{-4\pi(m_1+m_2)(\Im(z))^2}{\Im(\tau)}}\, dV_J.
\end{aligned}$$

Inserting the Fourier expansions of ϕ and ψ and using the iterated action of the heat operators L_{m_1}, L_{m_2} :

$$\begin{aligned}
L_{m_1}^\alpha(e^{(2\pi i(n\tau+rz))}) &= (4m_1n-r^2)^\alpha e^{(2\pi i(n\tau+rz))}, \\
L_{m_2}^\beta\psi(\tau,z) &= \sum_{\substack{n_1,r_1\in\mathbb{Z},\\ 4m_2n_1-r_1^2>0}} (4m_2n_1-r_1^2)^\beta a(n_1,r_1)e^{(2\pi i(n_1\tau+r_1z))},
\end{aligned}$$

and $L_{m_1+m_2}$ using (2.3), $\langle \phi, [P_{k_1,m_1;(n,r)}, \psi]_{X,\ 2\nu}^{k_1,k_2,m_1,m_2}\rangle$ equals

$$\sum_{\substack{\alpha,\beta,\gamma\in\mathbb{N}\cup\{0\}\\ \alpha+\beta+\gamma=\nu}} C_{\alpha,\beta,\gamma}(k_1,k_2)\,(1+m_1X)^{\beta}\,(2-m_2X)^{\alpha}\,(4m_1n-r^2)^{\alpha}$$
$$\times \sum_{\substack{n_2,r_2\in\mathbb{Z},\\ 4(m_1+m_2)n_2-r_2^2>0}} b(n_2,r_2) \sum_{\substack{n_1,r_1\in\mathbb{Z}\\ 4m_2n_1-r_1^2>0}} (4m_2n_1-r_1^2)^{\beta}(4(m_1+m_2)(n+n_1)$$
$$-(r+r_1)^2)^{\gamma}\,\overline{a(n_1,r_1)}$$
$$\times \int_{\Gamma_\infty^J\backslash\mathcal{H}\times\mathbb{C}} e^{2\pi i(n_2\tau+r_2z)}\,\overline{e^{2\pi i((n+n_1)\tau+(r+r_1)z)}}\,\Im(\tau)^{k_1+k_2+2\nu}\,e^{\frac{-4\pi(m_1+m_2)(\Im(z))^2}{\Im(\tau)}}\,dV_J.$$

A fundamental domain for the action of Γ_∞^J on $\mathcal{H}\times\mathbb{C}$ is given by the set

$$\{(u,v,x,y) : u\in[0,1], v\in[0,\infty], x\in[0,1], y\in\mathbb{R}\}.$$

Integrating in this region, $\langle \phi, [P_{k_1,m_1;(n,r)}, \psi]_{X,\ 2\nu}^{k_1,k_2,m_1,m_2}\rangle$ equals

$$\frac{(m_1+m_2)^{k_1+k_2+2\nu-2}\Gamma(k_1+k_2+2\nu-\frac{3}{2})}{2\pi^{k_1+k_2+2\nu-\frac{3}{2}}} \sum_{\substack{\alpha,\beta,\gamma\in\mathbb{N}\cup\{0\}\\ \alpha+\beta+\gamma=\nu}} C_{\alpha,\beta,\gamma}(k_1,k_2)(1+m_1X)^{\beta}(2-m_2X)^{\alpha}$$
$$\times\ (4m_1n-r^2)^{\alpha} \sum_{\substack{n_1,r_1\in\mathbb{Z}\\ 4m_2n_1-r_1^2>0\\ 4(m_1+m_2)(n+n_1)-(r+r_1)^2>0}} \frac{(4m_2n_1-r_1^2)^{\beta}\,\overline{a(n_1,r_1)}\,b(n+n_1,r+r_1)}{(4(m_1+m_2)(n+n_1)-(r+r_1)^2)^{k_1+k_2+2\nu-(\gamma+\frac{3}{2})}}.$$

In the above, we use the orthogonality relations for the exponential function to compute the integrals in x and u, the Gaussian integral to compute the integral in y and the integral representation of the Gamma function to compute the integral in v. Inserting the expression for $\langle \phi, [P_{k_1,m_1;(n,r)}, \psi]_{X,\ 2\nu}^{k_1,k_2,m_1,m_2}\rangle$ in (4.1), we obtain the required expression for $c_{\nu,X}(n,r)$ given in Theorem 3.1.

The proof of Theorem 3.2 is analogous to that of Theorem 3.1.

Acknowledgements. We thank the organizers for giving us the opportunity to participate in "Building Bridges: 3rd EU/US Summer School and Workshop on Automorphic Forms and Related Topics" during July 11-22, 2016 at the University of Sarajevo, Bosnia and Herzegovina. The first author would like to thank the National Board of Higher Mathematics (NBHM), India for providing travel support to attend the workshop. The first author is financially supported by the Council of Scientific and Industrial Research (CSIR), India. The second author is partially funded by the SERB grant SR/FTP/MS-053/2012. Finally, the authors thank the referee for careful reading of the paper and many helpful suggestions.

References

[1] S. Böcherer, *Bilinear holomorphic differential operators for the Jacobi group*, Comment. Math. Univ. St. Paul. **47** (1998), no. 2, 135–154. MR1660369

[2] Y. Choie, *Jacobi forms and the heat operator*, Math. Z. **225** (1997), no. 1, 95–101, DOI 10.1007/PL00004603. MR1451334

[3] Y. Choie, *Jacobi forms and the heat operator. II*, Illinois J. Math. **42** (1998), no. 2, 179–186. MR1612731

[4] Y. Choie and W. Eholzer, *Rankin-Cohen operators for Jacobi and Siegel forms*, J. Number Theory **68** (1998), no. 2, 160–177, DOI 10.1006/jnth.1997.2203. MR1605899

[5] Y. Choie, H. K. Kim, and M. Knopp, *Construction of Jacobi forms*, Math. Z. **219** (1995), no. 1, 71–76, DOI 10.1007/BF02572350. MR1340849

[6] Y. Choie and W. Kohnen, *Rankin's method and Jacobi forms*, Abh. Math. Sem. Univ. Hamburg **67** (1997), 307–314, DOI 10.1007/BF02940838. MR1481545

[7] H. Cohen, *Sums involving the values at negative integers of L-functions of quadratic characters*, Math. Ann. **217** (1975), no. 3, 271–285, DOI 10.1007/BF01436180. MR0382192

[8] M. Eichler and D. Zagier, *The theory of Jacobi forms*, Progress in Mathematics, vol. 55, Birkhäuser Boston, Inc., Boston, MA, 1985. MR781735

[9] B. Gross, W. Kohnen, and D. Zagier, *Heegner points and derivatives of L-series. II*, Math. Ann. **278** (1987), no. 1-4, 497–562, DOI 10.1007/BF01458081. MR909238

[10] S. D. Herrero, *The adjoint of some linear maps constructed with the Rankin-Cohen brackets*, Ramanujan J. **36** (2015), no. 3, 529–536, DOI 10.1007/s11139-013-9536-5. MR3317870

[11] A. K. Jha and B. Sahu, *Rankin-Cohen brackets on Jacobi forms and the adjoint of some linear maps*, Ramanujan J. **39** (2016), no. 3, 533–544, DOI 10.1007/s11139-015-9683-y. MR3472123

[12] W. Kohnen, *Cusp forms and special values of certain Dirichlet series*, Math. Z. **207** (1991), no. 4, 657–660, DOI 10.1007/BF02571414. MR1119963

[13] R. A. Rankin, *The construction of automorphic forms from the derivatives of a given form*, J. Indian Math. Soc. (N.S.) **20** (1956), 103–116. MR0082563

[14] R. A. Rankin, *The construction of automorphic forms from the derivatives of given forms*, Michigan Math. J. **4** (1957), 181–186. MR0092870

[15] H. Sakata, *Construction of Jacobi cusp forms*, Proc. Japan Acad. Ser. A Math. Sci. **74** (1998), no. 7, 117–119. MR1658858

[16] D. Zagier, *Modular forms and differential operators*, Proc. Indian Acad. Sci. Math. Sci. **104** (1994), no. 1, 57–75, DOI 10.1007/BF02830874. K. G. Ramanathan memorial issue. MR1280058

Department of Mathematics, Indian Institute of Science, Bangalore-560012, India
Email address: abhashkumarjha@gmail.com, abhashjha@iisc.ac.in

School of Mathematical Sciences, National Institute of Science Education and Research, HBNI, Bhubaneswar, PO: Jatni, Khurdha, Odisha- 752 050, India
Email address: brundaban.sahu@niser.ac.in

Contemporary Mathematics
Volume **732**, 2019
https://doi.org/10.1090/conm/732/14786

Some results in study of Kronecker limit formula and Dedekind sums

Jay Jorgenson and Lejla Smajlović

ABSTRACT. This article was prepared for the Research School component of the *Building Bridges: 3rd EU/US Summer School and Workshop on Automorphic Forms and Related Topics*, which was held from 11 July 2016 to 22 July 2016 at the University of Sarajevo in Sarajevo, Bosnia and Herzegovina. The aim of the lectures is to describe aspects of the ongoing research projects related to Kronecker limit formula in different settings.

1. Introduction

The purpose of these lecture notes is to report on work of the authors in the following three different but related directions: In the study of holomorphic functions associated to the function field of Atkin-Lehner groups, which is joint with Holger Then; in the study of Kronecker limit formulas for elliptic Eisenstein series, which is joint with Anna-Maria von Pippich; and in the study of Dedekind sums stemming from higher order non-holomorphic Eisenstein series, which is joint with Cormac O'Sullivan. Each of these topics will be described in a separate section. To begin, let us introduce the classical theory which serves to motivate our work.

1.1. Kronecker's limit formula. The classical Kronecker limit formula refers to evaluation of the first two terms in the Laurent series expansion at $s = 1$ of the non-holomorphic Eisenstein series $E_\infty(s, z)$ associated to the full modular group $\mathrm{PSL}(2, \mathbb{Z})$. Specifically, let

$$E_\infty(z, s) = \frac{1}{2} \sum_{(c,d)\in\mathbb{Z}^2,(c,d)=1} \frac{\mathrm{Im}(z)^s}{|cz + d|^{2s}}$$

be the non-holomorphic (parabolic) Eisenstein series associated to $\mathrm{PSL}(2, \mathbb{Z})$. The first Kronecker limit formula states that

$$E_\infty(z, s) = \frac{1}{\mathrm{vol}(\mathcal{F}_{\mathrm{PSL}(2,\mathbb{Z})})} \cdot \frac{1}{s-1} - \frac{1}{2\pi} \log\left(|\Delta(z)| \,\mathrm{Im}(z)^6\right) + \frac{1}{\pi}(6 - 72\zeta'(-1) - 6\log(4\pi)) + O(s-1)$$

as $s \to 1$. In this expansion, $\mathrm{vol}(\mathcal{F}_{\mathrm{PSL}(2,\mathbb{Z})})$ denotes the volume of a fundamental domain of the modular group $\mathrm{PSL}(2, \mathbb{Z})$, which is equal to $\pi/3$, and $\Delta(z)$ is Delta function. Recall that $\Delta(z)$ can be written as $\Delta(z) = \eta(z)^{24}$ where

$$\eta(z) = q_z^{1/24} \prod_{n=1}^{\infty} (1 - q_z^n) \tag{1.1}$$

and $q_z = \exp(2\pi i z) = e(z)$.

The first Kronecker limit formula can be written in a more succinct form as Taylor series expansion at $s = 0$, namely the formula

$$E_\infty(z, s) = 1 + \log\left(|\Delta(z)|^{1/6}\,\mathrm{Im}(z)\right) \cdot s + O(s^2) \quad \text{as } s \to 0. \tag{1.2}$$

The second Kronecker limit formula refers to evaluation of the constant term in the Taylor series expansion at $s = 1$ of the series

$$E_{u,v}(z, s) = \frac{1}{2} \sum_{(c,d)\in\mathbb{Z}^2,(c,d)=1} \frac{\mathrm{Im}(z)^s}{|cz + d|^{2s}} \exp 2\pi i(cu + dv),$$

where u and v are not both integers. The result is that

$$E_{u,v}(z, s) = -\pi \log |g_{-v,u}(z)| + O(s - 1),$$

as $s \to 1$ and $g_{-v,u}(z)$ is the Siegel function. One can also express $g_{-v,u}$ in terms of the Riemann theta function; see [**23**].

More on classical aspects of Kronecker limit formulas as well as proofs of the above statements can be found in [**23**]. We would also like to refer an interested reader to a beautiful paper by Ramachandra [**28**] where the second Kronecker limit formula is used in various applications, such as to construct ray class fields over imaginary quadratic fields.

Historically, the first Kronecker limit formula was used to study imaginary quadratic number fields. Namely, if K is an imaginary quadratic number field with discriminant $-D < 0$, then the zeta function $\zeta_K(s)$ can be expressed as

$$\zeta_K(s) = \frac{\zeta(2s)}{w} \left(\frac{4}{D}\right)^{s/2} \sum_{(a,b,c)\in Q(D)} E_\infty\left(\frac{-b + \sqrt{-D}}{2a}, s\right)$$

where $Q(D) = \{(a, b, c) \in \mathbb{Z}^3 : b^2 - 4ac = -D, -a < b \le a < c \text{ or } 0 < b \le a = c\}$ and the points $\frac{-b+\sqrt{-D}}{2a}$ are Heegner points. Combining the above equation with (1.2) one can evaluate $\zeta_K(s)$ at $s = 0$.

Somewhat generally and vaguely, one considers a Kronecker limit formula to mean the determination of the second term in the Laurent expansion of a non-holomorphic Eisenstein series $E(z, s)$ at some point, usually taken to be the real-value at the edge of a right half-plane of convergence. As we and others have noted, perhaps it would be better to consider the Laurent expansion at $s = 0$ and take advantage of a functional equation, in case it exists.

1.2. Kronecker's limit formula and the Laplacian. Let Γ_∞ denote the set of all elements of $\mathrm{PSL}(2, \mathbb{Z})$ fixing ∞ (i.e. the stabilizer of the point ∞). Then, one can prove that

$$E_\infty(z, s) = \sum_{\gamma\in\Gamma_\infty\backslash \mathrm{PSL}(2,\mathbb{Z})} \mathrm{Im}(\gamma z)^s.$$

There are various techniques which establish the meromorphic continuation and functional equation for the Eisenstein series. Provided that $E_\infty(z, s)$ admits a meromorphic continuation which is holomorphic at $s = 0$, we can write its Taylor series expansion at $s = 0$ as

$$E_\infty(z, s) = c + F(z) \cdot s + O(s^2)$$

for some constant c, from which we have that

$$F(z) = \frac{d}{ds}(E_\infty(z, s))\Big|_{s=0}.$$

Since $E_\infty(z, s)$ is $\mathrm{PSL}(2, \mathbb{Z})$ invariant, then so is $F(z)$. Let Δ_{hyp} denote the hyperbolic Laplacian. In the usual coordinate $z = x + iy$ of the upper half plane, one has that

$$\Delta_{\mathrm{hyp}} = -y^2 \left(\frac{\partial^2}{\partial x^2} + \frac{\partial^2}{\partial y^2} \right).$$

For $\mathrm{Re}(s) > 1$, the series defining $E_\infty(z, s)$ converges, and one can show that

$$\Delta_{\mathrm{hyp}} E_\infty(z, s) = s(1 - s) E_\infty(z, s).$$

Formally, one gets from this equation that $\Delta_{\mathrm{hyp}} F = c$. Trivially, $\Delta_{\mathrm{hyp}} \log(y)$ is also a constant, but clearly $\log y$ is not $\mathrm{PSL}(2, \mathbb{Z})$-invariant; however, since $y = \mathrm{Im}(z)$, it is elementary to relate $\mathrm{Im}(\gamma z)$ to $\mathrm{Im}(z)$. The evaluation of $E_\infty(z, 0) = 1$ becomes important, since it allows us to conclude that, at least locally, one can write $F(z) = \mathrm{Im}(z) \log |H(z)|$ for some non-vanishing, holomorphic function $H(z)$, which necessarily has some type of modularity with respect to $\mathrm{PSL}(2, \mathbb{Z})$. The series $E_\infty(z, s)$ is invariant under $x \mapsto x + 1$; hence the series admits a Fourier expansion. After significant computations, one can compute the Fourier expansion and prove that $H(z)$ is a cusp form. The group $\mathrm{PSL}(2, \mathbb{Z})$ has many specific properties, one of which being that it is the free product of the finite groups $\mathbb{Z}/2\mathbb{Z}$ and $\mathbb{Z}/3\mathbb{Z}$. Therefore, even though $H(z)$ may not be $\mathrm{PSL}(2, \mathbb{Z})$-invariant (and it is not), $H(z)^6$ is $\mathrm{PSL}(2, \mathbb{Z})$ modular. Using the computation for $\mathrm{Im}(\gamma z)/\mathrm{Im}(z)$, we have that

$$H(\gamma z)^6 = H(z)^6 (cz + d)^{12} \quad \text{for any} \quad \gamma = \begin{pmatrix} * & * \\ c & d \end{pmatrix} \in \mathrm{PSL}(2, \mathbb{Z}).$$

Through a separate argument, one shows that up to multiplicative constant, $\Delta(z)$ is the unique weight 12 cusp form for $\mathrm{PSL}(2, \mathbb{Z})$, so then $H(z)^6 = A\Delta(z)$ for some constant A.

Loosely speaking, we may say that the Laurent series expansion of (parabolic) Eisenstein series may be used to construct cusp forms. In [**22**], the authors defined and studied a non-holomorphic Eisenstein series associated to hyperbolic subgroups of a general Fuchsian group Γ of the first kind. Building on this approach, A.-M. von Pippich studied a non-holomorphic Eisenstein series associated to any elliptic subgroup of Γ. From a different point of view, the elliptic Eisenstein series is associated to any point w of the quotient space $\Gamma \backslash \mathbb{H}$, where $\mathbb{H}$ denotes the hyperbolic upper half space. In [**27**], the elliptic Eisenstein series is shown to have a meromorphic continuation and a Kronecker limit formula. These results were reproved and extended in [**14**] and [**15**] and are the subject of one of the lectures.

1.3. Dedekind sums arising from Kronecker's limit formula. Since $\Delta(z)$ is a weight twelve form, we have that

$$\eta(\gamma z) = \epsilon(\gamma)(cz + d)^{1/2} \eta(z),$$

for $\gamma = \begin{pmatrix} a & b \\ c & d \end{pmatrix} \in \mathrm{PSL}(2,\mathbb{Z})$ and $\epsilon(\gamma)$ a complex number of absolute value one. Dedekind proved that when taking all branches of the logarithm to be principal, one has the relation

$$\log(\eta(\gamma z)) = \log(\eta(z)) + \frac{1}{2}\log(cz+d) + \pi i S(a,b,c,d) \tag{1.3}$$

where for $c \neq 0$

$$S(a,b,c,d) = \frac{a+d}{12c} - \frac{1}{4}\frac{c}{|c|} - \sum_{m=0}^{|c|-1}\left(\left(\frac{dm}{|c|}\right)\right)\left(\left(\frac{m}{|c|}\right)\right),$$

with $((x)) = x - \lfloor x \rfloor - 1/2$, and $S(a,b,0,d) = b/(12d)$. The sum in the definition of $S(a,b,c,d)$ is called the *Dedekind sum* and is denoted by $s(c,d)$.

Generalizations of Dedekind sums are considered in different settings, see for example articles by Goldstein [**6**] or Takada [**30**] where Dedekind sums were computed for principal congruence groups $\Gamma(N)$.

In essence, Dedekind sums appear when dropping the absolute value which occurs within the Kronecker limit formula, though there are many other places where Dedekind sums manifest themselves; see [**1**]. In joint work with C. O'Sullivan, the authors defined and studied Dedekind sums associated to higher order non-holomorphic Eisenstein series; see [**13**]. The presentation of this work formed one of the lectures given.

1.4. Kronecker's limit function and holomorphic function theory. Returning to the setting of $\mathrm{PSL}(2,\mathbb{Z})$, one can use $\Delta(z)$ in the construction of holomorphic functions on $\mathbb{H}$ which are $\mathrm{PSL}(2,\mathbb{Z})$ invariant. Let $F(z)$ denote a product of holomorphic Eisenstein series and assume the weight of F is $12n$. Then the quotient $F(z)/\Delta^n(z)$ is a weight zero form, meaning a holomorphic function, whose only pole is at $i\infty$. In particular, if $F(z)$ has weight twelve, then the quotient is equal to $aj(z)+b$ for constants a and b and where $j(z)$ is the Hauptmodul associated to $\mathrm{PSL}(2,\mathbb{Z})$. The generalization of this construction for Atkin-Lehner groups was studied by the authors in collaboration with Holger Then, and the results of their study are given in one of the lectures.

1.5. The contents of these notes. The three sections below summarize the three different lectures given during the *Building Bridges* Summer School in Sarajevo. In brief, the sections are the following: The study of function field generators for Atkin-Lehner groups (joint with Holger Then); Kronecker's limit formula for elliptic Eisenstein series (joint with A.-M. von Pippich); and Dedekind sums associated to higher order Eisenstein series (joint with C. O'Sullivan). Each section is a summary of the results obtained with the aforementioned co-author, and we refer to the various articles for details of proofs and further discussion.

The mathematical literature contains a large, and growing, number of fascinating articles which, in some way or another, are related to Kronecker's limit formula, Eisenstein series, Dedekind sums, and holomorphic function theory. This article outlines recent research involving the authors. By no means does the paper give these important topics their full and exhaustive presentation; who knows if that is even possible. We are reminded of the first paragraph of S. Lang's book *Fundamentals of Differential Geometry* where Lang writes the following:

The present book aims to give a fairly comprehensive account of the fundamentals of differential manifolds and differential geometry. The size of the book influenced where to stop, and there would be enough material for a second volume (this is not a threat).

The lectures given at the *Building Bridges* summer school served to introduce the audience to certain areas of mathematics by summarizing the recent research of the authors, and this article reports the presentation at that time. We have purposefully avoided expanding the article to include additional research after the summer school, and we did not include any attempt to summarize the existing literature. In other words, we kept true to the lectures and were not tempted to carry out a version of Lang's so-called "threat".

Naturally, there is overlap between the discussion in these lecture notes and the research papers from which we report. This is unavoidable. However, we hope the reader will see that this article is not simply a "cut and paste" production, but rather an abbreviated review of our findings with the aforementioned collaborators. In doing so, we believe that we will communicate the idea that the synthesis of our results can be viewed as one example of a more general study for which further examples will each have their own interest.

1.6. Acknowledgements. Finally, let us express our profound thanks to Cormac O'Sullivan, Anna-Maria von Pippich, and Holger Then for the many hours of enjoyable mathematical discovery. We hope that one of the points conveyed during the summer school is the joyful collaboration we had with each of these individuals, and we hope that these notes properly communicate our appreciation to them. We have learned a lot from Cormac, Anna, and Holger. Additionally, the first named author acknowledges grant support from PSC-CUNY.

2. Generators of function fields for Atkin-Lehner groups

The results in this section stem from joint work with Holger Then. The reader is referred to the following papers and references therein for proofs and background material: [**16**], [**17**], [**18**], [**19**], [**20**], [**21**]. Specifically, we call attention to the following publications which are particularly important to the context of this section: [**4**] ,[**7**], [**8**], [**10**], [**29**] and [**31**].

2.1. Atkin-Lehner type groups. Let $N = p_1 \cdots p_r$ be a square-free, non-negative integer. The subset of $\mathrm{SL}(2,\mathbb{R})$, defined by

$$\Gamma_0(N)^+ := \left\{ e^{-1/2}\begin{pmatrix} a & b \\ c & d \end{pmatrix} \in \mathrm{SL}(2,\mathbb{R}) : ad-bc=e, \;\; a,b,c,d,e \in \mathbb{Z}, \;\; e \mid N, \;\; e \mid a, \;\; e \mid d, \;\; N \mid c \right\}$$

is an arithmetic subgroup of $\mathrm{SL}(2,\mathbb{R})$. We use the terminology "Atkin-Lehner group" of level N to describe $\Gamma_0(N)^+$ because the the groups $\Gamma_0(N)^+$ are obtained by adjoining to $\Gamma_0(N)$ all Atkin-Lehner involutions W_e, for $e \mid N$. Note that Atkin-Lehner groups of level N contain the Fricke involutions τ_N.

The groups $\Gamma_0(N)^+$ have the property that if a subgroup $G \subseteq \mathrm{SL}(2,\mathbb{R})$ is commensurable with $\mathrm{SL}(2,\mathbb{Z})$, then there exists a square-free, non-negative integer N such that G is a subgroup of $\Gamma_0(N)^+$. Moreover, the groups $\Gamma_0(N)^+$ are examples of groups which are commensurable with $\mathrm{SL}(2,\mathbb{Z})$ but not necessarily conjugate to a subgroup of $\mathrm{SL}(2,\mathbb{Z})$.

Let $\{\pm \mathrm{Id}\}$ denote the set of two elements consisting of the identity matrix Id and its product with -1. In general, if Γ is a subgroup of $\mathrm{SL}(2,\mathbb{R})$, we let $\overline{\Gamma} := \Gamma/\{\pm \mathrm{Id}\}$ denote its projection into $\mathrm{PSL}(2,\mathbb{R})$.

By X_N we denote the Riemann surface $\overline{\Gamma_0(N)^+} \setminus \mathbb{H}$. The surface X_N always has one cusp at ∞. By Γ_∞ we denote the stabilizer group of ∞, which consists of all translations from $\Gamma_0(N)^+$. Let f be a modular form on X_N of weight $2k$, not identically zero. Let $\mathcal{F}_N$ denote a fundamental domain of $\Gamma_0(N)^+$, and let $v_z(f)$ denote the order of zero z of f. Then,

$$k\frac{\mathrm{vol}(X_N)}{2\pi} = v_\infty(f) + \sum_{e\in\mathcal{E}_N} \frac{1}{n_e} v_e(f) + \sum_{z\in\mathcal{F}_N\setminus\mathcal{E}_N} v_z(f), \tag{2.1}$$

where $\mathcal{E}_N$ denotes the set of elliptic points in $\mathcal{F}_N$ and n_e is the order of the elliptic point $e \in \mathcal{E}_N$.

2.2. Kronecker limit formula for $\overline{\Gamma_0(N)^+}$. We start with the Fourier expansion of the parabolic Eisenstein series

$$\mathcal{E}_\infty(z,s) = \sum_{\gamma\in\Gamma_\infty\setminus\overline{\Gamma_0(N)^+}} \mathrm{Im}(\gamma z)^s,$$

which is

$$\mathcal{E}_\infty(z,s) = y^s + \varphi_N(s) y^{1-s} + \sum_{m\neq 0} \varphi_N(m,s) W_s(mz). \tag{2.2}$$

In this expression, $W_s(mz)$ is the Whittaker function given by

$$W_s(mz) = 2\sqrt{|m|\,y} K_{s-1/2}(2\pi\,|m|\,y) e(mx)$$

and $K_{s-1/2}$ is the K-Bessel function. The coefficients of the Fourier expansion (2.2) are given by

$$\varphi_N(s) = \frac{s}{s-1}\frac{\xi(2s-1)}{\xi(2s)} \cdot \prod_{\nu=1}^{r} \frac{p_\nu^{1-s}+1}{p_\nu^s+1},$$

where $\xi(s) = \frac{1}{2}s(s-1)\pi^{-s/2}\Gamma(s/2)\zeta_{\mathbb{Q}}(s)$ is the completed Riemann zeta function and

$$\varphi_N(m,s) = \pi^s \frac{m^{s-1}}{\Gamma(s)} \sum_{v|N} v^s N^{-2s} \sum_{n\in\mathbb{N}:(v,n)=1} \frac{a_m((N/v)n)}{n^{2s}}, \tag{2.3}$$

with

$$a_m(n) = \mu\left(\frac{n}{(|m|\,,n)}\right)\frac{\varphi(n)}{\varphi(n/(|m|\,,n))},$$

where $\varphi(n)$ denotes the Euler totient function and $\mu(n)$ classically stands for the Möbius function.

After a series of computations (see [**17**]), one derives the Kronecker limit formula:

$$\mathcal{E}_\infty(z,s) = \frac{C_{-1,N}}{(s-1)} + C_{0,N} - \frac{1}{\mathrm{vol}\,X_N}\log\left(\sqrt[2^r]{\prod_{v|N}|\eta(vz)|^4}\cdot \mathrm{Im}(z)\right) + O(s-1),$$

as $s \to 1$, where

$$C_{-1,N} = -C_{1,N} = \frac{3\cdot 2^r}{\pi\sigma(N)} = \frac{1}{\mathrm{vol}\,X_N},$$

$$C_{0,N} = \frac{3 \cdot 2^r}{\pi\sigma(N)} \left(-\frac{1}{2} \sum_{\nu=1}^{r} \frac{1+3p_\nu}{1+p_\nu} \log p_\nu + 2 - 24\zeta'(-1) - 2\log 4\pi \right),$$

and $\sigma(N)$ denotes the sum of divisors function.

The Kronecker limit formula at $s = 1$ becomes more elegant as the following statement which uses the functional equation of the Eisenstein series:

$$\mathcal{E}_\infty(z,s) = 1 + (\log \left(\sqrt[2^r]{\prod_{v|N} |\eta(vz)|^4} \cdot \operatorname{Im}(z) \right)) \cdot s + O(s^2), \text{ as } s \to 0. \tag{2.4}$$

2.3. The Kronecker limit function as a cusp form. From (2.4) we immediately conclude that $\sqrt[2^r]{\prod_{v|N} |\eta(vz)|^4} \cdot \operatorname{Im}(z)$ is $\Gamma_0(N)^+$-invariant. Therefore, there exists a character $\varepsilon_N(\gamma)$ on $\Gamma_0(N)^+$ such that

$$\prod_{v|N} \eta(v\gamma z) = \varepsilon_N(\gamma)(cz+d)^{2^{r-1}} \prod_{v|N} \eta(vz), \tag{2.5}$$

for all $\gamma \in \Gamma_0(N)^+$. Let us define the constant ℓ_N by

$$\ell_N = 2^{1-r}\operatorname{lcm}\Big(4,\ 2^{r-1}\frac{24}{(24,\sigma(N))}\Big).$$

Then the function

$$\Delta_N(z) := \left(\prod_{v|N} \eta(vz) \right)^{\ell_N}$$

is a weight $k_N = 2^{r-1}\ell_N$ holomorphic form on $\Gamma_0(N)^+$ vanishing only at the cusp.

Finally, it has been shown that the Kronecker limit function $\Delta_N(z)$ is *the minimal weight cusp form on* $\Gamma_0(N)^+$ *vanishing only at the cusp.* This property is important for two reasons: One is the minimality of weight, and the other is the fact that $\Delta_N(z)$ is non-vanishing on $\mathbb{H}$. By dividing holomorphic forms by powers of $\Delta_N(z)$ we will obtain holomorphic modular functions, by which we mean weight zero forms whose only pole is at infinity. In particular, one can divide linear combinations of products of holomorphic Eisenstein series $E_{2k}^{(N)}(z)$ with appropriate weights in order to construct holomorphic modular functions on $\Gamma_0(N)^+$ of weight zero. This point will be discussed in further detail below.

2.4. Construction of generators of function fields.

2.4.1. *The j-invariant.* From the uniformization theorem, one has the existence of a single-valued meromorphic function on $\mathrm{PSL}(2,\mathbb{Z})\backslash\mathbb{H}$ which has a simple pole at ∞ and which maps the one-point compactification of the quotient space $\mathrm{PSL}(2,\mathbb{Z})\backslash\mathbb{H}$ onto the projective space $\mathbb{P}^1$. The bi-holomorphic map f from the compactification of $\mathrm{PSL}(2,\mathbb{Z})\backslash\mathbb{H}$ onto $\mathbb{P}^1$ is uniquely determined by specifying constants $c_{-1} \neq 0$ and c_0 such that the local expansion of f near ∞ is of the form $f(q) = c_{-1}q^{-1} + c_0 + O(q)$ as $q \to 0$. For reasons coming from the theory of automorphic forms and relations with elliptic curves, one chooses $c_{-1} = 1$ and $c_0 = 744$. The unique function obtained by setting of $c_{-1} = 1$ and $c_0 = 744$ is known as the j-invariant, which we denote by $j(z)$.

Any point $\tau \in \mathrm{PSL}(2,\mathbb{Z})\backslash\mathbb{H}$ can be viewed as being the moduli point of an ellitpic curve $\mathcal{E}_L$ with lattice $L = [1, \tau]$ by

$$j(\tau) = 1728E_4^3(\tau)/(E_4^3(\tau) - E_6^2(\tau)) = \frac{E_4^3(\tau)}{\Delta(\tau)}. \tag{2.6}$$

The j-invariant admits the series expansion

$$j(z) = \frac{1}{q_z} + 744 + 196884q_z + 21493760q_z^2 + O(q_z^3) \quad \text{as } q_z \to 0. \tag{2.7}$$

An explicit evaluation of the coefficients in the expansion (2.7) was established by Rademacher.

For quite some time it has been known that the j-invariant has importance far beyond the setting of automorphic forms and the uniformization theorem as applied to $\mathrm{PSL}(2,\mathbb{Z})\backslash\mathbb{H}$. For example, in 1937 T. Schneider initiated the study of the transcendence properties of $j(z)$. Specifically, Schneider proved that if z is a quadratic irrational number in the upper half plane then $j(z)$ is an algebraic integer, and if z is an algebraic number but not imaginary quadratic then $j(z)$ is transcendental. More specifically, if z is any element of an imaginary quadratic extension of $\mathbb{Q}$, then $j(z)$ is an algebraic integer, and the field extension $\mathbb{Q}[j(z), z]/\mathbb{Q}(z)$ is abelian.

In a different direction, the coefficients of (2.7) are related to the dimensions of the components of a graded module admitting action by the monster group $\mathbb{M}$. Specifically, J. Conway and S. Norton established the "monstrous moonshine" conjectures which more precisely formulated relations between $\mathbb{M}$ and the j-invariant (2.7). R. Borcherds proved a significant portion of the Conway-Norton "monstrous moonshine" conjectures in his celebrated work in 1992. More recently, additional work by many authors (too numerous to list here) has extended "moonshine" to other simple groups and other j-invariants associated to certain genus zero Fuchsian groups.

Our work in [**17**] was motivated by the following statement by T. Gannon:

In genus > 0, two functions are needed to generate the function field. A complication facing the development of a higher-genus Moonshine is that, unlike the situation in genus 0 considered here, there is no canonical choice for these generators.

In other words, one does not know the analogue of the j-invariant for moonshine groups of genus greater than zero from which one can begin the quest for "higher genus moonshine".

2.4.2. *Construction of generators of the function fields.* Let us now describe the algorithms from [**17**] and [**18**] whose implementation constructs the sought-after generators of the holomorphic function fields of Atkin-Lehner groups of genus less than or equal to three.

For general square-free N, one can define a holomorphic Eisenstein series as in the case of $\mathrm{PSL}(2,\mathbb{Z})$. Specifically, for any integer $k \geq 2$, set

$$E_{2k}^{(N)}(z) := \sum_{\left(\begin{smallmatrix} * & * \\ c & d \end{smallmatrix}\right) \in \Gamma_\infty \backslash \Gamma_0(N)^+} (cz+d)^{-2k}.$$

Assuming the series converges uniformly and absolutely, it follows that $E_{2k}^{(N)}(z)$ are modular forms of weight $2k$. As shown in [**17**], the holomorphic Eisenstein series for the Atkin-Lehner group $\Gamma_0(N)^+$ can be written in terms of the holomorphic

Eisenstein series $E_{2k}(z)$ for $\mathrm{SL}(2,\mathbb{Z})$; namely, one has the relation

$$E_{2m}^{(N)}(z) = \frac{1}{\sigma_m(N)} \sum_{v|N} v^m E_{2m}(vz);$$

where $\sigma_m(N) = \sum_{d|N} d^m$ and $E_{2m}(z) = \sum_{\left(\begin{smallmatrix} * & * \\ c & d\end{smallmatrix}\right)\in\Gamma_\infty\backslash\mathrm{SL}(2,\mathbb{Z})}(cz+d)^{-2m}$.

For any positive integer M, it is evident that the function
(2.8)

$$F_b(z)=\prod_\nu\left(E_{2m_\nu}^{(N)}(z)\right)^{b_\nu}\Big/\left(\Delta_N(z)\right)^M \quad \text{where} \quad \sum_\nu b_\nu\cdot 2m_\nu=Mk_N \quad \text{and} \quad b=(b_1,\ldots)$$

is a holomorphic modular function on $\overline{\Gamma_0(N)^+\backslash\mathbb{H}}$, meaning a weight zero modular form with polynomial growth in q_z as $z\to\infty$. Its q-expansion follows from substituting $E_{2k}^{(N)}$ and Δ_N by their q-expansions. Note that the q-expansion of $E_{2k}^{(N)}$ follows from the q-expansion of E_{2k} and the relation

$$E_{2m}^{(N)}(z) = \frac{1}{\sigma_m(N)} \sum_{v|N} v^m E_{2m}(vz).$$

Let $\mathcal{S}_M$ denote the set of all rational functions defined in (2.8) for all possible vectors $b=(b_\nu)$ and $m=(2m_\nu)$ with fixed M. Our rationale for finding a set of generators for the function field of the smooth, compact algebraic curve associated to X_N is based on the assumption that a finite span of $\mathcal{S}=\cup_{M=0}^K\mathcal{S}_M$ contains the set of generators for the function field. Subject to this assumption, the set of generators follows from a base change in the sense of Gauss elimination. The algorithm is as follows, which we describe in the setting that X_N has genus g which is not necessarily zero.

Choose a non-negative integer κ. Let $M=1$ and set $\mathcal{S}=\mathcal{S}_1\cup\mathcal{S}_0$.

(1) Form the matrix $A_\mathcal{S}$ of coefficients from the q-expansions of all elements of $\mathcal{S}$, where each element in $\mathcal{S}$ is expanded along a row with each column containing the coefficient of a power, negative, zero or positive, of q. The expansion is recorded out to order q^κ.
(2) Apply Gauss elimination to $A_\mathcal{S}$, thus producing a matrix $B_\mathcal{S}$ which is in row-reduced echelon form.
(3) Implement the following decision to determine if the algorithm is complete: If the g lowest non-trivial rows at the bottom of $B_\mathcal{S}$ correspond to q-expansions whose lead terms have precisely g gaps in the set $\{q^{-1},\ldots,q^{-2g}\}$, then the algorithm is completed. If the indicator to stop fails, then replace M by $M+1$, $\mathcal{S}$ by $\mathcal{S}_M\cup\mathcal{S}$ and reiterate the algorithm.

The rationale for the stopping decision in Step 3 above is based on two ideas, one factual and one hopeful. First, the Weierstrass gap theorem states that for any point P on a compact Riemann surface there are precisely g gaps in the set of possible orders from 1 to $2g$ of functions whose only pole is at P. For instance, when $g=1$ there occurs exactly one gap which for topological reasons is always $\{q^{-1}\}$. Second, the assumption which is hopeful is that the function field is generated by the set of holomorphic modular functions defined in (2.8). The latter assumption is not obvious, but it has turned out to be true for all groups $\Gamma_0(N)^+$ that we studied so far. This includes in particular all groups $\Gamma_0(N)^+$ of genus zero, genus one, genus two, and genus three.

The following assertions are shown to be true.

1. In genus zero case if the Hauptmodul j_N with q-expansion normalized to begin with $1/q$ is such that all coefficients up to q^κ are integers, then all further coefficients are integers. As it turns out, this result was known.

2. In genus $g \geq 1$, if the two generators $j_{1,N}$ and $j_{2,N}$ of the function field of meromorphic functions on X_N are normalized to begin with q^{-a_1} and q^{-a_2} with $a_1 < a_2$ and if all coefficients of $j_{1,N}$ and $j_{2,N}$ up to q^κ are integers, then all further coefficients are integers.

This value of κ can be expressed in terms of a_1, a_2 and the level N explicitly and this is the value we refer to as *desired value*.

The proof is somewhat technical, but the main feature is that one uses the (unscaled) Hecke operators

$$T_p(f)(z) = f(pz) + \sum_{b=0}^{p-1} f\left(\frac{z+b}{p}\right)$$

associated to a suitably chosen prime p. Actually, one shows that all further coefficients are rational numbers whose denominators divide a power of p. By repeating the argument with the Hecke operator T_q for a different prime q, one concludes that all coefficients are integers.

After an implementation of the algorithm described above, we obtained generators of the function fields in genus $g \in \{1, 2, 3\}$ as well as the polynomial equation satisfied by generators. Furthermore, the algorithm shows that all coefficients up to κ are integral; hence, we conclude that the generators we construct have integral q-expansions. Except for the genus zero setting, we do not know if there is a connection to analogues of "monstrous moonshine" or any analogue.

2.4.3. *A slight modification of the algorithm in $g = 0$ case.* The algorithm explained above can be slightly modified when $g = 0$ to produce a representation of the Hauptmodul j_N as a rational function in $E_{2k}^{(N)}$, analogous to representation (2.6) of the j-invariant. Namely, in case when $g = 0$ the last row in the row-reduced echelon form of final output $B_{\mathcal{S}}$ is 1, the next to the last row consists of coefficients of j_N, the Hauptmodul for X_N. If we keep record of the linear transformations needed to deduce $B_{\mathcal{S}}$ from $A_{\mathcal{S}}$ and slightly change $A_{\mathcal{S}}$ to consist of coefficients with constant value of M in (2.8), from the last row, we are then able to represent $(\Delta_N)^M$ as a polynomial in finitely many $E_{2k}^{(N)}$ and then, from the next to the last row, we get j_N as a rational function of $E_{2k}^{(N)}$. The difference in the algorithm consists in replacing $\mathcal{S}$ by $\mathcal{S}_M$, not the union of all $\mathcal{S}_k$ for $k \leq M$. If this algorithm concludes successfully, then the equations for 1 and the Hauptmodul will be rational functions whose denominator is a power of the the Kronecker limit function and whose numerator is a polynomial in holomorphic Eisenstein series.

The successful conclusion of the algorithm will yield a set of generators of the ring of holomorphic modular forms on X_N as a rational functions in a finite set of holomorphic Eisenstein series; see [**18**]. This is a weak generalization of the classical result saying that the ring of holomorphic modular forms associated to $\mathrm{PSL}(2,\mathbb{Z})$ is generated by E_4 and E_6. We used the term "weak" generalization since we conclude that any holomorphic modular form is a rational function in a set of Eisenstein series rather than a polynomial.

2.5. Application to explicit class field theory. By definition, the singular moduli of the j-invariant are the values of j at imaginary quadratic arguments. Such values play a very important role in the class field theory of imaginary quadratic fields. Namely, let K be an imaginary quadratic field over $\mathbb{Q}$, of discriminant d_K and let $\mathcal{O}$ be an order in K. Then, for an imaginary quadratic argument $\tau \in \mathbb{H} \cap \mathcal{O}$ the singular moduli $j(\tau)$ is an algebraic integer. Moreover, the extension $K[j(\tau)]$ is the ring class field of $\mathcal{O}$, which is the Hilbert class field if $\mathcal{O}$ is the maximal order of K, and can be realized constructively as the splitting field over $\mathbb{Q}$ of the class polynomial $H_{d_K}(X)$, which itself is explicitly computable.

More precisely, the numbers $j(\tau)$, where τ ranges over the elements of the fundamental domain of $\mathrm{PSL}(2,\mathbb{Z})\backslash\mathbb{H}$ which satisfy a quadratic equation of discriminant d_K, are exactly the roots of the class polynomial. The class polynomial can be computed using the fact that it can be represented as a modular function, which then must be a polynomial in j. By examining the coefficients in a q-expansion, one obtains the coefficients of the polynomial.

Theoretically, the above approach to computing the class polynomial is satisfactory because one deduces the existence and the form of the polynomial which has also some nice reduction properties modulo primes which divide the discriminant. However, computationally the approach has considerable limits due to the fact that coefficients in the q-expansion of the Hauptmodul j grow rapidly even when the discriminant is modest. For example, when $d_K = -55$, the coefficients of the polynomial $H_{d_K}(X)$ are described in [**32**] as being of "astronomic size". Specifically, the class polynomial is

$$\begin{aligned} H_{-55}(X) = X^4 &+ 3^3 \cdot 5^3 \cdot 29 \cdot 134219 X^3 - 3^7 \cdot 5^3 \cdot 23 \cdot 101 \cdot 32987 X^2 \\ &+ 3^9 \cdot 5^7 \cdot 11^2 \cdot 83 \cdot 101 \cdot 110641 X - 3^{12} \cdot 5^6 \cdot 11^3 \cdot 29^3 \cdot 41^3, \end{aligned}$$

whose discriminant is approximately -4.7×10^{91}. Therefore, it is not really convenient to work with such a polynomial as a generating polynomial.

In [**20**], it is shown that the generators of the functions fields of $\Gamma_0(N)^+$ when evaluated at CM-points also generate the corresponding Hilbert class field. In addition, the minimial polynomials are much more tractable. For example, the minimal polynomial for the Hilbert class field of $\mathbb{Q}(\sqrt{-74})$ using the generators for $\Gamma_0(37)^+$ has coefficients which are less than 2.2×10^4, whereas the minimal polynomial obtained from the Hauptmodul of $\mathrm{PSL}(2,\mathbb{Z})$ has coefficients as large as 6.6×10^{73} and its discriminant approximately 5.7×10^{893}.

2.6. The Schwarzian derivative. For any holomorphic function f, consider the Schwarzian derivative which is defined by

$$S(f)(z) = \left(\frac{f''(z)}{f'(z)}\right)' - \frac{1}{2}\left(\frac{f''(z)}{f'(z)}\right)^2,$$

Using the q-expansion of the Hauptmoduli j_N for genus zero surfaces X_N, we were able to explicitly compute a rational function R_N with integer coefficients such that

$$S(j_N)(z) + R(j_N(z))(j_N'(z))^2 = 0.$$

In [**19**], the authors were able to derive the expression for the Schwarzian derivative of the Hauptmodul of genus zero Atkin-Lehner groups. In [**21**], the Schwarzian derivative of the generator x_N for the function fields of genus one and genus two Atkin-Lehner groups was investigated. As a result, the authors were able to evaluate the generators of the function fields at elliptic points of the Atkin-Lehner groups and relate them to subfields of the ring class fields of orders in imaginary quadratic extensions of the field $\mathbb{Q}$ corresponding to those elliptic points.

We refer the reader to [**19**] and [**21**] for a discussion of the applications of the Schwarzian derivative for the level one j-invariant to differential algebra and questions in logic, which leads one to hope that the results of [**19**] and [**21**], besides the obvious applications in explicit class field theory may find similar applications in other areas of mathematics.

3. Elliptic Kronecker limit formula and a Weierstrass-type factorization theorem

In this section, we discuss joint work with Anna-Maria von Pippich; see [**14**] and [**15**]. We begin with a natural question which will be addressed by the discussion below: Can one realize any holomorphic function, or holomorphic form, as a type of Kronecker limit function for a series similar to a non-holomorphic parabolic Eisenstein series?

3.1. Basic hyperbolic geometry. Let $\Gamma \subseteq \mathrm{PSL}(2,\mathbb{R})$ denote a general Fuchsian group of the first kind acting by fractional linear transformations on the hyperbolic upper half-plane $\mathbb{H} := \{z = x + iy \in \mathbb{C} \,|\, x, y \in \mathbb{R};\ y > 0\}$. We let $M := \Gamma\backslash\mathbb{H}$, which is a finite volume hyperbolic Riemann surface, and denote by $p : \mathbb{H} \longrightarrow M$ the natural projection. We assume that M has e_Γ elliptic fixed points and p_Γ cusps. We identify M locally with its universal cover $\mathbb{H}$.

We let μ_{hyp} denote the hyperbolic metric on M, which is compatible with the complex structure of M, and has constant negative curvature equal to minus one. The hyperbolic line element ds^2_{hyp}, resp. the hyperbolic Laplacian Δ_{hyp}, are given as

$$ds^2_{\mathrm{hyp}} := \frac{dx^2 + dy^2}{y^2}, \quad \text{resp.} \quad \Delta_{\mathrm{hyp}} := -y^2\left(\frac{\partial^2}{\partial x^2} + \frac{\partial^2}{\partial y^2}\right).$$

By $d_{\mathrm{hyp}}(z, w)$ we denote the hyperbolic distance from $z \in \mathbb{H}$ to $w \in \mathbb{H}$. Under the change of coordinates

$$x := e^\rho \cos(\theta)\,, \quad y := e^\rho \sin(\theta)\,,$$

the hyperbolic line element, resp. the hyperbolic Laplacian are rewritten as

$$ds^2_{\mathrm{hyp}} = \frac{d\rho^2 + d\theta^2}{\sin^2(\theta)}, \quad \text{resp.} \quad \Delta_{\mathrm{hyp}} = -\sin^2(\theta)\left(\frac{\partial^2}{\partial \rho^2} + \frac{\partial^2}{\partial \theta^2}\right).$$

The coordinates (ρ, θ) are called Fermi coordinates.

For $z = x + iy \in \mathbb{H}$, we define the hyperbolic polar coordinates $\varrho = \varrho(z), \vartheta = \vartheta(z)$ centered at $i \in \mathbb{H}$ by

$$\varrho(z) := d_{\mathrm{hyp}}(i, z)\,, \quad \vartheta(z) := \measuredangle(\mathcal{L}, T_z)\,,$$

where $\mathcal{L} := \{z \in \mathbb{H} \,|\, x = \mathrm{Re}(z) = 0\}$ denotes the positive y-axis and T_z is the euclidean tangent at the unique geodesic passing through i and z at the point i. In terms of the hyperbolic polar coordinates, the hyperbolic line element, resp. the hyperbolic Laplacian take the form

$$ds^2_{\mathrm{hyp}} = \sinh^2(\varrho)d\vartheta^2 + d\varrho^2, \quad \text{resp.} \quad \Delta_{\mathrm{hyp}} = -\frac{\partial^2}{\partial \varrho^2} - \frac{1}{\tanh(\varrho)}\frac{\partial}{\partial \varrho} - \frac{1}{\sinh^2(\varrho)}\frac{\partial^2}{\partial \vartheta^2}.$$

3.2. Hyperbolic and elliptic Eisenstein series. Let L_γ be the closed geodesic on M in the homotopy class determined by a hyperbolic element $\gamma \in \Gamma$ with associated stabilizer subgroup $\Gamma_\gamma = \langle \gamma \rangle$. There is a scaling-matrix $\sigma_\gamma \in \mathrm{PSL}(2, \mathbb{R})$ such that

$$\sigma_\gamma^{-1} \gamma \sigma_\gamma = \begin{pmatrix} e^{\ell_\gamma/2} & 0 \\ 0 & e^{-\ell_\gamma/2} \end{pmatrix},$$

where ℓ_γ denotes the hyperbolic length of L_γ. We note that $L_\gamma = p(\mathcal{L}_\gamma)$, where $\mathcal{L}_\gamma := \sigma_\gamma \mathcal{L}$ with $\mathcal{L}$ denoting the positive y-axis. Using the coordinates $\rho = \rho(z)$ and $\theta = \theta(z)$, the *hyperbolic Eisenstein series associated to the closed geodesic* L_γ is defined as a function of $z \in M$ and $s \in \mathbb{C}$ with $\mathrm{Re}(s) > 1$ by the series

$$\mathcal{E}_\gamma^{\mathrm{hyp}}(z, s) = \sum_{\eta \in \Gamma_\gamma \backslash \Gamma} \sin(\theta(\sigma_\gamma^{-1} \eta z))^s.$$

From elementary hyperbolic geometry, one can show that the hyperbolic distance $d_{\mathrm{hyp}}(z, \mathcal{L})$ from z to the geodesic line $\mathcal{L}$ satisfies the formula

$$\sin(\theta(z)) \cosh(d_{\mathrm{hyp}}(z, \mathcal{L})) = 1\,.$$

Therefore, we can re-write the series which defines the *hyperbolic Eisenstein series* by

$$\mathcal{E}_\gamma^{\mathrm{hyp}}(z, s) = \sum_{\eta \in \Gamma_\gamma \backslash \Gamma} \cosh(d_{\mathrm{hyp}}(\eta z, \mathcal{L}_\gamma))^{-s}\,. \tag{3.1}$$

The series (3.1) converges absolutely and locally uniformly in the half-plane $\mathrm{Re}(s) > 1$.

Now, let $w \in M$ be an arbitrary point with scaling matrix $\sigma_w \in \mathrm{PSL}(2, \mathbb{R})$ and stabilizer subgroup $\Gamma_w = \langle \gamma_w \rangle$. The group Γ_w is trivial unless $w = e_j$, where e_j $(j = 1, \ldots, e_\Gamma)$ is an elliptic fixed point of M, and, hence, γ_w is a primitive, elliptic element of Γ. Note that

$$\sigma_w^{-1} \gamma_w \sigma_w = \begin{pmatrix} \cos(\pi/\mathrm{ord}(w)) & \sin(\pi/\mathrm{ord}(w)) \\ -\sin(\pi/\mathrm{ord}(w)) & \cos(\pi/\mathrm{ord}(w)) \end{pmatrix},$$

where $\mathrm{ord}(w) = \mathrm{ord}(\Gamma_w)$ denotes the order of w. Note that σ_w is elliptic scaling matrix, such that $\sigma_w(i) = w$.

Using the hyperbolic polar coordinates $\varrho = \varrho(z)$ and $\vartheta = \vartheta(z)$, the *elliptic Eisenstein series* $\mathcal{E}_w^{\mathrm{ell}}(z, s)$ *associated to the point* w is defined for $\mathrm{Re}(s) > 1$ by the absolutely convergent series

$$\mathcal{E}_w^{\mathrm{ell}}(z, s) = \sum_{\eta \in \Gamma_w \backslash \Gamma} \sinh(\varrho(\sigma_w^{-1} \eta z))^{-s}\,. \tag{3.2}$$

The above definition is equivalent to

$$\mathcal{E}_w^{\mathrm{ell}}(z, s) = \frac{1}{\mathrm{ord}(w)} \sum_{\eta \in \Gamma} \sinh(d_{\mathrm{hyp}}(w, \eta z))^{-s}\,. \tag{3.3}$$

3.3. Relations amongst the non-holomorphic Eisenstein series. There are some basic similarities amongst the series. Each series admits a meromorphic continuation to all $s \in \mathbb{C}$. The poles of the meromorphic continuations have been identified and are closely related, in all cases involving data associated to the continuous and non-cuspidal discrete spectrum of the hyperbolic Laplacian and, for hyperbolic and elliptic series, involving data associated to the cuspidal spectrum as well. Finally, and most importantly for this article, the hyperbolic and elliptic Eisenstein series are holomorphic at $s = 0$, and for all known instances, the parabolic Eisenstein series also is holomorphic at $s = 0$. In all these cases, the value of each Eisenstein series at $s = 0$ is a constant as a function of z. The coefficient of s in the Taylor series expansion about $s = 0$ shall be called the Kronecker limit function.

There are some fundamental differences between the various types of Eisenstein series. Hyperbolic Eisenstein series are in $L^2(M)$, whereas parabolic and elliptic series are not. Elliptic Eisenstein series are defined as a sum over a finite index subset of Γ, and indeed the series (3.2) can be extended to all Γ by equation (3.3). However, hyperbolic and parabolic series are necessarily formed by sums over infinite index subsets of Γ. Parabolic Eisenstein series are eigenfunctions of the hyperbolic Laplacian; however, elliptic and hyperbolic Eisenstein series satisfy a differential-difference equation which involves the value of the series at $s + 2$.

Additionally, despite their differences, there are several intriguing ways in which the Eisenstein series interact. The hyperbolic Eisenstein series are in $L^2(M)$ and admit a spectral expansion which involves the parabolic Eisenstein series. If one considers a degenerating sequence of Riemann surfaces obtained by pinching a geodesic, then the associated hyperbolic Eisenstein series converges to parabolic Eisenstein series on the limit surface. If one studies a family of elliptically degenerating surfaces obtained by re-uniformizing at a point with increasing order, then the corresponding elliptic Eisenstein series converge to parabolic Eisenstein series on the limit surface; see [**5**] and references therein.

3.4. Kronecker limit formula for elliptic Eisenstein series. The analogue of Kronecker's limit formula for elliptic Eisenstein series was first proved in [**27**]. Specifically, it is shown that at $s = 0$, the series (3.2) admits the Laurent expansion

$$\operatorname{ord}(w)\,\mathcal{E}^{\mathrm{ell}}_{w}(z,s) - \frac{2^s\sqrt{\pi}\,\Gamma(s-\frac{1}{2})}{\Gamma(s)} \sum_{k=1}^{p_\Gamma} \mathcal{E}^{\mathrm{par}}_{\mathfrak{a}_k}(w,1-s)\,\mathcal{E}^{\mathrm{par}}_{\mathfrak{a}_k}(z,s) =$$

$$-c - \log\left(|H_\Gamma(z,w)|^{\operatorname{ord}(w)}(\operatorname{Im}(z))^c\right)\cdot s + O(s^2) \quad \text{as } s \to 0, \tag{3.4}$$

where $\mathfrak{a}_k$, $k = 1, \ldots, p_\Gamma$, are cusps of M, $c = 2\pi/\mathrm{vol}_{\mathrm{hyp}}(M)$ and $H_\Gamma(z,w)$ is a holomorphic automorphic function with respect to Γ and which vanishes only when $z = \eta w$ for some $\eta \in \Gamma$.

The function $H_\Gamma(z,w)$, defined in (3.4) is called the *elliptic Kronecker limit function at* w. It satisfies the transformation rule

$$H_\Gamma(\gamma z, w) = \varepsilon_w(\gamma)(cz+d)^{2C_w} H_\Gamma(z,w), \quad \text{for any } \gamma = \begin{pmatrix} * & * \\ c & d \end{pmatrix} \in \Gamma, \tag{3.5}$$

where $\varepsilon_w(\gamma) \in \mathbb{C}$ is a constant of absolute value 1, independent of z and

$$C_w = 2\pi/(\operatorname{ord}(w)\,\mathrm{vol}_{\mathrm{hyp}}(M)), \tag{3.6}$$

as proved in [**27**]. Since $H_\Gamma(z,w)$, as a function of z, is finite and non-zero at the cusp $\mathfrak{a}_1 = \infty$, we may re-scale the function and assume, without lost of generality, that $H_\Gamma(z,w)$ is real at the cusp ∞. Also, it is obvious that the above definition may be extended to any point w which is not elliptic by putting $\mathrm{ord}(w) = 1$ and noting that Γ_w consists solely of the identity element.

Two explicit computations are given in [**27**] for $\Gamma = \mathrm{PSL}(2,\mathbb{Z})$ when considering the elliptic Eisenstein series $E_w^{\mathrm{ell}}(z,s)$ associated to the points $w = i$ and $w = \rho = (1+i\sqrt{3})/2$. In these cases, the elliptic Kronecker limit function $H_\Gamma(z,w)$ at points $w = i$ and $w = \rho$ is such that

(3.7)

$$|H_\Gamma(z,i)| = \exp(-B_i)\,|E_6(z)| \text{ where } B_i = -3(24\zeta'(-1) - \log(2\pi) + 4\log\Gamma(1/4))$$

and

(3.8)

$$|H_\Gamma(z,\rho)| = \exp(-B_\rho)\,|E_4(z)| \text{ where } B_\rho = -2(24\zeta'(-1) - 2\log(2\pi/\sqrt{3}) + 6\log\Gamma(1/3)).$$

The Kronecker limit formulas for elliptic Eisenstein series at these points are the asymptotic formulas

$$\mathcal{E}_i^{\mathrm{ell}}(z,s) = -\log(|E_6(z)||\Delta(z)|^{-1/2})\cdot s + O(s^2) \quad \text{as } s \to 0, \tag{3.9}$$

and

$$\mathcal{E}_\rho^{\mathrm{ell}}(z,s) = -\log(|E_4(z)||\Delta(z)|^{-1/3})\cdot s + O(s^2) \quad \text{as } s \to 0, \tag{3.10}$$

where E_4 and E_6 are classical holomorphic Eisenstein series on $\mathrm{PSL}(2,\mathbb{Z})$ of weight four and six, respectively.

3.5. A factorization theorem. As stated in [**15**], it seems not be known in general if the parabolic Eisenstein series $\mathcal{E}_{\mathfrak{a}}^{\mathrm{par}}(z,s)$ is holomorphic at $s = 0$. In any specific examples, such as congruence subgroups or Atkin-Lehner groups, it can be shown that each parabolic Eisenstein series is holomorphic at $s = 0$. *As stated in* [**15**], *we must assume that each* $\mathcal{E}_{\mathfrak{a}}^{\mathrm{par}}(z,s)$ *is holomorphic at* $s = 0$. With this assumption, the following factorization theorem is proved in [**15**].

THEOREM. *Let $M = \Gamma \backslash \mathbb{H}$ be a finite volume Riemann surface with at least one cusp, which we assume to be at ∞ with identity scaling matrix. Let k be a fixed positive integer such that there exists a weight $2k$ holomorphic form f_{2k} on M which is non-vanishing in all cusps and with q-expansion at ∞ given by*

$$f_{2k}(z) = b_{f_{2k}} + \sum_{n=1}^{\infty} b_{f_{2k}}(n) q_z^n. \tag{3.11}$$

Let $Z(f_{2k})$ denote the set of all zeros f_{2k} counted according to their multiplicities and let us define the function

$$H_{f_{2k}}(z) := \prod_{w \in Z(f_{2k})} H_\Gamma(z,w),$$

where, as above, $H_\Gamma(z,w)$ is the elliptic Kronecker limit function. Then there exists a complex constant $c_{f_{2k}}$ such that

$$f_{2k}(z) = c_{f_{2k}} H_{f_{2k}}(z) \tag{3.12}$$

and

$$|c_{f_{2k}}| = |b_{f_{2k}}| \exp \left(\sum_{w \in Z(f_{2k})} B_{w,\infty} \right),$$

where $B_{w,\infty}$ *is defined by*

$$B_{w,\infty} = -C_w \left(2 - \log 2 + \log \left| \eta_\infty^4(w) \operatorname{Im}(w) \right| - \beta_{ll} \operatorname{vol}_{\mathrm{hyp}}(M) \right).$$

In other words, the form f_{2k} is a product of elliptic Kronecker limit functions over $w \in Z(f_{2k})$, the zeros of f_{2k}, modulo a constant $c_{f_{2k}}$ determined up to a factor of modulus 1. Here, $\eta_\infty(w)$ denotes the Kronecker limit function of the parabolic non-holomorphic Eisenstein series, introduced in [**6**] and which coincides with the classical eta function (1.1) in the case $\Gamma = \mathrm{PSL}(2, \mathbb{Z})$. The constants β_{ll} arise in the Fourier expansion of the elements of the hyperbolic scattering matrix.

A similar factorization theorem holds if M is compact, meaning there are no cusps. We refer the reader to [**15**] for the statement and proof.

3.6. Further examples. In the case when the surface has one cusp, we have the following, interesting formula:

$$\sum_{w \in Z(f_{2k})} \mathcal{E}_w^{\mathrm{ell}}(z, s) = -s \log \left(|f_{2k}(z)| |\eta_\infty^4(z)|^{-k} \right) + s \log |b_{f_{2k}}| + O(s^2)$$

as $s \to 0$. In other words, any holomorphic function or form can be realized as a Kronecker limit function, up to a factor involving the Kronecker limit function of the parabolic non-holomorphic Eisenstein series.

In the case when $\Gamma = \mathrm{PSL}(2, \mathbb{Z})$, the parabolic Kronecker limit function is given by $\eta_\infty(z) = \eta(z) = \Delta(z)^{1/12}$. If we take $k = 3$ and $f_{2k} = E_6$, hence $b_{E_6} = 1$ and $Z(E_6) = \{i\}$, so then we have another proof of (3.9). For $k = 2$ and $f_{2k} = E_4$, we have $b_{E_4} = 1$ and $Z(E_4) = \{\rho\}$, when then yields (3.10).

Consider the surface X_2 associated to the Atkin-Lehner group of level $N = 2$. There exists one elliptic point of order two, $e_1 = i/\sqrt{2}$, and one elliptic point of order four, $e_2 = 1/2 + i/2$. The surface has genus zero and one cusp, hence $\operatorname{vol}_{\mathrm{hyp}}(X_2) = \pi/2$. The following examples of the above transformation theorem are proved in [**15**]. First, corresponding to e_1, we have that

$$\mathcal{E}_{e_1}^{\mathrm{ell}}(z, s) = -s \log \left(|E_6^{(2)}(z)| |E_4^{(2)}(z)|^{-1/2} |\Delta(z)\Delta(2z)|^{-1/6} \right) + O(s^2) \quad \text{as } s \to 0.$$

Second, associated to the elliptic point e_2, we have that

$$\mathcal{E}_{e_2}^{\mathrm{ell}}(z, s) = -\log \left(|E_4^{(2)}(z)|^{1/2} |\Delta(z)\Delta(2z)|^{-1/12} \right) \cdot s + O(s^2) \quad \text{as } s \to 0,$$

or, equivalently, that

$$\mathcal{E}_{e_2}^{\mathrm{ell}}(z, s) = -\log \left(\frac{1}{\sqrt{5}} |E_4(z) + 4E_4(2z)|^{1/2} |\Delta(z)\Delta(2z)|^{-1/12} \right) \cdot s + O(s^2) \quad \text{as } s \to 0.$$

The proof of these formulae from [**15**] is quite general, so much so that it is evident one can compute many additional Kronecker limit formulas associated to elliptic points for Atkin-Lehner groups.

4. Kronecker limit formula for higher order Eisenstein series and generalized Dedekind sums

In this section, we discuss joint work with Cormac O'Sullivan; see [**13**]. The problem studied is to obtain a Kronecker limit formula associated to higher order Eisenstein series, thus generalizing the results from [**11**], from which we develop the theory of the associated Dedekind symbols.

4.1. Modular symbols. Let $\Gamma \subseteq \mathrm{PSL}(2,\mathbb{R})$ be a Fuchsian group of the first kind which possesses at least one parabolic element and denote by $S_k(\Gamma)$ be the space of holomorphic weight k cusp forms for Γ, meaning the vector space of holomorphic functions g on $\mathbb{H}$ which satisfy the transformation property

$$g(\gamma z) = j(\gamma, z)^k g(z) \quad \text{with} \quad j((\begin{smallmatrix} a & b \\ c & d \end{smallmatrix}), z) = cz + d \quad \text{for} \quad (\begin{smallmatrix} a & b \\ c & d \end{smallmatrix}) \in \Gamma,$$

and decay rapidly in each cusp in the quotient space $\Gamma\backslash\mathbb{H}$. As usual, we equip the vector space $S_k(\Gamma)$ with the well-known Petersson inner product. For $f \in S_k(\Gamma)$ we set $F_\mathfrak{a}$ to be

$$F_\mathfrak{a}(z) = 2\pi i \int_\mathfrak{a}^z f(w)\, dw.$$

We have the Fourier expansions at the cusp $\mathfrak{b}$

$$j(\sigma_\mathfrak{b}, z)^{-k} f(\sigma_\mathfrak{b} z) = \sum_{n=1}^{\infty} a_\mathfrak{b}(n) e(nz), \quad \text{where} \quad e(z) = e^{2\pi i z},$$

$$F_\mathfrak{a}(\sigma_\mathfrak{b} z) = 2\pi i \int_\mathfrak{a}^\mathfrak{b} f(w)\, dw + \sum_{n=1}^{\infty} \frac{a_\mathfrak{b}(n)}{n} e(nz).$$

Recall that $\sigma_\mathfrak{b}$ is the scaling matrix at the cusp $\mathfrak{b}$.

For the remainder of this lectures we set f to have weight two, meaning $f \in S_2(\Gamma)$. The *modular symbol* $\langle \cdot, f \rangle$ associated to f is the homomorphism from Γ to $\mathbb{C}$ given by

$$\langle \gamma, f \rangle = 2\pi i \int_z^{\gamma z} f(w)\, dw = F_\mathfrak{a}(\gamma z) - F_\mathfrak{a}(z).$$

It is necessarily zero on parabolic and elliptic elements γ.

Before continuing, let us have a small digression. Assume we are dealing with the groups $\Gamma = \Gamma_0(N)^+$ of genus 1, and assume that $f \in S_2(\Gamma)$ is normalized so its q-expansion begins with q. Then the mapping $\phi(\tau) = -F_\infty(\tau) = -2\pi i \int_{i\infty}^{\tau} f(w) dw$ is such that $\phi(\gamma\tau) - \phi(\tau) = -\langle \gamma, f \rangle$, hence is a constant function of τ for all $\gamma \in \Gamma$. Then the image $\Lambda_N = \{-\langle \gamma, f \rangle : \gamma \in \Gamma\}$ is a lattice in $\mathbb{C}$ and ϕ is an isomorphism between the compactified quotient of $\Gamma\backslash\mathbb{H}$ and the elliptic curve generated by the lattice Λ_N such that the pull-back of the canonical differential dx/y is exactly $-2\pi i f(\tau) d\tau$.

4.2. Higher order non-holomorphic Eisenstein series. The *higher order non-holomorphic Eisenstein series associated to f and the cusp $\mathfrak{a}$ of* $\Gamma\backslash\mathbb{H}$ are defined for integers $m, n \geq 0$ by

$$E_\mathfrak{a}^{m,n}(z, s) = \sum_{\gamma \in \Gamma_\mathfrak{a} \backslash \Gamma} \langle \gamma, f \rangle^m \overline{\langle \gamma, f \rangle}^n \, \mathrm{Im}(\sigma_\mathfrak{a}^{-1} \gamma z)^s. \tag{4.1}$$

Then $E_{\mathfrak{a}}^{m,n}(z,s)$ converges for $\mathrm{Re}(s) > 1$ and it has a meromorphic continuation to all s in $\mathbb{C}$; see [**11**] and [**26**]. Going further, it is shown in [**26**] that, in the case $m = n \geq 0$, the series (4.1) has a pole at $s = 1$ of order $m+1$ and

$$\text{(4.2)}\quad \frac{1}{A_m}E_{\mathfrak{a}}^{m,m}(z,s) = \frac{1}{(s-1)^{m+1}} + \frac{B_{\mathfrak{a}}^{(m)}(z)}{(s-1)^m} + \frac{C_{\mathfrak{a}}^{(m)}(z)}{(s-1)^{m-1}} + O\left(\frac{1}{(s-1)^{m-2}}\right),$$

as $s \to 1$ for

$$A_m := \frac{(16\pi^2)^m}{V^{m+1}} m!^2 \|f\|^{2m},$$

where V denotes the volume of the corresponding surface $M = \Gamma\backslash\mathbb{H}$.

We call the function $B_{\mathfrak{a}}^{(m)}(z)$ the *Kronecker limit function associated to the series* $E_{\mathfrak{a}}^{m,m}(z,s)$ or the *higher order Kronecker limit function*. Define

$$\text{(4.3)}\quad V_f := \frac{V}{16\pi^2\|f\|^2} = \frac{m^2 A_{m-1}}{A_m} \quad \text{for} \quad m \geq 1.$$

We can now give the basic properties of $B_{\mathfrak{a}}^{(m)}$. For m a non-negative integer:

(1) $B_{\mathfrak{a}}^{(m)}(z)$ is real valued;
(2) $\Delta_{\text{hyp}} B_{\mathfrak{a}}^{(m)}(z) = -1$;
(3) $B_{\mathfrak{a}}^{(0)}(z)$ is Γ-invariant and $B_{\mathfrak{a}}^{(m)}(z) - V_f|F_{\mathfrak{a}}(z)|^2$ is Γ-invariant if $m \geq 1$.

For any two cusps $\mathfrak{a}, \mathfrak{b}$ of Γ, the series $E_{\mathfrak{a}}^{m,m}(\sigma_{\mathfrak{b}}z, s)$, $m \geq 1$ admits the Fourier expansion

$$\text{(4.4)}\quad E_{\mathfrak{a}}^{m,m}(\sigma_{\mathfrak{b}}z,s) = \phi_{\mathfrak{a}\mathfrak{b}}^{m,m}(0,s)y^{1-s} + \sum_{k\neq 0}\phi_{\mathfrak{a}\mathfrak{b}}^{m,m}(k,s)W_s(kz),$$

were $W_s(z)$ denotes the classical Whittaker function, defined for $z = x+iy \in \mathbb{H}$ by

$$W_s(z) = 2\sqrt{y}K_{s-1/2}(2\pi y)e(x),$$

and where K stands for the K-Bessel function; [**11**]. As is often assumed, the Whittaker function is extended to the lower half-plane by imposing the symmetry $W_s(z) = W_s(\overline{z})$. The coefficients $\phi_{\mathfrak{a}\mathfrak{b}}^{m,m}(k,s)$, for $k \in \mathbb{Z}$, admit meromorphic continuations to the whole complex plane with poles at $s = 1$ of order at most $m+1$. Let

$$\text{(4.5)}\quad \frac{\phi_{\mathfrak{a}\mathfrak{b}}^{m,m}(k,s)}{A_m} = \frac{a_{\mathfrak{a}\mathfrak{b}}^{(m)}(k)}{(s-1)^{m+1}} + \frac{b_{\mathfrak{a}\mathfrak{b}}^{(m)}(k)}{(s-1)^m} + O\left(\frac{1}{(s-1)^{m-1}}\right)$$

be the Laurent series expansion of $\phi_{\mathfrak{a}\mathfrak{b}}^{m,m}(k,s)$ at $s=1$. Then one can prove that

$$\text{(4.6)}\quad B_{\mathfrak{a}}^{(m)}(\sigma_{\mathfrak{b}}z) = -\log y + b_{\mathfrak{a}\mathfrak{b}}^{(m)}(0) + 2\,\mathrm{Re}(H_{\mathfrak{a}\mathfrak{b}}^{(m)}(z)),$$

for any $z = x+iy \in \mathbb{H}$, where $H_{\mathfrak{a}\mathfrak{b}}^{(m)}(z) := \sum_{k>0} b_{\mathfrak{a}\mathfrak{b}}^{(m)}(k)e(kz)$.

Consider the difference $B_{\mathfrak{a}}^{(m)}(z) - B_{\mathfrak{a}}^{(1)}(z)$. By property 3 of the limit function $B_{\mathfrak{a}}^{(m)}(z)$ this difference is Γ-invariant. Moreover, we have that

$$\text{(4.7)}\quad B_{\mathfrak{a}}^{(m)}(\sigma_{\mathfrak{b}}z) - B_{\mathfrak{a}}^{(1)}(\sigma_{\mathfrak{b}}z) = b_{\mathfrak{a}\mathfrak{b}}^{(m)}(0) - b_{\mathfrak{a}\mathfrak{b}}^{(1)}(0) + 2\,\mathrm{Re}\left(H_{\mathfrak{a}\mathfrak{b}}^{(m)}(z) - H_{\mathfrak{a}\mathfrak{b}}^{(1)}(z)\right)$$

which shows that $B_{\mathfrak{a}}^{(m)}(z) - B_{\mathfrak{a}}^{(1)}(z)$ is bounded in cuspidal zones. Therefore it is an L^2 eigenfunction of the Laplacian, with eigenvalue zero. Part (2) then yields that the difference is a constant, say, $\mathcal{C}^{(m)}$. This constant is real by property (1). In summary, we conclude the following for $m > 0$:

(a) There exists a constant $\mathcal{C}^{(m)} \in \mathbb{R}$ so that $B_{\mathfrak{a}}^{(m)}(z) = B_{\mathfrak{a}}^{(1)}(z) + \mathcal{C}^{(m)}$.
(b) For any pairs of cusps $\mathfrak{a}$, $\mathfrak{b}$ we have $\mathcal{C}^{(m)} = b_{\mathfrak{ab}}^{(m)}(0) - b_{\mathfrak{ab}}^{(1)}(0)$.
(c) We have that $H_{\mathfrak{ab}}^{(m)}(z)$ is independent of m and so denote it just by $H_{\mathfrak{ab}}(z)$.

4.3. Higher order Dedekind sums. Let $\gamma \in \Gamma$ set $\gamma_{\mathfrak{b}} = \sigma_{\mathfrak{b}}{}^{-1}\gamma\sigma_{\mathfrak{b}}$. Then one can show that

(4.8)
$$H_{\mathfrak{ab}}(\gamma_{\mathfrak{b}}z) - H_{\mathfrak{ab}}(z) = -\log j(\gamma_{\mathfrak{b}}, z) + V_f\left(F_{\mathfrak{a}}(\sigma_{\mathfrak{b}}z)\overline{\langle\gamma, f\rangle} + |\langle\gamma, f\rangle|^2/2\right) + i\theta_{\mathfrak{ab}}(\gamma),$$

where we are taking the principal branch of the logarithm.

We will call the function $\theta_{\mathfrak{ab}}(\gamma)$ the *higher order Dedekind sum*. In order to justify this, let us define the function

$$\eta_{\mathfrak{ab}}(z, f) = \eta_{\mathfrak{ab}}(z) := \exp\left(-\frac{1}{2}H_{\mathfrak{ab}}(z) + \frac{V_f}{4}|F_{\mathfrak{a}}(\sigma_{\mathfrak{b}}z)|^2\right).$$

The function $\eta_{\mathfrak{ab}}(z)$ is a generalization of the classical Dedekind eta function. Analogously as above, it is easy to deduce that $\eta_{\mathfrak{ab}}(z)$ satisfies the transformation formula

$$\log \eta_{\mathfrak{ab}}(\gamma_{\mathfrak{b}}z) = \frac{1}{2}\log j(\gamma_{\mathfrak{b}}, z) + \log \eta_{\mathfrak{ab}}(z) + \pi i S_{\mathfrak{ab}}(\gamma), \tag{4.9}$$

where $S_{\mathfrak{ab}}(\gamma) = -\frac{1}{2\pi}\theta_{\mathfrak{ab}}(\gamma)$.

The transformation formula (4.9) is an analogue of the transformation formula (1.3) for the classical Dedekind eta function, hence it justifies the name higher order Dedekind sum for the function $\theta_{\mathfrak{ab}}(\gamma)$.

4.4. Properties of higher order Dedekind sums. Recall that in our setting, we consider surfaces with at least one cusp, of which one is assumed to be at ∞ with identity scaling matrix. The cusp at ∞ is always the most interesting one, since in that case the shape of the higher order and classical Kronecker limit functions are somehow most natural (see, e.g. computations of the Kronecker limit function for congruence groups or for principal congruence groups).

For any two matrices M, N in $\mathrm{SL}_2(\mathbb{R})$ set

$$\omega(M, N) = (-\log j(MN, z) + \log j(M, Nz) + \log j(N, z))/(2\pi i), \tag{4.10}$$

where we use the principal branch with argument from $(-\pi, \pi]$. Actually, the right side of (4.10) is independent of z and takes only the values $\{-1, 0, 1\}$. We call the function ω the *phase factor.*

We will use the notation $\theta_{\mathfrak{a}}$ for $\theta_{\mathfrak{a}\infty}$. Then, for all γ in Γ and any cusp $\mathfrak{b}$

$$\theta_{\mathfrak{a}}(\gamma) = \theta_{\mathfrak{ab}}(\gamma) + 2\pi\omega(\sigma_{\mathfrak{b}}{}^{-1}, \gamma) - 2\pi\omega(\sigma_{\mathfrak{b}}{}^{-1}\gamma\sigma_{\mathfrak{b}}, \sigma_{\mathfrak{b}}{}^{-1}).$$

This shows that the action of $\theta_{\mathfrak{ab}}$ on elements $\gamma \in \Gamma$ is reduced to the action of $\theta_{\mathfrak{a}}$ on γ, modulo an additive factor $2\pi k$, where k takes values from the set $\{-2, -1, 0, 1, 2\}$.

Therefore, it is actually sufficient to investigate properties of $\theta_{\mathfrak{a}}(\gamma)$.

First, one can prove that for all γ and τ in Γ we have

$$\theta_{\mathfrak{a}}(\gamma\tau) = \theta_{\mathfrak{a}}(\gamma) + \theta_{\mathfrak{a}}(\tau) - V_f \operatorname{Im}\left(\langle\gamma, f\rangle\overline{\langle\tau, f\rangle}\right) - 2\pi\omega(\gamma, \tau),$$

meaning that $\theta_{\mathfrak{a}}$ is not homomorphism. However, it satisfies an interesting relation. Namely, for all γ_1, γ_2 and γ_3 in Γ

$$\theta_{\mathfrak{a}}(\gamma_1\gamma_2\gamma_3) - \theta_{\mathfrak{a}}(\gamma_1\gamma_2) - \theta_{\mathfrak{a}}(\gamma_1\gamma_3) - \theta_{\mathfrak{a}}(\gamma_2\gamma_3) + \theta_{\mathfrak{a}}(\gamma_1) + \theta_{\mathfrak{a}}(\gamma_2) + \theta_{\mathfrak{a}}(\gamma_3)$$
$$= 2\pi\Big[\omega(\gamma_1,\gamma_3) + \omega(\gamma_2,\gamma_3) - \omega(\gamma_1\gamma_2,\gamma_3)\Big].$$

The action of $\theta_{\mathfrak{a}}$ on parabolic and elliptic elements of Γ is particularly simple. Namely, the following statements hold true:

i) Let $p_{\mathfrak{b}}$ be the parabolic element generating $\Gamma_{\mathfrak{b}}$, the elements of Γ fixing the cusp $\mathfrak{b}$. Then

$$\theta_{\mathfrak{a}}(p_{\mathfrak{b}}) = 2\pi\omega({\sigma_{\mathfrak{b}}}^{-1}, p_{\mathfrak{b}}).$$

ii) For every elliptic element ε in Γ of order r and fixing $z_0 \in \mathbb{H}$ we have

$$\theta_{\infty}(\varepsilon) = -i\log j(\varepsilon, z_0) = \frac{2\pi}{r}\sum_{k=1}^{r-1}\omega(\varepsilon^k,\varepsilon).$$

4.5. The difference between the first order and higher order multiplier systems at ∞. Consider, in general, a group with an arbitrary positive number of cusps. For each pair of cusps, we have defined the associated Dedekind sums. We then establish certain results, which can be considered a type of reduction formula, which shows the relations of Dedekind sums. Without loss of generality, we may assume one of the cusps is at infinity. Then we have related the Dedekind sums associated to the pair $(\mathfrak{a},\mathfrak{b})$ to the pair $(\mathfrak{a},\infty)$. Also, we have related the Dedekind sums of the pair $(\mathfrak{a},\infty)$ to the pair $(\mathfrak{b},\infty)$. With this, we can focus our attention to the Dedekind sums associated to the pair (∞,∞).

Also, note that our early analysis showed that for any $m > 0$, all Dedekind sums for the higher order Eisenstein series $E_{\mathfrak{a}}^{m,m}(z,s)$ agreed, so there are actually two types of Dedekind sums to consider: Those for $m = 1$ and those for $m = 0$. Define Ψ to be the function on Γ equal to the difference of the first order $m = 1$ Dedekind sums and the zeroth order $m = 0$ sums. As it turns out, we have the following formula: For any triple of elements γ_1, γ_2 and γ_3 of Γ, then

$$\Psi(\gamma_1\gamma_2\gamma_3) = \Psi(\gamma_1\gamma_2) + \Psi(\gamma_1\gamma_3) + \Psi(\gamma_2\gamma_3) - \Psi(\gamma_1) - \Psi(\gamma_2) - \Psi(\gamma_3) \tag{4.11}$$

This identity admits a cohomological interpretation, which we will not include in this discussion. For further details, we refer the reader to [**13**] and references therein.

4.6. An example: $\Gamma_0(37)^+$. The smallest level group $\Gamma_0(N)^+$ with squarefree N which possesses a unique normalized weight two cusp form, i.e. a weight two cusp form f whose lead coefficient in the q-expansion is 1 is $N = 37$. The compactified quotient $\Gamma_0(37)^+\backslash\mathbb{H}$ is isomorphic to the genus one curve $y^2 = 4x^3 - 4x + 1$ over $\mathbb{C}$ in the sense that there is a unique isomorphism between $\Gamma\backslash\mathbb{H}\cup\{\infty\}$ and the curve which sends ∞ to 0 and such that the pull-back of the canonic differential dx/dy is $-2\pi i f(z)dz$. The q-expansion of f is given by

$$f(z) = q_z - 2q_z^2 - 3q_z^3 + 2q_z^4 - 5q_z^5 + 6q_z^6 - q_z^7 + 6q_z^9 + \ldots$$

and its Petterson norm is $\|f\|^2 = \omega_1\omega_2/(2\pi^2 i)$ where ω_1 and ω_2 are real and complex periods of the curve $y^2 = 4x^3 - 4x + 1$. The numerical values of those periods are approximately $\omega_1 \approx 2.993458644$ and $\omega_2 \approx 2.451389381i$.

The set of generators of the group $\Gamma_0(37)^+$ is given by

$$g_1 = \begin{pmatrix} 1 & 1 \\ 0 & 1 \end{pmatrix}, \qquad g_2 = \frac{1}{\sqrt{37}} \begin{pmatrix} 37 & -1 \\ 37 & 0 \end{pmatrix},$$
$$g_3 = \frac{1}{\sqrt{37}} \begin{pmatrix} 185 & -31 \\ 222 & -37 \end{pmatrix}, \quad g_4 = \frac{1}{\sqrt{37}} \begin{pmatrix} 111 & -28 \\ 148 & -37 \end{pmatrix},$$
$$g_5 = \frac{1}{\sqrt{37}} \begin{pmatrix} 74 & -25 \\ 111 & -37 \end{pmatrix}, \quad g_6 = \frac{1}{\sqrt{37}} \begin{pmatrix} 37 & -19 \\ 74 & -37 \end{pmatrix},$$
$$g_7 = \frac{1}{\sqrt{37}} \begin{pmatrix} 74 & -15 \\ 185 & -37 \end{pmatrix}, \quad g_8 = \frac{1}{\sqrt{37}} \begin{pmatrix} 148 & -89 \\ 185 & -111 \end{pmatrix},$$

The sides of the fundamental polygon for $\Gamma_0(37)^+$ are given by

$$s_1 = \{z \mid \mathrm{Re}(z) = 0\},\ s_2 = \{z \mid |z| = \frac{1}{\sqrt{37}}\},\ s_3 = \{z \mid |z - \frac{1}{6}| = \frac{1}{\sqrt{1332}}\},$$
$$s_4 = \{z \mid |z - \frac{1}{5}| = \frac{1}{\sqrt{925}}\},\ s_5 = \{z \mid |z - \frac{1}{4}| = \frac{1}{\sqrt{592}}\},$$
$$s_6 = \{z \mid |z - \frac{1}{3}| = \frac{1}{\sqrt{333}}\},\ s_7 = \{z \mid |z - \frac{2}{5}| = \frac{1}{\sqrt{925}}\},$$
$$s_8 = \{z \mid |z - \frac{1}{2}| = \frac{1}{\sqrt{148}}\},\ s_{8+k} = \{z \mid (1 - \overline{z}) \in s_{9-k}\} \text{ for } k = 1, \ldots, 8.$$

with vertices

$$v_1 = \infty,\ v_2 = i\frac{1}{\sqrt{37}},\ v_3 = \frac{6}{37} + i\frac{1}{37},\ v_4 = \frac{33}{185} + i\frac{\sqrt{21}}{185},$$
$$v_5 = \frac{81}{370} + i\frac{\sqrt{99}}{370},\ v_6 = \frac{21}{74} + i\frac{\sqrt{3}}{74},\ v_7 = \frac{141}{370} + i\frac{\sqrt{99}}{370},$$
$$v_8 = \frac{78}{185} + i\frac{\sqrt{21}}{185},\ v_9 = \frac{1}{2} + i\frac{1}{2\sqrt{37}},\ v_{9+k} = 1 - \overline{v_{9-k}} \text{ for } k = 1, \ldots, 7.$$

The identification of sides of the fundamental polygon is given by the following mappings

$$g_1 : s_1 \mapsto s_{16},\ g_2 : s_2 \mapsto s_{15},\ g_3 : s_3 \mapsto s_{14},\ g_4 : s_5 \mapsto s_{12},$$
$$g_5 : s_6 \mapsto s_{11},\ g_6 : s_8 \mapsto s_9,\ g_7 : s_4 \mapsto s_7,\ g_8 : s_{10} \mapsto s_{13}.$$

The corresponding Riemann surface has genus 1, one cusp and four elliptic points, one of order three and three points of order two. The volume of the surface is $19\pi/3$. The points v_2, v_3, v_9, v_{16}, v_{15} and v_{12} are elliptic points of order two, while points v_6 and v_{12} have order three. The corresponding maps are

$$g_1^{-1}g_2 : v_2 \mapsto v_2 \quad \& \quad g_2 g_1^{-1} : v_{16} \mapsto v_{16} \quad \text{elliptic of order 2},$$
$$g_2^{-1}g_3 : v_3 \mapsto v_3 \quad \& \quad g_3 g_2^{-1} : v_{15} \mapsto v_{15} \quad \text{elliptic of order 2},$$
$$g_5^{-1}g_4 : v_6 \mapsto v_6 \quad \& \quad g_4 g_5^{-1} : v_{12} \mapsto v_{12} \quad \text{elliptic of order 3},$$
$$g_6 : v_9 \mapsto v_9 \quad \text{elliptic of order 2}.$$

Actually, the set of generators can be reduced to six mappings, since g_3 and g_4 can be expressed in terms of g_i, $i = 5, 6, 7, 8$; namely $g_3 = g_8 g_6 g_7$ and $g_4 = g_8 g_5 g_7$; see [**13**] for further information.

Let us put $\tau_{37} = E_1 = g_1^{-1}g_2$, $E_2 = g_2^{-1}g_3$, $E_3 = g_5^{-1}g_4$ and $E_4 = g_6$ to be elliptic elements of $\Gamma_0(37)^+$. We denote by $\widetilde{E}_i$, $i = 1,2,3$ their commutators.

In the sequel we will show how to compute values of $\Psi_{37}^+ = \theta_{37}^+ - \xi_{37}^+$, the difference of the first order and of the zeroth order Dedekind sum on some generators of the group $\Gamma_0(37)^+$.

It is easy to compute values of $\xi_{37}^+(\gamma)$, for any $\gamma \in \Gamma_0(37)^+$; one simply applies results from [**13**], Proposition 3.5. for $p = 37$ and $\xi_{37}^+(\gamma) = -2\pi S_{37}^+(\gamma)$. For example, a simple computation shows that $\xi_{37}^+(g_1) = -\pi/6$, $\xi_{37}^+(E_i) = \xi_{37}^+(\widetilde{E}_i) = \pi/2$, for $i = 1,2,4$ and $\xi_{37}^+(E_3) = -2\pi/3$.

The values of the higher order Dedekind sum θ_{37}^+ on parabolic and elliptic elements are easily computed using Proposition 3.5 and Proposition 3.6 of [**13**]. Namely, we get $\theta_{37}^+(g_1) = 0$, $\theta_{37}^+(E_i) = \theta_{37}^+(\widetilde{E}_i) = \pi/2$, for $i = 1,2,4$ and $\theta_{37}^+(E_3) = -2\pi/3 = \theta_{37}^+(\widetilde{E}_3) = -2\pi/3$. Therefore,

$$\Psi_{37}^+(E_i) = \Psi_{37}^+(\widetilde{E}_i) = 0, \text{ for } i = 1,2,3,4 \text{ and } \Psi_{37}^+(g_1) = \pi/6.$$

Now, we show how to compute $\Psi_{37}^+(g_i)$, for $i = 2,3,4,5$. Since g_1 is parabolic, it is easy to see that

$$0 = \Psi_{37}^+(E_1) = -\Psi_{37}^+(g_1) + \Psi_{37}^+(g_2),$$

therefore $\Psi_{37}^+(g_2) = \pi/6$. Now, we compute $\Psi_{37}^+(g_3)$. Applying the properties of Ψ from [**13**], we have

$$0 = \Psi_{37}^+(E_2) = \Psi_{37}^+(g_2^{-1}g_3) = -\Psi_{37}^+(g_2) + \Psi_{37}^+(g_3) + V_f \operatorname{Im}(\langle g_2, f\rangle \overline{\langle g_3, f\rangle}) \tag{4.12}$$

Moreover,

$$0 = \Psi_{37}^+(E_2) - \Psi_{37}^+(\widetilde{E}_2) = \Psi_{37}^+(g_2^{-1}g_3) + \Psi_{37}^+(g_2 g_3^{-1}) = 2V_f \operatorname{Im}(\langle g_2, f\rangle \overline{\langle g_3, f\rangle}).$$

This, together with (4.12) yields that $\Psi_{37}^+(g_2) = \Psi_{37}^+(g_3) = \pi/6$ and $\Psi_{37}^+(g_2g_3) = \Psi_{37}^+(g_3g_2) = \pi/3$.

Proceeding analogously, starting with $0 = \Psi_{37}^+(E_3) - \Psi_{37}^+(\widetilde{E}_3)$ we deduce that $\operatorname{Im}(\langle g_4, f\rangle \overline{\langle g_5, f\rangle}) = 0$ and $\Psi_{37}^+(g_4) = \Psi_{37}^+(g_5)$.

Exploiting this relation, together with results from [**13**] and representation of g_3 and g_4 in terms of g_i, $i = 5,6,7,8$ we get that

$$V_f \operatorname{Im}(\langle g_8, f\rangle \overline{\langle g_5, f\rangle}) - \operatorname{Im}(\langle g_7, f\rangle \overline{\langle g_5, f\rangle}) = \pi/6 \text{ and } \Psi_{37}^+(g_8g_7) = \pi/6,$$

where $V_f = 19/(48\pi\|f\|^2)$.

Proceeding analogously, computation of $\Psi_{37}^+(g_4g_5) = 2\Psi_{37}^+(g_5)$ yields that $\operatorname{Im}(\langle g_3, f\rangle \overline{\langle g_4, f\rangle}) = 0$ and that

$$V_f \operatorname{Im}(\langle g_7, f\rangle \overline{\langle g_5, f\rangle}) = -V_f \operatorname{Im}(\langle g_8, f\rangle \overline{\langle g_5, f\rangle}) = \pi/12.$$

References

[1] M. Atiyah, *The logarithm of the Dedekind η-function*, Math. Ann. **278** (1987), no. 1-4, 335–380, DOI 10.1007/BF01458075. MR909232

[2] I. Chen and N. Yui, *Singular values of Thompson series*, Groups, difference sets, and the Monster (Columbus, OH, 1993), Ohio State Univ. Math. Res. Inst. Publ., vol. 4, de Gruyter, Berlin, 1996, pp. 255–326. MR1400423

[3] D. Cox, J. McKay, and P. Stevenhagen, *Principal moduli and class fields*, Bull. London Math. Soc. **36** (2004), no. 1, 3–12, DOI 10.1112/S0024609303002583. MR2011972

[4] C. J. Cummins, *Congruence subgroups of groups commensurable with* $\mathrm{PSL}(2,\mathbb{Z})$ *of genus 0 and 1*, Experiment. Math. **13** (2004), no. 3, 361–382. MR2103333

[5] D. Garbin and A.-M. v. Pippich, *On the behavior of Eisenstein series through elliptic degeneration*, Comm. Math. Phys. **292** (2009), no. 2, 511–528, DOI 10.1007/s00220-009-0892-3. MR2544741

[6] L. J. Goldstein, *Dedekind sums for a Fuchsian group. I*, Nagoya Math. J. **50** (1973), 21–47. MR0318065

[7] D. A. Hejhal, *The Selberg trace formula for* PSL(2, **R**). *Vol. 2*, Lecture Notes in Mathematics, vol. 1001, Springer-Verlag, Berlin, 1983. MR711197

[8] M. N. Huxley, *Scattering matrices for congruence subgroups*, Modular forms (Durham, 1983), Ellis Horwood Ser. Math. Appl.: Statist. Oper. Res., Horwood, Chichester, 1984, pp. 141–156. MR803366

[9] Ö. Imamoḡlu and C. O'Sullivan, *Parabolic, hyperbolic and elliptic Poincaré series*, Acta Arith. **139** (2009), no. 3, 199–228, DOI 10.4064/aa139-3-1. MR2545926

[10] H. Iwaniec, *Spectral methods of automorphic forms*, 2nd ed., Graduate Studies in Mathematics, vol. 53, American Mathematical Society, Providence, RI; Revista Matemática Iberoamericana, Madrid, 2002. MR1942691

[11] J. Jorgenson and C. O'Sullivan, *Convolution Dirichlet series and a Kronecker limit formula for second-order Eisenstein series*, Nagoya Math. J. **179** (2005), 47–102, DOI 10.1017/S0027763000025605. MR2164401

[12] J. Jorgenson and C. O'Sullivan, *Unipotent vector bundles and higher-order non-holomorphic Eisenstein series* (English, with English and French summaries), J. Théor. Nombres Bordeaux **20** (2008), no. 1, 131–163. MR2434161

[13] J. Jorgenson, C. O'Sullivan, and L. Smajlović, Modular Dedekind symbols associated to Fuchsian groups and higher-order Eisenstein series. Submitted for publication.

[14] J. Jorgenson, A.-M. von Pippich, and L. Smajlović, *On the wave representation of hyperbolic, elliptic, and parabolic Eisenstein series*, Adv. Math. **288** (2016), 887–921, DOI 10.1016/j.aim.2015.10.019. MR3436401

[15] J. Jorgenson, A.-M. von Pippich, and L. Smajlović, *Applications of Kronecker's limit formula for elliptic Eisenstein series*, Ann. Math. Qué. **43** (2019), no. 1, 99–124, DOI 10.1007/s40316-017-0094-x. MR3925139

[16] J. Jorgenson, L. Smajlović, and H. Then, *On the distribution of eigenvalues of Maass forms on certain moonshine groups*, Math. Comp. **83** (2014), no. 290, 3039–3070, DOI 10.1090/S0025-5718-2014-02823-8. MR3246823

[17] J. Jorgenson, L. Smajlović, and H. Then, *Kronecker's limit formula, holomorphic modular functions, and q-expansions on certain arithmetic groups*, Exp. Math. **25** (2016), no. 3, 295–319, DOI 10.1080/10586458.2015.1053161. MR3478329

[18] J. Jorgenson, L. Smajlović, and H. Then, *Certain aspects of holomorphic function theory on some genus-zero arithmetic groups*, LMS J. Comput. Math. **19** (2016), no. 2, 360–381, DOI 10.1112/S1461157016000425. MR3621645

[19] J. Jorgenson, L. Smajlović, and H. Then, The Hauptmodul at elliptic points of certain arithmetic groups, accepted for publication in J. Number Theory, arXiv:1602.07426.

[20] J. Jorgenson, L. Smajlović, and H. Then, On the evaluation of singular invariants for canonical generators of certain genus one arithmetic groups, to appear in Exp. Math., doi: 10.1080/10586458.2017.1422161.

[21] J. Jorgenson, L. Smajlović, and H. Then, An investigation of the Schwarzian derivative for Atkin-Lehner groups of genus one and two. In preparation.

[22] S. S. Kudla and J. J. Millson, *Harmonic differentials and closed geodesics on a Riemann surface*, Invent. Math. **54** (1979), no. 3, 193–211, DOI 10.1007/BF01390229. MR553218

[23] S. Lang, Elliptic functions 2nd ed., Springer Verlag, New York, Berlin, Heidelberg, 1987.

[24] S. Lang, *Fundamentals of differential geometry*, Graduate Texts in Mathematics, vol. 191, Springer-Verlag, New York, 1999. MR1666820

[25] C. O'Sullivan, *Properties of Eisenstein series formed with modular symbols*, J. Reine Angew. Math. **518** (2000), 163–186, DOI 10.1515/crll.2000.003. MR1739405

[26] Y. N. Petridis and M. S. Risager, *Modular symbols have a normal distribution*, Geom. Funct. Anal. **14** (2004), no. 5, 1013–1043, DOI 10.1007/s00039-004-0481-8. MR2105951

[27] A.-M. von Pippich, The arithmetic of elliptic Eisenstein series, PhD thesis, Humboldt-Universität zu Berlin, 2010.

[28] K. Ramachandra, *Some applications of Kronecker's limit formulas*, Ann. of Math. (2) **80** (1964), 104–148, DOI 10.2307/1970494. MR0164950

[29] J.-P. Serre, *A course in arithmetic*, Springer-Verlag, New York-Heidelberg, 1973. Translated from the French; Graduate Texts in Mathematics, No. 7. MR0344216

[30] I. Takada, *Dedekind sums of* $\Gamma(N)$, Japan. J. Math. (N.S.) **12** (1986), no. 2, 401–411, DOI 10.4099/math1924.12.401. MR914303

[31] A. B. Venkov, *Spectral theory of automorphic functions*, Proc. Steklov Inst. Math. **4(153)** (1982), ix+163 pp. (1983). A translation of Trudy Mat. Inst. Steklov. **153** (1981). MR692019

[32] N. Yui and D. Zagier, *On the singular values of Weber modular functions*, Math. Comp. **66** (1997), no. 220, 1645–1662, DOI 10.1090/S0025-5718-97-00854-5. MR1415803

Department of Mathematics, The City College of New York, Convent Avenue at 138th Street, New York, New York 10031

Email address: jjorgenson@mindspring.com

Department of Mathematics, University of Sarajevo, Zmaja od Bosne 35, 71 000 Sarajevo, Bosnia and Herzegovina

Email address: lejlas@pmf.unsa.ba

Contemporary Mathematics
Volume **732**, 2019
https://doi.org/10.1090/conm/732/14792

Equidistribution of shears and their arithmetic applications

Dubi Kelmer

ABSTRACT. These are lecture notes from a talk given in the 3rd EU/US workshop on Automorphic Forms, presenting some arithmetic applications of results in homogenous dynamics. Explicitly, the dynamical result is an effective equidistribution result for shears of cuspidal geodesics. All new results presented here are based on joint work with Alex Kontorovich.

1. Introduction

In many cases, it is possible to encode an arithmetic problem (such as counting integer solutions to algebraic equations and understanding their distribution) by looking at the homogenous space constructed from the group of symmetries of the problem. This process gives rise to interesting dynamical, geometric, and analytic problems involving various group actions on homogenous spaces. It is then insightful to try and understand such problems more generally, forgetting the underlying arithmetic structure and approaching the problem from a purely analytic perspective. By doing this, it is possible to understand which features of the problem come from the arithmetic structure and which hold in greater generality. Moreover, in some cases pursuing this approach can lead to sharper results and better understanding of the original arithmetic problem.

In these notes I will consider one example of this general principle, presenting joint work with Kontorovich [**KK15**] where we show how a good understanding of the distribution of cuspidal geodesics and their shears can lead to interesting arithmetic applications. Before describing the dynamical problem in more detail, we introduce the arithmetic applications.

1.1. Application 1: Moments of GL_2 automorphic L-functions. The L-function $L(f,s)$ corresponding to an automorphic form, f, is an analytic function encoding a lot of its arithmetic information. Examples of GL_1 automorphic L-functions are given by the Riemann zeta function $\zeta(s)$ and Dirichlet L-functions, $L(s,\chi)$, and examples of GL_2-automorphic L-function are given by L-functions corresponding to weight-k holomorphic cusp forms. Explicitly, given such a form, f, with Fourier expansion $f=\sum_{n\geq 1} a_f(n)e^{2\pi i n z}$, the corresponding L-function is

The work presented here was supported by NSF grant DMS-1401747.

given by

$$L(f,s) = \sum_{n\geq 1} \frac{a_f(n)}{n^{s+\frac{k-1}{2}}}. \tag{1.1}$$

Such automorphic L-functions (originally defined by a converging series on the half plane $\mathfrak{Re}(s) > 1$) have an analytic continuation to the complex plane and satisfy a functional equation relating s and $1-s$. Much of the arithmetic information is then hiding in the behavior of these functions on the critical strip $0 < \mathfrak{Re}(s) < 1$, or even the critical line $\mathfrak{Re}(s) = 1/2$.

Our first application is for estimating the second moments along the critical line

$$\int_{-T}^{T} |L(f, \frac{1}{2} + it)|^2 dt.$$

Such moments for GL_2-forms are pretty well understood. In particular, a result of Good [**Goo83**] shows that as $T \to \infty$

$$\int_{-T}^{T} |L(f, \frac{1}{2} + it)|^2 dt = AT\log(T) + BT + O(T^{2/3}\log(T), \tag{1.2}$$

with A, B explicitly given in terms of f. The main point of our results is to give another, soft approach, to estimating these moments that could potentially be generalized to deal with other GL_n-automorphic forms.

1.2. Application 2: Counting quadratic forms. Another arithmetic problem, of a different nature, involves counting integer solutions to algebraic equations. Explicitly we will study the asymptotic behavior for the counting function, $N_d(T)$, counting the number of integral binary quadratic forms $ax^2+bxy+cy^2$ with a fixed discriminant d and bounded coefficients $\|(a,b,c)\| \leq T$.

This problem is amenable to other methods of attack, including classical methods of exponential sums, half-integral weight automorphic forms, Poincaré series and shifted convolutions, or multiple Dirichlet series [**Hoo63, Sar84, Blo08, TT13, HKKL16**]. In particular, [**HKKL16**] show that as $T \to \infty$

$$N_T(d) = T(C_1(d)\log T + C_2(d) + O_d(T^{-5/39})) \tag{1.3}$$

where $C_1(d) \neq 0$ if and only if d is a square.

Applying a dynamical approach to this problem allows us treat similar counting problems having no arithmetic structure, for which other methods do not apply. Moreover, we will see that even for this arithmetic problem, our soft methods can give results that are sharper in some aspects than the ones obtained by classical methods.

REMARK 1.4. Another remarkable consequence of this approach is the observation that the different behavior of $N_T(d)$ when d is a square can be traced back to the different geometry of the corresponding dynamical system.

2. The dynamical problem

After giving some motivation we describe the actual dynamical problem at hand.

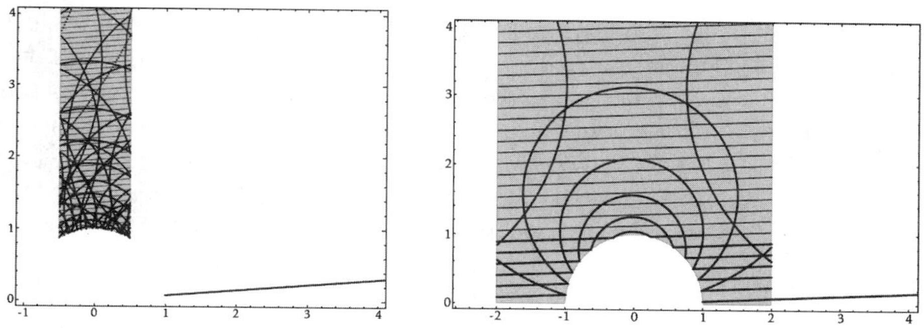

(a) Lattice case: $\Gamma = \mathrm{PSL}_2(\mathbb{Z})$ (b) Thin case: $\Gamma = \langle \left(\begin{smallmatrix} 1 & 4 \\ 0 & 1 \end{smallmatrix}\right), \left(\begin{smallmatrix} 0 & 1 \\ -1 & 0 \end{smallmatrix}\right) \rangle$

FIGURE 1. A shear of the cuspidal geodesic ray

2.1. Shears of cuspidal geodesics. Let $\Gamma \leq \mathrm{SL}_2(\mathbb{R})$ be discrete, geometrically finite, Zariski dense with at least one cusp (that we will assume is at infinity). We sill mostly be interested in the case when Γ is a lattice (i.e., that $\Gamma\backslash G$ has finite measure), however, our method works also for the case when Γ is a thin group (i.e., when $\Gamma\backslash G$ is of infinite measure). Key examples to keep in mind are the lattice given by the modular group $\Gamma = \mathrm{SL}_2(\mathbb{Z})$, and the thin group $\Gamma = \langle \left(\begin{smallmatrix} 1 & 4 \\ 0 & 1 \end{smallmatrix}\right), \left(\begin{smallmatrix} 0 & 1 \\ -1 & 0 \end{smallmatrix}\right) \rangle$.

Let $\mathbb{H} = \{z : \mathfrak{Im}(z) > 0\}$ denote the hyperbolic plane, with the hyperbolic area given by $d\mathrm{vol} = \frac{dxdy}{y^2}$. Recalling that the group Γ acts by isometries, we let $X = \Gamma\backslash\mathbb{H}$ denote corresponding hyperbolic surface, and identify its unit tangent bundle T^1X with the homogenous space $\Gamma\backslash G$. The geodesic flow on T^1X can then be realized as the right action of the diagonal group on $\Gamma\backslash G$, and closed geodesics are closed orbits under this flow. We note that closed geodesics can be either bounded (periodic) orbits, or unbounded orbits shooting out into a cusp. We call the latter cuspidal geodesics. A typical example of a cuspidal geodesic ray is the ray

$$\mathcal{C} = \{iy : y > 1\} \subseteq \Gamma\backslash\mathbb{H},$$

going out to the cusp at infinity (note that this ray is contained in a single fundamental domain for Γ).

Given a closed geodesic or geodesic ray, $\mathcal{C} \subset T^1X$, a shear of $\mathcal{C}$ is any translate by some element $g \in G$ (acting on the right). That is $\mathcal{C}g = \{xg : x \in \mathcal{C}\}$, note that this translate is no longer necessarily a geodesic. For example, a short reflection shows that (the projection to X of) the shears of the cuspidal geodesic rays given above are the rays

$$\mathcal{C}_\theta = \{re^{i\theta} : r > 1\},$$

folded back to a fundamental domain of $\Gamma\backslash\mathbb{H}$ (see Figures 1a 1b).

2.2. Equidistribution. Looking the pictures above, the sheared geodesic $\mathcal{C}_\theta$ for small θ, looks roughly evenly distributed in X. To make this notion more precise, consider the measure μ_θ on $\Gamma\backslash\mathbb{H}$ defined by

$$\mu_\theta(\psi) = \int_1^\infty \psi(re^{i\theta})\frac{dr}{r},$$

and study the weak limit of μ_θ as $\theta \to 0$.

Let μ denote the probability measure given by $d\mu = \frac{1}{\mathrm{vol}(X)}\frac{dxdy}{y^2}$. We say that the shears $\mathcal{C}_\theta$ become equidistributed as $\theta \to 0$ if, after multiplying by an appropriate normalizing factor c_θ, the measures $c_\theta\mu_\theta \to \mu$ (weakly). Since the measure μ_θ is an infinite measures it is not a priory obvious what the correct normalization c_θ should be. Nevertheless, it turns out that the correct normalization is $c_\theta = \frac{1}{\log(\cot(\theta))}$, and indeed, the results of Oh and Shah [**OS14**] imply that $c_\theta\mu_\theta \to \mu$ as $\theta \to 0$. More precisely, they show that for any smooth test functions $\psi \in C_c^\infty(\Gamma\backslash\mathbb{H})$

$$\mu_\theta(\psi) = c_\theta^{-1}\mu(\psi) + O_\psi(1). \tag{2.1}$$

One of the points of difficulty for this problem is coming from the fact that μ_θ is an infinite measure. For the analogous problem of shears of periodic geodesics, the corresponding measures are finite and can be normalized to be probability measures. In this setting the classical results of Duke Rudnick and Sarnak [**DRS93**] and Eskin and McMullen [**EM93**] show (in much greater generality) that the shears of periodic geodesics become equidistributed and with an effective rate. That is, for shears $\mathcal{C}g$ of a periodic geodesic $\mathcal{C}$, let μ_g denote the probability measure obtained by integrating along $\mathcal{C}g$ then there is some $\eta > 0$ such that

$$\mu_g(\psi) = \mu(\psi) + O_\psi(\|g\|^{-\eta}),$$

for an appropriate norm $\|g\|$.

Having such effective bounds for the error is crucial for many applications. In particular, for the application discussed in the introduction we will need such bounds for shears of cuspidal geodesics. However, in this case it turns out the rate $O_\psi(1)$ in (2.1) is actually sharp, and hence, in order to get an effective power saving estimate for the error one needs to first identify a secondary main term. In recent joint work with Kontorovich we were able to identify the appropriate secondary term as a certain Eisenstein measure we now describe.

2.3. The Eisenstein measure. Recall that the Eisenstein series (corresponding to a cusp at infinity of a lattice Γ) is defined by the series

$$E(s,z) = \sum_{\Gamma_\infty\backslash\Gamma} \mathfrak{Im}(\gamma z)^s,$$

converging for $\mathfrak{Re}(s) > 1$ and has an analytic continuation with a pole at $s = 1$ with residue $\frac{1}{v_\Gamma}$ with $v_\Gamma = \mathrm{vol}(\Gamma\backslash\mathbb{H})$. After subtracting the pole, the regularized function $\widetilde{E}(s,z) = E(s,z) - \frac{1}{v_\Gamma(s-1)}$ is analytic at $s = 1$ and we define the regularized Eisenstein measure as the infinite measure given by

$$\mu_{\mathrm{Eis}}(\psi) = \left\langle \psi, \widetilde{E}(1,\cdot)\right\rangle_{\Gamma\backslash\mathbb{H}}.$$

When Γ is a thin group the series

$$E(s,z) = \sum_{\Gamma_\infty\backslash\Gamma} \mathfrak{Im}(\gamma z)^s,$$

converges at $s = 1$ and the Eisenstein measure is simply given by the inner product

$$\mu_{\mathrm{Eis}}(\psi) = \langle \psi, E(1,\cdot)\rangle_{\Gamma\backslash\mathbb{H}}.$$

REMARK 2.2. For the special case of $\Gamma = \mathrm{SL}_2(\mathbb{Z})$ the Kronecker limit formula takes the explicit form

$$\widetilde{E}(1,z) = \frac{3}{\pi}\left(2\gamma - 2\frac{\zeta'}{\zeta}(2) - \log(4y|\eta(z)|^4)\right),$$

where $\gamma = 0.577\cdots$ is Euler's constant, $\zeta(s)$ is the Riemann zeta function, and $\eta(z)$ is the Dedekind eta function. There are similar explicit formulas also for some congruence groups, but for a general lattice the Eisenstein series and hence also the Eisenstein measures are more mysterious.

2.4. Effective equidistribution. The following is a special case of [**KK15**, Theorem 1.3] for the specific case of shears of the cuspidal geodesics $\mathcal{C}_\theta$ described above. To simplify notation we denote by $T = \cot(\theta)$ so that $c_\theta = \frac{1}{\log(T)}$.

THEOREM 2.3. *(*[**KK15**, Theorem 1.3]*) There is $\eta > 0$ depending on the first non zero Laplace eigenvalue in $L^2(\Gamma\backslash\mathbb{H})$ such that for any $\psi \in C_c^\infty(\Gamma\backslash\mathbb{H})$*

(1) *For Γ a lattice*

$$\mu_\theta(\psi) = \log(T)\mu(\psi) + \mu_{\mathrm{Eis}}(\psi) + O_\psi(T^{-\eta}),$$

(2) *for Γ a thin group*

$$\mu_\theta(\psi) = \mu_{\mathrm{Eis}}(\psi) + O_\psi(T^{-\eta})$$

REMARK 2.4. The explicit value of η obtained by this method can be computed explicitly, and it is an interesting problem to try and get the optimal value, however, for many applications the important point is that there is some explicit power saving for the error.

REMARK 2.5. It is worth noting that, while in the lattice case the Eisenstein term appears as a secondary term, for thin groups it is actually the main term so that in this case the measures $\mu_\theta \to \mu_{\mathrm{Eis}}$ as $\theta \to 0$.

We will not discuss the proof of these results here and refer to [**KK15**] for details. Instead, we show how to use these effective equidistribution results to attack the two arithmetic problems described in the introduction.

3. Moments of GL_2 automorphic L-functions

The L-function $L(f,s)$ given in (1.1) (multiplied by some Γ-factors) can be obtained as a Melin transform of f. More generally, the two are related by the following modified Hecke integral: for any $T > 0$

$$\int_0^\infty f(Ty+iy)y^{s+\frac{k-1}{2}}\frac{dy}{y} = L(f,s)\mathcal{W}_k(s,T),$$

where $\mathcal{W}_k(s,T)$ is a twisted Γ-factor

$$\mathcal{W}_k(s,T) = (2\pi)^{-(s+\frac{k-1}{2})}\Gamma(s+\frac{k-1}{2})(1-iT)^{-(s+\frac{k-1}{2})}.$$

By using this formula for $s = \frac{1}{2}+it$ and applying Parseval we get the identity

$$\int_0^\infty |f(Ty+iy)|^2y^k\frac{dy}{y} = \frac{1}{2\pi}\int_{\mathbb{R}}|L(f,\tfrac{1}{2}+it)|^2|\mathcal{W}_k(\tfrac{1}{2}+it,T)|^2dy.$$

For a weight-k modular form f the function $\psi = |f|^2y^k$ is Γ invariant and the left hand side is precisely $\mu_\theta(\psi)$, with $T = \cot(\theta)$. The right hand side on the other hand is a good approximation for the second moment, in the sense that the function $|\mathcal{W}_k(\frac{1}{2}+it,T)|^2$ has the bulk of its mass for $-T<t<T$ and decays exponentially for $t\geq T$. Using this observation together with the results on effective equidistribution, specializing to the arithmetic case of $\Gamma=\mathrm{SL}_2(\mathbb{Z})$, gives the following formula for the (smoothed) second moment

THEOREM 3.1. *([**KK15**, Theorem 1.17]) There is some $\eta>0$ such that for any weight k holomorphic Hecke form, f, as $T\to\infty$*

$$\frac{1}{2\pi}\int_{\mathbb{R}}|L(f,\tfrac{1}{2}+it)|^2F_{k,T}(t)dt=$$
$$2T\frac{\|f\|^2}{v_\Gamma}\left(\log T+\frac{\Lambda'}{\Lambda}(\mathrm{sym}^2f,1)+\gamma-2\frac{\zeta'}{\zeta}(2)+O_f(T^{-\eta})\right),$$

where $\Lambda(\mathrm{sym}^2f,s)$ is the completed symmetric-square L-function and $F_{k,T}(t)=c_k(\frac{t}{T})^ke^{-t/T}$ is a smoothed step function.

REMARK 3.2. Using the same method it is possible to obtain similar results also for a general lattice, however, without the arithmetic information it is not possible to identify the secondary term as the special value of the symmetric-square L-function.

REMARK 3.3. So far we were not able to remove the smoothing to get a power saving error term with a sharp cutoff. Doing so will give another soft proof of Good's result (1.2), leading to sub convex bounds in the t-aspect.

4. Counting orbit points on affine quadrics

For our second application (counting quadratic forms with a fixed discriminant) the results are more satisfying. We note that this is a special case of the more general problem of counting orbit points in affine quadrics, we now describe.

Given a ternary indefinite binary quadratic form Q (e.g. $Q(a,b,c)=b^2-4ac$) and $d\in\mathbb{R}$ let $V_{Q,d}$ denote the affine quadric $Q(x)=d$. The quadric $V_{Q,d}$ is either a one sheeted hyperboloid (if $d<0$), a two sheeted hyperboloid (when $d>0$) or a cone (when $d=0$). Let $G=\mathrm{SO}(Q)$ denote the group of symmetries, let $\Gamma\leq G$ be a discrete subgroup. Given a real a point $x_0\in V_{Q,d}(\mathbb{R})$ with a discrete orbit $\mathcal{O}=x_0\Gamma$ our goal is to count orbit points in an Archimedean ball $\mathcal{N}_{\mathcal{O}}(T)=\#B_T\cap\mathcal{O}$ with

$$B_T=\{x\in\mathbb{R}^3:\|x\|\leq T\}$$

(for an appropriate norm $\|\cdot\|$ on $\mathbb{R}^3$).

4.1. Relation to equidistribution. In order to relate this counting problem to equidistribution, we first need to reformulate it as a lattice point counting problem. Given the orbit $\mathcal{O}=x_0\Gamma$ let $H=\mathrm{Stab}_{x_0}(G)$ denote the stabilizer of x_0 in G and let $\Gamma_H=\Gamma\cap H$. We can then naturally identify the quadric $V_{Q,d}$ with the quotient $H\backslash G$ and the Archimedean norm $\|\cdot\|$ on $\mathbb{R}^3$ restricted to $V_{Q,d}$ gives a corresponding norm on $H\backslash G$. Let $\tilde{B}_T\subseteq H\backslash G$ denote the corresponding norm balls, so that $\mathcal{N}_{\mathcal{O}}(T)=\#(\Gamma_H\backslash\Gamma)\cap\tilde{B}_T$.

This lattice point counting problem is intimately related to behavior of the closed orbit $\Gamma_H\backslash H\subseteq\Gamma\backslash G$ and the distribution of its translates $\Gamma_H\backslash Hg$ as $\|g\|\to\infty$ (in $H\backslash G$). This relation goes back to the work of Duke Rudnick and Sarnak

[**DRS93**] and Eskin and McMullen [**EM93**], showing that, assuming (i) $\mu(\Gamma\backslash G) < \infty$ and (ii) $\mu_H(\Gamma_H\backslash H) < \infty$ implies that the translates $\Gamma_H\backslash Hg$ become equidistributed in $\Gamma\backslash G$ (with an effective rate) and as a consequence, that there is $\eta > 0$ such that as $T \to \infty$

$$\mathcal{N}_{\mathcal{O}}(T) = \frac{\mu(\Gamma_H\backslash H)}{\mu(\Gamma\backslash G)}T + O(T^{1-\eta}).$$

Using the new results on shears of cuspidal geodesics allows us to generalize this method and obtain similar results also for cases where the assumptions (i) and (ii) are not satisfied. In particular, we get the following.

THEOREM 4.1. *([**KK15**, Theorem 1.24]) There are constants C_1, C_2, and $\eta > 0$ so that as $T \to \infty$:*
For Γ a lattice

$$\mathcal{N}_{\mathcal{O}}(T) = \big(C_1 T\log T + C_2 T\big)\big(1 + O(T^{-\eta})\big), \tag{4.2}$$

and for Γ thin:

$$\mathcal{N}_{\mathcal{O}}(T) = \big(C_1 T + C_2 T^{\delta}\big)\big(1 + O(T^{-\eta})\big), \tag{4.3}$$

where the constants C_1, C_2 can be determined explicitly in terms of volumes, special values of Eisenstein series, and Patterson-Sullivan measures.

REMARK 4.4. For thin groups all previously known cases where ones with $C_1 = 0$. For the lattice case, similar results for cases with $C_1 \neq 0$ were also obtained by Oh and Shah in an unpublished manuscript.

4.2. Counting quadratic forms. We now go back to the problem of counting quadratic forms discussed in the introduction. Specializing to the form $Q(a,b,c) = b^2 - 4ac$, let $N_d(T)$ denote the number of primitive[1] points in $V_{Q,d}(\mathbb{Z})$ (i.e., the points with co-prime entries). The action of $\Gamma = \mathrm{SL}_2(\mathbb{Z})$ on binary quadratic forms preserves the discriminant and hence gives a natural action on $V_{Q,d}(\mathbb{Z})$, splitting it into $h(d)$ disjoint orbits[2]

$$V_{Q,d} = \bigcup_{j=1}^{h(d)} x_j\Gamma$$

We can thus reduce the problem of estimating $N_d(T)$ to the orbit counting problem $\mathcal{N}_{x_j\Gamma}(T)$ for each of the $h(d)$ orbits. We note that for $d > 0$ the stabilizer $H = \mathrm{Stab}_{x_j}(G)$ is conjugate to the diagonal group so $\Gamma_H\backslash H$ is a closed geodesic. Moreover, this closed geodesic is periodic if Γ_H is non-trivial and cuspidal otherwise.

REMARK 4.5. The condition that Γ_H is non-trivial is equivalent to the existence of integer solutions to a certain Pellian equation, and hence one can show that the closed geodesic is cuspidal exactly when d is a perfect square. It is this geometric feature of the dynamical problem that causes the difference in the asymptotics between the two cases.

[1] It is not hard to deduce the formula for counting non-primitive points from results on primitive points by summing over all divisors of d

[2] When d is a square free fundamental discriminant, the number of orbits $h(d)$ is the class number of the quadratic field $Q(\sqrt{d})$ which changes wildly with d. On the other hand when d is a perfect square $h(d)$, can be computed explicitly.

Applying the general results (4.2) to this case recovers the the result (1.3) from[**HKKL16**] (with perhaps a different exponent). Moreover, by using the extra structure available for $\Gamma = \mathrm{SL}_2(\mathbb{Z})$ (such as an optimal bounds on the spectral gap, good bounds on decay of Fourier coefficients of Maass forms and an understanding of the behavior of $h(d)$), in [**KK17**] we use this method to get the following sharper result.

THEOREM 4.6. *Let $\theta = 7/64$ be the bound towards the Ramanujan Conjecture (that $\theta = 0$). For any $\eta < \frac{3}{40+72\theta}$ and $\beta > \frac{3}{2} + 2\theta$ for any square d and $T > d^\beta$*

$$N_d(T) = \frac{\sqrt{72}\phi(d)}{\pi d} T\Big(\log(T) + C + O((\tfrac{T}{d^\beta})^{-\eta})\Big)$$

where $\phi(d)$ is Euler's function and $C = 2\gamma - 1 - 2\frac{\zeta'}{\zeta}(2) - \log(\frac{\Gamma(1/4)^4}{2\sqrt{2\pi^3}})$ is an explicit computable constant, independent of d.

REMARK 4.7. A nice feature of this result is that the bounds are uniform in both T and d. It is an interesting problem to obtain such uniform bunds that apply for larger values of d going as large as T.

References

[Blo08] Valentin Blomer, *Sums of Hecke eigenvalues over values of quadratic polynomials*, Int. Math. Res. Not. IMRN **16** (2008), Art. ID rnn059. 29, DOI 10.1093/imrn/rnn059. MR2435749

[DRS93] W. Duke, Z. Rudnick, and P. Sarnak, *Density of integer points on affine homogeneous varieties*, Duke Math. J. **71** (1993), no. 1, 143–179, DOI 10.1215/S0012-7094-93-07107-4. MR1230289

[EM93] Alex Eskin and Curt McMullen, *Mixing, counting, and equidistribution in Lie groups*, Duke Math. J. **71** (1993), no. 1, 181–209, DOI 10.1215/S0012-7094-93-07108-6. MR1230290

[Goo83] Anton Good, *On various means involving the Fourier coefficients of cusp forms*, Math. Z. **183** (1983), no. 1, 95–129, DOI 10.1007/BF01187218. MR701361

[HKKL16] Thomas A. Hulse, Chan Ieong Kuan, Eren Mehmet Kıral, and Li-Mei Lim, *Counting square discriminants*, J. Number Theory **162** (2016), 255–274, DOI 10.1016/j.jnt.2015.10.015. MR3448270

[Hoo63] Christopher Hooley, *On the number of divisors of a quadratic polynomial*, Acta Math. **110** (1963), 97–114, DOI 10.1007/BF02391856. MR0153648

[KK15] Dubi Kelmer and Alex Kontorovich, *Effective equidistribution of shears and applications*, Math. Ann. **370** (2018), no. 1-2, 381–421, DOI 10.1007/s00208-017-1580-9. MR3747491

[KK17] D. Kelmer and A. Kontorovich. Exponents for the equidistribution of shears and applications, 2018, preprint.

[OS14] Hee Oh and Nimish A. Shah, *Limits of translates of divergent geodesics and integral points on one-sheeted hyperboloids*, Israel J. Math. **199** (2014), no. 2, 915–931, DOI 10.1007/s11856-013-0063-2. MR3219562

[Sar84] Peter Sarnak, *Additive number theory and Maass forms*, Number theory (New York, 1982), Lecture Notes in Math., vol. 1052, Springer, Berlin, 1984, pp. 286–309, DOI 10.1007/BFb0071548. MR750670

[TT13] Nicolas Templier and Jacob Tsimerman, *Non-split sums of coefficients of GL(2)-automorphic forms*, Israel J. Math. **195** (2013), no. 2, 677–723, DOI 10.1007/s11856-012-0112-2. MR3096570

DEPARTMENT OF MATHEMATICS, BOSTON COLLEGE, CHESTNUT HILL, MASSACHUSETTS 02467
Email address: `kelmer@bc.edu`

Contemporary Mathematics
Volume **732**, 2019
https://doi.org/10.1090/conm/732/14790

Fake proofs for identities involving products of Eisenstein series

Kamal Khuri-Makdisi

ABSTRACT. In the workshop of the July 2016 Building Bridges 3 conference in Sarajevo, I presented the results from a joint article with W. Raji (Mathematische Annalen, 2017). That article gave a proof of various linear relations between products of two Eisenstein series on $\Gamma(N)$, including an interesting identity related to the action of a Hecke operator on such a product. The real proofs involve some care to deal with issues of convergence. In this note we give "fake" proofs for these identities, ignoring the convergence issues; some of these fake proofs appeared in the workshop lecture as an amusing side note before I sketched the real proofs. Something in these fake proofs is quite suggestive, even though the proofs themselves are clearly invalid (and even produce wrong results). It would be interesting to understand what exactly is going on here.

1. Introduction

The basic object of study in this note is the Eisenstein series of weight $\ell \geq 1$ on $\Gamma(N)$, with parameter $\lambda \in N^{-1}\mathbf{Z}^2/\mathbf{Z}^2 \subset \mathbf{Q}^2/\mathbf{Z}^2$:

$$E_{\ell,\lambda}(z) = \sum_{\substack{(a,b)\equiv\lambda \pmod{\mathbf{Z}^2} \\ (a,b)\neq(0,0)}} (az+b)^{-\ell}. \tag{1.1}$$

The above is not quite right if $\ell \in \{1,2\}$, as the above series does not converge. In that case, one evaluates the sum, following Hecke, by replacing $(az+b)^{-\ell}$ by $(az+b)^{-\ell}|az+b|^{-s}$ for a complex parameter s, and then setting $s=0$ after one has analytically continued the resulting sum in s. In the convergent case, when $\ell \geq 3$, the Eisenstein series is a holomorphic function of z, and this holomorphy fortuitously still holds for $\ell = 1$; however, for $\ell = 2$, Hecke's summation procedure yields that $E_{2,\lambda}(z)$ is the sum of a nonholomorphic expression $-\pi/\mathrm{Im}\, z$ (which is the same for all λ) with a holomorphic function (which of course depends on λ).

The product $E_{\ell,\lambda}E_{m,\mu}$ of two of these Eisenstein series is then a form of weight $\ell+m$ on $\Gamma(N)$. The articles [**BG01, BG03**] prove a number of linear relations between such products — more precisely, they show that certain linear combinations of such products belong to the space $Eis_{\ell+m}$ of Eisenstein series of weight $\ell+m$. These linear relations modulo $Eis_{\ell+m}$ have a structure that is reminiscent of the Manin relations between periods of cusp forms; this was further

2010 *Mathematics Subject Classification.* Primary 11F11, 33E05.

codified in [**Paş06**]. For example, in weight 2, we have the following relations:

$$\begin{aligned} &E_{1,\lambda}E_{1,\mu} + E_{1,\mu}E_{1,-\lambda} = 0, \\ &E_{1,\lambda}E_{1,\mu} + E_{1,\mu}E_{1,-\lambda-\mu} + E_{1,-\lambda-\mu}E_{1,\lambda} \equiv 0 \pmod{Eis_2}. \end{aligned} \tag{1.2}$$

The first identity, which is very simple, is analogous to the two-term Manin relation involving $\left(\begin{smallmatrix} 0 & -1 \\ 1 & 0 \end{smallmatrix}\right)$, while the second, more interesting, identity, is analogous to the three-term Manin relation involving $\left(\begin{smallmatrix} 0 & 1 \\ -1 & -1 \end{smallmatrix}\right)$. The explanation for this parallelism can be found in [**KMR17**], where we show that the Petersson inner product of any cusp form $f \in \mathcal{S}_2$ with $E_{1,\lambda}E_{1,\mu} + E_{1,\mu}E_{1,-\lambda-\mu} + E_{1,-\lambda-\mu}E_{1,\lambda}$ expands to a combination of periods of f and its transforms $f|_2 M$ for certain $M \in GL_2^+(\mathbf{Q})$, and this combination of periods vanishes precisely by the Manin relations. The proof in that article, which works for a similar identity in arbitrary weight, involves carrying throughout the parameter s in the Eisenstein series $E(z,s)$, and controlling the analysis fairly carefully. Our goal in this note is to provide fake proofs of that and other results from [**KMR17**], relying on intriguing identities between rational functions, but with no attention paid to convergence. It would be interesting to find the connection between these intriguing identities and the structure of the Manin relations, and to see what parts of the fake proofs can be salvaged.

Fake proof of the second identity in (1.2). Let us write $\nu = -\lambda - \mu$; hence we can assume given a triple (λ, μ, ν) for which $\lambda + \mu + \nu = (0,0)$. Similarly, consider the set $T = T_{(\lambda,\mu,\nu)}$ of all triples $((a,b),(c,d),(e,f)) \in (\mathbf{Q}^2)^3$ with

$$\begin{aligned} (a,b) &\equiv \lambda \pmod{\mathbf{Z}^2}, \\ (c,d) &\equiv \mu \pmod{\mathbf{Z}^2}, \\ (e,f) &\equiv \nu \pmod{\mathbf{Z}^2}, \\ (a,b) + (c,d) &+ (e,f) = (0,0), \\ \text{None of } (a,b), (c,d),& \text{ or } (e,f) \text{ equals } (0,0). \end{aligned} \tag{1.3}$$

(The set T is nonempty precisely because $\lambda + \mu + \nu = (0,0)$, and the last condition of nonvanishing only matters if one of λ, μ, ν is zero in $\mathbf{Q}^2/\mathbf{Z}^2$.) Then, ignoring all issues of convergence, we formally have

$$E_{1,\lambda}E_{1,\mu} \equiv \sum_{((a,b),(c,d),(e,f)) \in T} \frac{1}{az+b} \cdot \frac{1}{cz+d} \pmod{Eis_2}. \tag{1.4}$$

The reason is that once one chooses (a,b) and (c,d) arbitrarily in the congruence classes of λ and μ, respectively, the pair $(e,f) = -(a,b) - (c,d)$ is uniquely determined. In the event that $(e,f) = (0,0)$, which anyhow only occurs when $\lambda = -\mu$, the pairs that we omit are those with $(c,d) = (-a,-b)$, which corresponds to being off by $-E_{2,\lambda}$, which (ignoring its nonholomorphy) "is" an element of Eis_2.

We obtain similar expressions (modulo Eis_2) for $E_{1,\mu}E_{1,\nu}$ and for $E_{1,\nu}E_{1,\lambda}$. Thus, working modulo Eis_2, we obtain

$$\begin{aligned} &E_{1,\lambda}E_{1,\mu} + E_{1,\mu}E_{1,\nu} + E_{1,\nu}E_{1,\lambda} \\ \equiv &\sum_{((a,b),(c,d),(e,f)) \in T} \left[\frac{1}{(az+b)(cz+d)} + \frac{1}{(cz+d)(ez+f)} + \frac{1}{(ez+f)(az+b)} \right], \end{aligned} \tag{1.5}$$

and this last sum vanishes, thanks to the identity

$$p+q+r=0 \quad\Longrightarrow\quad \frac{1}{pq}+\frac{1}{qr}+\frac{1}{rp}=0, \tag{1.6}$$

with $p = az+b$, $q = cz+d$, and $r = ez+f$. □

It is not apparent to me how to salvage the above fake proof, for instance by summing over elements of T in a particular order so as to obtain convergence, in the style of Eisenstein [**Wei99**]. Indeed, when λ, μ, ν are all nonzero, then our fake proof would imply that $E_{1,\lambda}E_{1,\mu} + E_{1,\mu}E_{1,\nu} + E_{1,\nu}E_{1,\lambda}$ is actually zero, which is not the case; its expression as an explicit weight 2 Eisenstein series is known.

2. The identity in higher weight

We now turn to the case of general weight $k \geq 2$. The analog of the two-term Manin relation is again simple, and can be found in equation (2.23) of [**KMR17**]. The interesting three-term relation in higher weight amounts to the following.

PROPOSITION 2.1 (Theorem 2.8 of [**KMR17**]). *Let λ, μ, ν satisfy $\lambda+\mu+\nu = 0$, as before. Let $\alpha, \beta, \gamma \in \mathbf{C}$ satisfy $\alpha+\beta+\gamma = 0$ (these can be thought of as formal variables). Let $k \geq 2$. Then the following expression is orthogonal to all cusp forms $f \in \mathcal{S}_k$:*

$$\sum_{\substack{\ell+m=k\\ \ell,m\geq 1}} \alpha^{\ell-1}\beta^{m-1}E_{\ell,\lambda}E_{m,\mu} + \sum_{\substack{\ell+m=k\\ \ell,m\geq 1}} \beta^{\ell-1}\gamma^{m-1}E_{\ell,\mu}E_{m,\nu} + \sum_{\substack{\ell+m=k\\ \ell,m\geq 1}} \gamma^{\ell-1}\alpha^{m-1}E_{\ell,\nu}E_{m,\lambda}. \tag{2.1}$$

Morally speaking, this means that the expression (2.1) should belong to Eis_k, but the presence of nonholomorphic E_2 terms complicates the statement somewhat.

FAKE PROOF. We again ignore all issues of convergence, and work modulo "Eis_k" (ignoring the nonholomorphy coming from any E_2). Analogously to (1.4), we formally write the first term $\sum_{\ell+m=k} \alpha^{\ell-1}\beta^{m-1}E_{\ell,\lambda}(z)E_{m,\mu}(z)$ as the following sum over $((a,b),(c,d),(e,f)) \in T$ and $\ell, m \geq 1$ with $\ell+m = k$:

$$\sum_{((a,b),(c,d),(e,f))}\sum_{\ell,m} \frac{\alpha^{\ell-1}\beta^{m-1}}{(az+b)^{\ell}(cz+d)^{m}} = \sum_{((a,b),(c,d),(e,f))}\sum_{\ell,m} \frac{\alpha^{\ell-1}\beta^{m-1}}{p^{\ell}q^{m}}, \tag{2.2}$$

using the notation of p, q, r as in (1.6) and the sentence that follows it. Some manipulation with the finite geometric series over ℓ, m then gives us the following congruence modulo "Eis_k":

$$\sum_{\substack{\ell+m=k\\ \ell,m\geq 1}} \alpha^{\ell-1}\beta^{m-1}E_{\ell,\lambda}E_{m,\mu} \equiv \sum_{((a,b),(c,d),(e,f))\in T} \frac{\left(\frac{\alpha}{p}\right)^{k-1} - \left(\frac{\beta}{q}\right)^{k-1}}{\alpha q - \beta p}. \tag{2.3}$$

Taking simultaneous cyclic permutations of (α, β, γ) and (p, q, r), we obtain similar (purely formal) expressions for the second and third terms of (2.1). But now a miracle occurs: the identities $\alpha+\beta+\gamma = 0$ and $p+q+r = 0$ imply that

$$\alpha q - \beta p = \beta r - \gamma q = \gamma p - \alpha r. \tag{2.4}$$

(An amusing way to avoid verifying the above fact algebraically is to stick to real variables, and use the cross product in $\mathbf{R}^3$: in that case, the vectors (α, β, γ) and (p, q, r) both lie in the plane orthogonal to $(1, 1, 1)$, so their cross product is parallel to $(1,1,1)$, which gives precisely the equalities in (2.4).) Now adding up the cyclic permutations of the expressions on the right hand side of (2.3) gives a sum of the cyclic permutations of $\left(\frac{\alpha}{p}\right)^{k-1} - \left(\frac{\beta}{q}\right)^{k-1}$ over the **same** common denominator, and this immediately gives the desired sum of zero! $\square$

The identities between rational functions that we have used in the above two fake proofs, namely $1/pq + 1/qr + 1/rp = 0$ and the analogous cyclic sum involving also α, β, γ in higher weight, date back to Eisenstein; a good reference for this is Chapter II, section 2 and Chapter IV, section 1 of [**Wei99**] (but note that Weil uses $r = p + q$, whereas we use $r = -p - q$). The identities there, which are proved by a partial fraction decomposition and/or successive differentiation, may look more complicated than ours, particularly since they involve various sums and binomial coefficients. The relation to the identities we used above with auxiliary variables α, β, γ are however straightforward: write $\gamma = -\alpha - \beta$ in our identities, expand everything into a polynomial in α and β, then equate the coefficients of the same monomial $\alpha^\ell \beta^m$ on both sides to obtain the identities in [**Wei99**]. However, those identities, just like ours, always end up involving some terms with only a first power or square of p (or q, or r) in the denominator. This appears to make the convergence difficult to control, even if one sums both sides in Eisenstein style, with a sum $\sum_{(m,n)\in\mathbf{Z}^2}$ being carried out as $\lim_{M\to\infty}\sum_{m=-M}^{M} \lim_{N\to\infty}\sum_{n=-N}^{N}$. For that reason, the treatment in Chapter IV of [**Wei99**], following Eisenstein, proceeds by a somewhat different route.

3. An example related to Hecke operators

Another result of interest relates to the trace from $\Gamma(NM)$ to a lower level $\Gamma(M)$ of certain products of Eisenstein series. The basic expression for which we derive an identity in [**KMR17**] is a sum of the form $\sum_{\tau\in N^{-1}\mathbf{Z}^2/\mathbf{Z}^2} E_{\ell,\lambda+\tau} E_{m,\mu-S\tau}$, as explained in Section 4 of that article. Here $\lambda, \mu \in M^{-1}\mathbf{Z}^2/\mathbf{Z}^2$, and $S \in \mathbf{Z}$. The variable $\tau \in N^{-1}\mathbf{Z}^2/\mathbf{Z}^2$ in the sum can be thought of as a sum over all N-torsion points of the elliptic curve given analytically as $\mathbf{C}/(\mathbf{Z}z + \mathbf{Z})$. In this note, we will only deal with one example, but the result holds generally, as does the fake proof, in whatever sense fake proofs can be said to hold. The combinatorics are again reminiscent of the combinatorics one obtains when one computes Hecke operators on spaces of modular symbols, and involve sublattices of $\mathbf{Z}^2$ and the convex hull of the lattice points in the first quadrant; the references for this are Theorem 3.16 of [**BG01**] and its proof, Lemma 7.3 of [**BG03**], and Subsection 2.3 and Section 3 of [**Mer94**]. Here we will just illustrate these phenomena for the case $N = 5$ and $S = 3$, and make the connection with fake proofs based on interesting identities of rational functions.

We thus fix $\lambda, \mu \in \mathbf{Q}^2$ (where usually only their image in $\mathbf{Q}^2/\mathbf{Z}^2$ matters). The exact level of λ, μ, i.e., their denominator, M is immaterial, since the formulas we obtain are insensitive to M. We then consider just the following identity, which is a special case of Proposition 4.1 of [**KMR17**] (and the notation $L_{\lambda,\mu,\alpha,\beta}$ is taken from there as well).

PROPOSITION 3.1. *Write* $L_{\lambda,\mu,\alpha,\beta} = \sum_{\substack{\ell+m=k\\ \ell,m\geq 1}} \alpha^{\ell-1}\beta^{m-1} E_{\ell,\lambda} E_{m,\mu}$ *for the expression in weight* k *that has appeared repeatedly in Section* 2. *Then, modulo "*Eis_k*"*

as usual, we have

$$(3.1)\qquad \begin{aligned}(1/5)&\sum_{\tau\in 5^{-1}\mathbf{Z}^2/\mathbf{Z}^2} L_{\lambda+\tau,\mu-3\tau,\alpha,\beta}\\ &\equiv L_{5\lambda,3\lambda+\mu,5\alpha,3\alpha+\beta}+L_{3\lambda+\mu,\lambda+2\mu,3\alpha+\beta,\alpha+2\beta}+L_{\lambda+2\mu,5\mu,\alpha+2\beta,5\beta}.\end{aligned}$$

REMARK 3.2. The vectors $(5,0)$, $(3,1)$, $(1,2)$, and $(0,5)$ in sequence are obtained by taking the convex hull of the nonzero points in the first quadrant of the sublattice $\{(x,y)\mid x-3y\equiv 0 \pmod 5\}$ of $\mathbf{Z}^2$, as in the figure below. Any pair of consecutive vectors has a determinant of 5, the index. This is described further in the references mentioned above.

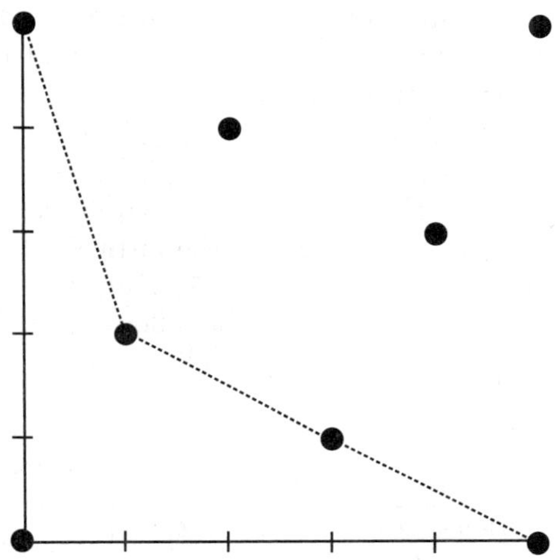

FAKE PROOF OF PROPOSITION 3.1. It is clearer if we first restrict to $k=2$, in which case everything related to α,β can be omitted, and $L_{\lambda,\mu,\text{anything}}=E_{1,\lambda}E_{1,\mu}$. We formally expand $\sum_\tau E_{1,\lambda+\tau}E_{1,\mu-3\tau}$ using as usual $v=(a,b)\equiv\lambda+\tau$ and $w=(c,d)\equiv\mu-3\tau \pmod{\mathbf{Z}^2}$, taking into account also the sum over τ. Write as usual $p=az+b$ and $q=cz+d$; these depend linearly on v and w. Our left hand side is thus the sum of all terms $1/(5pq)$, as $(v,w)=(\lambda+v',\mu+w')\in\mathbf{Q}^4$ ranges over all possible shifts of (λ,μ) by the lattice $\Lambda=\{(v',w')\in 5^{-1}\mathbf{Z}^4\mid 3v'+w'\in\mathbf{Z}^4\}$.

For each two consecutive vectors in the list $\{(5,0),(3,1),(1,2),(0,5)\}$, say for example the vectors $(3,1)$ and $(1,2)$, one can see that as (v',w') varies over Λ, the resulting combination $(3v'+w',v'+2w')$ (made using the coefficients of the two consecutive vectors) varies precisely over all of $\mathbf{Z}^4$. It follows that the pair of values $(3p+q,p+2q)$ varies over the terms in such a way that the (nonconvergent, as always) sum of all the products $(3p+q)^{-1}(p+2q)^{-1}$ yields $E_{1,3\lambda+\mu}E_{1,\lambda+2\mu}$. It now remains to make use of the hopefully impressive identity

$$(3.2)\qquad \frac{1}{5pq}=\frac{1}{(5p)(3p+q)}+\frac{1}{(3p+q)(p+2q)}+\frac{1}{(p+2q)(5q)}$$

to conclude the fake proof for $k=2$. In all this, we have blithely ignored the fact that in all our sums, we omitted any terms that look like $1/0$, which may have introduced correction terms that with luck will belong to Eis_2; as mentioned at the end of Section 1, however, the issues with convergence seem to produce further unavoidable corrections from Eis_2, even if the above formal argument has not omitted any terms (e.g., if all of $5\lambda, 3\lambda+\mu, \lambda+2\mu, 5\mu$ are nonzero in $\mathbf{Q}^2/\mathbf{Z}^2$).

We note that the identity (3.2) may become more apparent if we observe that

$$
\begin{aligned}
\frac{1}{(5p)(3p+q)} &= \frac{1/q}{5p} - \frac{3/(5q)}{3p+q}, \\
\frac{1}{(3p+q)(p+2q)} &= \frac{3/(5q)}{3p+q} - \frac{1/(5q)}{p+2q}, \\
\frac{1}{(p+2q)(5q)} &= \frac{1/(5q)}{p+2q} - \frac{0}{5q}, \\
\text{generally, } \frac{1}{(ap+bq)(cp+dq)} &= \frac{a/((ad-bc)q)}{(ap+bq)} - \frac{c/((ad-bc)q)}{(cp+dq)}.
\end{aligned}
\tag{3.3}
$$

Thus when we add the terms coming from each consecutive pair of vectors in the list $\{(N,0),\dots,(0,N)\}$, all the middle terms cancel, and we are left with $1/(Npq)$.

We now turn to the fake proof for arbitrary weight k. For general k, the sum giving the left hand side of (3.1), $(1/5)\sum_\tau L_{\lambda+\tau,\mu-3\tau,\alpha,\beta}$, is a sum not merely of $1/(5pq)$, but rather, as in (2.3), of $[(\alpha/p)^{k-1}-(\beta/q)^{k-1}]/[5(\alpha q-\beta p)]$. The sum runs over the same collection of p,q as before, and the same type of combinations (using consecutive vectors in the list $\{(5,0),\dots\}$) relate the lattice Λ to $\mathbf{Z}^4$, specifically for the terms appearing on the right hand side of (3.1). One obtains that the three L expressions there amount to summing each of the following three terms:

$$
\begin{aligned}
&\frac{(5\alpha/5p)^{k-1} - \big((3\alpha+\beta)/(3p+q)\big)^{k-1}}{(5\alpha)(3p+q)-(3\alpha+\beta)(5p)}, \\
&\frac{\big((3\alpha+\beta)/(3p+q)\big)^{k-1} - \big((\alpha+2\beta)/(p+2q)\big)^{k-1}}{(3\alpha+\beta)(p+2q)-(\alpha+2\beta)(3p+q)}, \\
&\frac{\big((\alpha+2\beta)/(p+2q)\big)^{k-1} - (5\beta/5q)^{k-1}}{(\alpha+2\beta)(5q)-(5\beta))((p+2q))}.
\end{aligned}
\tag{3.4}
$$

Once again, a minor miracle occurs in that the denominators of the terms are all equal to the same expression, namely $5(\alpha q-\beta p)$, so we formally get the desired left hand side. The same phenomenon happens in general, not just for $N=5$. □

References

[BG01] Lev A. Borisov and Paul E. Gunnells, *Toric modular forms and nonvanishing of L-functions*, J. Reine Angew. Math. **539** (2001), 149–165, DOI 10.1515/crll.2001.071. MR1863857

[BG03] Lev A. Borisov and Paul E. Gunnells, *Toric modular forms of higher weight*, J. Reine Angew. Math. **560** (2003), 43–64, DOI 10.1515/crll.2003.060. MR1992801

[KMR17] Kamal Khuri-Makdisi and Wissam Raji, *Periods of modular forms and identities between Eisenstein series*, Math. Ann. **367** (2017), no. 1-2, 165–183, DOI 10.1007/s00208-016-1380-7. MR3606438

[Mer94] Loïc Merel, *Universal Fourier expansions of modular forms*, On Artin's conjecture for odd 2-dimensional representations, Lecture Notes in Math., vol. 1585, Springer, Berlin, 1994, pp. 59–94, DOI 10.1007/BFb0074110. MR1322319

[Paş06] Vicenţiu Paşol, *A modular symbol with values in cusp forms*, electronically available from `http://arxiv.org/abs/math/0611704`, preprint, 2006.

[Wei99] André Weil, *Elliptic functions according to Eisenstein and Kronecker*, Classics in Mathematics, Springer-Verlag, Berlin, 1999. Reprint of the 1976 original. MR1723749

Mathematics Department, American University of Beirut, Bliss Street, 1107 2020 Beirut, Lebanon

Email address: `kmakdisi@aub.edu.lb`

Contemporary Mathematics
Volume **732**, 2019
https://doi.org/10.1090/conm/732/14791

Modular forms constructed from moduli of elliptic curves, with applications to explicit models of modular curves

Kamal Khuri-Makdisi

ABSTRACT. These are the lecture notes from my portion of a mini-course for the summer school "Building Bridges 3" that was held in Sarajevo during July 2016. My lectures covered the Katz definition of modular forms, a family of forms defined from this perspective and their relation to Eisenstein series, and methods of finding explicit models of modular curves. The treatment is purely expository, and the results are mostly standard, although a few points of view may not be as widely known as they deserve to be.

1. Lecture 1

In the previous lectures in this summer school, we have considered modular forms as holomorphic functions $f(\tau)$ for $\tau \in \mathcal{H}$, with the q-expansion (when f is a newform) encoding an associated Galois representation that we have used as a black box.

We now want to describe the connection between modular forms and the modular curves, such as $X(N)$, parametrizing elliptic curves with level structure. A word of caution: this parametrization of elliptic curves by the points of the modular curve $X(N)$ is completely different from the arithmetic parametrization of a single elliptic curve over $\mathbf{Q}$ as a quotient of the Jacobian of $X_0(N)$.

Recall our fundamental congruence subgroups of $\mathrm{SL}(2,\mathbf{Z})$:

$$
\begin{aligned}
&\mathrm{SL}(2,\mathbf{Z}) = \Gamma(1) > \Gamma_0(N) > \Gamma_1(N) > \Gamma(N),\\
&\Gamma_0(N) = \{\begin{pmatrix} a & b \\ c & d \end{pmatrix} \in \Gamma(1) \mid c \equiv 0 \pmod{N}\},\\
&\Gamma_1(N) = \{\begin{pmatrix} a & b \\ c & d \end{pmatrix} \mid c \equiv 0, \quad a \equiv d \equiv 1 \pmod{N}\},\\
&\Gamma(N) = \{\begin{pmatrix} a & b \\ c & d \end{pmatrix} \mid b \equiv c \equiv 0, \quad a \equiv d \equiv 1 \pmod{N}\}.
\end{aligned}
\tag{1.1}
$$

The connection between modular forms and elliptic curves arises by associating, to each value $\tau \in \mathcal{H}$, an elliptic curve E_τ, which is analytically $E_\tau = \mathbf{C}/(\mathbf{Z} + \mathbf{Z}\tau)$.

2010 *Mathematics Subject Classification.* Primary 11F11, 14G35.

Suppose $\tau, \tau' \in \mathcal{H}$ are related via an element $\gamma = \left(\begin{smallmatrix} a & b \\ c & d \end{smallmatrix}\right) \in \Gamma(1)$, so that $\tau' = \gamma\tau = (a\tau + b)/(c\tau + d)$. It then follows that $E_\tau \cong E_{\tau'}$, via

$$z + \mathbf{Z} + \mathbf{Z}\tau \in E_\tau \longleftrightarrow z' = \frac{z}{c\tau + d} + \mathbf{Z} + \mathbf{Z}\tau' \in E_{\tau'}. \tag{1.2}$$

This basically expresses, in terms of the elliptic curves E_τ and $E_{\tau'}$, the standard fact that the lattices $\mathbf{Z} + \mathbf{Z}\tau$ and $\mathbf{Z} + \mathbf{Z}\tau'$ are homothetic if and only if τ' and τ are related by an element $\gamma \in \Gamma(1)$. For equivalence under the smaller congruence subgroup $\Gamma(N)$, we have the following more precise statement.

PROPOSITION 1.1. *Let $\tau, \tau' \in \mathcal{H}$. Then τ and τ' are related by an element $\gamma \in \Gamma(N)$ if and only if there is an isomorphism $\phi : E_\tau \to E_{\tau'}$ between the corresponding elliptic curves, such that*

$$\phi(1/N) = 1/N, \qquad \phi(\tau/N) = \tau'/N. \tag{1.3}$$

The above is of course shorthand for saying $\phi(1/N + \mathbf{Z} + \mathbf{Z}\tau) = 1/N + \mathbf{Z} + \mathbf{Z}\tau'$ and $\phi(\tau/N + \mathbf{Z} + \mathbf{Z}\tau) = \tau'/N + \mathbf{Z} + \mathbf{Z}\tau'$.

From the above, we deduce that the quotients $Y(1) = \Gamma(1)\backslash\mathcal{H}$ and $Y(N) = \Gamma(N)\backslash\mathcal{H}$ parametrize, respectively, isomorphism classes of elliptic curves over $\mathbf{C}$, and isomorphism classes of triples (E, P, Q) where E is an elliptic curve over $\mathbf{C}$, and $\{P, Q\}$ is a basis for the N-torsion $E[N]$ that is symplectic, in the sense that $e_N(P, Q) = \exp(2\pi i/N)$ for the Weil pairing. The $\Gamma(N)$-orbit of a point $\tau \in \mathcal{H}$ corresponds to the isomorphism class of the triple $(\mathbf{C}/(\mathbf{Z} + \mathbf{Z}\tau), 1/N, \tau/N)$.

One says that $Y(1)$ and $Y(N)$ are *moduli spaces* parametrizing the moduli of elliptic curves, respectively without or with a basis for the N-torsion. We leave it to the reader to look up or determine as an exercise the moduli problems that are parametrized by the quotients $Y_0(N) = \Gamma_0(N)\backslash\mathcal{H}$ and $Y_1(N) = \Gamma_1(N)\backslash\mathcal{H}$.

It turns out to be much better to work with a compactification, the modular curve $X(N)$, of $Y(N)$, which one can think of as adding the cusps to $Y(N)$[1]. With the cusps included, one can view the space of modular forms $\mathcal{M}_k(\Gamma(N)) = H^0(X(N), \mathcal{L}^k)$ as being the space of holomorphic sections of the kth power of a line bundle $\mathcal{L}$ on $X(N)$, at least for $N \geq 3$ to avoid issues of elliptic elements (which would arise if we used, say, $X_0(N)$). This point of view makes the required behavior of modular forms at cusps automatic, once we require holomorphy at the cusps that have been added to obtain $X(N)$. We will follow up on the interpretation of modular forms as sections of line bundles in Lecture 2.

In this lecture, we will focus instead on interpreting modular forms based on the moduli of elliptic curves parametrized by $X(N)$. The precise algebraic formulation is due to N. Katz; see for example Section 2.1 of [**Kat76**]. We shall be somewhat cavalier with the precise definition, and simply state the following.

DEFINITION 1.2. Let $k \geq 1$ be an integer. A Katz modular form of weight k on $\Gamma(N)$ is a "nice" function $f(E, P, Q, \omega)$ satisfying the homogeneity property

$$f(E, P, Q, c\omega) = c^{-k} f(E, P, Q, \omega). \tag{1.4}$$

[1]There is an extensive theory of these models of modular curves, not just over $\mathbf{C}$, but over number fields and in characteristic p. We will not have the space to touch directly on the arithmetic aspects in these lectures, but the reader is encouraged to think at least about the way in which our discussion over $\mathbf{C}$ in fact takes place over a number field (which is usually $\mathbf{Q}$ or the Nth cyclotomic field), viewed as a subfield of $\mathbf{C}$.

Informally, the domain of definition of f is tuples (E, P, Q, ω), where E is an elliptic curve, the pair (P, Q) is a symplectic basis for the N-torsion $E[N]$, and $\omega \in H^0(E, \Omega^1)$ is a global 1-form on E. The precise definition allows the argument E in the tuple to be a generalized elliptic curve scheme E/S over a base S, and brings in compatibility conditions under change of base; over $\mathbf{C}$, these compatibility conditions amount to holomorphy on $\mathcal{H}$ and at the cusps (this is what we mean by a "nice" function). In these lectures, however, we will pretend to consider only tuples defined over $\mathbf{C}$. In this context, the choice of ω, up to a complex scalar, determines the homothety class of the lattice L of periods of E, namely $L = \{\int_\gamma \omega \mid \gamma \in H_1(E, \mathbf{Z})\}$, and $E \cong \mathbf{C}/L$. This ties in with the perspective seen elsewhere, of modular forms on $\Gamma(1)$ as functions of lattices. In the context of these lectures, we will simply pass between modular forms $f(\tau)$ with $\tau \in \mathcal{H}$, and the corresponding Katz modular form which we evaluate on the tuple $(E, P, Q, \omega) = (\mathbf{C}/(\mathbf{Z} + \mathbf{Z}\tau), 1/N, \tau/N, dz)$; here z is the coordinate on $\mathbf{C}$ when we view $E = \mathbf{C}/(\mathbf{Z} + \mathbf{Z}\tau)$.

EXAMPLE 1.3. In case of level $N = 1$, we can dispense with specifying the P and Q in the tuples above. Then the Eisenstein series (restricted in this example to even weight $k \geq 4$) is given in Katz and traditional form as

$$(1.5)\qquad \begin{aligned} G_k(E, \omega) &= \sum_{0 \neq \gamma \in H_1(E, \mathbf{Z})} \left[\int_\gamma \omega\right]^{-k}, \\ G_k(\tau) &= \sum_{0 \neq m + n\tau \in \mathbf{Z} + \mathbf{Z}\tau} (m + n\tau)^{-k} = \sum{}' (m + n\tau)^{-k}, \end{aligned}$$

where as usual $\sum'$ denotes a sum where we omit terms that look like 0^{-k} and are hence meaningless.

We admit that calling (1.5) a Katz form is not quite fair, because one really wants to have an algebraic construction of the value on the tuple. So we point out that G_4 and G_6 can be obtained from the following algebraic construction: starting from the elliptic curve E, its short Weierstrass form $y^2 = x^3 + ax + b$ (over, say, a field not of characteristic 2 or 3) is uniquely determined up to changing (x, y) into $(x', y') = (c^{-2}x, c^{-3}y)$ with $c \neq 0$; this transforms (a, b) into $(a', b') = (c^{-4}a, c^{-6}b)$. However, if one has access not only to E but also to the global differential ω on E, then one can normalize the choice of coordinates to obtain $\omega = dx/2y$. In this context, the coefficients a and b in the normalized Weierstrass equation are Katz modular forms of level $\Gamma(1)$ of weights 4 and 6, and are in fact simple multiples of G_4 and G_6, respectively. This can be seen via the parametrization of a complex elliptic curve by $x = \wp(z) = z^{-2} + \cdots, y = (1/2)\wp'(x) = -z^{-3} + \cdots$, and the usual differential equation for $\wp$. We can also identify the higher weight Eisenstein series on $\Gamma(1)$ as coefficients in the Laurent expansion of $\wp$, which can be defined purely algebraically over a field of characteristic zero (in terms of the completion of the local ring of E at the origin O, which allows us to integrate the formal power series of ω and obtain an "analytic" uniformizer z in this completed local ring).

We wish to generalize the abovementioned principle to $\Gamma(N)$, and to define a wide family of Katz-style modular forms, which when evaluated on a tuple (E, P, Q, ω) are given as coefficients in the Laurent series of certain elements in the function field of E, which analytically can be viewed as elliptic functions with respect to $\mathbf{Z}+\mathbf{Z}\tau$. This family of modular forms will include all the Eisenstein series on $\Gamma(N)$. We therefore begin by recalling the definition of the relevant Eisenstein series.

DEFINITION 1.4. For $i, j \in \mathbf{Z}$, let $\alpha = (i/N, j/N)$, which we will usually view as an element of $\mathbf{Q}^2/\mathbf{Z}^2$; sometimes, by abuse of notation, we will identify α with the torsion point $(i+j\tau)/N = iP + jQ \in E_\tau[N]$, all of which depends of course on a varying $\tau \in \mathcal{H}$, or equivalently on the corresponding tuple $(E = E_\tau, P, Q, \omega)$. We then define the Eisenstein series of arbitrary weight k, with parameter α, by

$$G_{k,\alpha}(\tau) = \sum_{m,n\in\mathbf{Z}}{}' (m + n\tau + i/N + (j/N)\tau)^{-k} = \sum_{\ell\in\mathbf{Z}+\mathbf{Z}\tau}{}' (\ell + \alpha)^{-k}. \tag{1.6}$$

In the above, the notation $\sum'$ means we omit the term with $\ell + \alpha = 0$, if it is present in the sum (which is essentially only when $i = j = 0$). For $k > 2$, the sum in (1.6) converges absolutely and uniformly for τ in any compact set, and yields a modular form of weight k on $\Gamma(N)$.

It is traditional to modify the definition to make sense of the lower weights $k \in \{1, 2\}$ by Hecke's summation method [**Hec27**]:

$$\begin{aligned} G_{k,\alpha}(\tau, s) &= \sum_{\ell\in\mathbf{Z}+\mathbf{Z}\tau}{}' (\ell+\alpha)^{-k}|\ell+\alpha|^{-2s}, \\ G_{k,\alpha}(\tau) &= G_{k,\alpha}(\tau, 0) \text{ after analytic continuation in } s. \end{aligned} \tag{1.7}$$

This yields the same Eisenstein series as before for $k > 2$. For $k = 2$, it turns out that $G_{2,\alpha}(\tau)$ is not quite holomorphic, but is the sum of $-\pi/(\operatorname{Im}\tau)$ and a holomorphic function of τ; so to obtain a holomorphic weight 2 Eisenstein series one must consider a difference such as $G_{2,\alpha} - G_{2,0}$. Reassuringly, for weight $k = 1$, $G_{1,\alpha}$ is indeed holomorphic. We will not consider the case $k = 0$ in these lectures, but elsewhere in this summer school the series $G_{0,0}(\tau, s)$ plays a major role via the Kronecker limit formula.

We need a few more preliminaries before we can construct the general family of modular forms that we promised above. In the meantime, to whet the reader's appetite, let us give an *ad hoc* Katz-style interpretation of the Eisenstein series of weights 2 and 3. Take a nonzero α, and view it also as a nonzero N-torsion point on the elliptic curve E_τ. Then its coordinates on the Weierstrass model are $(x_\alpha, y_\alpha) = (\wp(\alpha), (1/2)\wp'(\alpha))$, so we obtain from the series for $\wp$ and $\wp'$ that

$$\begin{aligned} \wp(\alpha) &= \sum_{m,n}{}' \left[(\alpha + m + n\tau)^{-2} - (m + n\tau)^{-2}\right] = G_{2,\alpha}(\tau) - G_{2,0}(\tau), \\ \wp'(\alpha) &= -2\sum_{m,n}{}' (\alpha + m + n\tau)^{-3} = -2G_{3,\alpha}(\tau). \end{aligned} \tag{1.8}$$

This interprets the holomorphic weight 2 form $G_{2,\alpha} - G_{2,0}$ as the x-coordinate of a torsion point on the Weierstrass model (once normalized by the choice of global differential); similarly, the form $G_{3,\alpha}$ is essentially the y-coordinate. The identification $\wp(\alpha) = G_{2,\alpha}(\tau) - G_{2,0}(\tau)$ is however slightly more delicate than written above, since convergence issues prevent us from simply expanding the sum over (m, n) in the first line above. The end result is correct, however, using techniques similar to the construction in Definition 1.5 below.

It is also interesting to consider the slope $\lambda = (y_\beta - y_\alpha)/(x_\beta - x_\alpha)$ through two torsion points in the Weierstrass model; this has a natural interpretation as a Katz modular form. Here we assume that $\alpha, \beta \neq 0$ and that $\alpha + \beta \neq 0$. The line in question through the points (x_α, y_α) and (x_β, y_β) also passes through the point (x_γ, y_γ) on the elliptic curve, with $\alpha + \beta + \gamma = 0$, by the addition law on the elliptic curve. It is immediate that this slope λ is essentially the ratio $(G_{3,\beta} - G_{3,\alpha})/(G_{2,\beta} - G_{2,\alpha})$, so it transforms under $\Gamma(N)$ the same way as a modular form of weight 1. The question is whether the quotient λ (when viewed as a function of τ) is holomorphic or merely meromorphic on $\mathcal{H}$ and the cusps. However, it is known from the formulas for the addition law on a Weierstrass curve that $\lambda^2 = x_\alpha + x_\beta + x_\gamma$, and this last sum is a genuine modular form (being a linear combination of G_2's), so λ cannot have any poles. We will show later in this lecture that λ itself is essentially the sum $G_{1,\alpha} + G_{1,\beta} + G_{1,\gamma}$; this is roughly equivalent to a classical formula for the Weierstrass ζ function, but we will present the argument differently below.

As our final preparatory comment on Eisenstein series, we point out that one can bypass Hecke's analytic continuation in s by adopting a different point of view on Eisenstein series of weights 1 and 2. Instead of defining a single $G_{k,\alpha}$, it turns out that one can write down convergent series for certain linear combinations $\sum_\alpha m_\alpha G_{k,\alpha}$, with good convergence for all $k \geq 1$. The technique is as follows [**KM12**].

DEFINITION 1.5. Consider a finite number of $\alpha \in (1/N)\mathbf{Z}^2$, and attach to each α a coefficient $m_\alpha \in \mathbf{C}$, with the properties

$$\sum_\alpha m_\alpha = 0, \qquad \sum_\alpha m_\alpha \alpha = 0. \tag{1.9}$$

Write $D = \sum_\alpha m_\alpha[\alpha]$ for the formal linear combinations of symbols $[\alpha]$. We then define

$$G_{k,D}(\tau) = \left[\sum_\alpha m_\alpha G_{k,\alpha}(\tau, s)\right]_{s=0} = \sum_{\ell \in \mathbf{Z}+\mathbf{Z}\tau} \left(\sum_\alpha{}' m_\alpha(\ell+\alpha)^{-k}\right). \tag{1.10}$$

The latter sum converges in the given order $\sum_\ell(\sum'_\alpha)$, since condition (1.9) implies that $\sum'_\alpha m_\alpha(\ell+\alpha)^{-k} = O(\ell^{-k-2})$, which implies good convergence of the sum over ℓ for all $k > 0$. Note that if the coefficients m_α are integers, we can view the formal sum D as a divisor on the elliptic curve E_τ. In that setting, (1.9) says that D is a principal divisor on E_τ, and that the preimages α of the points $\alpha + \mathbf{Z} + \mathbf{Z}\tau \in E_\tau$ are chosen so that their sum (in $\mathbf{C}$, not just in $\mathbf{C}/(\mathbf{Z}+\mathbf{Z}\tau)$) is exactly zero.

We can now describe the construction of our family of Katz modular forms.

DEFINITION 1.6. Take a finite number of $\alpha \in N^{-1}\mathbf{Z}^2$ and coefficients $m_\alpha \in \mathbf{Z}$, satisfying condition (1.9) above. Write $D = \sum_\alpha m_\alpha[\alpha]$ for the formal sum. For $k \geq 1$, we define functions $f_{k,D}$ and $g_{k,D}$ of the tuple (E, P, Q, ω) by the following procedure (which works only if E is defined over a field K of characteristic zero):

(1) Associate to each $\alpha = (i/N, j/N)$ the point $P_\alpha = iP + jQ$ on the elliptic curve E, as usual, and form the associated divisor (which by abuse of notation will also be called D) as above. Thus $D = \sum_\alpha m_\alpha P_\alpha$, and we will ignore the distinction between α and P_α whenever it suits us for the exposition. It follows that D is a principal divisor on E, and there exists

an element $\phi_D \in K(E)$ (the function field of E) with $\operatorname{div} \phi_D = D$. The values $f_{k,D}(E,P,Q,\omega)$ and $g_{k,D}(E,P,Q,\omega)$ will be constructed out of the Laurent expansion of ϕ_D and its logarithmic derivative $d\phi_D/\phi_D$ at the origin O of the elliptic curve E.

(2) The Laurent expansion of ϕ_D at O needs to be expressed in terms of a uniformizer, i.e., a local coordinate near O which vanishes there. Since we are in characteristic zero, we can integrate the global form ω to define an "analytic uniformizer" z at O, with $dz = \omega$. More precisely, start with an "algebraic uniformizer" $t \in K(E)$ at the origin O; for example, if a Weierstrass form of E is $y^2 = x^3 + ax + b$, then one choice of algebraic uniformizer is $t = x/y$. Then the completed local ring of E at O is $\widehat{\mathcal{O}}_{E,O} = K[[t]]$. In this situation, the expansion of ω at O can be written as $\omega = (c_0 + c_1 t + c_2 t^2 + \cdots)dt$, with $c_0 \neq 0$. Then define $z \in \widehat{\mathcal{O}}_{E,O}$ by $z = \int \omega = c_0 t + c_1 t^2/2 + c_2 t^3/3 + \cdots$. It follows that in fact $\widehat{\mathcal{O}}_{E,O} = k[[z]]$ as well.

(3) Normalize ϕ_D, which is unique up to a factor in K^*, by requiring that its Laurent expansion at the origin O be of the form $\phi_D = z^n(1 + f_1 z + f_2 z^2 + f_3 z^3 + \cdots)$, where n is the multiplicity of O in D (this is the sum of m_α over all those α that map to O in the curve). Similarly, and without need for normalizing ϕ_D, consider the logarithmic differential $d\phi_D/\phi_D$, and its expansion in terms of z at O: $d\phi_D/\phi_D = (n/z + g_1 + g_2 z + g_3 z^2 + \cdots)dz$.

(4) The value of $f_{k,D}$ at our tuple is then the coefficient f_k above, while the value of $g_{k,D}$ is the coefficient g_k above.

PROPOSITION 1.7. *The functions $f_{k,D}$ and $g_{k,D}$ defined above are Katz modular forms, whose value on the standard tuple $(E_\tau, 1/N, \tau/N, dz)$ give modular forms in $\mathcal{M}_k(\Gamma(N))$. Moreover, when we evaluate $g_{k,D}$ at the standard tuple associated to $\tau \in \mathcal{H}$, the value is $g_{k,D}(E_\tau, 1/N, \tau/N, dz) = -G_{k,D}$.*

PROOF. For the full proof, see Sections 2 and 3 of [**KM12**]. We note that in arbitrary characteristic, it is still possible to define Katz modular forms from the Laurent series coefficients of Φ_D and its logarithmic differential, when expanded in terms of the algebraic uniformizer t, which can be chosen sufficiently canonically.

Let us sketch a proof of the assertion that $g_{k,D} = -G_{k,D}$, thereby exhibiting Eisenstein series as special cases of this construction. We first identify the function ϕ_D (up to a constant factor, which disappears in the logarithmic differential). Morally, we would like to write down directly the desired function with zeros and poles as predicted by the (translates of the) $\{\alpha\}$:

$$\phi_D(z) = \prod_{\ell \in \mathbf{Z}+\mathbf{Z}\tau} \left[\prod_\alpha{}' \left(1 - \frac{z}{\ell + \alpha}\right)^{m_\alpha} \right], \tag{1.11}$$

where in the above, the notation $\prod'$ means that if $\ell + \alpha = 0$, then we include the factor z^{m_α} instead. (The above product is somewhat different from the usual construction of $\phi_D = \prod_\alpha \sigma(z-\alpha)^{m_\alpha}$ as a product of shifted Weierstrass σ-functions.)

The question is whether the product in (1.11) is really doubly periodic with respect to $\mathbf{Z} + \mathbf{Z}\tau$. The conditions (1.9) ensure that the above product over ℓ converges well. Then the identity $1 - \frac{z+w}{\ell+\alpha} = (1 - \frac{w}{\ell+\alpha})(1 - \frac{z}{\ell-w+\alpha})$, plus the good convergence of the products $\prod_\ell \prod_\alpha$ of each factor on the right hand side, tells us that $\phi_D(z+w) = C_w \phi_D(z)$ for $w \in \mathbf{Z} + \mathbf{Z}\tau$, and some appropriate constant C_w for each w. It turns out however that the C_w are 1, which follows by comparing the logarithmic differential of the product in (1.11):

$$\frac{d\phi_D}{\phi_D} = \left[\sum_{\ell \in \mathbf{Z}+\mathbf{Z}\tau} \left(\sum_{\alpha} \frac{m_\alpha}{z - \ell - \alpha} \right) \right] dz, \tag{1.12}$$

which agrees with the logarithmic derivative from the "correct" product of σ-functions, as presented in Theorem 2.8 of [**KM12**]. Once again, the sum over ℓ has good convergence, and we can expand $m_\alpha/(z-\ell-\alpha) = -m_\alpha((\ell+\alpha)^{-1} + (\ell+\alpha)^{-2}z + \cdots)$ for the pairs with $\ell+\alpha \neq 0$; the terms with $\ell+\alpha = 0$ contribute n/z. Combining this, we get the expansion of $d\phi_D/\phi_D$ in terms of the coefficients $\{g_k\}$, and we immediately identify each g_k as the negation of the desired Eisenstein series. □

2. Lecture 2

This lecture will discuss explicit models for modular curves. At first, we will work primarily with the function field of $X_0(N)$ when we discuss the modular equation, but in the second part of the lecture, we will view modular forms on $\Gamma(N)$ primarily as sections of line bundles on the modular curve $X(N)$. We assume some familiarity with Riemann-Roch spaces, but not necessarily with line bundles and their connection with projective embeddings, which we will discuss informally once we start using those concepts.

We will work entirely over $\mathbf{C}$, but, as in the first lecture, the reader is encouraged to picture how most of our constructions actually take place over a number field, viewed as a subfield of $\mathbf{C}$. This is the main, but not the only, source of the arithmetic subtlety captured by modular forms. The background to this is that modular curves such as $X(N)$ and $X_0(N)$ have a rich structure in arithmetic geometry, so that rational points on these curves (over a number field K) correspond to interesing elliptic curves defined over K. Having access to good models of modular curves is also useful in a number of algorithmic applications, such as the Schoof–Elkies–Atkin algorithm for counting points on an elliptic curve over a finite field. The supreme arithmetic application of modular curves is in their relation to the Galois representations attached to Hecke eigenforms, as extensively illustrated in other lectures from this summer school. For a cuspidal eigenform f of weight 2 on $\Gamma_1(N)$, say, the mod ℓ Galois representation $\rho_{f,\ell}$ can be realized inside the ℓ-torsion points of the Jacobian variety of the modular curve $X_1(N)$, and the ℓ-adic Galois representation can be assembled out of the ℓ^n torsion points of the Jacobian, for varying n. It is thus of interest to be able to find explicit algebraic equations for modular curves and their Jacobians.

These explicit equations can even help with finding explicit models for the Galois representations $\rho_{f,\ell}$ if the weight of f is greater than 2: in that case, the Galois representation is realized in an étale cohomology group of a modular curve with respect to a nonconstant system of coefficients. This is less amenable to direct computation, but it turns out that our given f is in fact congruent modulo ℓ to a Hecke eigenform g of weight 2 but of level $\Gamma_1(N\ell)$, so that the mod ℓ representations of f and g are the same. So, subject to increasing the level, this reinforces the usefulness of having access to explicit models for modular curves and for working with their Jacobians. This approach is used in the work of Couveignes-Edixhoven [**EC11**] and their students to give algorithms to compute explicit Galois representations attached to forms of higher weight.

Now that the reader is, we hope, sufficiently motivated to find models of modular curves, we address the issue of precisely how we can represent a smooth projective algebraic curve X, such as $X(N)$. Broadly speaking, one can view such a curve algebraically, via a model for its function field $\mathbf{C}(X)$, or geometrically, via an embedding of the curve X in some projective (or other explicit) space $\mathbf{P}^n$. From the algebraic point of view, the field $\mathbf{C}(X)$ is of transcendence degree 1 over $\mathbf{C}$, so one chooses a transcendental element $x \in \mathbf{C}(X)$, and considers the finite extension $\mathbf{C}(X)$ of $\mathbf{C}(x)$. This is the same as considering a finite map of curves $X \to \mathbf{P}^1$, where x is the coordinate on $\mathbf{P}^1$, so $\mathbf{C}(\mathbf{P}^1) = \mathbf{C}(x)$. Let $y \in \mathbf{C}(X)$ be a primitive element for the field extension, so $\mathbf{C}(X) = \mathbf{C}(\mathbf{P}^1)[y]$. Then the elements x, y generate the function field $\mathbf{C}(X)$, and they satisfy a polynomial equation $f(x, y) = 0$. The geometric meaning of this is that X is birationally equivalent to the plane curve with affine equation $f(x, y) = 0$. However, the plane curve in question will usually have singularities (including at infinity, once one moves to the projective plane), and working directly with that plane model can be delicate. What one usually does is to work with the extension $\mathbf{C}(X)/\mathbf{C}(\mathbf{P}^1)$ using algorithms analogous to those used for computing in a number field $\mathbf{Q}(\alpha)/\mathbf{Q}$: there, one computes the integral closure R of $\mathbf{Z}$ in $\mathbf{Q}(\alpha)$ to find the ring of integers, and represents fractional ideals of R as $\mathbf{Z}$-lattices of rank $[\mathbf{Q}(\alpha) : \mathbf{Q}]$. In the function field case, one has to consider integral closures over both $\mathbf{C}[x]$ (the analog of $\mathbf{Z}$ here) and a ring such as $\mathbf{C}[1/x]$, in order to get a handle on the points of X lying above $\infty \in \mathbf{P}^1$. In this lecture, we will use the algebraic point of view to describe a model for $X_0(N)$, where the map to $\mathbf{P}^1$ is the natural projection to $X(1)$, the transcendental element generating $\mathbf{C}(X(1))$ is the usual j-function, and the polynomial that we called $f(x, y)$ above is in fact the modular polynomial $\Phi_N(j, j')$.

As for the geometric point of view, one can run the range between two extremes. On the one hand, one can represent X as a curve in $\mathbf{P}^2$ or $\mathbf{P}^3$, so the equations of X involve few variables, but can be of high degree. On the other hand, one can take an embedding of X arising from a line bundle of moderately large degree; this yields an embedding of X into a projective space $\mathbf{P}^n$ with n moderately large (but still comparable to the genus g of X), however with the benefit that the equations for X now have low degree and a simpler structure. We will illustrate this second approach later for $X(N)$, where the line bundle in question is the one whose sections are modular forms of a given weight on $\Gamma(N)$. That will require us to review a few constructions in algebraic geometry, so we will postpone it, and start with the more concrete approach of using the modular equation to get models for $X_0(N)$.

We thus proceed to study the modular curve $X_0(N)$, which parametrizes pairs of elliptic curves connected by an isogeny whose kernel is cyclic of order N. Over $\mathbf{C}$, one can always analytically bring this situation to the map $\mathbf{C}/(\mathbf{Z} + \mathbf{Z}N\tau) \to \mathbf{C}/(\mathbf{Z} + \mathbf{Z}\tau)$. Equivalently, there are two maps from $X_0(N)$ to $X(1)$, the first sending $\tau \in \Gamma_0(N)\backslash\mathcal{H}$ to $\tau \in \Gamma(1)\backslash\mathcal{H}$, and the second sending τ to $N\tau \in \Gamma(1)\backslash\mathcal{H}$. The resulting map $X_0(N) \to X(1) \times X(1)$, induced by the map $\tau \mapsto (\tau, N\tau)$ from $\mathcal{H}$ to $\mathcal{H} \times \mathcal{H}$, is a birational equivalence between $X_0(N)$ and its image. Concretely, we can use the j-function as a coordinate on $X(1)$ to identify $X(1)$ with $\mathbf{P}^1$. We then obtain that the function field of $X_0(N)$ is generated by the two modular functions

$j(\tau)$ and $j' = j(N\tau)$. These play the role of the elements x, y in our above discussion of function fields of general curves. We thus have the following classical result.

PROPOSITION 2.1. *The function field* $\mathbf{C}(X_0(N))$ *is generated by the two elements* $j(\tau)$ *and* $j(N\tau)$*, with a single polynomial relation between them, called the modular equation:*

$$\Phi_N(j(\tau), j(N\tau)) = 0, \qquad \Phi_N(x,y) \in \mathbf{Z}[x,y]. \tag{2.1}$$

The polynomial $\Phi_N(x,y)$ *is called the* N*th modular polynomial; we assert that its coefficients actually belong to* $\mathbf{Z}$*. It has the property that if* E *and* E' *are elliptic curves (a priori, over* $\mathbf{C}$*, but this works more generally) with* j*-invariants* $j(E)$ *and* $j(E')$*, then there exists a cyclic* N*-isogeny between* E *and* E' *if and only if* $\Phi_N(j(E), j(E')) = 0$*.*

For the proof, see for example Chapter 5 of [**Lan87**]. Let us sketch in these notes one way to compute the polynomial $\Phi_N(x,y) = \sum_{k,\ell} c_{k,\ell}x^k y^\ell$. Similarly to the situation with Hecke operators, one considers the decomposition of a double coset into single cosets, where we know each $\Gamma(1)$-coset contains an upper triangular representative:

$$\Gamma(1)\begin{pmatrix} N & \\ & 1\end{pmatrix}\Gamma(1) = \bigsqcup_{\text{certain } a,b,d} \Gamma(1)\begin{pmatrix} a & b \\ & d\end{pmatrix}. \tag{2.2}$$

This means effectively that if one fixes τ and hence a value $j(\tau)$, then the roots in y of the polynomial $\Phi_N(j(\tau), y)$ are the values $y = j((a\tau+b)/d)$ for those (a,b,d) that appear (parametrizing different sublattices of $\mathbf{Z} + \mathbf{Z}\tau$ of cyclic index N); note that one of the values of (a,b,d) is $(N,0,1)$, which corresponds to the root $y = j(N\tau)$. We thus conclude that

$$\sum_{k,\ell} c_{k,\ell}\, j(\tau)^k y^\ell = \Phi_N(j(\tau), y) = \prod_{\text{the same } a,b,d} \big[y - j((a\tau+b)/d)\big]. \tag{2.3}$$

It follows that the coefficients $c_{k,\ell}$ for fixed ℓ and varying k are obtained when one expresses the ℓth symmetric polynomial in the $\{j((a\tau+b)/d)\}$ as a polynomial in $j(\tau)$. This is possible because this symmetric polynomial is a modular function that is $\Gamma(1)$-invariant (due to the double coset in (2.2)), and its only pole is at the cusp ∞. We can compute the q-expansion of this ℓth symmetric polynomial from the q-expansion of $j(\tau) = q^{-1} + 744 + \cdots \in \mathbf{Z}[[q]]$, and then identify the resulting series in q as a polynomial $\sum_k c_{k,\ell}\, j(\tau)^k$. In carrying out this calculation, one uses $j((a\tau+b)/d) = \zeta_d^{-b} q^{-a/d} + 744 + \cdots$, where $\zeta_d = \exp((2\pi i)/d)$. In fact, the calculation takes place over $\mathbf{Z}[\zeta_N]$, which contains all the ζ_d. The invariance of everything under $\Gamma(1)$ implies however that the final result is invariant under any Galois automorphism of $\mathbf{Q}(\zeta_N)$, which shows that the coefficients in the final answer all belong to $\mathbf{Z}$.

The coefficients of Φ_N are notoriously large, and the birational plane model for $X_0(N)$ given by the equation $\Phi_N(j,j') = 0$ is rather singular, but this model is still quite useful in explicit computations. It should be pointed out that there are now better ways to compute the modular polynomial, namely, by interpolation. The degree of Φ_N is known (e.g., for N prime, it is $N+1$), and one knows that $\Phi_N(x,y) = \Phi_N(y,x)$, because the dual morphism to a cyclic N-isogeny is again cyclic of degree N. It follows that it is enough to generate enough points (j_α, j'_α) on the curve $\Phi_N(j,j') = 0$, in order to obtain enough values to solve for the coefficients

$c_{k,\ell} = c_{\ell,k}$. The articles [**Eng09**] and [**BLS12**] do this respectively for collections of points $(j_\alpha, j'_\alpha) \in \mathbf{C}^2$ or $(j_\alpha, j'_\alpha) \in \overline{\mathbf{F}}_p^2$, by taking a suitable collection of isogenous pairs of elliptic curves over $\mathbf{C}$ or $\overline{\mathbf{F}}_p$. In the latter setting, one gets equations for the $c_{k,\ell}$ mod p, which one can combine for various p to obtain the true value over $\mathbf{Z}$.

We now move on to the second approach outlined in the introduction of finding equations for modular curves. As promised, we begin with an informal overview of the needed prerequisites from algebraic geometry: line bundles on (as always, smooth projective) algebraic curves. For pedagogical reasons, we continue to work over $\mathbf{C}$, to allow the reader to visualize the situation in the analytic category, not just algebraically.

DEFINITION 2.2. A complex line bundle $\mathcal{L}$ on an algebraic curve X is a choice, for each point $p \in X$, of a one-dimensional complex vector space $\mathcal{L}_p$, in a way that "varies holomorphically" with p.

Concretely, this means that one can cover X by open sets $\{U_i\}$ such that, for each U_i, the totality of vector spaces $\mathcal{L}_{U_i} = \{\mathcal{L}_p \mid p \in U_i\}$ is isomorphic to the product $U_i \times \mathbf{C}$. This means that there is an isomorphism (of two-dimensional complex manifolds) $\psi_i : \mathcal{L}_{U_i} \to U_i \times \mathbf{C}$, where a vector $v \in \mathcal{L}_p$ is mapped to $\psi_i(v) = (p, c)$ for some $c \in \mathbf{C}$, and the map sending v to c is a $\mathbf{C}$-linear isomorphism between $\mathcal{L}_p$ and $\mathbf{C}$; hence the set $\mathcal{L}_p \subset \mathcal{L}_{U_i}$ is identified with $\{p\} \times \mathbf{C} \subset U_i \times \mathbf{C}$. Whenever $U_i \cap U_j \neq \emptyset$, these two different identifications of $\mathcal{L}_p$ can be compared via a homomorphic nowhere vanishing transition function $\varphi_{i,j} : U_i \cap U_j \to \mathbf{C}^*$, where $\psi_j(\psi_i^{-1}(p, c)) = (p, \varphi_{i,j}(p)c)$.

Conversely, given a covering $\{U_i\}$ of X by open sets, and a collection of transition functions $\varphi_{i,j}$ (which need to be compatible on intersections $U_i \cap U_j \cap U_k$), then one can glue the line bundles $\{U_i \times \mathbf{C}\}$ together, using the $\varphi_{i,j}$, to obtain a line bundle $\mathcal{L}$ on X.

The key concept that will matter to us is that of a holomorphic section of a line bundle $\mathcal{L}$ on X. This generalizes holomorphic functions on X, which are sections of the trivial line bundle $X \times \mathbf{C}$.

DEFINITION 2.3. Let $\mathcal{L}$ be a line bundle on X. A (holomorphic) section s of $\mathcal{L}$ is a function $s : X \to \mathcal{L}$, such that for every $p \in X$, we have $s(p) \in \mathcal{L}_p$. In terms of the local isomorphisms $\{\psi_i : \mathcal{L}_{U_i} \to U_i \times \mathbf{C}\}$, requiring s to be holomorphic means that for $p \in U_i$, $\psi_i(s(p)) = (p, f_i(p))$ with $f_i : U_i \to \mathbf{C}$ a holomorphic function. The resulting "values in local coordinates" f_i of the section s will then be compatible in the sense that for $p \in U_i \cap U_j$, we have $f_j(p) = \varphi_{i,j}(p) f_i(p)$.

The set of holomorphic sections is written $H^0(X, \mathcal{L})$; it is a finite-dimensional vector space, that we can always identify with a Riemann-Roch space, as we will discuss presently. We can also consider meromorphic sections of $\mathcal{L}$, which the reader should have no trouble defining. Although $H^0(X, \mathcal{L})$ can be zero, there are always nonzero meromorphic sections of $\mathcal{L}$.

We now describe the relation with Riemann-Roch spaces. Recall that for a divisor $D = \sum n_p p$ on X, the Riemann-Roch space $L(D)$ is the set of function field elements $f \in \mathbf{C}(X)$ that satisfy $\operatorname{div} f + D \geq 0$; in other words, for each of the (finitely many) p in the support of D, we have $v_p(f) \geq -n_p = -v_p(D)$. Here the valuation v_p gives the order of the zero of f at p (or of the pole, if $v_p(f) < 0$); for convenience, we set $v_p(0) = +\infty$. We remark that the valuation v_p also makes sense for a meromorphic section s of a line bundle $\mathcal{L}$, where it will be written $v_{p.\mathcal{L}}(s)$. Namely, suppose $p \in U_i$ for one of the open sets of the cover, where s is represented by the function f_i. Then $v_{p,\mathcal{L}}(s) = v_p(f_i)$. This is independent of the choice of U_i containing p, since $\varphi_{i,j}(p) \neq 0$ whenever $p \in U_i \cap U_j$.

PROPOSITION 2.4. *Let D be a divisor on X. Then there exists a line bundle $\mathcal{L}_D$ with the property that a meromorphic section s of $\mathcal{L}_D$ can be identified with a meromorphic function $\phi_s \in \mathbf{C}(X)$ on X, but with a modified valuation: $v_{p,\mathcal{L}_D}(s) = v_p(\phi_s) + v_p(D)$. Thus $s \in H^0(X, \mathcal{L}_D)$ if and only if for every $p \in X$, we have $v_{p,\mathcal{L}_D}(s) \geq 0$, which corresponds precisely to $\phi_s \in L(D)$; note that it is possible to have $H^0(X, \mathcal{L}_D) = L(D) = 0$.*

Conversely, every line bundle $\mathcal{L}$ on X is isomorphic to a line bundle $\mathcal{L}_D$ for some D, which is unique up to equivalence of divisors (by principal divisors of rational functions in $\mathbf{C}(X)$).

PROOF. Every divisor D is locally principal, in the sense that there exists an open cover $\{U_i\}$ of X (in either the analytic or Zariski topology), with a nonzero meromorphic function u_i on each U_i satisfying $(\operatorname{div} u_i)|_{U_i} = D|_{U_i}$. (The restriction $D|_{U_i}$ of a divisor D can be thought of as the intersection $D \cap U_i$, i.e., the restriction includes only those points of D that belong to U_i.) Then construct $\mathcal{L}_D$ by gluing the $U_i \times \mathbf{C}$ along the transition functions $\varphi_{i,j} = u_j/u_i$. A holomorphic (respectively, meromorphic) section s of $\mathcal{L}_D$ thus corresponds to a collection $\{f_i\}$ of holomorphic (respectively, meromorphic) functions on each U_i, satisfying $f_j = (u_j/u_i)f_i$. Every section s corresponds to the unique ϕ_s that is obtained by gluing together the functions f_i/u_i. So locally, $f_i = \phi_s \cdot u_i$. Recall that for $p \in U_i$, we have $v_{p,\mathcal{L}_D}(s) = v_p(f_i)$; this yields the desired relation between the valuations of s and ϕ_s.

For the converse, let $\mathcal{L}$ be given, choose any nonzero meromorphic section s_0 of $\mathcal{L}$, and let $D = \sum_p v_{p,\mathcal{L}}(s_0) \cdot p = \operatorname{div}_{(\mathcal{L})} s_0$ be the divisor of s_0, viewed as a section of $\mathcal{L}$. Then we can identify any other meromorphic section s of $\mathcal{L}$ with the meromorphic function $\phi_s = s/s_0 \in \mathbf{C}(X)$; note that although the values of s and s_0 at a point p belong to $\mathcal{L}_p$, their ratio is canonically an element of $\mathbf{C}$ (at least, away from the poles of ϕ_s). This identifies $\mathcal{L}$ with $\mathcal{L}_D$; incidentally, the section s_0 of $\mathcal{L}$ corresponds to the collection of functions $\{f_i\} = \{u_i\}$ which give a section of $\mathcal{L}_D$. In terms of $\mathcal{L}$, the bijection between $L(D)$ and $H^0(X, \mathcal{L})$ identifies $\phi \in L(D)$ with $\phi \cdot s_0 \in H^0(X, \mathcal{L})$. Finally, if we make a different choice of meromorphic section s_1 instead of s_0 at the start, this modifies D by the principal divisor $\operatorname{div}(s_1/s_0)$. □

The next important notion in our overview is the degree of a line bundle.

DEFINITION 2.5. Let $\mathcal{L}$ be a line bundle on the Riemann surface X. We say that $\deg \mathcal{L} = d$ if one meromorphic section s of $\mathcal{L}$ vanishes at exactly d points, counting multiplicities, and subtracting any multiplicities of poles. Thus $\deg \mathcal{L} = \deg \operatorname{div}_{(\mathcal{L})} s = \sum_p v_{p,\mathcal{L}}(s)$, and this degree does not depend on the choice of s, since all other choices are of the form sf with $f \in \mathbf{C}(X)$, with moreover $\deg \operatorname{div} f = 0$.

Equivalently, if $\mathcal{L} \cong \mathcal{L}_D$, then $\deg \mathcal{L} = \deg D$.

A basic consequence of Riemann-Roch is that if X has genus g, and $\deg \mathcal{L} \geq 2g-1$, then $\dim H^0(X, \mathcal{L}) = \deg \mathcal{L} + 1 - g$. Another consequence is that if $\deg \mathcal{L} \geq 2g$, then $\mathcal{L}$ is base point free, which means that for every $p \in X$, there exists a holomorphic section $s \in H^0(X, \mathcal{L})$ with $s(p) \neq 0$.

We are now ready to discuss some aspects of the relation between line bundles on a curve X, and maps from X to a projective space.

DEFINITION 2.6. Let $\mathcal{L}$ be a base point free line bundle on X. Take a basis $\{s_0, s_1, \ldots, s_n\}$ for $H^0(X, \mathcal{L})$ (more generally, we only need a basis for a base point free subspace of $H^0(X, \mathcal{L})$). The associated map from X to the projective space $\mathbf{P}^n$ is given by

$$\varphi : X \to \mathbf{P}^n, \qquad \varphi(p) = [s_0(p) : s_1(p) : \cdots : s_n(p)]. \tag{2.4}$$

Note in the above that, as usual, the values $s_i(p)$ all belong to the one-dimensional vector space $\mathcal{L}_p$, but that the proportions between their values make enough sense for us to get the projective coordinates of a point in $\mathbf{P}^n$. The reason for requiring $\mathcal{L}$ to be base point free is to ensure that we never map a point p to the invalid projective point $[0 : 0 : \cdots : 0]$.

EXAMPLE 2.7. Let X be an elliptic curve, say for definiteness with affine equation $y^2 = x^3 + 3141x + 5926$, and let $O \in X$ be the point at infinity. Consider the line bundles $\mathcal{L}_{3O}$ and $\mathcal{L}_{4O}$. We can identify $H^0(X, \mathcal{L}_{3O})$ with the Riemann-Roch space L_{3O}, which has the basis $\{1, x, y\}$. The resulting map from X to the projective plane is the usual one; it sends the affine point p to the projective point $[1 : x(p) : y(p)]$, while the point O is sent to $[0 : 0 : 1]$. One can see this by "continuity", because of the Laurent series $x = t^{-2} + \cdots$ and $y = t^{-3} + \cdots$ in terms of a uniformizer t at O, so as our point "approaches" O, its projective coordinates $[1 : t^{-2} + \cdots : t^{-3} + \cdots] = [t^3 : t + \cdots : 1 + \cdots]$ "approach" $[0 : 0 : 1]$. A less informal way to see this is to remember that the sections in $H^0(X, \mathcal{L}_{3O})$ corresponding to $1, x, y$ are in fact everywhere holomorphic, when viewed as sections of the line bundle, and to work with a trivialization of $\mathcal{L}_{3O}$ in a neighborhood of O; this corresponds to the transition function y^{-1} sending the local coordinate functions $1, x, y$ to $1/y, x/y, 1$ near O.

As for the line bundle $\mathcal{L}_{4O}$, take the basis $\{s_0, s_1, s_2, s_3\}$ of $H^0(X, \mathcal{L}_{4O})$, corresponding to the basis $\{1, x, y, x^2\}$ of $L(4O)$. The resulting map $p \mapsto [s_0(p) : s_1(p) : s_2(p) : s_3(p)] \in \mathbf{P}^3$ embeds X as the intersection of the two quadric surfaces $s_1^2 - s_0 s_3 = 0$ and $s_2^2 - s_1 s_3 - 3141 s_0 s_1 - 5926 s_0^2 = 0$.

In the above example, the image of the genus 1 curve X under the projective embedding given by $\mathcal{L}_{4O}$ is described by quadrics (i.e., by polynomials of degree 2). This is a special case of the following general theorem, due independently to Fujita [**Fuj77**] and Saint-Donat [**SD72a, SD72b**], building on results of Castelnuovo and Mumford:

THEOREM 2.8. *If X has genus g, and $\deg \mathcal{L} \geq 2g+2$, then the map to projective space given by $\mathcal{L}$ is an embedding of X, and the image is defined by quadrics; more precisely, the homogeneous ideal of vanishing of the image of X in projective space is generated by its degree 2 elements.*

We now finally come to the application of all this to modular curves. We first review how modular forms of weight k on $\Gamma(N)$, with $N \geq 3$, are sections of a particular line bundle $\mathcal{L}_k$ on $X(N)$; this result holds in fact for any subgroup of $\Gamma(1)$, but the advantage of the group $\Gamma(N)$ is that it has no elliptic points for $N \geq 3$, and all its cusps are moreover regular; it follows that $\mathcal{L}_k \cong \mathcal{L}_1^{\otimes k}$, so that $\deg \mathcal{L}_k = k \deg \mathcal{L}_1$. In the presence of elliptic points or irregular cusps, the degree of $\mathcal{L}_k$ is slightly more delicate; see for example the discussion of the divisor of a modular form in Chapter 2 of [**Shi94**].

To define the line bundle $\mathcal{L}_k$ on $X(N)$, we depart from our previous description in terms of an open cover, and instead obtain $\mathcal{L}_k$ as the quotient of the trivial bundle on $\mathcal{H}$ by a nontrivial action of $\Gamma(N)$. To be precise, we need to consider the extended upper half plane $\mathcal{H}^* = \mathcal{H} \cup \mathbf{Q} \cup \{\infty\}$, with the topology given in Chapter I of [**Shi94**], so as to correctly deal with the cusps; we will however ask for the reader's indulgence, and gloss over this important point from here on. The idea is that (holomorphic) sections of the trivial bundle $\mathcal{H} \times \mathbf{C}$ are precisely holomorphic functions $f : \mathcal{H} \to \mathbf{C}$. We define an action of $\Gamma(1)$ on the line bundle $\mathcal{H} \times \mathbf{C}$, in such a way that sections invariant under a subgroup Γ of $\Gamma(1)$ are precisely the modular forms of weight k on Γ. One can then see that the desired action of $\gamma = \left(\begin{smallmatrix} a & b \\ c & d \end{smallmatrix}\right) \in \Gamma(1)$ on a pair $(\tau, z) \in \mathcal{H} \times \mathbf{C}$ is given by

$$\gamma \cdot (\tau, z) = \Big(\frac{a\tau + b}{c\tau + d}, (c\tau + d)^k \cdot z\Big). \tag{2.5}$$

We then define $\mathcal{L}_k$ to be the resulting line bundle on $X(N) = \Gamma(N)\backslash\mathcal{H}$ (where we apologize one last time about the cusps) whose total space is $\Gamma(N)\backslash(\mathcal{H} \times \mathbf{C})$. It follows that

$$H^0(X(N), \mathcal{L}_k) = \mathcal{M}_k(\Gamma(N)); \tag{2.6}$$

it turns out to be slightly more convenient to use the full space of modular forms than to restrict to cusp forms, which would correspond to sections of $\mathcal{L}_k$ that vanish at all cusps.

We can now use a basis $\{f_0, \dots, f_n\}$ for $\mathcal{M}_k(\Gamma(N))$ to obtain a projective embedding of $X(N)$ into $\mathbf{P}^n$, at least when $\deg \mathcal{L}_k \geq 2g + 2$, with g the genus of $X(N)$. Similarly to (2.4), this sends $\tau \in \Gamma(N)\backslash\mathcal{H}$ to $[f_0(\tau) : \cdots : f_n(\tau)] \in \mathbf{P}^n$; the value in projective space is independent of the representative τ chosen, provided all the f_i are evaluated at the same τ.

PROPOSITION 2.9. *Let $N \geq 3$. For the resulting curve $X(N)$, the line bundle $\mathcal{L}_2$ has degree* $\deg \mathcal{L}_2 \geq 2g + 2$, *and hence gives rise to a projective embedding of the modular curve with image described by quadrics. Knowing the equations of the resulting modular curve is equivalent to knowing the multiplication map $\mathcal{M}_2(\Gamma(N)) \times \mathcal{M}_2(\Gamma(N)) \to \mathcal{M}_4(\Gamma(N))$.*

PROOF. We can relate $\mathcal{L}_2$ to the canonical line bundle Ω^1 of holomorphic 1-forms on $X(N)$. It is standard that $H^0(X(N), \Omega^1) \cong \mathcal{S}_2(\Gamma(N))$ (in fact, this works for any group Γ), by identifying a cusp form $f(z)$ with the differential form $f(z)dz$, which is now invariant under $\Gamma(N)$. The reason that the corresponding modular forms are cuspidal can be seen in terms of the local coordinate at infinity $q^{1/N} = \exp(2\pi i z/N)$. Since $2\pi i N^{-1}\, dz = q^{-1/N}\, d(q^{1/N})$, the q-expansion of f must start with $c_1 q^{1/N} + c_2 q^{2/N} + \cdots$ for $f(z)dz$ to avoid a pole at $q = 0$, i.e., at $z = \infty$; a similar condition holds at the other cusps. A modular form which does not have to be cuspidal thus corresponds to a meromorphic section of Ω^1, with a possible simple pole at each cusp of $X(N)$. This means that $\mathcal{L}_2 \cong \Omega^1 \otimes \mathcal{L}_{\mathbf{cusps}}$, where **cusps** is the divisor of the cusps. In particular,

$$\deg \mathcal{L}_2 = \deg \Omega^1 + \deg(\mathbf{cusps}) = (2g - 2) + c, \tag{2.7}$$

where c is the number of cusps of $X(N)$. But one knows that $c \geq 4$ once $N \geq 3$. Thus $\deg \mathcal{L}_2 \geq 2g + 2$, as desired, and the projective embedding given by a basis of $\mathcal{M}_2(\Gamma(N))$ is described by quadrics.

Let $\{f_0, \ldots, f_n\}$ be a basis for $\mathcal{M}_2(\Gamma(N))$. Suppose we wish to determine the quadrics that vanish on the image of $X(N)$, since these generate the homogeneous ideal of the projective curve. The presence of such a quadric of the form $q(T_0, \ldots, T_n) = \sum_{i,j} c_{i,j} T_i T_j$, in terms of the homogeneous coordinates $[T_0 : \cdots : T_n]$, corresponds to the identity of modular forms $\sum_{i,j} c_{i,j} f_i f_j = 0 \in \mathcal{M}_4(\Gamma(N))$. Hence, to find the generators of our homogeneous ideal, it is enough to know how to multiply $f_i f_j$ for every pair (i,j), and how to find linear relations between these elements of $\mathcal{M}_4(\Gamma(N))$. □

We remark that the same result holds for $X_1(N)$ with $N \geq 5$. Let us therefore compute equations for $X_1(5)$ as an example. This is not extremely interesting, since the genus of $X_1(5)$ is zero, but it illustrates the above theorem. In this situation, we have $\dim \mathcal{M}_2(\Gamma_1(5)) = 3$, and a basis is $\{f, g, h\}$ with

$$\begin{aligned} f &= 1 \quad + 60q^3 - 120q^4 + \cdots, \\ g &= \quad q \quad + 6q^3 - 9q^4 + \cdots, \\ h &= \qquad q^2 - 4q^3 + 12q^4 + \cdots. \end{aligned} \tag{2.8}$$

One also knows that $\mathcal{M}_4(\Gamma_1(5))$ is 5-dimensional, and its elements are determined by knowing their q-expansions up to and including the q^4 term. One then computes $f^2 = 1 + 120q^3 - 240q^4 + \cdots$, $fg = q + 6q^3 + 51q^4 + \cdots$, and so forth, all of which belong to $\mathcal{M}_4(\Gamma_1(5))$. One then obtains the equation

$$g^2 - fh - 4gh - 16h^2 = 0, \tag{2.9}$$

which means that we have identified $X_1(5)$ with the conic in the projective plane given by the equation $U^2 - TV - 4UV - 16V^2 = 0$. The conic contains the rational point $[T : U : V] = [1 : 0 : 0]$ (namely, the point $q = 0$ corresponding to the cusp ∞), so we can identify the curve $X_1(5)$ with $\mathbf{P}^1$ over $\mathbf{Q}$, not just over $\mathbf{C}$.

We point out that describing the multiplication map $\mathcal{M}_2(\Gamma(N)) \times \mathcal{M}_2(\Gamma(N)) \to \mathcal{M}_4(\Gamma(N))$ can be done by interpolation, which ties in with our earlier description of finding the modular equation Φ_N by interpolation. Namely, suppose that we take a large number ("L") of points $\tau_1, \ldots, \tau_L \in X(N)$, where we require $L > \deg \mathcal{L}_4$. Then a modular form $f \in \mathcal{M}_4(\Gamma(N))$ is completely determined by its values at these L points, because if g were a different form agreeing with f at $\tau_1, \ldots, \tau_L$, then the difference $f - g$ would be a nonzero section of the line bundle $\mathcal{L}_4$, so $f - g$ could only vanish at $\deg \mathcal{L}_4$ points, contradicting the fact that it vanishes at $\tau_1, \ldots, \tau_L$. In that case, the basis $\{f_0, \ldots, f_n\}$ of $\mathcal{M}_2(\Gamma(N))$, as well as all products $f_i f_j$, can be represented by their values at $\tau_1, \ldots, \tau_L$. Thus we can identify f_i by its vector of values $(f_i(\tau_1), \ldots, f_i(\tau_L))$, and carry out multiplication into $\mathcal{M}_4$ componentwise in order to find the quadrics that vanish on the image of $X(N)$. This is equivalent to finding the quadrics on $\mathbf{P}^n$ that vanish on (i.e., interpolate through) the projective points $P_1, \ldots, P_L$, which are the images of $\tau_1, \ldots, \tau_L$ under the projective embedding. Concretely,

$$P_j = [f_0(\tau_j) : \cdots : f_n(\tau_j)] \tag{2.10}$$

so we have reversed our viewpoint. Whereas we previously fixed f_i and represented it as a function by its values at varying τ_j, we now fix τ_j and represent it as a projective point by the values of the various f_i at that point. It turns out [**KM07**] that this is an effective way to represent the curve if one is interested in computing with its Jacobian; this is called "Representation B" in that article. We also note that the points $\tau_1, \dots, \tau_L$ do not actually have to be distinct; representing modular forms by their q-expansions up to degree q^L, as we did in the example above, is a way of evaluating the forms at the divisor $L \cdot \infty$. We note that the family of Katz modular forms from Lecture 1 gives an ample supply of modular forms that can be easily evaluated at points, since evaluating at a τ_j can be carried out algebraically by evaluating on a tuple (E, P, Q, ω).

3. Exercises

The following exercises were distributed to students at the summer school.

Exercise 0. Give the argument that evaluating a weight k Katz modular form on the tuple $(E_\tau, 1/N, \tau/N, dz)$ defines a function $f(\tau)$ that transforms like a usual weight k modular form.

Exercise 1. a) Use SageMath or Magma (or any other software) to find the modular polynomial $\Phi_2(X, Y)$ from the identity

$$\Phi_2(X, j(\tau)) = (X - j(2\tau))(X - j(\tau/2))(X - j((\tau+1)/2)),$$

by comparing q-expansions.

b) Find $\Phi_2(X, Y)$ in a different way, by finding enough pairs (j, j') of j-invariants of elliptic curves that are 2-isogenous to interpolate Φ_2 through these points. (What is the degree of Φ_2 in each of X and Y, and how many points are needed? I suggest taking curves $E : y^2 = x(x-1)(x-\lambda)$ for a few values of $\lambda \in \mathbf{Q}$, and all their quotients by cyclic 2-torsion subgroups.)

c) Question to think about later: can you find, e.g., $\Phi_5(X, Y)$ by finding its reduction modulo many primes? This involves finding for each p a number of pairs of 5-isogenous elliptic curves over $\mathbf{F}_p$ and interpolating through the corresponding $(j, j') \bmod p$.

Exercise 2. In this exercise, you may assume that $N \geq 3$ is prime if you like, but see if you can do the general case too.

a) What is the index $[\Gamma(1) : \Gamma(N)]$? What is the degree d of the map $\pi : X(N) \to X(1)$ between modular curves?

b) What is the ramification of π at the cusps? Use this to find the number c of cusps of $X(N)$.

c) Let $\mathcal{L}$ be the line bundle on $X(N)$ whose sections give $\mathcal{M}_1(\Gamma(N))$. Show that $\deg \mathcal{L} = d/12$. (Hint: $\Delta(z) \in \mathcal{S}_{12}(\Gamma(1)) \subset \mathcal{S}_{12}(\Gamma(N))$.)

d) Find the genus of $X(N)$ in terms of d and c. (Hint: consider the line bundles $\mathcal{L}^2$, whose sections are $\mathcal{M}_2(\Gamma(N))$, and $\Omega^1 \cong \mathcal{L}^2(-\mathbf{cusps})$, whose sections are $\mathcal{S}_2(\Gamma(N))$.)

Exercise 3. Let X be the (projective model of the) curve $y^3 = x^4 + x + 2$. Its points are the affine points satisfying the above equation, plus one point P_0 at infinity where the rational function x has a pole of order 3, and y has a pole of order 4. This is a $C_{3,4}$ curve; generally, $C_{a,b}$ curves are a nice source of examples. (However, if $|a - b| \geq 2$, then the plane model of a $C_{a,b}$ curve is singular at infinity.)

a) Compute the Riemann-Roch spaces $L(kP_0)$ for $k \leq 15$, and deduce that X has genus 3, either from Riemann-Roch or by any other method you like.

b) For each of $k = 6, 7, 8$, consider the resulting map of X to projective space, and give generators for the ideal describing the image.

Exercise 4. Our goal is to find equations for $X = X_1(11)$, which has genus 1.

a) Use SageMath or Magma to find q-expansions of a basis for $\mathcal{M}_2(\Gamma_1(11))$, ordered as an echelon basis in terms of the order of vanishing at the cusp ∞ (i.e., $q = 0$).

b) The projective embedding of X given by $\mathcal{M}_2(\Gamma_1(11))$ has too large dimension for a human-readable model of the curve. Instead, obtain a smaller embedding by restricting to a subset $V \subset \mathcal{M}_2(\Gamma_1(11))$ defined by imposing a certain order of vanishing at the cusp ∞. This means that, viewing $\mathcal{M}_2(\Gamma_1(11)) = H^0(X, \mathcal{L})$ for a suitable line bundle $\mathcal{L}$ on X, your space V will be $H^0(X, \mathcal{L}(-k\infty))$ for some k. Thus the line bundle you will consider will have degree $(\deg \mathcal{L}) - k$.

Suggestion: Take $(\deg \mathcal{L}) - k = 3$ or 4. This will produce either one cubic equation in $\mathbf{P}^2$, or two quadric equations in $\mathbf{P}^3$.

References

[BLS12] Reinier Bröker, Kristin Lauter, and Andrew V. Sutherland, *Modular polynomials via isogeny volcanoes*, Math. Comp. **81** (2012), no. 278, 1201–1231, DOI 10.1090/S0025-5718-2011-02508-1. MR2869057

[EC11] Bas Edixhoven and Jean-Marc Couveignes (eds.), *Computational aspects of modular forms and Galois representations*, Annals of Mathematics Studies, vol. 176, Princeton University Press, Princeton, NJ, 2011. How one can compute in polynomial time the value of Ramanujan's tau at a prime. MR2849700

[Eng09] Andreas Enge, *Computing modular polynomials in quasi-linear time*, Math. Comp. **78** (2009), no. 267, 1809–1824, DOI 10.1090/S0025-5718-09-02199-1. MR2501077

[Fuj77] T. Fujita, *Defining equations for certain types of polarized varieties*, Complex analysis and algebraic geometry, Iwanami Shoten, Tokyo, 1977, pp. 165–173. MR0437533

[Hec27] E. Hecke, *Theorie der Eisensteinschen Reihen höherer Stufe und ihre Anwendung auf Funktionentheorie und Arithmetik* (German), Abh. Math. Sem. Univ. Hamburg **5** (1927), no. 1, 199–224, DOI 10.1007/BF02952521. MR3069476

[Kat76] Nicholas M. Katz, *p-adic interpolation of real analytic Eisenstein series*, Ann. of Math. (2) **104** (1976), no. 3, 459–571, DOI 10.2307/1970966. MR0506271

[KM07] Kamal Khuri-Makdisi, *Asymptotically fast group operations on Jacobians of general curves*, Math. Comp. **76** (2007), no. 260, 2213–2239, DOI 10.1090/S0025-5718-07-01989-8. MR2336292

[KM12] Kamal Khuri-Makdisi, *Moduli interpretation of Eisenstein series*, Int. J. Number Theory **8** (2012), no. 3, 715–748, DOI 10.1142/S1793042112500418. MR2904927

[Lan87] Serge Lang, *Elliptic functions*, 2nd ed., Graduate Texts in Mathematics, vol. 112, Springer-Verlag, New York, 1987. With an appendix by J. Tate. MR890960

[SD72a] Bernard Saint-Donat, *Sur les équations définissant une courbe algébrique* (French), C. R. Acad. Sci. Paris Sér. A-B **274** (1972), A324–A327. MR0289516

[SD72b] Bernard Saint-Donat, *Sur les équations définissant une courbe algébrique* (French), C. R. Acad. Sci. Paris Sér. A-B **274** (1972), A487–A489. MR0289517

[Shi94] Goro Shimura, *Introduction to the arithmetic theory of automorphic functions*, Publications of the Mathematical Society of Japan, vol. 11, Princeton University Press, Princeton, NJ, 1994. Reprint of the 1971 original; Kanô Memorial Lectures, 1. MR1291394

Mathematics Department, American University of Beirut, Bliss Street, 1107 2020 Beirut, Lebanon

Email address: `kmakdisi@aub.edu.lb`

Contemporary Mathematics
Volume **732**, 2019
https://doi.org/10.1090/conm/732/14814

Some remarks on the coefficients of symmetric power L-functions

Balesh Kumar, Jaban Meher, and Sudhir Pujahari

ABSTRACT. In this article, we study the coefficients of symmetric power L-functions attached to holomorphic cusp forms for the full modular group $SL_2(\mathbb{Z})$. We find the abscissa of absolute convergence of the coefficients of symmetric power L-functions attached to holomorphic cusp forms for $SL_2(\mathbb{Z})$. We also study the oscillation in signs of the coefficients of symmetric power L-functions attached to cusp forms.

1. Introduction

Let S_k be the space of cusp forms of integral weight k for the full modular group $SL_2(\mathbb{Z})$. Suppose that

$$f(z) = \sum_{n=1}^{\infty} a_n q^n \in S_k$$

is a normalized eigenform, where $q = e^{2\pi i z}$ and z is in complex upper-half plane $\mathcal{H}$. It is well known that for $n \geq 1$, a_n are real algebraic numbers [**12**, Corollary 7.3.2, pp. 91]. In fact, the coefficients a_n lie in a number field K_f, where the number field depends only on the form f. Let

$$a(n) = \frac{a_n}{n^{\frac{k-1}{2}}}.$$

The Ramanujan-Petersson conjecture which was proved by Deligne [**2**], is the fact that

$$|a(n)| \leq d(n),$$

where $d(n) = \sum_{d|n} 1$ is the divisor function. For $\Re(s) > 1$, the L-function attached to the normalized eigenform f is given by

$$L(s, f) = \sum_{n=1}^{\infty} \frac{a(n)}{n^s} = \prod_p \left(1 - \frac{\alpha(p)}{p^s}\right)^{-1} \left(1 - \frac{\beta(p)}{p^s}\right)^{-1},$$

where p in the product expansion is over all prime numbers. According to Deligne [**2**], for any prime p, we have

$$|\alpha(p)| = |\beta(p)| = \alpha(p)\beta(p) = 1. \tag{1}$$

2010 *Mathematics Subject Classification.* Primary 11F66, 11F11.
Key words and phrases. Cusp forms, symmetric power L-functions.

It is well known that the L-function $L(s, f)$ is analytically continued to the whole complex plane having a certain functional equation. For any positive integer r, the r-th symmetric power L-function attached to f is defined as

$$L(s, sym^r f) = \prod_p \prod_{m=0}^{r} \left(1 - \frac{\alpha(p)^{r-m}\beta(p)^m}{p^s}\right)^{-1} \tag{2}$$

for $\Re(s) > 1$. For $\Re(s) > 1$, we write the Dirichlet series expansion as

$$L(s, sym^r f) = \sum_{n=1}^{\infty} \frac{a_{sym^r f}(n)}{n^s}. \tag{3}$$

The analytic properties of symmetric power L-functions are related to the Sato-Tate conjecture. It is also known [**13**, Section 4] that for each integer $r \geq 1$, the series $L(s, sym^r f)$ can be analytically continued to the region $\Re(s) \geq 1$ and it is non-vanishing in that region. The Rankin-Selberg convolution of L-functions attached to $sym^r f$ and $sym^t f$ is defined as

$$L(s, sym^r f \times sym^t f) := \prod_p \prod_{m=0}^{r} \prod_{l=0}^{t} \left(1 - \frac{\alpha(p)^{r-m}\beta(p)^m \alpha(p)^{t-l}\beta(p)^l}{p^s}\right)^{-1} \tag{4}$$

for $\Re(s) > 1$. For $\Re(s) > 1$, we write the Dirichlet series expansion as

$$L(s, sym^r f \times sym^t f) = \sum_{n=1}^{\infty} \frac{a_{sym^r f \times sym^t f}(n)}{n^s}. \tag{5}$$

Based on the work of Cogdell and Michel [**1**], Lau and Wu [**8**] have shown that for $r = 2, 3, 4$, the Rankin-Selberg convolution $L(s, sym^r f \times sym^r f)$ has analytic continuation to the whole complex plane with a simple pole at $s = 1$ and it satisfies a functional equation, as predicted by the Langlands program. Our first result of the paper is about the abscissa of absolute convergence of symmetric power L-series attached to any eigenform $f \in S_k$. More precisely, we prove the following result.

THEOREM 1.1. *The series*

$$L(s, sym^r f) = \sum_{n=1}^{\infty} \frac{a_{sym^r f}(n)}{n^s}$$

has abscissa of absolute convergence 1.

To prove the above result, we prove the following more general result on Dirichlet series.

THEOREM 1.2. *Let $a(n)(n \geq 1)$ be a sequence of complex numbers such that $a(n) = O_\epsilon(n^{\alpha+\epsilon})$ for any $\epsilon > 0$, and the series $\sum_{n=1}^{\infty} \frac{|a(n)|^2}{n^s}$ has a singularity at $s = \beta \geq 0$, where α, β are real numbers such that $2\alpha + 1 \leq \beta$. Then the series $\sum_{n=1}^{\infty} \frac{a(n)}{n^s}$ has abscissa of absolute convergence $\alpha + 1$.*

Our next result is on the sign change of the coefficients of symmetric square L-function attached to a modular form. More precisely, we prove the following result.

THEOREM 1.3. *There exists δ with $0 < \delta < 1$ such that the number sign changes of the sequence $\{a_{sym^2 f}(p)\}$ for $p \in [x, 2x]$ is at least ax^δ for some $a > 0$ and sufficiently large x, where p runs over prime numbers. In particular, the sequence $\{a_{sym^2 f}(p)\}$ changes signs infinitely often.*

2. Hoheisel phenomenon

In this section, we will briefly describe the Hoheisel property for Dirichlet series. We follow the presentations and notations as those given in [**10**]. Let

$$\phi(s) = \sum_{n=1}^{\infty} \frac{a_n}{n^s}, \quad s = \sigma + it,$$

be a Dirichlet series with non-negative coefficients. Let $N_\phi(\sigma, T)$ denote the number of zeros $\rho = \beta + i\gamma$ of $\phi(s)$ with $\beta \geq \sigma$ and $|\gamma| \leq T$. We say that $\phi(s)$ has Hoheisel Property if the following properties hold.

(1) **Explicit formula:**

$$\psi(x) = \sum_p a_p \log p = rx - \sum_{|\gamma| \leq T} \frac{x^\rho}{\rho} + O\left(\frac{x}{T}(\log xT)^2\right),$$

with $r = 0$ or 1 and where the first sum is over primes p and the second sum is over the zeros $\rho = \beta + i\gamma$ of $\phi(s)$ with $|\gamma| \leq T \leq x^{1/2}$ and $\beta \geq 0$.

(2) **Zero free region:** $\phi(s) \neq 0$ in the region $\sigma \geq 1 - a/\log(2 + |t|)$, $a > 0$.

(3) **Log free zero density estimate:**

$$N_\phi(\sigma, T) \ll T^{c(1-\sigma)}, \ (c > 0)$$

uniformly for $\frac{1}{2} \leq \sigma \leq 1$, as $t \to \infty$.

(4) **Zero density estimate:**

$$N_\phi(0, T) \ll T \log T.$$

We state the following main theorem of [**10**].

THEOREM 2.1. *Assume the Dirichlet series $\phi(s) = \sum_{n=1}^{\infty} \frac{a_n}{n^s}$ has Hoheisel Property with $r = 1$. Then there exists a positive constant $\nu < 1$ such that for any $h = x^\theta$ with $\nu < \theta < 1$, one has*

$$\sum_{x \leq p < x+h} a_p \log p \geq c_0 h$$

for some $c_0 > 0$.

In fact, we can relax the condition on the non-negativity of the coefficients a_n in the above theorem. By following the methods given in [**10**], we have the following result.

THEOREM 2.2. *Assume the Dirichlet series $\phi(s) = \sum_{n=1}^{\infty} \frac{a_n}{n^s}$ has Hoheisel Property with $r = 0$ with the property that all the coefficients a_n need not be non-negative. Then there exists a positive constant ν such that if $h = x^\theta$, $\nu < \theta < 1$, we have*

$$\sum_{x \leq p < x+h} a_p \log p = o(h).$$

The above result for Hecke cusp forms has been stated in [**11**, Lemma 4].

3. Proof of Theorem 1.1

We apply Theorem 1.2 to prove the result. By the Euler product expansion, we see that

$$L(s, sym^r f \times sym^r f) = \zeta(s) \prod_{l=1}^{r} L(s, sym^{2l} f). \tag{6}$$

Since for each positive integer l, $L(s, sym^l f)$ is analytically continued to $\Re(s) \geq 1$, from the above identity we deduce that $L(s, sym^r f \times sym^r f)$ is analytically continued to $\Re(s) \geq 1$ whose only singularity a simple pole at $s = 1$. Since Ramanujan-Petersson conjecture is true for symmetric power L-functions attached to a holomorphic cusp form, we have

$$a_{sym^r f}(n) \ll n^{\epsilon}.$$

for any $\epsilon > 0$. To proceed further, we need the following result.

PROPOSITION 3.1. *The series*

$$\sum_{n=1}^{\infty} \frac{|a_{sym^r f}(n)|^2}{n^s}$$

has a singularity at $s = 1$.

Assuming the above proposition we first prove Theorem 1.1. In the setting of Theorem 1.2, we have $\alpha = 0$ and $\beta = 1$. Applying Theorem 1.2, we get Theorem 1.1. Now we prove Proposition 3.1. To prove Proposition 3.1, we state the following proposition the proof of which is exactly the same as [**4**, Theorem 6] since in the proof of [**4**, Theorem 6], automorphic properties of L-functions are not used. Only thing used in that proof is the Deligne's bound which is known to be true in the cases of symmetric power L-functions attached to holomorphic cusp forms.

PROPOSITION 3.2. *For positive integers* r *and* t *if*

$$L(s, sym^r f) = \sum_{n=1}^{\infty} \frac{a_{sym^r f}(n)}{n^s}$$

and

$$L(s, sym^t f) = \sum_{n=1}^{\infty} \frac{a_{sym^t f}(n)}{n^s},$$

then

$$\sum_{n=1}^{\infty} \frac{a_{sym^r f}(n) a_{sym^t f}(n)}{n^s} = L(s, sym^r f \times sym^t f) g(s),$$

where $g(s)$ *is a Dirichlet series absolutely convergent for* $\Re(s) > \frac{1}{2}$.

We also prove the following lemma. The lemma is similar to Lemma 2.5 of [**9**]. The result of following lemma is about the coefficients of Rankin-Selberg L-functions attached to symmetric power L-functions whereas Lemma 2.5 of [**9**] is about the coefficients of Rankin-Selberg L-functions attached to irreducible cuspidal automorphic representations of GL_m. The automorphicity of symmetric power L-functions is not known in general. However, we use the known analytic properties of symmetric power L-functions to prove our result.

LEMMA 3.3. *The series*

$$\sum_{p \text{ prime}} \frac{a_{sym^r f \times sym^r f}(p)}{p}$$

is divergent.

PROOF. To prove the lemma, we prove that

$$\sum_{p \le x} a_{sym^r f \times sym^r f}(p) \sim \frac{x}{\log x}, \tag{7}$$

where p varies over all primes less than or equal to x. To prove (7), we first show that

$$\sum_{n \le x} a_{sym^r f \times sym^r f}(n) \Lambda(n) \sim x, \tag{8}$$

where $\Lambda(n)$ is the von Mangoldt function. Then by partial summation, we get (7). We apply Tauberian theorem to the Dirichlet series

$$-\frac{L'(s, sym^r f \times sym^r f)}{L(s, sym^r f \times sym^r f)} = \sum_{n=1}^{\infty} \frac{a_{sym^r f \times sym^r f}(n) \Lambda(n)}{n^s}.$$

By [**13**, Section 4], we know that for each positive integer l, $L(s, sym^l f)$ is analytically continued to $\Re(s) \ge 1$ and it is non-vanishing in that region. Thus from (6), we deduce that $L(s, sym^r f \times sym^r f)$ is analytically continued to $\Re(s) \ge 1$ with only singularity a simple pole at $s = 1$ and it is non-vanishing for $\Re(s) \ge 1$. Thus the series

$$-\frac{L'(s, sym^r f \times sym^r f)}{L(s, sym^r f \times sym^r f)} = \sum_{n=1}^{\infty} \frac{a_{sym^r f \times sym^r f}(n) \Lambda(n)}{n^s}.$$

converges absolutely for $\Re(s) > 1$ and it is analytically continued for $\Re(s) \ge 1$ except for a simple pole at $s = 1$. Thus applying the Tauberian theorem, we get (8) and hence (7).

Now by partial summation, we have

$$\sum_{p \le x} \frac{a_{sym^r f \times sym^r f}(p)}{p} = \frac{A(x)}{x} + \int_2^x \frac{A(t)}{t^2} dt,$$

where $A(x) = \sum_{p \le x} a_{sym^r f \times sym^r f}(p)$. Now using (7), we get

$$\sum_{p \le x} \frac{a_{sym^r f \times sym^r f}(p)}{p} = \frac{1}{\log x} + o\left(\frac{1}{\log x}\right) + \int_2^x \left(\frac{1}{t \log t} + o\left(\frac{1}{t \log t}\right)\right) dt.$$

By simplifying, we deduce that

$$\sum_{p \le x} \frac{a_{sym^r f \times sym^r f}(p)}{p} = \log \log x + o(\log \log x).$$

This proves the result. □

Now we prove Proposition 3.1. By Proposition 3.2, for any positive integer r, we have

$$\sum_{n=1}^{\infty} \frac{|a_{sym^r f}(n)|^2}{n^s} = L(s, sym^r f \times sym^r f) h(s), \tag{9}$$

where $h(s)$ is a Dirichlet series absolutely convergent for $\Re(s) > \frac{1}{2}$.

Assume on the contrary that the series

$$\sum_{n=1}^{\infty} \frac{|a_{sym^r f}(n)|^2}{n^s}$$

does not have any singularity at $s = 1$. By Landau's theorem on Dirichlet series with non-negative coefficients, there exists a real number σ such that the Dirichlet series

$$\sum_{n=1}^{\infty} \frac{|a_{sym^r f}(n)|^2}{n^s}$$

is convergent for $\Re(s) > \sigma$ and it has a singularity at $s = \sigma$. Since the series

$$L(s, sym^r f \times sym^r f) = \sum_{n=1}^{\infty} \frac{a_{sym^r f \times sym^r f}(n)}{n^s}$$

is absolutely convergent for $\Re(s) > 1$ and the series $h(s)$ is absolutely convergent for $\Re(s) > 1/2$, we deduce from (9) that $\sigma < 1$. But we will prove that the series

$$\sum_{n=1}^{\infty} \frac{|a_{sym^r f}(n)|^2}{n}$$

is divergent. This will lead to a contradiction. We know that for any prime p, we have

$$|a_{sym^r f}(p)|^2 = a_{sym^r f \times sym^r f}(p).$$

Thus

$$\sum_{p \text{ prime}} \frac{a_{sym^r f \times sym^r f}(p)}{p} \leq \sum_{n=1}^{\infty} \frac{|a_{sym^r f}(n)|^2}{n},$$

and thus Proposition 3.1 follows from Lemma 3.3.

4. Proof of Theorem 1.2

We follow the idea of Kohnen [7] to prove the theorem. We recall the basic fact that if

$$L(s) = \sum_{n=1}^{\infty} \frac{b(n)}{n^s}$$

is a Dirichlet series such that $\sum_{n=1}^{\infty} b(n)$ is divergent, then the abscissa of convergence σ of $L(s)$ is given by

$$\sigma = \inf\{t \in \mathbb{R} | \sum_{n \leq x} b(n) = O_t(x^t)\}. \tag{10}$$

We first observe that the series

$$\sum_{n=1}^{\infty} |a(n)|$$

is divergent. If it were not divergent, then since

$$\sum_{n=1}^{\infty} |a(n)|^2 \leq \left(\sum_{n=1}^{\infty} |a(n)| \right)^2,$$

it would imply that the series

$$\sum_{n=1}^{\infty} \frac{|a(n)|^2}{n^s}$$

is convergent for $\Re(s) \geq 0$. This contradicts the hypothesis of the theorem since the above series has a singularity at $s = \beta > 0$. Now from the first condition of the hypothesis, the series

$$\sum_{n=1}^{\infty} \frac{a(n)}{n^s}$$

is absolutely convergent for $s > \alpha + 1$. Thus the abscissa of absolute convergence σ_a of the series

$$\sum_{n=1}^{\infty} \frac{a(n)}{n^s}$$

is less than or equal to $\alpha + 1$. Suppose that $\sigma_a < \alpha + 1$. Then by (10), there exists a positive real number c such that

$$\sum_{n \leq x} |a(n)| = O(x^{\alpha+1-c}). \tag{11}$$

Now by Cauchy-Schwarz inequality, we have

$$\sum_{n\leq x} |a(n)|^2 \leq \left(\sum_{n\leq x} |a(n)| \right)^{\frac{1}{2}} \left(\sum_{n\leq x} |a(n)|^3 \right)^{\frac{1}{2}}.$$

Using (11) and the hypothesis of the theorem, we see that

$$\sum_{n\leq x} |a(n)|^2 \ll_\epsilon x^{\frac{\alpha+1-c}{2}} x^{\frac{3\alpha+1+3\epsilon}{2}} \ll_\epsilon x^{2\alpha+1-\frac{c}{2}+\frac{3\epsilon}{2}}$$

is true for any $\epsilon > 0$. Thus by (10), the abscissa of convergence of the series

$$\sum_{n=1}^{\infty} \frac{|a(n)|^2}{n^s}$$

is less than $2\alpha + 1$. This gives a contradiction to the hypothesis that it has a singularity at $\beta \geq 2\alpha + 1$. This proves the theorem.

5. Proof of Theorem 1.3

We apply Theorem 2.1 and Theorem 2.2 respectively for the symmetric square L-function and the Rankin-Selberg convolution of symmetric square L-functions attached to an Hecke eigenform to get the corresponding bounds and then we compare both the bounds to get the required result. Let us first see that the hypothesis of Theorem 2.1 are satisfied by $L(s, sym^2 f)$. The required zero free region of $L(s, sym^2 f)$ follows by the classical method in analytic number theory. This has been also stated in [**11**, pp. 438]. Also the log free zero density estimate for $L(s, sym^2 f)$ follows from Corollary 1.2 of [**14**]. The zero density estimate for $L(s, sym^2 f)$ follows from [**6**, Theorem 5.8].

Since $L(s, sym^2 f)$ satisfies Ramanujan-Petersson conjecture, by [**6**, Exercise 7, pp. 112], we have

(12)

$$\psi(sym^2 f, x) = \sum_{p\leq x} a_{sym^2 f}(p) \log p = - \sum_{|\gamma|\leq T} \frac{x^\rho - 1}{\rho} + O\left(\frac{x}{T}(\log x) \log(x^3 q_1)\right),$$

where $\rho = \beta + i\gamma$ runs over the zeros of $L(s, sym^2 f)$ in the critical strip of height up to T, with any $1 \le T \le \sqrt{x}$ and the implied constant is absolute, q_1 is the analytic conductor of $L(s, sym^2 f)$. Now we estimate the sum $\sum_{|\gamma| \le T} \frac{1}{\rho}$. Since

$$|\sum_{|\gamma| \le T} \frac{1}{\rho}| \le \sum_{|\gamma| \le T} \frac{1}{|\rho|} \le \sum_{|\gamma| \le T} \frac{1}{|\gamma|},$$

it is enough to estimate $\sum_{0<\gamma \le T} \frac{1}{\gamma}$. If $N(t)$ denotes the number of zeros of $L(s, sym^2 f)$ in the critical strip with ordinates less than t, as it was done in [**3**, pp. 111] for Riemann zeta function, here we have

$$\sum_{0<\gamma \le T} \frac{1}{\gamma} = \int_0^T t^{-1} dN(t) = \frac{1}{T} N(T) + \int_0^T t^{-2} dN(t).$$

By the zero density estimate for $L(s, sym^2 f)$, we have $N(t) \ll t \log t$. Thus from the above identity, we deduce that

$$\sum_{0<\gamma \le T} \frac{1}{\gamma} \ll (\log T)^2,$$

and therefore

$$\sum_{|\gamma| \le T} \frac{1}{\rho} \ll (\log T)^2.$$

Substituting the above estimate in (12), we get the required explicit formula:

$$\psi(sym^2 f, x) = \sum_{p \le x} a_{sym^2 f}(p) \log p = - \sum_{|\gamma| \le T} \frac{x^\rho}{\rho} + O\left(\frac{x}{T} (\log x) \log (x^3 q_1)\right). \tag{13}$$

Thus $L(s, sym^2 f)$ satisfies all the hypothesis of Theorem 2.2. Therefore by Theorem 2.2 there exists a positive integer $\nu_1 < 1$ such that for any α with $\nu_1 < \alpha < 1$, we have

$$\sum_{x \le p < x + x^\alpha} a_{sym^2 f}(p) \log p = o(x^\alpha). \tag{14}$$

Now to get a bound for $\sum_{p \le x} |a_{sym^2 f}(p)|^2 \log p$, we note that since

$$|a_{sym^2 f}(p)|^2 = a_{sym^2 f \times sym^2 f}(p)$$

is true for any prime p, we consider the Dirichlet series $L(s, sym^2 f \times sym^2 f)$. By [**15**, Theorem B], $L(s, sym^2 f \times sym^2 f)$ does not have any exceptional zero. Thus by [**5**, Theorem A.1], we get the required zero free region for $L(s, sym^2 f \times sym^2 f)$. The required log free zero density estimate follows from [**14**, Theorem 1.1] since $sym^2 f$ satisfies generalised Ramanujan conjecture. The zero density estimate follows from [**6**, Theorem 5.8]. Now as we have done in the case of $L(s, sym^2 f)$ in (13), we get the following explicit formula for $L(s, sym^2 f \times sym^2 f)$:

$$\begin{aligned} \psi(sym^2 f \times sym^2 f, x) &= \sum_{p \le x} a_{sym^2 f \times sym^2 f}(p) \log p \\ &= x - \sum_{|\gamma| \le T} \frac{x^\rho}{\rho} + O\left(\frac{x}{T} (\log x) \log (x^9 q_2)\right), \end{aligned}$$

where $\rho = \beta + i\gamma$ runs over the zeros of $L(s, sym^2 f \times sym^2 f)$ in the critical strip of height up to T, with any $1 \le T \le \sqrt{x}$ and the implied constant is absolute,

q_2 is the analytic conductor of $L(s, sym^2 f \times sym^2 f)$. Therefore we have shown that $L(s, sym^2 f \times sym^2 f)$ satisfies all the hypothesis of Theorem 2.1 and hence by Theorem 2.1, there exists a positive number $\nu_2 < 1$ such that for any $\nu_2 < \beta < 1$, we have

$$\sum_{x \le p < x + x^\beta} |a_{sym^2 f}(p)|^2 \log p \gg x^\beta. \tag{15}$$

Choose $\nu = \max\{\nu_1, \nu_2\}$. Now assume on the contrary that there is no sign change for the sequence $a_{sym^2 f}(p)$ for $p \in [x, x + x^\theta)$, where $\nu < \theta < 1$. Without loss of any generality, we assume that all coefficients are positive in that interval. Since

$$|a_{sym^2 f}(p)| \le 3,$$

by comparing (14) and (15), we deduce that

$$x^\theta \ll \sum_{x \le p < x + x^\theta} |a_{sym^2 f}(p)|^2 \log p \ll \sum_{x \le p < x + x^\theta} a_{sym^2 f}(p) \log p = o(x^\theta).$$

The above identity gives a contradiction. This proves the theorem.

Acknowledgments. We would like to thank Prof. M. Ram Murty, Prof. A. Raghuram and Dr. G. K. Viswanadham for useful discussions. We also thank Prof. R. Thangadurai for giving useful comments on an earlier version of the paper. The research of the second author is partially supported by the DST-SERB grant MTR/2017/000022.

References

[1] J. Cogdell and P. Michel, *On the complex moments of symmetric power L-functions at $s = 1$*, Int. Math. Res. Not. **31** (2004), 1561–1617, DOI 10.1155/S1073792804132455. MR2035301

[2] P. Deligne, *La conjecture de Weil. I* (French), Inst. Hautes Études Sci. Publ. Math. **43** (1974), 273–307. MR0340258

[3] H. Davenport, *Multiplicative number theory*, 3rd ed., Graduate Texts in Mathematics, vol. 74, Springer-Verlag, New York, 2000. Revised and with a preface by Hugh L. Montgomery. MR1790423

[4] S. Gun and M. Ram Murty, *Generalization of an identity of Ramanujan*, J. Ramanujan Math. Soc. **31** (2016), no. 2, 125–135. MR3518178

[5] P. Humphries and F. Brumley, *Standard zero-free regions for Rankin-Selberg L-functions via sieve theory*, Math. Z., https://doi.org/10.1007/s00209-018-2136-8.

[6] H. Iwaniec and E. Kowalski, *Analytic number theory*, American Mathematical Society Colloquium Publications, vol. 53, American Mathematical Society, Providence, RI, 2004. MR2061214

[7] W. Kohnen, *A short note on Fourier coefficients of half-integral weight modular forms*, Int. J. Number Theory **6** (2010), no. 6, 1255–1259, DOI 10.1142/S1793042110003484. MR2726580

[8] Y.-K. Lau and J. Wu, *A density theorem on automorphic L-functions and some applications*, Trans. Amer. Math. Soc. **358** (2006), no. 1, 441–472, DOI 10.1090/S0002-9947-05-03774-8. MR2171241

[9] J. Meher and M. R. Murty, *Oscillations of coefficients of Dirichlet series attached to automorphic forms*, Proc. Amer. Math. Soc. **145** (2017), no. 2, 563–575, DOI 10.1090/proc/13264. MR3577861

[10] C. J. Moreno, *The Hoheisel phenomenon for generalized Dirichlet series*, Proc. Amer. Math. Soc. **40** (1973), 47–51, DOI 10.2307/2038629. MR0327682

[11] M. R. Murty, *Oscillations of Fourier coefficients of modular forms*, Math. Ann. **262** (1983), no. 4, 431–446, DOI 10.1007/BF01456059. MR696516

[12] M. R. Murty, M. Dewar, and H. Graves, *Problems in the theory of modular forms*, Institute of Mathematical Sciences Lecture Notes, vol. 1, Hindustan Book Agency, New Delhi, 2015. MR3330491

[13] M. R. Murty and S. Pujahari, *Distinguishing Hecke eigenforms*, Proc. Amer. Math. Soc. **145** (2017), no. 5, 1899–1904, DOI 10.1090/proc/13446. MR3611306

[14] R. J. L. Oliver and J. Thorner, *Effective log-free zero density estimates for automorphic L-functions and the Sato-Tate conjecture*, Int. Math. Res. Not. IMRN, https://doi.org/10.1093/imrn/rnx309.

[15] D. Ramakrishnan and S. Wang, *On the exceptional zeros of Rankin-Selberg L-functions*, Compositio Math. **135** (2003), no. 2, 211–244, DOI 10.1023/A:1021761421232. MR1955318

Institute of Mathematical Sciences, HBNI, CIT Campus, Taramani, Chennai 600113, India

Email address: baleshk@imsc.res.in

School of Mathematical Sciences, National Institute of Science Education and Research, Bhubaneswar, HBNI, P. O. Jatni, Khurda 752050, Odisha, India

Email address: jaban@niser.ac.in

The University of Hong Kong, Pokfulam, Hong Kong

Email address: sudhir@hku.hk

Contemporary Mathematics
Volume **732**, 2019
https://doi.org/10.1090/conm/732/14789

On primes in arithmetic progressions

Junxian Li

ABSTRACT. Fix k a positive integer, the least prime modulo k is defined as $P(k) = \max_{(\ell,k)=1} p(k,\ell)$, where $p(k,\ell)$ is the least prime that is equivalent to ℓ modulo k. The purpose of the paper is to give a summery on recent progress on $P(k)$. Based on probabilistic models of primes, it is expected that $P(k) \asymp \phi(k)\log^2 k$. While upper bounds for $P(k)$ have been studied since Linnik, results on lower bounds are less known. The recent work on large gaps between primes allows an improvement of a lower bound for $P(k)$ due to Pomerance. In particular, for almost all k we have $P(k) \gg \phi(k)\log k \log_2 k \log_4 k/\log_3 k$, answering a question of Ford, Green, Konyangin, Maynard, and Tao.

Fix a positive integer k and let ℓ be coprime to k. Let $p(k,\ell)$ denote the smallest prime equivalent to ℓ modulo k. We are interested in finding the size of

$$P(k) := \max_{(\ell,k)=1} p(k,\ell).$$

If we assume that primes behave like random variables, we can then interpret this problem as a variant of the coupon collector's problem. For a fixed integer k, there are $\phi(k)$ distinct coupons to collect, which are represented by the residue classes coprime to k. Each prime p can be identified as a coupon for the class $p \pmod k$. It is known that if each coupon is equally likely to be collected and comes independently, then the expected waiting time T_k to collect a full set of all $\phi(k)$ coupons is $\phi(k)\log\phi(k)$. Furthermore, using a result of Erdős and Rényi [**3**], one would expect the limiting distribution of T_k is given by

$$\lim_{k\to\infty} \mathbb{P}\left(\frac{T_k}{\phi(k)} - \log\phi(k) < x\right) = \exp(-e^{-x}),$$

which is the known as the standard Gumbel distribution. Indeed, we computed the distribution of the normalized waiting time $r_k := \frac{\pi(P(k))}{\phi(k)} - \log(\phi(k))$, and compared it with the standard Gumbel distribution in Figure 1. The discrepancy may due to the week dependency between primes. In particular, two primes that are within k distance are not in the same residue class modulo k. On the other hand, one should still expect

$$\liminf_k \frac{P(k)}{\phi(k)\log^2 k} = 1, \qquad \limsup_k \frac{P(k)}{\phi(k)\log^2 k} = 2,$$

as is shown in [**8**].

2010 *Mathematics Subject Classification.* Primary 11N13, 11N05; Secondary 11N36.
Key words and phrases. Primes in progressions, gaps between primes.

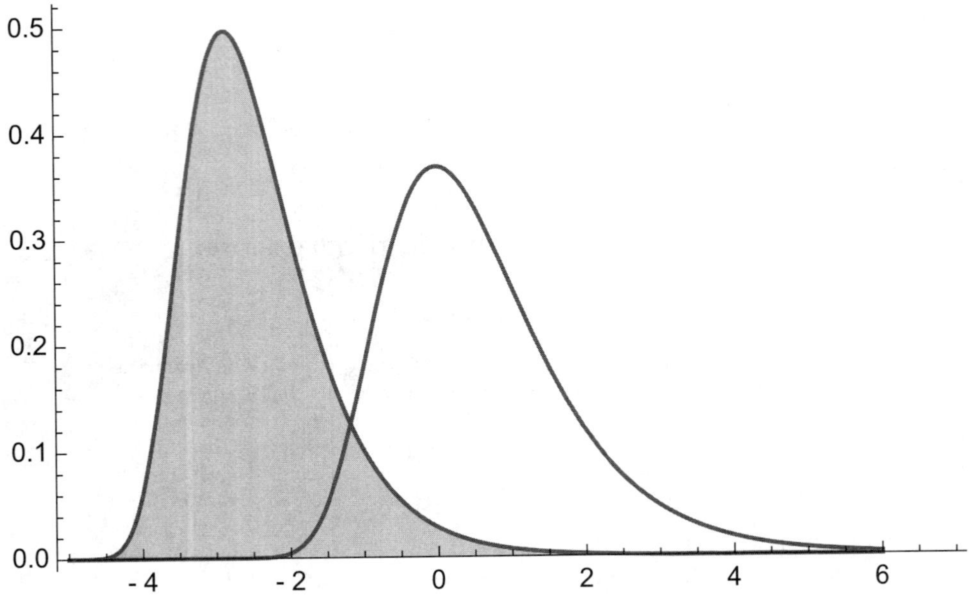

FIGURE 1. Histogram of $r_k = \frac{\pi(P(k))}{\phi(k)} - \log(\phi(k))$ for $k \leq 10^6$, and the density function of the distribution $e^{-e^{c-bx}}$, with $b \approx 1.35$, $c \approx -3.86$ and $b = 1$, $c = 0$ respectively.

For upper bounds for $P(k)$, it is Linnik [**7**] who first proved that $P(k) \ll k^L$, where $L > 0$ is a fixed constant. Subsequent authors improved upon the value of L, including Chen [**1**], Graham [**5**], Heath-Brown [**6**], Jutila [**9**], Pan [**10**], and Wang [**12**]. Recently Xylouris [**13**] showed that $L \leq 5.18$. Chowla [**2**] observed that the Generalized Riemann Hypothesis implies $P(k) \ll_\epsilon k^{2+\epsilon}$ for any fixed $\epsilon > 0$, and conjectured that $P(k) \ll_\epsilon k^{1+\epsilon}$. Less is known for lower bounds for $P(k)$. Pomerance [**11**] showed that $P(k) \geq (e^\gamma + o(1))\phi(k)\log k$, improving a constant factor e^γ of the trivial bound given by the prime number theorem. He also showed that $P(k)/(\phi(k)\log k)$ tends to infinity for almost every k. Specifically, let Q be the set of integers k with more than $\exp(\log_2 k/\log_3 k)^1$ distinct prime factors. Pomerance [**11**] showed that

$$P(k) \geq (e^\gamma + o(1))\phi(k)\log k \log_2 k \frac{\log_4 k}{(\log_3 k)^2}$$

for every $k \notin Q$. Recent breakthroughs in large gaps between primes by the work of Ford, Green, Konyangin, Maynard, and Tao [**4**] allows one to improve this result by a factor of $\log_3 k$ for a thinner exceptional set. Precisely, it is shown in [**8**] that for any $\epsilon > 0$, there exists $k_0(\epsilon)$ such that for all integers $k > k_0(\epsilon)$ with no more than $\exp((\frac{1}{2} - \epsilon)\log_2 k \log_4 k/\log_3 k)$ distinct prime factors, we have

$$P(k) \gg \phi(k)\log k \log_2 k \frac{\log_4 k}{\log_3 k},$$

which answers the question in [**4**] asking for an improvement for $P(k)$.

[1] As usual in the subject, $\log_2 x = \log\log x, \log_3 x = \log\ \log\log x$, and so on.

References

[1] J. R. Chen, *On the least prime in an arithmetical progression and two theorems concerning the zeros of Dirichlet's L-functions*, Sci. Sinica **20** (1977), no. 5, 529–562. MR0476668

[2] S. Chowla, *On the least prime in an arithmetical progression*, J. Indian Math. Soc. **1** (2) (1934), 1-3.

[3] P. Erdős and A. Rényi, *On a classical problem of probability theory* (English, with Russian summary), Magyar Tud. Akad. Mat. Kutató Int. Közl. **6** (1961), 215–220. MR0150807

[4] K. Ford, B. Green, S. Konyagin, J. Maynard, and T. Tao, *Long gaps between primes*, J. Amer. Math. Soc. **31** (2018), no. 1, 65–105, DOI 10.1090/jams/876. MR3718451

[5] S. Graham, *On Linnik's constant*, Acta Arith. **39** (1981), no. 2, 163–179, DOI 10.4064/aa-39-2-163-179. MR639625

[6] D. R. Heath-Brown, *Zero-free regions for Dirichlet L-functions, and the least prime in an arithmetic progression*, Proc. London Math. Soc. (3) **64** (1992), no. 2, 265–338, DOI 10.1112/plms/s3-64.2.265. MR1143227

[7] U. V. Linnik, *On the least prime in an arithmetic progression. I. The basic theorem* (English, with Russian summary), Rec. Math. [Mat. Sbornik] N.S. **15(57)** (1944), 139–178. MR0012111

[8] J. Li, K. Pratt, and G. Shakan, *A lower bound for the least prime in an arithmetic progression*, Q. J. Math. **68** (2017), no. 3, 729–758, DOI 10.1093/qmath/hax001. MR3698292

[9] M. Jutila, *On Linnik's constant*, Math. Scand. **41** (1977), no. 1, 45–62, DOI 10.7146/math.scand.a-11701. MR0476671

[10] C. D. Pan, *On the least prime in an arithmetical progression*, Sci. Record (N.S.) **1** (1957), 311–313. MR0105398

[11] C. Pomerance, *A note on the least prime in an arithmetic progression*, J. Number Theory **12** (1980), no. 2, 218–223, DOI 10.1016/0022-314X(80)90056-6. MR578815

[12] W. Wang, *On the least prime in an arithmetic progression*, Acta Math. Sinica (N.S.) **7** (1991), no. 3, 279–289, DOI 10.1007/BF02583005. A Chinese summary appears in Acta Math. Sinica **35** (1992), no. 4, 575. MR1141242

[13] T. Xylouris, *On the least prime in an arithmetic progression and estimates for the zeros of Dirichlet L-functions*, Acta Arith. **150** (2011), no. 1, 65–91, DOI 10.4064/aa150-1-4. MR2825574

University of Illinois Urbana-Champaign, Urbana, Illinois 61801
Email address: `jli135@illinois.edu`

Contemporary Mathematics
Volume **732**, 2019
https://doi.org/10.1090/conm/732/14788

The Fourier coefficients of Eisenstein series newforms

Benjamin Linowitz and Lola Thompson

Abstract. In this article, we study the Fourier coefficients of Eisenstein series newforms. We obtain a sharp refinement of the strong multiplicity-one theorem by showing that the density of primes p for which the pth Hecke eigenvalues of two distinct Eisenstein series newforms differ is of the form $1/n$ for some $n \geq 2$. Additionally, we show that if f is an Eisenstein series newform whose Fourier coefficients $a_f(n)$ are real then there is a constant $\delta > 0$ such that the sequence $(a_f(n))_{n \leq x}$ has at least δx sign changes.

1. Introduction

An important problem in the theory of modular forms is to determine the extent to which a newform is determined by its Fourier coefficients. A similar problem asks for characterizations of a newform that enable one to distinguish it from other newforms. The classical strong multiplicity-one theorem, for instance, shows that if the pth Fourier coefficients of two (normalized) cuspidal newforms agree for all but finitely many primes p then the two newforms are equal. Statistical refinements of results such as this are of great interest. With respect to the strong multiplicity-one theorem, Ramakrishnan [**9**] has shown that, if the pth Fourier coefficients of two cuspidal newforms agree for any set of primes having density greater than 7/8, then the two newforms are equal. See Walji [**10**] for a further refinement.

A related problem that has attracted a great deal of attention in recent years is to consider a newform whose Hecke eigenvalues are real and determine the extent to which the newform is determined by the *signs* of its Hecke eigenvalues (see [**3, 4, 6, 7**]). Some of the many questions that have been addressed in this context are bounds on the first sign change in the sequence of Hecke eigenvalues (a variant of the classical least quadratic non-residue problem in analytic number theory), determining how often the sequence of signs of Fourier coefficients of a newform agree with a fixed sequence of signs, and counting the number of sign changes in the sequence of Fourier coefficients up to x.

In a previous paper of the authors [**5**], the aforementioned problems were considered for the Fourier coefficients of Eisenstein series newforms. It was shown in [**5**, Theorem 5.4], for instance, that an Eisenstein series newform is determined by the signs of its pth Fourier coefficients for any set of primes having density greater than 1/2. (Note that by using quadratic twists the authors were able to produce examples which show that this bound is best possible.) In this paper we continue

2010 *Mathematics Subject Classification.* Primary 11F41.

our study of the Fourier coefficients of Eisenstein series newforms. With respect to the strong multiplicity-one theorem, we prove a new result (Theorem 3.4) that implies:

THEOREM 1. *Let f, g be Eisenstein series newforms of weight k and levels N, M. Then the density of the set of primes p for which the pth Fourier coefficients of f and g agree is of the form $1/d$ for some divisor d of $\varphi(\mathrm{lcm}[N, M])$, where φ is the Euler totient function. Moreover, as g varies over all Eisenstein series newforms of weight k, every fraction of the form $1/n$ occurs as the density of primes for which the pth Fourier coefficients of f and g agree.*

Let f be a cuspidal newform for the full modular group with Hecke eigenvalues $\lambda_f(n)$. Matomäki and Radziwiłł have recently proven [**7**] that there is a positive constant δ such that, for x sufficiently large, the sequence $(\lambda_f(n))_{n\leq x}$ has at least δx sign changes. In Theorem 4.6 we prove an analogous result for the Fourier coefficients of Eisenstein series newforms with real coefficients. Our result can be reformulated in the following manner:

THEOREM 2. *Let f be an Eisenstein series newform whose Fourier coefficients $a_f(n)$ are real numbers. There exists a positive constant δ such that, for x sufficiently large, the sequence $(a_f(n))_{n\leq x}$ has at least δx sign changes. Moreover, half of the non-zero $a_f(n)$ are positive and the other half are negative.*

We now discuss the organization of the paper. In Section 2 we review the newform theory of Eisenstein series. This theory was developed by Weisinger [**12**] in his (unpublished) thesis and extended to Hilbert modular forms by the first author and Atwill [**1**]. In Section 3 we prove a refinement of the strong multiplicity-one theorem for Eisenstein series newforms. In Section 4 we prove a result concerning the number of sign changes of the Fourier coefficients of an Eisenstein series newform.

2. A newform theory for Eisenstein series

Given a positive integer N, a Dirichlet character χ modulo N and an integer $k \geq 3$, we let $M_k(N, \chi)$ be the complex vector space of modular forms for $\Gamma_0(N)$ with weight k and character χ. Let $E_k(N, \chi)$ denote the subspace of Eisenstein series and $S_k(N, \chi)$ denote the subspace of cusp forms. For a prime p, we will denote by T_p the pth Hecke operator and recall that T_p preserves both $E_k(N, \chi)$ and $S_k(N, \chi)$. Additionally, for a positive integer d, we denote by B_d the shift operator defined by

$$f \mid B_d = d^{-k/2} f \mid \begin{pmatrix} d & 0 \\ 0 & 1 \end{pmatrix}$$

for $f \in M_k(N, \chi)$. In terms of Fourier coefficients, if

$$f = \sum_{n=0}^{\infty} a_f(n) q^n,$$

where $q = e^{2\pi i z}$, then

$$f \mid B_d = \sum_{n=0}^{\infty} a_f(n) q^{nd}.$$

This operator maps $M_k(N,\chi)$ to $M_k(dN,\chi)$ and takes Eisenstein series to Eisenstein series and cusp forms to cusp forms. We note that if $(d,p)=1$ then $B_dT_p = T_pB_d$.

Let N_1, N_2 be positive integers such that $N = N_1N_2$ and let χ_1, χ_2 be Dirichlet characters modulo N_1, N_2 for which $(\chi_1\chi_2)(-1) = (-1)^k$. We now define a modified sum-of-divisors function by

$$\sigma^{k-1}_{\chi_1,\chi_2}(n) = \sum_{d|n} \chi_1(n/d)\chi_2(d)d^{k-1}. \tag{2.1}$$

Associated to the triple (χ_1, χ_2, k) is a function

$$E(\chi_1,\chi_2,k) = \frac{\delta(\chi_1)}{2}L(1-k,\chi_2) + \sum_{n\geq 1} \sigma^{k-1}_{\chi_1,\chi_2}(n)q^n,$$

where $q = e^{2\pi i z}$, $L(s,\chi_2)$ is the Dirichlet L-function associated to χ_2, and $\delta(\chi_1) = 1$ if χ_1 is principal and equal to 0 otherwise. It is well-known that $E(\chi_1,\chi_2,k) \in E_k(N,\chi)$ for $\chi = \chi_1\chi_2$ (see [**2**, Chapter 5]).

Definition. We will say that the Eisenstein series $E(\chi_1,\chi_2,k)$ is a newform if the characters χ_1, χ_2 are primitive.

The following proposition shows that, just like the cuspidal setting, Eisenstein series newforms are eigenforms for all of the Hecke operators T_p and have their T_p^{th}-eigenvalue equal to their p^{th} Fourier coefficient.

Proposition 2.1. *Let $E(\chi_1,\chi_2,k)$ be as above. Then for every prime p we have*

$$E(\chi_1,\chi_2,k) \mid T_p = \sigma^{k-1}_{\chi_1,\chi_2}(p)\cdot E(\chi_1,\chi_2,k).$$

Proof. If $p \nmid N$ then the desired equality is easily verified by examining the action of T_p on the Fourier coefficients of $E(\chi_1,\chi_2,k)$. When $p \mid N$ the equality follows from Proposition 3.15 of [**1**]. ☐

Denote by $E_k^{new}(N,\chi)$ the subspace of $E_k(N,\chi)$ spanned by newforms of exact level N (i.e., newforms of level N which are not shifts of newforms of level $M < N$). The following result (Proposition 3.11 of [**1**]) shows that, as in the setting of cusp forms, every Eisenstein series can be expressed uniquely as a linear combination of shifts of newforms.

Theorem 2.2. *Let the notation be as above. We have the following decompsiton of $E_k(N,\chi)$:*

$$E_k(N,\chi) = \bigoplus_{\mathrm{cond}(\chi)|M|N} \ \bigoplus_{d|NM^{-1}} E_k^{new}(M,\chi) \mid B_d.$$

3. The strong multiplicity-one theorem and matching densities

In this section we will recall a refinement of the strong multiplicity-one theorem proven by the authors in [**5**, Section 5]. This refinement shows that if the set of primes at which the pth Hecke eigenvalues of two Eisenstein series newforms coincide has a sufficiently large Dirichlet density then the two newforms are equal. This result raises a number of questions which we will address in this section. For instance, if f and g are two Eisenstein series newforms, what can one say about the density of the set of primes for which their Hecke eigenvalues coincide? In order to address these questions we introduce some useful terminology developed by Walji [**11**].

DEFINITION. Let f, g be Eisenstein series newforms. The matching density of f and g is the Dirichlet density, if it exists, of the set of primes p for which

$$a_f(p) = a_g(p).$$

We will denote this matching density by $\delta(f, g)$.

The first strong multiplicity-one theorem in the setting of Eisenstein series was proven by Weisinger [**12**], who showed that if f and g are Eisenstein series newforms for which $a_f(p) = a_g(p)$ for all but finitely many primes p then $f = g$. One of the main results of [**5**, Theorem 5.1] is the following refinement of Weisinger's result.

THEOREM 3.1. *Let $f \in E_k(N, \chi_f)$ and $g \in E_{k'}(M, \chi_g)$ be normalized newforms such that $\delta(f, g) > 1/2$. Then $k = k'$, $N = M$, $\chi_f = \chi_g$ and $f = g$.*

We note that Theorem 3.1 is best possible in that one may easily obtain, using quadratic twists, pairs of newforms whose Hecke eigenvalues agree for a set of primes having density 1/2. Indeed, suppose that $f \in E_k(N, \chi_f)$ is a newform and that θ is a quadratic Dirichlet character whose conductor is coprime to N. Then by [**1**, Corollary 5.5], the character twist $f \otimes \theta$ is a newform. Moreover, the pth coefficient of $f \otimes \theta$ is equal to $a_f(p)$ if $\theta(p) = 1$ and is equal to $-a_f(p)$ if $\theta(p) = -1$. As each of these cases occurs for a set of primes having density 1/2 and the set of primes for which $a_f(p) = 0$ is finite, we have $\delta(f, f \otimes \theta) = 1/2$.

Note that the authors have also shown [**5**, Theorem 5.3] that, in essence, the above quadratic twist construction is the only way of producing newforms f, g for which $\delta(f, g) = 1/2$. This raises the interesting question of what the maximal value of $\delta(f, g)$ is for non-twist-equivalent f and g. In what follows, we will prove a theorem (Theorem 3.4) which has as an immediate consequence that the maximal such value is 1/3. The following example of non-twist-equivalent Eisenstein newforms with $\delta(f, g) = 1/3$ shows that this maximal value is indeed attained.

EXAMPLE 3.2. Let χ denote the Dirichlet character modulo 7 defined by $\chi(3) = \zeta_6$ and let ψ be the Dirichlet character modulo 7 defined by $\psi(3) = -1$. Then χ and ψ are primitive characters of conductor 7 with the property that, for a prime $p \neq 7$, one has $\chi(p) = \psi(p)$ if and only if $p \equiv \pm 1 \pmod 7$. In particular the density of primes p for which $\chi(p) = \psi(p)$ is exactly 1/3. Now let θ be the trivial character (primitive of conductor 1) and define $f = E(\theta, \chi, k)$ and $g = E(\theta, \psi, k)$ for any odd integer $k \geq 3$. Then f and g are Eisenstein series newforms of level 7 which are clearly not twist-equivalent and satisfy $\delta(f, g) = 1/3$.

In order to prove Theorem 3.4 we will employ the following lemma (for a proof, see [**5**, Lemma 5.2]).

LEMMA 3.3. *Let $\chi_1, \chi_2, \psi_1, \psi_2$ be Dirichlet characters and let $k \geq 2$ be an integer. There exists a positive constant p_0 such that for all primes $p > p_0$, if*

$$\chi_1(p) + \chi_2(p)p^{k-1} = \psi_1(p) + \psi_2(p)p^{k-1}$$

then $\chi_1(p) = \psi_1(p)$ and $\chi_2(p) = \psi_2(p)$.

THEOREM 3.4. *Let $f \in E_k(N, \chi_f)$ and $g \in E_{k'}(M, \chi_g)$ be newforms.*

(1) *$\delta(f, g) = 0$ if and only if $k \neq k'$.*

(2) *If* $k = k'$ *then* $\delta(f,g) = 1/d$ *for some divisor* d *of* $\varphi(\mathrm{lcm}[N,M])$.
(3) *For every positive integer* n, *there exists a newform* g *such that* $\delta(f,g) = 1/n$.

PROOF. Write $f = E(\chi_1,\chi_2,k)$ and $g = E(\psi_1,\psi_2,k')$ and note that $a_f(p) = \chi_1(p)+\chi_2(p)p^{k-1}$ and $a_g(p) = \psi_1(p)+\psi_2(p)p^{k'-1}$. Fix $\epsilon \in (0,1)$. For all sufficiently large primes p, we have $|a_f(p)| \in (p^{k-1-\epsilon}, p^{k-1+\epsilon})$ and $|a_g(p)| \in (p^{k'-1-\epsilon}, p^{k'-1+\epsilon})$. It follows that $k \neq k'$ implies $\delta(f,g) = 0$, as these two intervals intersect if and only if $k = k'$. To prove the converse, we must show that if $k = k'$ then $\delta(f,g) > 0$. Suppose therefore that $k = k'$. If $p \equiv 1 \pmod{\mathrm{lcm}[N,M]}$ then

$$a_f(p) = \chi_1(p) + \chi_2(p)p^{k-1} = 1 + p^{k-1} = \psi_1(p) + \psi_2(p)p^{k-1} = a_g(p).$$

Assertion (1) now follows from the fact that the set of primes congruent to 1 modulo $\mathrm{lcm}[N,M]$ has positive density.

Henceforth we will assume that $k = k'$. We now prove (2). Denote by $\chi_1', \chi_2', \psi_1', \psi_2'$ the characters modulo $\mathrm{lcm}[N,M]$ induced by $\chi_1, \chi_2, \psi_1, \psi_2$. Let $p_1, \dots, p_{\varphi(\mathrm{lcm}[N,M])}$ be primes representing the residue classes of $(\mathbf{Z}/\mathrm{lcm}[N,M]\mathbf{Z})^\times$ which are all large enough that Lemma 3.3 holds. In particular, if $a_f(p_i) = a_g(p_i)$ then Lemma 3.3 implies that $\chi_1'(p_i) = \psi_1'(p_i)$ and $\chi_2'(p_i) = \psi_2'(p_i)$, hence $\chi_1(p_i) = \psi_1(p_i)$ and $\chi_2(p_i) = \psi_2(p_i)$. Conversely, if $\chi_1(p_i) = \psi_1(p_i)$ and $\chi_2(p_i) = \psi_2(p_i)$ then we will have $a_f(p) = a_g(p)$ for every prime $p \equiv p_i \pmod{\mathrm{lcm}[N,M]}$. From this we conclude that if there are precisely s values of i for which $\chi_1(p_i) = \psi_1(p_i)$ and $\chi_2(p_i) = \psi_2(p_i)$, then $\delta(f,g) = s/\varphi(\mathrm{lcm}[N,M])$. In particular, to prove assertion (2), it suffices to show that s is a divisor of $\varphi(\mathrm{lcm}[N,M])$. But this is clear, as the set of such primes p_i comprise a subgroup of $(\mathbf{Z}/\mathrm{lcm}[N,M]\mathbf{Z})^\times$. Explicitly, this subgroup is $\ker(\chi_1'\overline{\psi_1'}) \cap \ker(\chi_2'\overline{\psi_2'})$, where $\chi_1'\overline{\psi_1'}$ and $\chi_2'\overline{\psi_2'}$ are regarded as homomorphisms from $(\mathbf{Z}/\mathrm{lcm}[N,M]\mathbf{Z})^\times$ to $\mathbf{C}^\times$.

We now prove (3). Fix a positive integer n. We will construct a newform g for which $\delta(f,g) = 1/n$. As we trivially have $\delta(f,f) = 1$, we may assume that $n \geq 2$. Let q be a prime which does not divide N and which satisfies $q \equiv 1 \pmod{n}$. As $(\mathbf{Z}/q\mathbf{Z})^\times$ is cyclic of order $q-1$, there exists a Dirichlet character θ_0 modulo q which has order $q-1$. Let $\theta = \theta_0^{(q-1)/n}$ and note that the density of primes p for which $\theta(p) = 1$ is $1/n$. By [**1**, Corollary 5.5], $f \otimes \theta$ is a newform, and by definition, $a_{g\otimes\theta}(p) = \theta(p)a_f(p)$. It follows that $\delta(f, f\otimes\theta) = 1/n$, proving assertion (3). □

REMARK. We note that the third assertion of Theorem 3.4 can alternatively be proven by associating to every Eisenstein series newform a reducible two-dimensional Galois representation (which would arise from the pair of Dirichlet characters defining the newform) and then using the fact that the matching densities of abelian representations are of the form $\{1/n : n \geq 2\}$ (see [**11**]).

4. Sign changes of Fourier coefficients

A series of recent papers studies when the first negative Fourier coefficient occurs. In the cusp form case, we have a well-known theorem of Kowalski, Lau, Soundararajan and Wu [**4**], which was recently improved by Matomäki [**6**]. In order to state these results, we denote by f a cuspidal newform of weight k, level N, and trivial nebentypus. Note that the restriction to trivial nebentypus ensures that all of the Fourier coefficients of f are real. We denote by $\lambda_f(n)$ the normalized Fourier coefficients of f.

THEOREM 4.1 (Kowalski, Lau, Soundararajan, Wu, 2010). *For k even and $(n,N)=1$, we have $\lambda_f(n)<0$ for some $n \ll (k^2N)^{9/20}$.*

THEOREM 4.2 (Matomäki, 2012). *For k even and $(n,N)=1$, we have $\lambda_f(n)<0$ for some $n \ll (k^2N)^{\mathbf{3/8}}$.*

Matomäki and Radziwiłł [**7**] went on to prove a stronger result in the special case where $N=1$.

THEOREM 4.3 (Matomäki and Radziwiłł, 2015). *For $N=1$, there exists a positive constant δ such that, for x sufficiently large, the sequence $(\lambda_f(n))_{n\le x}$ has at least δx sign changes. Furthermore, half of the non-zero $\lambda_f(n)$ are positive and half are negative.*

We can obtain analogous results for Eisenstein series newforms. Recall that if f is an Eisenstein series newform then f is associated to a triple (χ_1,χ_2,k) and satisfies

$$f = E(\chi_1,\chi_2,k) = \frac{\delta(\chi_1)}{2}L(1-k,\chi_2) + \sum_{n\ge 1}\sigma^{k-1}_{\chi_1,\chi_2}(n)q^n,$$

where $q=e^{2\pi i z}$, $L(s,\chi_2)$ is the Dirichlet L-function associated to χ_2 and $\delta(\chi_1)=1$ if χ_1 is principal and equal to 0 otherwise. In order to ensure that our Fourier coefficients are real, we will only consider Eisenstein series newforms associated to quadratic Dirichlet characters. We additionally assume that $k\ge 3$.

Our proofs all rely on a single key lemma, which shows that understanding the signs of Fourier coefficients of Eisenstein series newforms amounts to understanding the behavior of certain quadratic Dirichlet characters. The proof of the key lemma first appeared in [**5**]. We reproduce it here for the sake of completeness.

LEMMA 4.4. *If $(n,N)=1$ and χ_1,χ_2 are quadratic, the sign of $\sigma^{k-1}_{\chi_1,\chi_2}(n)$ is completely determined by the behavior of $\chi_2(n)$.*

PROOF. Starting with the definition of $\sigma^{k-1}_{\chi_1,\chi_2}(n)$ given in (2.1), we can separate off the $d=n$ term, which gives us

$$\begin{aligned}\sigma^{k-1}_{\chi_1,\chi_2}(n) &= \sum_{d\mid n}\chi_1(n/d)\chi_2(d)d^{k-1}\\ &= \chi_2(n)n^{k-1} + \sum_{\substack{d\mid n\\ d<n}}\chi_1(n/d)\chi_2(d)d^{k-1}\\ &= n^{k-1}\left(\chi_2(n) + \sum_{\substack{d\mid n\\ d<n}}\frac{\chi_1(n/d)\chi_2(d)}{(n/d)^{k-1}}\right). \qquad (4.1)\end{aligned}$$

Consider the summation over the proper divisors of n in (4.1). Since $|\chi_1(d)|, |\chi_2(d)| \le 1$ for all $d\in \mathbf{Z}^+$, we can bound each numerator from above by 1, which yields

$$\left|\sum_{\substack{d\mid n\\ d<n}}\frac{\chi_1(n/d)\chi_2(d)}{(n/d)^{k-1}}\right| \le \sum_{\substack{d\mid n\\ d<n}}\frac{1}{(n/d)^{k-1}} \le \sum_{m\ge 2}\frac{1}{m^{k-1}} = \zeta(k-1)-1<1.$$

Thus, the $\chi_2(n)n^{k-1}$ term dominates in (4.1) and so we have

$$\operatorname{sgn} \sigma^{k-1}_{\chi_1,\chi_2}(n) = \chi_2(n).$$

□

In [**5**], we used Lemma 4.4 to demonstrate when the first sign change occurs.

THEOREM 4.5. *Let p_0 represent the smallest prime for which $\sigma^{k-1}_{\chi_1,\chi_2}(p_0) < 0$. Then, for any fixed $\varepsilon > 0$, we have*

$$p_0 \ll_\varepsilon N^{\frac{1}{4\sqrt{e}}+\varepsilon}.$$

Note that the $1/4\sqrt{e}$ in the exponent comes from the classical Burgess bound for the least quadratic non-residue. We can use Theorem 4.5 to obtain an analogue of Theorem 4.3 for Eisenstein series:

THEOREM 4.6. *Let f be an Eisenstein series newform with Fourier coefficients $\sigma^{k-1}_{\chi_1,\chi_2}(n)$, where χ_1, χ_2 are quadratic Dirichlet characters. There exists a positive constant δ such that, for x sufficiently large, the sequence $(\sigma^{k-1}_{\chi_1,\chi_2}(n))_{n\leq x}$ has at least δx sign changes. Moreover, half of the non-zero $\sigma^{k-1}_{\chi_1,\chi_2}(n)$ are positive and the other half are negative.*

The proof of the first part of this theorem relies on a recent result of Matomäki and Radziwiłł that appeared in [**8**].

THEOREM 4.7 (Matomäki & Radziwiłł, 2016). *Let $f : \mathbf{N} \to \mathbf{R}$ be a multiplicative function. Then $f(n)$ has a positive proportion of sign changes if and only if $f(n) < 0$ for some integer $n > 0$ and $f(n) \neq 0$ for a positive proportion of integers n.*

PROOF OF THEOREM 4.6. For the first assertion, recall that we are assuming that χ_2 is a quadratic character (mod N_2). From Lemma 4.4, the sign of $\sigma^{k-1}_{\chi_1,\chi_2}$ is completely determined by the sign of χ_2. In particular, we can rewrite

$$\operatorname{sgn} \sigma^{k-1}_{\chi_1,\chi_2}(n) = \prod_{p^\ell || n} (\chi_2(p))^\ell.$$

Call this function $g(n)$. The function g is multiplicative because it is a product of $\chi_2(p)$'s and χ_2 is totally multiplicative. By Theorem 4.5, we know that $g(n) < 0$ for some integer $n > 0$. We now show that $g(n) \neq 0$ for a positive proportion of integers n. Indeed, if $n \equiv 1 \pmod{N_2}$ then $\chi_2(n) = 1$ since the conductor of χ_2 is N_2. It follows that, for $n \equiv 1 \pmod{N_2}$, we have $\operatorname{sgn}(\chi_2(n)) = 1$. Hence, by Lemma 4.4, $\operatorname{sgn}(\sigma^{k-1}_{\chi_1,\chi_2}) = 1$, so $g(n) = 1$. Because the density of integers n such that $n \equiv 1$ (mod N_2) is positive, we see that $g(n) \neq 0$ for a positive proportion of integers. Therefore the hypotheses for Theorem 4.7 are met and we have that g changes sign a positive proportion of the time. In other words, there exists a positive constant δ such that the sequence $(\sigma^{k-1}_{\chi_1,\chi_2}(n))_{n\leq x}$ has at least δx sign changes.

For the second assertion, we use the fact that the Dirichlet character $\chi_2(n) = -1$ half of the time and $\chi_2(n) = 1$ the other half of the time. By Lemma 4.4, it follows that half of the non-zero $\sigma^{k-1}_{\chi_1,\chi_2}(n)$ are positive and the other half are negative. (Note: The proof of this second claim originally appeared as Corollary 3.2 in [**5**].) □

Acknowledgments

The authors would like to thank the organizers of the 3rd Building Bridges Workshop on Automorphic Forms and Related Topics for providing them with the opportunity to participate in a wonderful conference. The second author was supported by an AMS Simons Travel Grant. She would like to thank the Max Planck Institute for Mathematics for hosting her while she was completing this work.

References

[1] Timothy W. Atwill and Benjamin Linowitz, *Newform theory for Hilbert Eisenstein series*, Ramanujan J. **30** (2013), no. 2, 257–278, DOI 10.1007/s11139-012-9418-2. MR3017929

[2] Fred Diamond and Jerry Shurman, *A first course in modular forms*, Graduate Texts in Mathematics, vol. 228, Springer-Verlag, New York, 2005. MR2112196

[3] H. Iwaniec, W. Kohnen, and J. Sengupta, *The first negative Hecke eigenvalue*, Int. J. Number Theory **3** (2007), no. 3, 355–363, DOI 10.1142/S1793042107001024. MR2352824

[4] E. Kowalski, Y.-K. Lau, K. Soundararajan, and J. Wu, *On modular signs*, Math. Proc. Cambridge Philos. Soc. **149** (2010), no. 3, 389–411, DOI 10.1017/S030500411000040X. MR2726725

[5] Benjamin Linowitz and Lola Thompson, *The sign changes of Fourier coefficients of Eisenstein series*, Ramanujan J. **37** (2015), no. 2, 223–241, DOI 10.1007/s11139-013-9552-5. MR3341687

[6] Kaisa Matomäki, *On signs of Fourier coefficients of cusp forms*, Math. Proc. Cambridge Philos. Soc. **152** (2012), no. 2, 207–222, DOI 10.1017/S030500411100034X. MR2887873

[7] Kaisa Matomäki and Maksym Radziwiłł, *Sign changes of Hecke eigenvalues*, Geom. Funct. Anal. **25** (2015), no. 6, 1937–1955, DOI 10.1007/s00039-015-0350-7. MR3432161

[8] Kaisa Matomäki and Maksym Radziwiłł, *Multiplicative functions in short intervals*, Ann. of Math. (2) **183** (2016), no. 3, 1015–1056, DOI 10.4007/annals.2016.183.3.6. MR3488742

[9] Dinakar Ramakrishnan, *A refinement of the strong multiplicity one theorem for* GL(2). *Appendix to: "l-adic representations associated to modular forms over imaginary quadratic fields. II" [Invent. Math.* **116** *(1994), no. 1-3, 619–643; MR1253207 (95h:11050a)] by R. Taylor*, Invent. Math. **116** (1994), no. 1-3, 645–649, DOI 10.1007/BF01231576. MR1253208

[10] Nahid Walji, *Further refinement of strong multiplicity one for* GL(2), Trans. Amer. Math. Soc. **366** (2014), no. 9, 4987–5007, DOI 10.1090/S0002-9947-2014-06103-5. MR3217707

[11] Nahid Walji, *Matching densities for Galois representations*, Proc. Amer. Math. Soc. **144** (2016), no. 8, 3309–3316, DOI 10.1090/proc/12996. MR3503699

[12] James Weisinger, *SOME RESULTS ON CLASSICAL EISENSTEIN SERIES AND MODULAR FORMS OVER FUNCTION FIELDS*, ProQuest LLC, Ann Arbor, MI, 1977. Thesis (Ph.D.)–Harvard University. MR2940742

Department of Mathematics, Oberlin College, Oberlin, Ohio 44074
Email address: benjamin.linowitz@oberlin.edu

Department of Mathematics, Oberlin College, Oberlin, Ohio 44074
Email address: lola.thompson@oberlin.edu

Contemporary Mathematics
Volume **732**, 2019
https://doi.org/10.1090/conm/732/14800

Properties of Sturm's formula

Kathrin Maurischat

ABSTRACT. In contrary to large weights, Sturm's operator fails to represent the holomorphic projection operator for small weights in arbitrary Siegel genus. We emphasize the roles of continuation of Poincaré series, holomorphic projection, Sturm's formula, and spectral analysis in this context, and explain their interaction.

CONTENTS

This article summarizes a series of papers [**5**]–[**8**] on the program of realizing holomorphic projection in the Siegel case of arbitrary genus and small weight. Parts of these results are joint with R. Weissauer ([**7**],[**8**]). We also contain a sketch of ongoing projects as well as some expectations we have on this subject. Each of the papers [**5**]–[**8**] has special aims on its own. But their common intention is to understand Sturm's operator and its connection with the holomorphic projection operator from a conceptional representation theoretic point of view. For this reason we enlarge the genus of the special linear group SL_2 and study symplectic groups instead, and we keep the weight small.

1. Holomorphic projection

In order to use the properties of holomorphic automorphic forms it is often necessary to reduce C^∞-forms to their holomorphic part. To have a precise notion of this holomorphic projection, we describe it as L^2-operator. Let $G_m = \mathrm{Sp}_m(\mathbb{R})$ be the real symplectic group of genus m. Let Γ be a subgroup of finite index in $\mathrm{Sp}_m(\mathbb{Z})$ which contains the subgroup of translations Γ_∞. This last condition is made to keep Fourier expansions as simple as possible. It can be removed by standard technics. Then $L^2(\Gamma\backslash G_m)$ is a Hilbert space with unitary action of G_m by right translations.

2010 *Mathematics Subject Classification.* Primary 11F46, 11F70, 46G20.
This work was supported by the European Social Fund.

Let ρ be an irreducible representation of the maximal compact group K_m of G_m. Let $L^2(\Gamma\backslash G_m)_\rho$ be the subspace of functions of K_m-type ρ. Then the orthogonal projection operator

$$S\,:\,L^2(\Gamma\backslash G_m)_\rho \xrightarrow{proj} L^2(\Gamma\backslash G_m)_{\rho,hol}$$

describes the holomorphic projection. Here $L^2(\Gamma\backslash G_m)_{\rho,hol}$ is the holomorphic part of the spectrum, where a C^∞-function is called holomorphic if under the action of the Lie algebra $\mathfrak{g}_m^{\mathbb{C}}$ it is annihilated by the minus part $\mathfrak{p}^-$ in the Cartan decomposition $\mathfrak{g}_m^{\mathbb{C}} = \mathfrak{k}\oplus\mathfrak{p}^+\oplus\mathfrak{p}^-$. The latter depends on a system of positive roots which is chosen to be compatible with the following.

It is often convenient to work with functions on the Siegel upper half space $\mathcal{H}_m$ rather than on G_m. Let $J : G_m \times \mathcal{H}_m \to \mathrm{GL}_m(\mathbb{C})$ be given by $J((\begin{smallmatrix} A & B \\ C & D\end{smallmatrix}), Z) = CZ + D$. Restricted to K_m, we get an isomorphism $J(\cdot, i) : K_m \xrightarrow{\sim} U(m)$ we rely on for the whole paper. Let $C^\infty(\mathcal{H}_m, V_\rho)$ be the space of C^∞-functions on $\mathcal{H}_m$ with values in V_ρ, and let $C^\infty(G_m, V_\rho) = C^\infty(G_m) \otimes V_\rho$. There is an isomorphism

$$\begin{aligned} C^\infty(\mathcal{H}_m, V_\rho) &\xrightarrow{\sim} C^\infty(G, V_\rho)_\tau\,, \\ f(Z) &\mapsto F(G) = \rho^{-1}(J(g,i))f(gK\cdot i)\,. \end{aligned}$$

Here $\tau = \rho^{-1}\circ J(\cdot, i)$, and we identify the $U(m)$-representation (ρ, V_ρ) with the corresponding representation of $\mathrm{GL}_m(\mathbb{C})$. Under this isomorphism, the action of $\mathfrak{p}^-$ corresponds to the anti-holomorphic differential operator $\partial_{\bar{Z}}$. Normalizing the measures dk and dg of K and G_m such that $dg = dV_{inv}dk$, where dV_{inv} is the invariant measure on $\mathcal{H}_m$, we have the following identity of scalar products. For C^∞-cusp forms f and h of weight ρ for Γ it holds

$$\langle f, h\rangle \;=\; \langle\!\langle F, H\rangle\!\rangle_{L^2(\Gamma\backslash G_m)}\,.$$

For simplicity we assume $\rho = \det^\kappa$ to be scalar for the rest of this paragraph. As the holomorphic projection S is orthogonal, it is characterized by

$$\langle f, h\rangle \;=\; \langle S(f), h\rangle$$

for all holomorphic cusp forms $h \in [\Gamma, \kappa]_0$ of weight κ. Now assume we have a system of Poincaré series p_τ in $[\Gamma, \kappa]_0$, where τ is half-integral and positive definite, so that the Fourier coefficients $a(\tau)$ of $S(f) \in [\Gamma, \kappa]_0$ are given by

$$\langle S(f), p_\tau\rangle \;=\; a(\tau)c(m, \kappa)\,.$$

Here $c(m, \kappa)$ is a constant independent of τ. As this must equal $\langle f, p_\tau\rangle$ for the non-holomorphic form f, we look at its Fourier expansion

$$f(Z) \;=\; \sum_\tau a(\tau, Y)e^{2\pi i\,\mathrm{tr}(\tau X)}$$

with $Z = X + iY$. By unfolding we find

$$\langle f, p_\tau\rangle \;=\; \det(\tau)^{-\frac{m+1}{2}} \int_{Y>0} a(\tau, Y)e^{-2\pi\,\mathrm{tr}(Y)} \det(Y)^\kappa\, \frac{dY_{inv}}{\det(Y)^{\frac{m+1}{2}}}\,,$$

where $dY_{inv} = \det(Y)^{-\frac{m+1}{2}}\prod_{j\le k} dy_{jk}$ for $Y = (y_{jk})_{jk}$. So turning things around, defining

$$(1)\quad a(\tau) := c(m, \kappa)^{-1}\det(\tau)^{-\frac{m+1}{2}} \int_{Y>0} a(\tau, Y)e^{-2\pi\,\mathrm{tr}(Y)} \det(Y)^\kappa\, \frac{dY_{inv}}{\det(Y)^{\frac{m+1}{2}}}\,,$$

we obtain the Fourier expansion $\sum_{\tau>0} a(\tau)e^{2\pi i\,\mathrm{tr}(\tau Z)}$ of the holomorphic projection $S(f)$. The resulting operator

$$St_\kappa \ : \sum_{\tau} a(\tau, Y)e^{2\pi i\,\mathrm{tr}(\tau X)} \mapsto \sum_{\tau>0} a(\tau)e^{2\pi i\,\mathrm{tr}(\tau Z)}\,,$$

where $a(\tau)$ is defined by (1), is called **Sturm's operator**. It is not only defined on L^2-functions, but also for all non-holomorphic modular forms of moderate growth, i.e. forms such that the integral (1) exists.

THEOREM 1.1. *In the following cases Sturm's operator realizes the holomorphic projection operator* $St_\kappa = S$.

(i) [**12**], [**13**] *In the classical case of genus* $m = 1$ *and large weight* $\kappa > 2$.
(ii) [**10**] *For arbitrary genus* m *and large weight* $\kappa > 2m$.
(iii) [**3**] *For genus* $m = 1$ *and weight* $\kappa = 2m = 2$.
(iv) [**5**] *For genus* $m = 2$ *and weight* $\kappa = 2m = 4$.

To illustrate the significance of holomorphic projection we shortly recall that first, in [**3**] it was an important tool to interpret certain convolution L-series. Second, it was used in [**14**] to reveal the true nature of mock modular forms as the holomorphic parts of weak harmonic Maass forms.

2. Poincaré series

The whole argument in Section 1 relies on a given system of Poincaré series in the space $[\Gamma, \kappa]_0$ of holomorphic cusp forms for Γ of weight κ. That is a system of functions generating $[\Gamma, \kappa]_0$ and producing the Fourier coefficients via the inner product. The theory of Poincaré series was studied systematically in [**11**]. For large weights $\kappa > 2m$ Panshichkin [**10**] introduced such systems tracking back to definitions by Neuenhöffer [**9**]. For matters of applications we prefer to work with Poincaré series of exponential type also tracking back to [**9**] and first introduced by Klingen [**4**],

$$P_\tau(g) \;=\; \sum_{\gamma\in\Gamma_\infty\backslash\Gamma} \det(J(\gamma g, i))^{-\kappa} e^{2\pi i\,\mathrm{tr}(\tau(\gamma g\cdot i))}\,.$$

These (as all) Poincaré series converge if and only if $\kappa > 2m$. From a purely analytic point of view this is the reason why things become interesting for small weights. For those the common procedure is to analytically continue these Poincaré series. We define for complex variables $s_1, \ldots, s_m$ and $Z = X + iY \in \mathcal{H}_m$ the non-holomorphic functions

$$H_\tau(Z, s_1, \ldots, s_m) \;=\; e^{2\pi i\,\mathrm{tr}(\tau Z)} \cdot \prod\nolimits_{q=1}^{m} \mathrm{tr}((\tau^{\frac{1}{2}} Y \tau^{\frac{1}{2}})^{[q]})^{s_q}\,,$$

and the corresponding operator valued non-holomorphic Poincaré series

$$P_\tau(g, s_1, \ldots, s_m) \;=\; \sum_{\gamma\in\Gamma_\infty\backslash\Gamma} \rho(J(\gamma g, i)^{-\kappa}) \cdot H_\tau(\gamma g\cdot i, s_1, \ldots, s_m)\,.$$

Here $Y^{[q]}$ denotes the q-th alternating power of the matrix Y. These Poincaré series converge if $s_q >> 0$ for $q = 1, \ldots, m$. We note this precisely for the special case where we restrict to two variables s_1, s_m.

THEOREM 2.1. [**5**], [**8**] *Let* $(l_1, \ldots, l_m)$, $l_1 \geq \cdots \geq l_m = \kappa$ *be the dominant highest weight of the representation* ρ. *The Poincaré series* $P_\tau(g, s_1, 0, \ldots, 0, s_m)$

converge absolutely and uniformly on compact sets with respect to the operator norm in the domain

$$\left\{(s_1, s_m) \in \mathbb{C}^2 \mid \operatorname{Re} s_2 > m - \frac{\kappa}{2} \text{ and } \operatorname{Re}(m s_2 + s_1) > m^2 - \frac{\sum_j l_j}{2}\right\}.$$

For fixed such (s_1, s_2) *and for all* $v \in V_\rho$ *the functions* $P_\tau(g, s_1, 0, \dots, 0, s_m) \cdot v$ *are bounded and belong to* $L^2(\Gamma\backslash G_m) \cap C^\infty(\Gamma\backslash G_m)$. *In particular, in case the weight* $\kappa > 2m$ *is large, at the critical point* $(s_1, s_m) = (0, 0)$ *the Poincaré series converge absolutely and are holomorphic as functions on* $\Gamma\backslash G_m$.

Continuing the Poincaré series analytically to the critical point $(s_1, \dots, s_m) = (0, \dots, 0)$ is a non-trivial problem in case $m > 1$ (for $m = 1$ see [**3**]).

For scalar K-type $\rho = \det^\kappa$ we have the following results in case of genus two and small κ. Here a weight κ is called small if $m < \kappa \leq 2m$.

THEOREM 2.2. [**5**] *Let the genus* $m = 2$ *equal two and the weight* $\kappa = 2m = 4$ *be four. Then the Poincaré series* $P_\tau(g, s_1, s_2)$ *have analytic continuation to the critical point* $(s_1, s_2) = (0, 0)$. *The limits* $P_\tau(g, 0, 0)$ *are holomorphic functions in* $L^2(\Gamma\backslash G_2)$.

For the proof we study differential operators D belonging to the center $\mathfrak{z}_2$ of the universal enveloping algebra of $\mathfrak{g}_2^{\mathbb{C}}$ such that in the equation

$$D\big(P_\tau(g, s_1, s_2)\big) = \mathcal{P}_\tau(g, s_1, s_2) \tag{2}$$

the auxiliary Poincaré series $\mathcal{P}_\tau(g, s_1, s_2)$ have better convergency properties. Applying the resolvents of $D_\pm$ then gives analytic continuations of the Poincaré series as functions in $L^2(\Gamma\backslash G_2)$. So we pick up the classical idea by Maass and Selberg for the trace formula. The main difference is that for trace formulas the input functions are Eisenstein series, which naturally are eigenfunctions of $\mathfrak{z}_2$, while the non-holomorphic Poincaré series are not.

The are some technically functional analytic arguments to make this work, together with arguments from Langlands' theory of Eisenstein series. We want to mention that there are exactly two differential operators D_+ and D_- of minimal degree four for which we obtain better convergency properties on the right hand side of (2). On Langland's Eisenstein series $E_B(g, \Lambda)$ with analytic spectral parameter $\Lambda \in \mathfrak{h}_{\mathbb{C}}^*$ they have the compelling shape

$$D_+(u) E_B(g, \Lambda) = \prod_{\alpha \text{ long root}} (\check{\alpha}(\Lambda) - u) \cdot E_B(g, \Lambda),$$

$$D_-(v) E_B(g, \Lambda) = \prod_{\alpha \text{ short root}} (\check{\alpha}(\Lambda) - v) \cdot E_B(g, \Lambda),$$

for some affine linear combinations u, v of the variables s_1, s_2 and weight κ, where $\check{\alpha}$ is the dual of the root α.

Theorem 2 implies Theorem 1.1(iv) on holomorphic projection by Sturm's operator above. For the second small weight $\kappa = m + 1 = 3$ we have jointly with R. Weissauer the following result.

THEOREM 2.3. [**7**] *Let the genus* $m = 2$ *equal two and the weight* $\kappa = 3$ *be three. Then the Poincaré series* $P_\tau(g, s_1, s_2)$ *have analytic continuation to the critical point* $(s_1, s_2) = (0, 0)$. *The limits* $P_\tau(g, 0, 0)$ *are* C^∞ *functions in* $L^2(\Gamma\backslash G_2)$. *The unique non-zero isotypical spectral components are discrete and given by the holomorphic* *discrete series representation* $\pi^{hol}_{(2,1)}$ *of minimal* K*-type* $(3,3)$*, and a holomorphic but non-discrete series representation* π^{hol}_1 *of minimal* K*-type* $(1,1)$.

We use the same methods as in [**5**], but employ much deeper insight to the unitary spectrum of $\mathrm{Sp}_2(\mathbb{R})$. Two surprising facts occur. First, the analytic continuations exist as a C^∞-function in $L^2(\Gamma\backslash G_2)$, but they are not holomorphic anymore in the critical point $(s_1, s_2) = (0,0)$. The reason for this is the occurrence of the spectral component π^{hol}_1 of K-type $(1,1)$, which carries the K-type $(3,3)$ (i.e. scalar weight $\kappa = 3$) non-trivially, but in which only functions of K-type $(1,1)$ are holomorphic. Second, the general expectation that the crucial problems will be located within the continuous spectrum (which happens in case $m = 1$ and $\kappa = 1$) does not hold. In weight $\kappa = 3$ the continuous spectral components are well-behaved and do not contribute to the continuation. They will indeed make non-trivial contributions in case $\kappa = 2$, so this is a problem of weight $\kappa = m$.

The articles [**5**] and [**7**] exhaust the scalar weight case in genus two. Because genus two in general holds for the pivot case for all higher genera, we expect all these phenomenons and more to occur for any higher genus.

3. Sturm's operator

The statement of Theorem 2.3 that for $\kappa = m+1$ the analytic continuation of the Poincaré series is non-holomorphic in the critical point (s_1, s_2) is surprising and unexpected. It suggests that Sturm's operator fails to realize the holomorphic projection operator in this case. We show that this is indeed the case.

Consider the spectral component π^{hol}_1 which produces the non-holomorphic part of the continued Poincaré series. The generating cusp form $h \in [\Gamma, 1]_0$ of this representation has minimal K-type $(1,1)$. (Here we identify representations of $\mathrm{GL}_m(\mathbb{C})$ with their highest weight.) The only possibility to obtain a function of K-type $(3,3)$ from h is by application of Maass' shift operator. This is defined by

$$\Delta_+^{[m]}(h)(Z) \;=\; (2i)^m(\tau\otimes\det^{\frac{1-m}{2}})(Y^{-1})\det(\partial_Z)\Big((\tau\otimes\det^{\frac{1-m}{2}})(Y)\cdot h(Z)\Big)\,,$$

which for any irreducible rational $\mathrm{GL}_m(\mathbb{C})$-representation τ sends $C^\infty(\mathcal{H}_m, V_\tau)$ to $C^\infty(\mathcal{H}_m, V_{\tau\otimes\det^2})$.

THEOREM 3.1. [**7**], [**6**] *Let the genus* $m \geq 2$ *be arbitrary. Let* $h \in [\Gamma, k]_0$ *be a non-zero holomorphic cusp form of weight* k*. Then the image of its Maass shift* $\Delta_+^{[m]}(h)$ *under Sturm's operator*

$$St_{k+2}\left(\Delta_+^{[m]}(h)\right)$$

is non-zero if and only if $k = m-1$.

On the other hand, the Maass shift $\Delta_+^{[m]}(h)$ is a non-holomorphic function in the spectral component generated by h. So the holomorphic projection $S(\Delta_+^{[m]}(h))$ vanishes. Accordingly, Sturm's operator fails to realize the holomorphic projection operator for weight $\kappa = m+1$.

THEOREM 3.2. [**7**], [**6**] *For genus* $m \geq 2$ *and* $\kappa = m+1$ *it holds*

$$St_{m+1}(f) \;=\; S(f) + Ph(f)\,,$$

where the phantom term $Ph(f)$ *is non-zero in general, as there is a contribution by Maass shifts from* $[\Gamma, m-1]$.

By virtue of Theorem 3.2, the process of obtaining holomorphic Poincaré series by analytic continuation of analogs of those for larger weight must be expected to fail in general. Because, such holomorphic continuations would make the argument of Section 1 for Sturm's operator work. This reveals once more the fact that the spaces $[\Gamma,\kappa]_0$ of holomorphic cusp forms are very difficult to describe when κ is small.

For genus two we are more precise, thereby ruling out the possibility that our choice of Poincaré series simply was unlucky.

THEOREM 3.3. [**7**] *For genus $m=2$ and weight $\kappa=3$ let p_T be the images on the Siegel half space of the limit series $P_T(\cdot,0,0)$. Then p_T decompose*

$$p_T \;=\; f_T+\Delta_+^{[2]}(h_T)$$

as sums of holomorphic cusp forms $f_T\in[\Gamma,3]_0$ and Maass derivatives $\Delta_+^{[2]}(h_T)$, where $h_T\in[\Gamma,1]$. In general, f_T and h_T are non-zero. The forms h_T can be recovered by the anti-holomorphic Maass operator $\Delta_-^{[2]}$

$$\Delta_-^{[2]}(p_T) \;=\; \frac{3}{4}\cdot h_T\,.$$

Here $\Delta_-^{[m]}$ is given by

$$\Delta_-^{[m]}(f) \;=\; (2i)^m(\tau\otimes\det(Y)^{\frac{1-m}{2}})\det(\bar\partial_Z)(\det(Y^{-1})^{\frac{1-m}{2}})f(Z))\,.$$

It is an interesting question how phantom terms decompose in general. Apart from $\Delta_+^{[m]}(h)$ for $h\in[\Gamma,\kappa-1]_0$ coming from the holomorphic representation π_{m-1}^{hol}, there can appear many other terms coming from unitary representations in which the K-type $\rho=(\kappa,\dots,\kappa)$ occurs non-trivially. In terms of cusp forms, this could be any $\Delta^\sigma(h)$ for a vector valued cusp form $h\in[\Gamma,\tau]_0$ and a differential operator Δ^σ such that $\tau+\sigma=\rho$.

A result in this direction is the notion of Sturm's operator for general K-type ρ (see [**8**]). Let τ run through the positive definite half-integral matrices, and let $f(Z)=\sum_\tau \rho(\tau^{\frac12})a(\tau,\tau^{\frac12}Y\tau^{\frac12})\cdot e^{2\pi\,\mathrm{tr}(\tau X)}$ be the Fourier expansion of a vector valued C^∞-modular form. Then, sending the vector valued Fourier coefficient $a(\tau,\tau^{\frac12}Y\tau^{\frac12})$ to the coefficient $a(\tau)^T$ defined by the vector valued integral

$$\det(\tau)^{-\frac{m+1}{2}}\int_{Y>0} a(\tau,Y)^T\rho(\tau^{\frac12})C(m,\rho)^{-1}\rho(Y)\rho(\tau^{-\frac12})e^{-2\pi\,\mathrm{tr}(Y)}\frac{dY_{inv}}{\det(Y)^{\frac{m+1}{2}}}\,.$$

we define Sturm's operator

$$St_\rho(f)(Z) \;=\; \sum_{\tau>0}\rho(\tau^{\frac12})a(\tau)e^{2\pi i\,\mathrm{tr}(\tau Z)}\,.$$

Here $C(m,\rho)$

$$C(m,\rho) \;=\; \int_{Y>0}\rho(Y)e^{-4\pi\,\mathrm{tr}(Y)}\frac{dY_{inv}}{\det(Y)^{\frac{m+1}{2}}}$$

is an operator-valued integral such that on holomorphic cusp forms Sturm's operator is the identity. It is classically known to be convergent if the dominant highest weight $(l_1,\dots,l_m)$ of ρ satisfies $l_m>\frac{m-1}{2}$ ([**2**]). In [**8**] we prove that $C(m,\rho)$ to be invertible for genus $m=2$ as well as for some general classes of representations. In particular, if $\rho = \det^\kappa$, then $C(m,\rho) = c(\kappa, m)$, and the notions of Sturm's operator coincide. We have the following result on holomorphic projection generalizing Theorem 1.1 to vector valued forms.

THEOREM 3.4. [**8**] *Let ρ be an irreducible representation of $\mathrm{GL}_m(\mathbb{C})$ of highest weight $(l_1, \dots, l_m)$ satisfying $l_m > 2m$. Then Sturm's operator realizes the holomorphic projection operator.*

But, generalizing Theorems 3.1 and 3.2, for small weight this does not hold true.

THEOREM 3.5. [**8**] *Let the genus m equal two. Let τ be an the irreducible representation of $\mathrm{GL}_2(\mathbb{C})$ of highest weight $(k+r, k)$ with $k \geq 1$ and $r \geq 0$. Let $C(m,\rho)$ be invertible. Let $h \in [\Gamma, \tau]_0$ be a vector valued holomorphic cusp form of weight τ. Then the image of its Maass shift $\Delta_+^{[m]}(h)$ under Sturm's operator*

$$St_{\tau\otimes\det^2}\left(\Delta_+^{[m]}(h)\right)$$

is non-zero if and only if $k = 1$.

In case ρ has highest weight $(3+r, 3)$ Sturm's operator does not realize the holomorphic projection but produces phantom terms

$$St_\rho(f) \;=\; S(f) + ph(f)\,.$$

4. Spectral point of view

An irreducible representation containing a non-zero holomorphic function H of weight ρ is generated by H and has minimal K-type ρ. Most of them are (limit of) holomorphic discrete series representations, but there are others like π_1^{hol} above. In particular, for $\rho = (m+1, \dots, m+1)$ the underlying representation is the holomorphic discrete series with Harish-Chandra parameter $\delta = (m, m-1, \dots, 1)$. In general, $\rho = (\kappa_1, \dots, \kappa_m)$ with $\kappa_m \geq m+1$ (and the natural condition $\kappa_{j+1} \geq \kappa_j$) has Harish-Chandra parameter $(\kappa_1 - 1, \kappa_2 - 2, \dots, \kappa_m - m)$ and belongs to the cone given by the holomorphic discrete series within the root space, that is the cone given by the Weyl chamber translated by δ, where δ equals half the sum of positive roots.

The wall orthogonal to the short root is characterized by the scalar minimal K-types $(\kappa, \dots, \kappa)$, where $\kappa \geq m+1$. For those with $\kappa > m+1$ Sturm's operator realizes the holomorphic projection correctly at least in case $m = 2$ by Theorem 1.1, i.e. its image belongs to the holomorphic discrete series in question. For $\kappa = m+1$, belonging to the apex δ of the cone, this is wrong by Theorem 3.2. Then, the image of Sturm's operator has a non-zero part in π_{m-1}^{hol}, which has Harish-Chandra parameter $(m-2, \dots, -1)$ (respectively, a Weyl-conjugate of this). By Theorem 3.5 the same phenomenon occurs for $\rho = (3+r, 3)$, $r > 0$, which are the minimal K-types of the discrete series on the wall of the cone perpendicular to the long root. So they have Harish Chandra parameters $(2+r, 1)$. There Sturm's operator fails. Whereas Sturm's operator acts as holomorphic projection everywhere else on the cone.

We expect that, for higher genera, Sturm's operator will fail to realize the holomorphic projection on all facets of the cone which are not perpendicular to every short root.

The exhaustive results in genus $m = 2$ ([**7**]) rely on the analysis of the unitary spectrum of $\mathrm{Sp}_m(\mathbb{R})$. For results on higher genera an extensive insight to the unitary spectrum of $\mathrm{Sp}_m(\mathbb{R})$ is necessary. At the moment, the spectral parts parametrized by parameters Λ belonging to the ball of radius $\|\delta\|$ are little understood in detail. An exact description of the truly arising small K-types within (unitary) representations of small minimal K-types, holomorphic as well as non-holomorphic ones, will be a good help for our program. Explicitly, we have to determine those spectral components which are zeros of the differential operators $D_{\pm}$ used for analytic continuation of the Poincaré series. Written with respect to Langlands' Eisenstein series, these elements of the center of the universal enveloping Lie algebra themselves depend on the weight ρ of the Poincaré series. Their zeros are certain affine lines perpendicular to the roots, which get the nearer to the origin the smaller the weight becomes. From the representation theoretic point of view, this is why things become interesting for small weight.

5. Perspectives

There are several projects resulting from the discussion above which are ongoing work. These include a general theory of analytic continuation of Poincaré series of exponential type for general genus and for vector valued weights. Also a comprehensive atlas of the state of the art of the representation theoretic questions is needed.

Another current project is the case of half-integral weight. Most of our methods apply to them as well. We have some evidence that there should be a positive result on holomorphic projection even in case of weight $\kappa = m + \frac{1}{2}$. This would have direct applications to theta correspondences for certain Hilbert modular surfaces (see [**1**]).

An even more appealing project is to transfer these questions to the dualized setting, the special orthogonal groups $\mathrm{SO}(2, n)$. Some parts of the theory have direct counterparts there, for example the concrete differential operators used for analytic continuation, while other parts must be developed, like systems of Poincaré series. It is the case of special orthogonal groups we think of for arithmetic applications of our results.

In the very focus of our work stands the interpretation of the phantom terms arising from Sturm's formula. They occur in an area where analytic, representation theoretic, and arithmetic objects deeply interact with each other. We expect the phantom terms produced by analysis to carry arithmetic impact. In a first step we study their influence within convolution L-series involving derivatives of Eisenstein series.

References

[1] J. Funke and J. Millson, *The geometric theta correspondence for Hilbert modular surfaces*, Duke Math. J. **163** (2014), no. 1, 65–116, DOI 10.1215/00127094-2405279. MR3161312

[2] Godement, R.: *Function holomorphes de carré sommable dans le demi-plan de Siegel,* Sém. Cartan, ENS (1957/58), 6-01–6-22

[3] B. H. Gross and D. B. Zagier, *Heegner points and derivatives of L-series*, Invent. Math. **84** (1986), no. 2, 225–320, DOI 10.1007/BF01388809. MR833192

[4] H. Klingen, *Über Poincarésche Reihen vom Exponentialtyp* (German), Math. Ann. **234** (1978), no. 2, 145–157, DOI 10.1007/BF01420965. MR0480357

[5] K. Maurischat, *On holomorphic projection for symplectic groups*, J. Number Theory **182** (2018), 131–178, DOI 10.1016/j.jnt.2017.06.005. MR3703935

[6] K. Maurischat, *Sturm's operator for scalar weight in arbitrary genus*, Int. J. Number Theory **13** (2017), no. 10, 2677–2686, DOI 10.1142/S1793042117501482. MR3713097
[7] K. Maurischat and R. Weissauer, *Phantom holomorphic projections arising from Sturm's formula*, Ramanujan J. **47** (2018), no. 1, 21–46, DOI 10.1007/s11139-018-0033-8. MR3857933
[8] Maurischat, K., Weissauer, R.: *Sturm's operator acting on vector valued K-types,* Preprint 2016/2018
[9] H. Neunhöffer, *Über die analytische Fortsetzung von Poincaréreihen* (German), S.-B. Heidelberger Akad. Wiss. Math.-Natur. Kl. (1973), 33–90. MR0352007
[10] M. Courtieu and A. Panchishkin, *Non-Archimedean L-functions and arithmetical Siegel modular forms*, 2nd ed., Lecture Notes in Mathematics, vol. 1471, Springer-Verlag, Berlin, 2004. MR2034949
[11] H. Petersson, *Einheitliche Begründung der Vollständigkeitssätze für die Poincaréschen Reihen von reeller Dimension bei beliebigen Grenzkreisgruppen von erster Art* (German), Abh. Math. Sem. Hansischen Univ. **14** (1941), 22–60, DOI 10.1007/BF02940741. MR0005769
[12] J. Sturm, *Projections of C^∞ automorphic forms*, Bull. Amer. Math. Soc. (N.S.) **2** (1980), no. 3, 435–439, DOI 10.1090/S0273-0979-1980-14757-6. MR561527
[13] J. Sturm, *The critical values of zeta functions associated to the symplectic group*, Duke Math. J. **48** (1981), no. 2, 327–350. MR620253
[14] Zwegers, S.P.: *Mock Theta Functions,* Utrecht University Repository (2003)

MATHEMATISCHES INSTITUT, HEIDELBERG UNIVERSITY, IM NEUENHEIMER FELD 205, 69120 HEIDELBERG, GERMANY

Email address: maurischat@mathi.uni-heidelberg.de

Contemporary Mathematics
Volume **732**, 2019
https://doi.org/10.1090/conm/732/14801

An application of a special form of a Tauberian theorem

Almasa Odžak and Lamija Šćeta

ABSTRACT. We are presenting an overview of the applications of a special form of a theorem of Tauberian type to derive new bounds for the remainder term in the Weyl law for the counting function of positive eigenvalues of the Laplacian in three different settings: in the setting of the non-compact, cofinite surfaces of dimension $d = 3$, in the setting of symmetric spaces of real rank one and in the setting of quantum graphs with general self adjoint boundary conditions.

1. Introduction

Historically, Tauberian theorems may be viewed as converses of the Abel second theorem for the power series with a finite circle of convergence. One of the first theorems of this type was proved by A. Tauber [**Ta**] in 1897. However, after thirty years of work, Hardy and Littlewood, see e.g. [**HL**], are considered the most influential contributors in the development of the Tauberian theory. The development of Tauberian theory is further accelerated by potential applications in the number theory and the theory of probability.

Roughly speaking, the Tauberian type theorems relate behavior of generalized function in a neighborhood of zero with the asymptotic behavior of its integral representation at some kind at infinity. In this paper we are presenting Tauberian type theorem with the remainder terms for Laplace integral representation.

The basic contribution in the study of Tauberian theorems with remainder terms for Laplace transforms comes from the works of G. Freud, J. Korevar, T. Ganelius and M. A. Subhankulov. This type of theorems have been used as a powerful tool in number theory. One of the most substantial applications is the study of asymptotic properties of the eigenvalues of differential operators. We will consider the Laplace-Beltrami operator Δ, which appears in different contexts and can be defined in different settings.

In this paper we are interested in the distribution of eigenvalues of the Laplacian in three different settings: in the setting of non-compact, cofinite surfaces of dimension $d = 3$, in the setting of symmetric spaces of rank one and in the setting of quantum graphs with general self-adjoint boundary conditions. The formula, which describes the asymptotic behavior of the counting function of the number of discrete eigenvalues of the Laplacian (plus the suitably normalized integral of the

2010 *Mathematics Subject Classification.* Primary 11M45, 11F72; Secondary 34L20, 81Q10.
Key words and phrases. Tauberian theorems, Weyl law, Kleinian groups, symmetric spaces, quantum graphs.

spectral measure, in the case when the continuous spectrum is present), is called the Weyl law. The main tool to derive the Weyl law with a good bound on the remainder term in these three different settings is the Tauberian theorem for the Laplace transform with an optimal bound for the remainder term proved in [**SS1**] applied on the indefinite integral of the counting function and suitably transformed trace formulas.

2. A Tauberian theorem with an optimal bound for the remainder term

L. Smajlović and L. Šćeta [**SS1**, Theorem 2.2] have proven a generalization of the classical Tauberian theorem, which gives an optimal bound for the remainder term under given conditions on the Laplace-Stieltjes transform.

In this paper we are using one special form of [**SS1**, Theorem 2.2] when the error term in the formula relating the two Laplace-Stieltjes integrals is $O(\exp(-c\sigma^{-\delta}))$, (for $c > 0$, $0 < \delta \le 1$), and are applying the special case when $f(t) \equiv 1$ and $m = \beta = 0$:

PROPOSITION 2.1. [**SS1**, Corollary 2.4.] *Let $\psi(u)$ and $\varphi(u)$ be defined for $u \ge 0$, bounded for $u \le h$, non-negative and non-decreasing for $u \ge h$. Assume also that the following conditions are fulfilled: $\psi(0) = \varphi(0) = 0$,*

$$\frac{\varphi(v)}{\varphi(u)} < \left(\frac{v}{u}\right)^b, \quad b > 0, \quad 0 \le h \le u < v$$

and that the integrals $\int_0^\infty e^{-\sigma u} d\varphi(u)$ and $\int_0^\infty e^{-\sigma u} d\psi(u)$ converge for $\sigma > 0$.

Then, the estimate

$$\int_0^\infty e^{-\sigma u} d\psi(u) = \int_0^\infty e^{-\sigma u} d\varphi(u) + O\left(\varphi\left(\frac{1}{\sigma}\right) e^{-c\sigma^{-\delta}}\right), \quad \text{as } \sigma \downarrow 0, \tag{2.1}$$

where $c > 0$, $0 < \delta \le 1$ implies that

$$\psi(T) = \varphi(T) \times \left\{1 + O\left(T^{-\frac{\delta}{\delta+1}}\right) + O\left(\frac{1}{\varphi(T)}\right)\right\}, \quad \text{as } T \to \infty. \tag{2.2}$$

Moreover, under the assumption (2.1), *the error term in* (2.2) *is optimal.*

3. The Weyl law

The asymptotic formula

$$N(T) = \frac{\omega_n \text{vol}(X)}{(2\pi)^n} T^{n/2} + o(T^{n/2}), \text{ as } T \to \infty$$

for the distribution of the counting function $N(T)$ of the eigenvalues of the Laplacian for the bounded region X in $\mathbb{R}^n$ (with reasonable boundary conditions), proven by H. Weyl in 1911 was the first general result describing the distribution of eigenvalues of a elliptic operator and showing that the leading term in asymptotics depends only on the dimension n and the volume of the region X. Therefore, the results of this type are often called the Weyl law. Since then many have obtained and published results connected to the Weyl law, in which they studied different settings and contexts. Weyl also assumed the existence of the second term in the asymptotics and conjectured the form of the optimal remainder term; both were proven to be significant in applications of the Weyl law.

The spectrum of Laplacian on non-compact manifolds is the union of a point spectrum and a continuous spectrum, hence the counting function $N(T)$ is defined to be the sum of the counting function of discrete eigenvalues and the integral (from $-T$ to T) of the spectral measure.

In this section we derive the Weyl law with the remainder term with the power saving exponent in three different settings. The Weyl law in all three settings is proven similarly: The trace formula in each setting is transformed in the way that its spectral side is written as the Laplace-Stieltjes integral of the function $e^{-\sigma t}$ with respect to the measure generated by the counting function and its geometric side is written as the sum of the Laplace-Stieltjes integral of a function $e^{-\sigma t}$ with respect to a certain explicit measure and an error term. Then, we applie Proposition 2.1 to the indefinite integral of the counting function and derive the Weyl law using a simple truncation argument.

3.1. Cofinite Kleinian groups with finite dimensional unitary representation. Suppose Γ is a cofinite Kleinian group and $\mathrm{Rep}\,(\Gamma, V)$ is the space of finite dimensional unitary representations of Γ in a $n-$dimensional complex vector space V. The Hilbert space of $\chi-$automorphic functions ($\chi \in \mathrm{Rep}\,(\Gamma, V)$) is defined by

$$\mathcal{H}(\Gamma,\chi) = \{f : \mathbb{H}^3 \to V \mid f(\gamma P) = \chi(\gamma) f(P) \quad \forall \gamma \in \Gamma, P \in \mathbb{H}^3$$
$$\text{and } \langle f, f\rangle = \int_{\mathcal{F}} \langle f(P), f(P)\rangle_V dv(P) < \infty\},$$

where $\mathcal{F}$ denotes a fundamental domain for Γ in the hyperbolic three-dimensional upper half space $\mathbb{H}^3 := \{(z,r) : z \in \mathbb{C}, r > 0\}$. Let the Δ be the positive self-adjoint Laplace-Beltrami operator on the Hilbert space $\mathcal{H}(\Gamma,\chi)$.

In this setting, the counting function of eigenvalues of the Laplacian less than u is defined to be the sum of the counting function of discrete eigenvalues

$$N_\Gamma(u) := \sum_{\lambda_m < u} 1$$

and the normalized integral of the spectral measure

$$M_\Gamma(u) := \begin{cases} \frac{1}{4\pi}\int_{-\sqrt{u-1}}^{\sqrt{u-1}}(-\frac{\phi_\Gamma'}{\phi_\Gamma}(it))dt, & u \geq 1; \\ 0, & 0 < u < 1, \end{cases}$$

where $\phi_\Gamma\,(s)$ denotes the determinant of the scattering matrix.

Using certain estimates for the loxodromic and elliptic terms that appear on the geometric side of the trace formula (see [**SS2**, p. 118] and [**SP**, p. 283]) one can prove that the special form of the Selberg trace formula from [**Fr**] with the test function $h(t) = \exp(-\sigma t), 0 < \sigma < 1$ (see [**SS2**, Lemma 2.1]) yields these asymptotics

$$\int_0^\infty e^{-\sigma u} d\xi\,(u) = \int_0^\infty e^{-\sigma u} d\zeta\,(u) + \tilde{c}_\Gamma e^{-\sigma} + O\left(\frac{e^{-\frac{c}{\sigma}}}{\sigma^{\frac{3}{2}}}\right), \text{ as } \sigma \downarrow 0 \tag{3.1}$$

for certain constants $\widetilde{c}_\Gamma$ depending solely upon the Γ and $c > 0$, where functions $\xi\,(u)$ and $\zeta\,(u)$, $u \geq 0$, are defined by

$$\xi\,(u) = N_\Gamma\,(u) + M_\Gamma\,(u) + d_\Gamma P_\Gamma\,(u)$$

and

$$\zeta(u) = \left(\frac{\mathrm{vol}\,(\mathcal{F})}{6\pi^2}\dim_{\mathbb{C}} V(u-1)^{3/2} + \frac{c_\Gamma}{2\pi}\sqrt{u-1}\right)\chi_{[1,\infty)},$$

where $P_\Gamma(u) := \begin{cases} \int_{-\sqrt{u-1}}^{\sqrt{u-1}} \frac{\Gamma'}{\Gamma}(1+it)dt, & u \geq 1; \\ 0, & 0 < u < 1 \end{cases}$ and c_Γ and d_Γ are certain constants depending upon Γ.

The direct application of the classic Wiener-Ikehara theorem (see, e.g. [**Ko**]) to the asymptotic relation (3.1) yields the Weyl law with a poor remainder term:

$$N_\Gamma(T) + M_\Gamma(T) = \frac{\mathrm{vol}(\mathcal{F})}{6\pi^2} \dim_{\mathbb{C}} V \cdot T^{\frac{3}{2}} + O\left(\frac{T^{\frac{3}{2}}}{\log T}\right), \quad \text{as } T \to \infty. \tag{3.2}$$

Therefore, one needs to transform the asymptotic relation (3.1), first by integrating by parts and then, by introducing indefinite integrals. Putting

$$\psi(u) = \int_0^u \left(\xi(t) + \frac{C}{2\pi}\sqrt{t-1}\chi_{[1,\infty)}\right) dt$$

and

$$\varphi(u) = \int_0^u \left(\zeta(t) + \left(\frac{C}{2\pi}\sqrt{t-1} + \tilde{c}_\Gamma\right)\chi_{[1,\infty)}\right) dt$$

with a suitably chosen constant C, the formula (3.1) becomes

$$\int_0^\infty e^{-\sigma u} d\psi(u) = \int_0^\infty e^{-\sigma u} d\varphi(u) + O\left(\frac{e^{-\frac{c}{\sigma}}}{\sigma^{\frac{5}{2}}}\right) \quad \text{as } \sigma \downarrow 0. \tag{3.3}$$

The asymptotic behavior of the logarithmic derivative of classical gamma function and certain growth property of the scattering determinant, together with some simple inequalities are sufficient to show that functions $\psi(u)$ and $\varphi(u)$ satisfy the conditions stated in Proposition 2.1.

The application of Proposition 2.1 to the asymptotic relation (3.3) together with a simple truncation argument yields the Weyl law with the remainder term for cofinite Kleinian groups with finite dimensional unitary representation, which can be presented as the following theorem.

THEOREM 3.1. [**SS2**, Theorem 3.1] *Utilizing the notation above, one has the following asymptotic relation*

$$N_\Gamma(T) + M_\Gamma(T) = \frac{\mathrm{vol}(\mathcal{F})}{6\pi^2} \dim_{\mathbb{C}} V \cdot T^{\frac{3}{2}} + O\left(T^{\frac{5}{4}}\right), \quad \textit{as } T \to \infty.$$

Notice that the error term in the Weyl law mentioned above is much better than the one in (3.2), which is obtained by the direct application of the classic Wiener-Ikehara Tauberian theorem to the same trace formula.

3.2. Symmetric spaces of rank one. Let G be a connected simple Lie group, with the maximal compact subgroup K. With Γ we denote a discrete subgroup of G such that G/Γ is non-compact and $\mathrm{vol}(G/\Gamma) < \infty$. We denote by $X = G/K$ the space of real rank one and let $n = \dim X$.

Let $\mathfrak{g} = \mathfrak{k} + \mathfrak{p}$ be the Cartan decomposition of $\mathfrak{g}$ with respect to the involution θ, where $\mathfrak{g}$ and $\mathfrak{k}$ are the Lie algebras of G and K, respectively. Denote the maximal abelian subspace of $\mathfrak{p}$ by $\mathfrak{a}_\mathfrak{p}$, and the maximal abelian θ-stable subalgebra of $\mathfrak{g}$ by $\mathfrak{a}$ and assume $dim\mathfrak{a}_\mathfrak{p} = 1$. Let $\mathfrak{g}^C$, $\mathfrak{k}^C$ and $\mathfrak{a}^C$ be complexification of $\mathfrak{g}$, $\mathfrak{k}$ and $\mathfrak{a}$, respectively. Then $\mathfrak{a}^C$ is a Cartan subalgebra of $\mathfrak{g}^C$. Let $P_+ := \{\alpha \in \Phi^+ | \alpha \neq 0 \; on\}$, where $\Phi^+ = \Phi^+(\mathfrak{g}, \mathfrak{a})$ is the set of the positive roots of $(\mathfrak{g}^C, \mathfrak{a}^C)$ (under the compatible orders on the dual spaces of $\mathfrak{a}_\mathfrak{p}$ and $\mathfrak{a}_\mathfrak{p} + i\mathfrak{a}_\mathfrak{k}$).

For the set Σ of the elements of P_+ restricted to $\mathfrak{a}_\mathfrak{p}$ one can find an element $\beta \in \Sigma$ such that 2β is the only other possible other element of Σ (see [**GW**], p. 23). With p (resp. q) we will denote the number of elements of P_+ whose restriction is β (resp. 2β).

The spectrum of the Laplace-Beltrami operator Δ on X is the union of a point spectrum and continuous spectrum. The point spectrum consist of the eigenvalues $0 = \lambda_0 < \lambda_1 \leq \lambda_2 \leq \dots$ with the finite multiplicities $n_0, n_1, n_2, \dots$

The continuous spectrum is parametrized by the generalized eigenfunctions of the Laplacian on $K \backslash G/\Gamma$ which are given by the Eisenstein series. So, the counting function (which counts the number of eigenvalues of Laplacian which are less than u) is defined as the sum of the counting function of the discrete eigenvalues:

$$N_\Gamma(u) = \sum_{\{j:(r_j{}^+)^2+\rho_0^2 \leq u\}} n_j$$

and the integral of the spectral measure

$$M_\Gamma(u) = \begin{cases} \frac{1}{4\pi} \int\limits_{-\sqrt{u-\rho_o{}^2}}^{\sqrt{u-\rho_o{}^2}} (-\frac{\phi_\Gamma'}{\phi_\Gamma}(it))dt, & u \geq \rho_o{}^2; \\ 0, & 0 < u < \rho_o{}^2, \end{cases}$$

where ϕ_Γ denotes the determinant of the scattering matrix and ρ_0^2 is the lower bound for the continuous spectrum. The trace formula in this setting relates the spectrum of the representation U of G on $L^2(G/\Gamma)$ with the geometry of underlying manifold $K \backslash G/\Gamma$.

The idea is now to study the trace formula in the special case when the Fourier transform of the test function h is equal to $\hat{h} = e^{-t\left(r^2(\nu)+\rho_0^2\right)}$ $(t \geq 0)$.

Using the properties of the classical Γ function and certain asymptotic relations of the Harish-Chandra c-function, it can be deduced that the trace formula [**GW**, Formula 2.2.] after suitable transformations yields the following asymptotics

$$\int\limits_0^\infty e^{-\sigma u} d\psi(u) = \int\limits_0^\infty e^{-\sigma u} d\varphi(u) + O\left(e^{-\frac{c_\Gamma}{\sigma}}\right), \text{ as } \sigma \downarrow 0 \tag{3.4}$$

where the functions $\psi(u)$ and $\varphi(u)$ satisfy the assumptions of Proposition 2.1 and $c_\Gamma > 0$.

Assuming the following inequality of the scattering determinant along the imaginary axis:

$$-\frac{\phi_\Gamma'}{\phi_\Gamma}(it) + C|t|^k \geq 0$$

for some constant $C \geq 0$, a real number $k \leq n-1$ and all $t \geq t_\Gamma \geq 0$, Proposition 2.1 can be applied to the asymptotic (3.4). A simple truncation argument then yields the following Weyl law for symmetric spaces of real rank one:

THEOREM 3.2. *With the notation as above, one has the following asymptotic relation*

$$N_\Gamma(T) + M_\Gamma(T) + \frac{d}{2\pi} P_\Gamma(T) = C_G \mathrm{vol}(G/\Gamma) T^{\frac{n}{2}} + O\left(T^{\frac{2n-1}{4}}\right) \text{ as } T \to \infty,$$

where

$$C_G = \frac{|Z(\Gamma)|}{n\pi 2^{q+1}} \left[\frac{\Gamma\left(\frac{p+q}{2}\right)}{\Gamma(p+q)} \right]^2,$$

$|Z(\Gamma)|$ *is the number of elements in the centralizer of* Γ, d *is the number of equivalence classes of the* Γ*-cuspidal minimal parabolic subgroups of* G *and*

$$P_\Gamma(T) = \begin{cases} 2 \int\limits_{-\sqrt{T-\rho_o{}^2}}^{\sqrt{T-\rho_o{}^2}} \frac{\Gamma'}{\Gamma}(1+it)dt, & T \geq \rho_o{}^2; \\ 0, & 0 < T < \rho_o{}^2. \end{cases}$$

3.3. Quantum graphs with general self adjoint boundary conditions. Let $\Gamma = (V, E, \mathbf{l})$ be a metric graph, where V is a finite set of vertices and E $= \{e_1, e_2, \dots e_E\}$ is a finite set of edges. The metric structure is introduced through the vector $\mathbf{l} = (l_1, l_2, \dots, l_E)$, where the interval $[0, l_i]$ is assigned to the edge e_i, $i = 1, \dots, E$. The numbers l_i are called edge lengths and the largest and the smallest edge length are denoted by $l_{\max}$ and $l_{\min}$, respectively.

We consider a *quantum graph* as a metric graph which is equipped with the self-adjoint realization of the Laplacian $-\Delta$ acting on a suitable Hilbert space. The Laplacian acts as the negative second derivative along the edges and is accompanied by appropriate vertex conditions; it is given by $-\Delta F = (-f_1'', \dots, -f_E'')$, where f_i are functions defined on the edges of the graph.

The distribution of eigenvalues of the Laplacian is one of the main questions related to its spectrum. As it turns out, it is more convenient to study the eigenvalue-problem in the form $-\Delta\psi = k^2\psi$, where $\lambda_j = k_j^2$ and $k_0 \leq k_1 \leq \dots$ The asymptotic behavior of the counting function $N(T) = \#\{j : k_j^2 \leq T^2\}$ as $T \to \infty$, where the eigenvalues are counted according to their multiplicities, is particularly important. J. Bolte and S. Andres [**BE**] have proven quantum graph version of the Weyl law stating that $N(T) \sim \frac{\mathcal{L}}{\pi}T$, as $T \to \infty$, in a general case of self-adjoined realization of the Laplacian on a compact metric graph, where $\mathcal{L} = l_1 + \dots + l_E$ is the total length of the graph. This result shows that the leading asymptotics in the Weyl law are independent on any graph details except its length.

It is of interest to obtain the remainder term in the Weyl law which improves the existing error term $O(T)$, as $T \to \infty$. This is done in two steps. First, we insert the heat kernel $h(k) = e^{-k^2 t}, (t > 0)$ as a test function into the trace formula derived in [**BE**] (which, in this setting, provides the connection between the spectrum of the Laplacian and the sum over all periodic orbits). Then, one transforms the formula obtained to include the indefinite integral of the counting function and applies Proposition 2.1 to this formula to deduce the following theorem.

THEOREM 3.3. [**OS**, Theorem 3.1] *Let* Γ *be the compact metric graph with the self-adjoint realisation* $-\Delta(A, B; \mathbf{l})$ *of Laplacian, such that* $l_{\min}$ *is suitably large. Then,*

$$N(T) = \frac{\mathcal{L}}{\pi}T + O\left(T^{\frac{2}{3}}\right) \quad \textit{as } T \to \infty.$$

References

[BE] J. Bolte and S. Endres, *The trace formula for quantum graphs with general self adjoint boundary conditions*, Ann. Henri Poincaré **10** (2009), no. 1, 189–223, DOI 10.1007/s00023-009-0399-7. MR2496307

[Fr] J. S. Friedman, *The Selberg trace formula and Selberg zeta-function for cofinite Kleinian groups with finite-dimensional unitary representations*, Math. Z. **250** (2005), no. 4, 939–965, DOI 10.1007/s00209-005-0806-9. MR2180383

[GW] R. Gangolli and G. Warner, *Zeta functions of Selberg's type for some noncompact quotients of symmetric spaces of rank one*, Nagoya Math. J. **78** (1980), 1–44. MR571435

[HL] G. H. Hardy and J. E. Littlewood, *Tauberian Theorems Concerning Power Series and Dirichlet's Series whose Coefficients are Positive*, Proc. London Math. Soc. (2) **13** (1914), 174–191, DOI 10.1112/plms/s2-13.1.174. MR1577498

[Ko] J. Korevaar, *A century of complex Tauberian theory*, Bull. Amer. Math. Soc. (N.S.) **39** (2002), no. 4, 475–531, DOI 10.1090/S0273-0979-02-00951-5. MR1920279

[OS] A. Odžak and L. Šćeta, *On the Weyl Law for Quantum Graphs*, Bull. Malays. Math. Sci. Soc. **42** (2019), no. 1, 119–131, DOI 10.1007/s40840-017-0469-9. MR3894619

[SP] P. Sarnak, *The arithmetic and geometry of some hyperbolic three-manifolds*, Acta Math. **151** (1983), no. 3-4, 253–295, DOI 10.1007/BF02393209. MR723012

[SS1] L. Smajlović and L. Šćeta, *On a Tauberian theorem with the remainder term and its application to the Weyl law*, J. Math. Anal. Appl. **401** (2013), no. 1, 317–335, DOI 10.1016/j.jmaa.2012.09.039. MR3011272

[SS2] L. Smajlović and L. Šćeta, *On the remainder term in the Weyl law for cofinite Kleinian groups with finite dimensional unitary representation*, Arch. Math. (Basel) **102** (2014), no. 2, 117–126, DOI 10.1007/s00013-014-0613-9. MR3169021

[Ta] A. Tauber, *Ein Satz aus der Theorie der unendlichen Reihen* (German), Monatsh. Math. Phys. **8** (1897), no. 1, 273–277, DOI 10.1007/BF01696278. MR1546472

[GW] R. Gangolli and G. Warner, *Zeta functions of Selberg's type for some noncompact quotients of symmetric spaces of rank one*, Nagoya Math. J. **78** (1980), 1–44. MR571435

Department of Mathematics, University of Sarajevo, Zmaja od Bosne 35, 71 000 Sarajevo, Bosnia and Herzegovina

Email address: almasa.odzak@pmf.unsa.ba

School of Economics and Business, University of Sarajevo, Alije Izetbegovića 1, 71 000 Sarajevo, Bosnia and Herzegovina

Email address: lamija.sceta@efsa.unsa.ba

Contemporary Mathematics
Volume **732**, 2019
https://doi.org/10.1090/conm/732/14783

On the zeros of some L functions from the extended Selberg class

Almasa Odžak and Lamija Šćeta

ABSTRACT. In this paper a family of functions from the extended Selberg class having zeros off the critical line is constructed. Some properties of such functions and their zeros are derived. Also very efficient numerical method that can be used to detect some half-planes containing zeros of such functions is derived from the generalized τ-Li criterion for $\mathcal{S}^\sharp_\mathbb{R}$ class.

1. Introduction

One of the classical problems when dealing with an L function is location of its zeros. Riemann hypothesis is an assertion that the zeros of the Riemann zeta function $\zeta(s)$ are located on the critical line $\mathrm{Re}s = \frac{1}{2}$. Analogous assertion for the other zeta or L functions yields to different kinds of generalization of the Riemann hypothesis. There are several concepts developed to characterize and classify all the L functions for which generalized Riemmann hypothesis is believed to be true. One of such concepts is the Selberg class $\mathcal{S}$ introduced in [**12**]. It is a class of functions which possess a Dirichlet series representation with coefficients satisfying Ramanujan bound, as well as the Euler product representation. Functions in the Selberg class may be meromorphically continued to the whole complex plane and satisfy a functional equation with gamma factors. Throughout the mathematical literature there exist many extensions and modifications of the Selberg class. One of the most important is the extended Selberg class $\mathcal{S}^\sharp$. It consists of Dirichlet series which possess a meromorphic continuation and satisfy appropriate functional equation, while the bound for coefficients in the Dirichlet series representation is not assumed, as well as an Euler product representation. It is believed that the extended Selberg class contains all functions of the number-theoretical interest. Location of zeros of a function from the extended Selberg class is of special interest, since there are known functions in $\mathcal{S}^\sharp$ violating generalized Riemann hypothesis. The most famous of such functions is the Davenport-Heilbronn function, but there are also some other families of such functions, for example those investigated in [**1**] and [**3**].

For the investigation of the location of zeros of a zeta or L function some criteria equivalent to the generalized Riemman hypothesis can be used. One of very simple criteria equivalent to the Riemann hypothesis is the Li criterion [**10**], which states that the Riemann hypothesis is equivalent to the non-negativity of a

2010 *Mathematics Subject Classification.* Primary 11M26, 11M36.

certain sequence of real numbers called the Li coefficients. Analogously, in [**13**] and [**11**], it is proved that the generalized Riemann hypothesis for $f \in \mathcal{S}$ is equivalent to non-negativity (and even an asymptotic relation) of the generalized Li coefficients.

Generalizing these coefficients by introducing an additional parameter τ yields the τ-Li criterion stating that non-negativity of the τ-Li coefficients is equivalent to the $\tau/2$-Riemman hypothesis. Here, δ-Riemman hypothesis is the statement that every nontrivial zero ρ of the Riemann zeta function satisfies $\mathrm{Re}\rho \leq \delta$. Generalization of the τ-Li criterion to different classes of functions containing $\mathcal{S}$ is obtained in [**6**], [**7**] and [**3**].

The main purpose of this paper is to construct a family of functions from the extended Selberg class having zeros off the critical line and to investigate position of those zeros. First, we prove existence of the zero in the half-plane $\mathrm{Re}s > 1$. Then, we present numerical criterion, that can be derived from the generalized τ-Li criterion deduced in [**3**] to detect some half-planes containing zeros of constructed functions.

We have focused on a special case to be able to illustrate numerical results, but presented methods can be applied to other similar functions and to analogously defined families.

2. Certain families of L functions with zeros in the half-plane $\mathrm{Re}s > 1$

The extended Selberg class, denoted by $\mathcal{S}^\sharp$, and introduced by Kaczorowski and Perelli in [**9**] consists of functions $L(s)$ satisfying the following axioms:

(i) (Dirichlet series representation) L possesses a Dirichlet series representation

$$L(s) = \sum_{n=1}^{\infty} \frac{a_L(n)}{n^s}, \tag{2.1}$$

that converges absolutely for $\mathrm{Re}s > 1$.

(ii) (Analytic continuation) There exists an integer $m \geq 0$ such that $(s-1)^m L(s)$ is an entire function of finite order. The smallest such number is denoted by m_L and called the *polar order of* L.

(iii) (Functional equation) The function

$$\Phi_L(s) = L(s) Q_L^s \prod_{j=1}^{r} \Gamma(\lambda_j s + \mu_j),$$

with $Q_L > 0$, $r \geq 0$, $\lambda_j > 0$, $|w| = 1$, $\mathrm{Re}\mu_j \geq 0$, $j = 1, \ldots, r$ satisfies the functional equation

$$\Phi_L(s) = w\overline{\Phi_L(1-\bar{s})}.$$

The numbers λ_j and μ_j are called the spectral parameters of L.

A well known example of a function in $\mathcal{S}^\sharp$ with zeros in the half plane $\mathrm{Re}s > 1$ is the Davenport-Heilbronn L function introduced in [**4**].

Additionally, Davenport-Heilbronn L function is contained in the subclass $\mathcal{S}^\sharp_\mathbb{R}$ of the extended Selberg class consisting of Dirichlet series (2.1) from $\mathcal{S}^\sharp$ with real coefficients $a_L(n)$ and real spectral parameters μ_j, $j = 1, \ldots, r$. Coefficients in the Dirichlet series representation of the Davenport-Heilbronn L function are periodic with period 5. In the following example we will define a family of L functions from the class $\mathcal{S}^\sharp_\mathbb{R}$ with coefficient periodic with period 13.

EXAMPLE 2.1. Let $\xi = (\xi_1, \xi_2, \xi_3, \xi_4, \xi_5, \xi_6)$ be a sextuple of real numbers where $\xi_1 = 1$ and which satisfy the system of equations

$$\sum_{k=1}^{6} \xi_k \sin \frac{2\pi jk}{13} = \xi_j \sum_{k=1}^{6} \xi_k \sin \frac{2\pi k}{13}, \quad (j = 2, \dots, 6) \tag{2.2}$$

and let

$$L_{13}(s, \xi) = \sum_{j=1}^{\infty} \frac{b(j, \xi)}{j^s}, \tag{2.3}$$

where

$$b(j, \xi) = \begin{cases} \xi_k, & j \equiv k \pmod{13}, \\ -\xi_k, & j \equiv 13 - k \pmod{13}, \\ 0, & j \equiv 0 \pmod{13}, \end{cases}$$

for some $k \in \{1, \dots, 6\}$.

REMARK 2.2. In the given example condition $\xi_1 = 1$ is a form of normalization and may be omitted without any significant changes. The condition that the components of ξ are real numbers is posed since our intention is to present application of the generalized τ-Li criterion proved for the class $\mathcal{S}^{\sharp}_{\mathbb{R}}$. However, the family of functions from the Example 2.1 can be analogously defined for sextuples ξ with complex components.

Performing analogous steps as those used to derive properties of the family of functions $L_7(s, \xi, \rho)$, introduced in [**3**], we derive properties of functions $L_{13}(s, \xi)$, summarized in the following proposition.

PROPOSITION 2.3. *For an arbitrary chosen but fixed ξ satisfying system (2.2), function $L_{13}(s, \xi)$ defined by (2.3) has the following properties.*

(i) *It can be represented in terms of the Hurwitz zeta functions*

$$L_{13}(s, \xi) = \frac{1}{13^s} \sum_{k=1}^{6} \xi_k \left(\zeta\left(s, \frac{k}{13}\right) - \zeta\left(s, \frac{13-k}{13}\right) \right).$$

(ii) *$L_{13}(s, \xi)$ belongs to the class $\mathcal{S}^{\sharp}_{\mathbb{R}}$, where function $\Phi_{L_{13}}(s)$ from the functional equation axiom is defined as*

$$\Phi_{L_{13}}(s) = \left(\frac{13}{\pi}\right)^{s/2} \Gamma\left(\frac{s+1}{2}\right) L_{13}(s, \xi). \tag{2.4}$$

PROOF. Here we will give main steps of the proof. For details we refer to analogous results in [**14**] and [**3**]. Assertion (i) immediately follows from the periodicity of coefficients in the Dirichlet series representation and the definition of the Hurwitz zeta functions. In order to prove assertion (ii) validity of the three axioms from definition of the extended Selberg class need to be proved. Axioms (i) and (ii) are trivially satisfied. The functional equation axiom can be derived from the part (i) of this proposition. Namely, the functional equation for the Hurwitz zeta function, properties of the gamma function and well known trigonometric identities imply that

$$L_{13}(s, \xi) = 2 \cdot 13^{-s} \pi^{s-1/2} \frac{\Gamma\left(1 - \frac{s}{2}\right)}{\Gamma\left(\frac{1}{2} + \frac{s}{2}\right)} \sum_{j=1}^{\infty} \frac{1}{j^{1-s}} \left(\sum_{k=1}^{6} \xi_k \sin \frac{2\pi jk}{13} \right).$$

Using the system (2.2) satisfied by the components of ξ and comparing the coefficients of the series appearing in the above equation with the coefficients of the series

$$\sum_{j=1}^{\infty} \frac{b(j,\xi)}{j^{1-s}} \left(\sum_{k=1}^{6} \xi_k \sin \frac{2\pi k}{13} \right),$$

together with the fact that

$$\left(\sum_{k=1}^{6} \xi_k \sin \frac{2\pi k}{13} \right)^2 = \frac{13}{4} \tag{2.5}$$

implies that the function (2.4) satisfies the functional equation given in the axiom (iii). The fact that the coefficients in the Dirichlet series representation of $L_{13}(s,\xi)$ are real by definition implies that $L_{13}(s,\xi) \in \mathcal{S}^{\sharp}_{\mathbb{R}}$. □

Next, we show that $L_{13}(s,\xi)$ possesses zero in the half-plane $\mathrm{Re} s > 1$. To do so, we introduce some notation. First, we recall that $\zeta\left(s, \frac{a}{13}\right)$ for $1 \le a \le 12$ can be written as a sum over Dirichlet characters χ of modulus 13 of associated Dirichlet L functions

$$\zeta\left(s, \frac{a}{13}\right) = \frac{13^s}{\phi(13)} \sum_{\chi} \bar{\chi}(a) L(s,\chi).$$

For any $1 \le a \le 12$ let

$$Z_a(s) = \frac{13^s}{c(a)\phi(13)} \sum_{\chi} \bar{\chi}(a) M(s,\chi), \tag{2.6}$$

where the sum is taken over all Dirichlet characters modulus 13;

$$M(s,\chi) = \sum_{n=1}^{\infty} \frac{c(n)\chi(n)}{n^s}$$

with multiplicative coefficients $c(n)$ defined for primes p as

$$c(p) = \begin{cases} 1, & \chi_1(p) = 0 \text{ or } \chi_1(p) = 1, \\ i, & \chi_1(p) = -1, \end{cases}$$

where χ_1 is an arbitrary fixed real non-principal character modulus 13. Additionally, denote by χ_0 principal character modulus 13. Functions $Z_a(s)$ i $M(s,\chi)$ can be defined for general modulus q and positive integers a relatively prime to q. Their properties are proved in [**4**] and [**5**] and summarized in the following lemma.

Lemma 2.4. *Let q be an odd prime number and a $(0 < a < q)$ an integer relatively prime to q. Then*

(i) $Z_a(s)$ *and* $\frac{1}{c(a)}\left(M(s,\chi_0) + \chi_1(a)M(s,\chi_1)\right) = 2\mathrm{Re}\left(\frac{M(s,\chi_0)}{c(a)}\right)$ *are real for* $s > 1$.

(ii) *For* $s > 1$

$$\log M(s,\chi_0) = -\frac{1}{2}(1+i)\log(s-1) + O(1),$$

$$\log M(s,\chi_1) = -\frac{1}{2}(1-i)\log(s-1) + O(1),$$

$$\log M(s,\chi) = O(1),$$

for $\chi \notin \{\chi_0, \chi_1\}$, *as* $s \downarrow 1$.

Next proposition gives us some properties of the zeros of functions $L_{13}(s,\xi)$.

PROPOSITION 2.5. *For an arbitrary chosen but fixed ξ satisfying properties given in Example 2.1, function $L_{13}(s,\xi)$ defined by (2.3) has the following properties.*

(i) *There exists a constant $\sigma_{L_{13}} > 0$ such that $L_{13}(s,\xi)$ has no zeros in the half-plane* $\mathrm{Re}s > \sigma_{L_{13}}$.
(ii) *$L_{13}(s,\xi)$ has a zero in the half plane* $\mathrm{Re}s > 1$.

PROOF. The proof of the claim (i) immediately follows from the periodicity of coefficients in the Dirichlet series representation of the function $L_{13}(s,\xi)$. Namely,

$$||L_{13}(s,\xi)| - \xi_1| \le |L_{13}(s,\xi) - \xi_1| \le \max_{k=1,\dots,6}\{|\xi_k|\}\sum_{n=2}^{\infty}\frac{1}{n^{\mathrm{Re}s}}$$
$$\le \max_{k=1,\dots,6}\{|\xi_k|\}\frac{1}{\mathrm{Re}s-1}\frac{1}{2^{\mathrm{Re}s-1}}.$$

For $\mathrm{Re}s > 1 + \frac{\max\limits_{k=1,\dots,6}\{|\xi_k|\}}{|\xi_1|}$ the above inequality reduces to

$$-\frac{1}{2^{\mathrm{Re}s-1}}\,|\xi_1| < |L_{13}(s,\xi)| - |\xi_1| < \frac{1}{2^{\mathrm{Re}s-1}}\,|\xi_1|,$$

i.e.

$$|\xi_1|\left(1-\frac{1}{2^{\mathrm{Re}s-1}}\right) < |L_{13}(s,\xi)| < |\xi_1|\left(1+\frac{1}{2^{\mathrm{Re}s-1}}\right).$$

The bound for the $\mathrm{Re}s$, the fact that $|\xi_1| \le \max\limits_{k=1,\dots,6}\{|\xi_k|\}$ and inequality $\mathrm{Re}s - 1 < 2^{\mathrm{Re}s-1}$ imply

$$|\xi_1|\left(1-\frac{1}{2^{\mathrm{Re}s-1}}\right) > 0,$$

so $|L_{13}(s,\xi)| > 0$ for all s such that $\mathrm{Re}s > 1 + \frac{\max\limits_{k=1,\dots,6}\{|\xi_k|\}}{|\xi_1|}$. Thus, we may take $\sigma_{L_{13}} = 1 + \frac{\max\limits_{k=1,\dots,6}\{|\xi_k|\}}{|\xi_1|}$, which proves part (i).

To prove (ii), representation (i) from Proposition 2.3 will be used combined with Lemma 2.4.

Let $Z_L(s)$ be a function associated to $L_{13}(s,\xi)$ defined by

$$Z_L(s) = \frac{1}{13^s}\sum_{j=1}^{6}\xi_j\left(Z_j(s) - Z_{13-j}(s)\right),$$

where $Z_j(s)$ is defined in (2.6). Lemma 2.4 enables us to reduce the sum over characters in the definition of $Z_j(s)$ to the sum of terms involving χ_0 and χ_1, up to a constant, and moreover to write it in the following form

$$Z_L(s) = \frac{2}{\phi(13)}\sum_{j=1}^{6}\xi_j\mathrm{Re}\left(\left(\frac{1}{c(j)} - \frac{1}{c(13-j)}\right)M(s,\chi_0)\right) + O(1),$$

for $s > 1$, as $s \downarrow 1$. Writing $\frac{1}{c(j)} - \frac{1}{c(13-j)}$ as $a_j + ib_j$, where a_j and b_j are real numbers, we get

$$\mathrm{Re}\left(\left(\frac{1}{c(j)} - \frac{1}{c(13-j)}\right)M(s,\chi_0)\right) = a_j\mathrm{Re}M(s,\chi_0) - b_j\mathrm{Im}M(s,\chi_0),$$

thus

$$Z_L(s) = \frac{2}{\phi(13)} \left(A\mathrm{Re}M(s,\chi_0) - B\mathrm{Im}M(s,\chi_0) \right) + O(1), \tag{2.7}$$

for $s > 1$, as $s \downarrow 1$; where $A = \sum_{j=1}^{6} \xi_j a_j$ and $B = \sum_{j=1}^{6} \xi_j b_j$.

The first expression in Lemma 2.4, part (ii) implies that

$$M(s,\chi_0) = \frac{O(1)}{\sqrt{s-1}} \left(\cos\left(\log \frac{1}{\sqrt{s-1}} \right) + i \sin\left(\log \frac{1}{\sqrt{s-1}} \right) \right),$$

for $s > 1$, as $s \downarrow 1$, thus

$$A\mathrm{Re}M(s,\chi_0) - B\mathrm{Im}M(s,\chi_0) = \frac{O(1)}{\sqrt{s-1}} \left(A\cos\left(\log \frac{1}{\sqrt{s-1}} \right) - B\sin\left(\log \frac{1}{\sqrt{s-1}} \right) \right).$$

For the values of s closed to 1 growth rate of the above expression is determined by the factor $1/\sqrt{s-1}$, while the sign is determined by other term. Therefore, for any $\delta > 0$ and $H > 0$ there exist s_1 and s_2 such that $1 < s_1 < s_2 < 1+\delta$, $A\mathrm{Re}M(s_1,\chi_0) - B\mathrm{Im}M(s_1,\chi_0) > H$ and $A\mathrm{Re}M(s_2,\chi_0) - B\mathrm{Im}M(s_2,\chi_0) < -H$.

If H is sufficiently large, relation (2.7) implies $Z_L(s_2) < 0 < Z_L(s_1)$, hence there exists $s \in (1, 1+\delta)$ such that $Z_L(s) = 0$.

In order to prove proposition we apply Kroneker's and Rouche's theorems to relate zeros of $Z_L(s)$ and $L_{13}(s,\xi)$. We start with

$$\begin{aligned} L_{13}(s+it,\xi) - Z_L(s) = \frac{1}{\phi(13)} \sum_{k=1}^{6} \xi_k \sum_{\chi} \Bigg(\bar{\chi}(k) \left(L(s+it,\chi) - \frac{M(s,\chi)}{c(k)} \right) \\ -\bar{\chi}(13-k) \left(L(s+it,\chi) - \frac{M(s,\chi)}{c(13-k)} \right) \Bigg). \end{aligned} \tag{2.8}$$

Euler product for the Dirichlet L function and the function $M(s,\chi)$ combined with some basic inequalities and the fact that $|c(n)| = 1$ for any n, implies that for $\mathrm{Re}s \geq 1+\delta > 1$ and $k \in \{1,\ldots,6\}$ we have the bound

$$\begin{aligned} \log L(s+it,\chi) - \log \frac{M(s,\chi)}{c(k)} &= \sum_p \left(\log\left(c(k) \left(1 - \frac{c(p)\chi(p)}{p^s} \right) \right) - \log\left(1 - \frac{\chi(p)}{p^{s+it}} \right) \right) \\ &= \sum_{p \leq P} \log\left(c(k) \left(\frac{p^s - c(p)\chi(p)}{p^s - p^{-it}\chi(p)} \right) \right) + O\left(\frac{1}{P^\delta} \right) \\ &= O\left(\sum_{p\leq P,\, p\neq 13} \frac{\left| c(p) - p^{-it} \right|}{p^{\mathrm{Re}s}} \right) + O\left(\frac{1}{P^\delta} \right). \end{aligned} \tag{2.9}$$

The above expression can be made arbitrary small by Kronecker's theorem. Namely, let $c(p) = e^{2\pi i d(p)}$; for given $q > 0$ and $p \leq P$, there is a real number t and an integer x_p such that

$$\left| t\frac{\log p}{2\pi} + d(p) - x_p \right| \leq \frac{1}{q},$$

i.e. $\left| c(p) - p^{-it} \right| \leq e^{2\pi/q} - 1$. Hence, by appropriately choosing P, and then q expression (2.9) can be made arbitrary small for any character χ of modulus 13. Now, from (2.8) follows that for given $\varepsilon > 0$ and $\delta > 0$ there is a real number t such that $|L_{13}(s+it,\xi) - Z_L(s)| < \varepsilon$ when $\mathrm{Re}s \geq 1+\delta$.

We have proved the existence of zero of $Z_L(s)$ such that $s > 1$, denote such zero by $\tilde{s}$. For any $\eta > 0$ there exists an $\eta_1 \in (0, \eta)$, $\eta_1 < \tilde{s} - 1$ such that $Z_L(s) \neq 0$ for $|s - \tilde{s}| = \eta_1$. Put $\varepsilon = \min_{|s-\tilde{s}|=\eta_1} |Z_L(s)|$ and choose δ such that $\delta < \tilde{s} - \eta_1 - 1$; then Rouche's theorem may be applied to deduce that $Z_L(s)$ and $Z_L(s) - (Z_L(s) - L_{13}(s+it, \xi))$ have the same number of zeros inside $|s - \tilde{s}| = \eta_1$. Thus, $L_{13}(s+it, \xi)$ has at least one zero in $|s - \tilde{s}| = \eta_1$, i.e. $L_{13}(s, \xi)$ has at least one zero inside $|s - \tilde{s} - it| = \eta_1$. This proves the proposition. □

3. Numerical criterion for detecting half-planes containing zeros of L function

In this part we will present criterion that can be used to derive information about position of zeros of function $L_{13}(s, \xi)$ by numerically evaluating quantities called τ-Li coefficients attached to the function under consideration. The method is based on the generalized τ-Li criterion for the class $\mathcal{S}_{\mathbb{R}}^{\sharp}$ derived in [**3**].

First, we give definition of τ-Li coefficients as a sum over non-trivial zeros of corresponding function.

DEFINITION 3.1. For $L(s) \in \mathcal{S}_{\mathbb{R}}^{\sharp}$, integer n and real parameter τ the nth τ-Li coefficient attached to the function L is defined by

$$\lambda_L(n, \tau) = \sum_{\rho \in Z(L)}^{*} \left(1 - \left(\frac{\rho}{\rho - \tau}\right)^n\right), \tag{3.1}$$

where $*$ indicates that the sum is interpreted as a limit $\lim_{T\to\infty} \sum_{\rho \in Z(L), |\mathrm{Im}\rho| < T}$.

Second, the existence of such coefficients is proved and alternative definition in terms of, so-called, completed L function defined by

$$\Phi_L^c(s) = s^{m_F}(s-1)^{m_F}\Phi_L(s)$$

is given. The zeros of Φ_L^c are called the non-trivial zeros of $L \in \mathcal{S}_{\mathbb{R}}^{\sharp}$. The set of non-trivial zeros of L is denoted by $Z(L)$.

PROPOSITION 3.2 ([**3**]). *Fix $\tau \in [1, +\infty)$ an arbitrary real number. Let $L \in \mathcal{S}_{\mathbb{R}}^{\sharp}$ with the property that $0, \tau \notin Z(L)$. Then, for every positive integer n*

(i) *the sum (3.1) defining $\lambda_F(n, \tau)$ is $*-$convergent,*
(ii)

$$\lambda_L(n, \tau) = \frac{\tau}{(n-1)!}\left[\frac{d^n}{ds^n}\left(s^{n-1}\log \Phi_L^c(s)\right)\right]_{s=\tau}. \tag{3.2}$$

Third, generalized τ-Li criterion that gives us possibility to relate information about τ-Li coefficients to the information about positions of the zeros of corresponding L function is formulated.

THEOREM 3.3 ([**3**]). *Let $L \in \mathcal{S}_{\mathbb{R}}^{\sharp}$ and let $\tau \in [1, \infty)$ such that $0, \tau \notin Z(L)$. The following statements are equivalent.*

(i) *All the non-trivial zeros of the function L lie in the strip $1 - \tau/2 \leq \mathrm{Re} s \leq \tau/2$.*
(ii) *The τ-Li coefficients $\lambda_L(n, \tau) \geq 0$ for all $n \geq 1$.*
(iii) *For every fixed $\delta > 0$ there exists a constant $c(\delta)$ such that $\lambda_L(n, \tau) \geq -c(\delta)\exp(\delta n)$.*

(iv) *The τ-Li coefficients of L have the property that*

$$\limsup_{n\to\infty} |\lambda_L(n+1,\tau)|^{1/n} \leq 1.$$

We end this section with a numerical example illustrating how to numerically prove existence of zeros of a specific function $L_{13}(s,\xi)$ in certain half-planes.

Obviously, the above result gives us possibility to determine whether all the zeros of some L function are located in the strip $1-\tau/2 \leq \mathrm{Re}s \leq \tau/2$ by investigating non-negativity of the $\tau-$Li coefficients attached to L. However, proving non-negativity of coefficients is not an easy task. Alternatively, numerically we can evaluate only finite number of τ-Li coefficients. Existence of a negative value of $\tau-$Li coefficient indicates existence of a zero of the function L in the half-plane $\mathrm{Re}s > \tau/2$. Thus, above results gives us numerical criterion to find half-planes (determined by τ) containing zeros of an L function.

Illustration of the above observation will be given in the following example. Numerical calculations are based on the formula (3.2) and are done using ARB, a C library for arbitrary-precision floating-point ball arithmetic, developed by F. Johansson [**8**]. The advantage gained by using ARB is the fact that ball-arithmetic provides automatic error control. Also, it supports efficient computation with power series and special functions over the real and complex numbers so it turns out to be very efficient for calculations of τ-Li coefficients attached to functions under consideration. Let us notice that the presented values are actually approximate values of τ-Li coefficients, i.e. mid-points of the balls representing real numbers in ball-arithmetic. However, the precision in calculation is set such that there is no noticeable difference between the mid-point and mid-point $\pm$ the error term, so graphs are completely reliable.

EXAMPLE 3.4. Let as consider $L_{13}(s,\xi_0)$ where

$$\xi_0 = (1, 0.8731494282703761, -4.294850574354864, 0, 1, 2.567015075752353).$$

In addition to normalization $\xi_1 = 1$, here ξ_4 and ξ_5 are set to be 0 and 1, respectively, while the other components are calculated such that ξ_0 is a solution of the system (2.2).

The assertion (*ii*) of Proposition 2.3 implies that $L_{13}(s,\xi_0) \in \mathcal{S}^\sharp_{\mathbb{R}}$, thus by Proposition 3.2, coefficients $\lambda_{L_{13}}(n,\tau)$ are well defined for real parameters $\tau \geq 1$ and can be calculated using (3.2).

On the Figure 1 τ-Li coefficients attached to function $L_{13}(s,\xi_0)$ for $\tau = 1$ and $\tau = 2$ and for n from 1 to 600 are presented. Existence of negative values of τ-Li

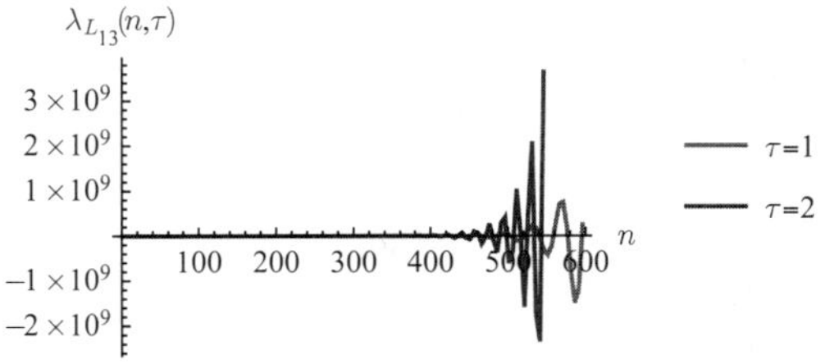

FIGURE 1. τ-Li coefficients attached to $L_{13}(s,\xi_0)$ for $\tau \in \{1,2\}$

coefficient implies by Theorem 3.3 existence of a zero ρ of function $L_{13}(s,\xi_0)$ such that $\mathrm{Re}\rho > \tau/2$, i.e. for $\tau = 1$, ρ is a zero off the critical line, and for $\tau = 2$ zero ρ is such that $\mathrm{Re}\rho > 1$. Notice that the above reasoning provides another proof of claim (ii) from Proposition 2.5 for the function $L_{13}(s, \xi_0)$. The same procedure can be applied for other choices of sextuple ξ.

The advantage of presented reasoning is the fact that we able to vary τ from 2 to $2\sigma_{L_{13}}$ (from the claim (i) in the Proposition 2.5) and thus obtain more information about position of zeros of the function under consideration. Let us point out that the expression for $\sigma_{L_{13}}$ is very rough, so better insight of position of zeros is of interest.

Figure 2 presents τ-Li coefficients attached to function $L_{13}(s, \xi_0)$ for $\tau = 3$ and $\tau = 4$ and for n from 1 to 600. Existence of oscillations in the case $\tau =$

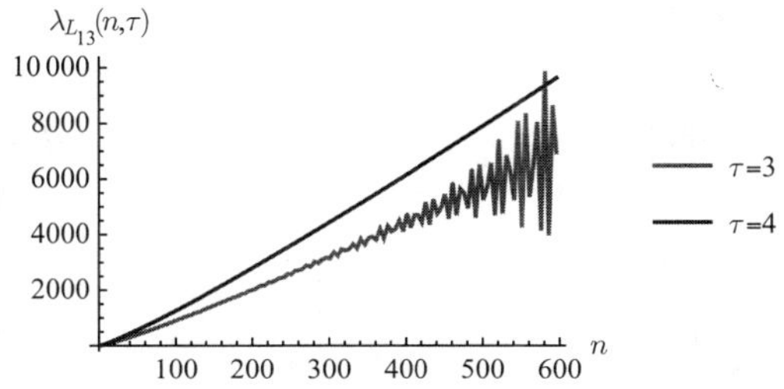

FIGURE 2. τ-Li coefficients attached to $L_{13}(s, \xi_0)$ for $\tau \in \{3, 4\}$

3 indicates that we need to expand range of n to attain negative value of τ-Li coefficients. This is due to the fact that existence of the zero outside the strip implied by τ is equivalent to the oscillations of τ-Li coefficients with amplitudes growing exponentially (see Theorem 3.3, claim (iii)). Existence of negative value is obtained for greater values of n as it is clearly visible from the Figure 3. This implies that there is a zero of the function $L_{13}(s, \xi_0)$ in the half-plane $\mathrm{Re}s > 3/2$. It

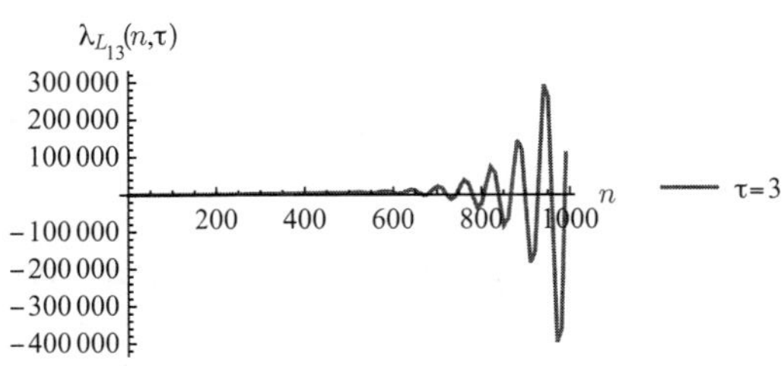

FIGURE 3. τ-Li coefficients attached to $L_{13}(s, \xi_0)$ for $\tau = 3$

is important to mention that the fact that we have monotonically increasing values of τ-Li coefficients for some range of n, as in the case $\tau = 4$ in the Figure 2, does not imply absence of zeros in the half-plane $\mathrm{Re}s > \tau/2$. It may happen that the oscillations develop later in the range of n and then become big enough to produce negative coefficients. Such situation is illustrated in Figures 4 and 5. Namely, to analyze zeros contained in half-plane $\mathrm{Re}s > 3/2$ values of τ-Li coefficients for $\tau \in \{3.1, 3.2, 3.3, 3.4\}$ are evaluated. Clearly visible oscillations in the case $\tau = 3.1$ indicate that we will have negative value for wider range of n, tiny oscillation for

$\tau = 3.2$ may be used to come to the same conclusion, but the range of n needs to be much wider. Monotonic increase of τ-Li coefficients for $\tau = 3.3$ and $\tau = 3.4$ does not allow us to conclude anything, specially, since Figure 5 shows existence of oscillations for $\tau = 3.3$ arising for large values of n.

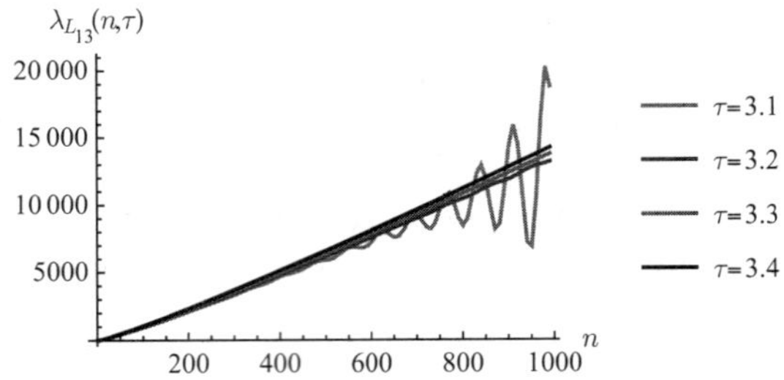

FIGURE 4. τ-Li coefficients attached to $L_{13}(s, \xi_0)$ for $\tau \in \{3.1, 3.2, 3.3, 3.4\}$

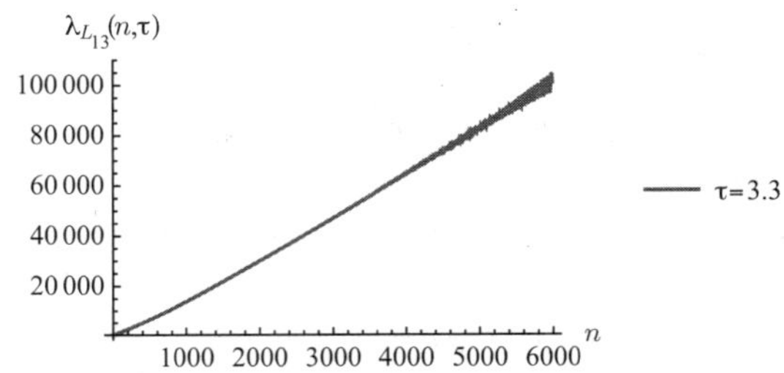

FIGURE 5. τ-Li coefficients attached to $L_{13}(s, \xi_0)$ for $\tau = 3.3$

Results presented in the above examples indicate that numerical criterion based on the τ-Li coefficients gives us possibility to obtain more information about position of zeros off the critical line of the function under consideration compared to information obtained in a classical way (e.g. Proposition 2.5). Moreover, though we are focused on family of functions constructed in Example 2.1, the method may be extended analogously to some other families of functions, defined in a similar way.

Additionally, we believe that values of τ-Li coefficients may be used to reveal some other properties of zeros of the corresponding L function. We believe that position of the first negative τ-Li coefficient, amplitudes of the oscillation, as well as values of n where $\lambda_L(n, \tau)$ starts to oscillate are related to overall distribution of zeros and proportion of zeros outside and inside the strip $1 - \tau/2 \leq \text{Re}s \leq \tau/2$.

References

[1] È. Bomb′eri and A. Gosh, *On the Davenport-Heilbronn function* (Russian, with Russian summary), Uspekhi Mat. Nauk **66** (2011), no. 2(398), 15–66, DOI 10.1070/RM2011v066n02ABEH004740; English transl., Russian Math. Surveys **66** (2011), no. 2, 221–270. MR2847789

[2] A. Bucur, A.-M. Ernvall-Hytönen, A. Odžak, E. Roditty-Gershon, and L. Smajlović, *On τ-Li coefficients for Rankin-Selberg L-functions*, Women in numbers Europe, Assoc. Women Math. Ser., vol. 2, Springer, Cham, 2015, pp. 167–190, DOI 10.1007/978-3-319-17987-2_7. MR3596605

[3] A. Bucur, A.-M. Ernvall-Hytönen, A. Odžak, and L. Smajlović, *On a Li-type criterion for zero-free regions of certain Dirichlet series with real coefficients*, LMS J. Comput. Math. **19** (2016), no. 1, 259–280, DOI 10.1112/S1461157016000115. MR3569320

[4] H. Davenport and H. Heilbronn, *On the Zeros of Certain Dirichlet Series*, J. London Math. Soc. **11** (1936), no. 3, 181–185, DOI 10.1112/jlms/s1-11.3.181. MR1574345

[5] H. Davenport and H. Heilbronn, *On the Zeros of Certain Dirichlet Series*, J. London Math. Soc. **11** (1936), no. 4, 307–312, DOI 10.1112/jlms/s1-11.4.307. MR1574931

[6] A. D. Droll, *Variations of Li's criterion for an extension of the Selberg class*, ProQuest LLC, Ann Arbor, MI, 2012. Thesis (Ph.D.)–Queen's University (Canada). MR3218185

[7] A.-M. Ernvall-Hytönen, A. Odžak, L. Smajlović, and M. Sušić, *On the modified Li criterion for a certain class of L-functions*, J. Number Theory **156** (2015), 340–367, DOI 10.1016/j.jnt.2015.03.019. MR3360344

[8] F. Johansson, 'Arb: a C library for ball arithmetic', *ACM Communications in Computer Algebra*, 47(4) (2013) 166–169.

[9] J. Kaczorowski and A. Perelli, *On the structure of the Selberg class. I.* $0 \leq d \leq 1$, Acta Math. **182** (1999), no. 2, 207–241, DOI 10.1007/BF02392574. MR1710182

[10] X.-J. Li, *The positivity of a sequence of numbers and the Riemann hypothesis*, J. Number Theory **65** (1997), no. 2, 325–333, DOI 10.1006/jnth.1997.2137. MR1462847

[11] A. Odžak and L. Smajlović, *On asymptotic behavior of generalized Li coefficients in the Selberg class*, J. Number Theory **131** (2011), no. 3, 519–535, DOI 10.1016/j.jnt.2010.08.009. MR2741223

[12] A. Selberg, *Old and new conjectures and results about a class of Dirichlet series*, Proceedings of the Amalfi Conference on Analytic Number Theory (Maiori, 1989), Univ. Salerno, Salerno, 1992, pp. 367–385. MR1220477

[13] L. Smajlović, *On Li's criterion for the Riemann hypothesis for the Selberg class*, J. Number Theory **130** (2010), no. 4, 828–851, DOI 10.1016/j.jnt.2009.10.012. MR2600405

[14] E. C. Titchmarsh, *The Theory of the Riemann Zeta-Function*, Oxford, at the Clarendon Press, 1951. MR0046485

Department of Mathematics, University of Sarajevo, Zmaja od Bosne 35, 71 000 Sarajevo, Bosnia and Herzegovina

Email address: almasa.odzak@pmf.unsa.ba

School of Economics and Business, University of Sarajevo, Alije Izetbegovića 1, 71 000 Sarajevo, Bosnia and Herzegovina

Email address: lamija.sceta@efsa.unsa.ba

Contemporary Mathematics
Volume **732**, 2019
https://doi.org/10.1090/conm/732/14782

Rational points on twisted modular curves

Ekin Ozman

ABSTRACT. In this survey we summarize some results about the rational points of twisted modular curves. We also raise some open questions and conjectures.

1. Introduction

Let N be a squarefree integer. The modular curve $X_0(N)$ is a moduli space of generalized elliptic curves. In other words, $X_0(N)$ is the compactification of $Y_0(N)$ which is the moduli space of tuples (E, C) where E is an elliptic curve and C is cyclic N-torsion subgroup of $E[N]$. In [**Maz78**], Mazur proved that the only $\mathbb{Q}$-rational points on $X_0(N)$ are cusps when N is prime and greater than 163. Later, the result was extended to composite levels by Kenku and Momose. Much work has been done regarding the study of L-rational points on modular curves for L a number field by Merel in [**Mer96**].

A special class of elliptic curves, namely, $\mathbb{Q}$-*curves*, are parametrized by rational points of certain twists of $X_0(N)$. $\mathbb{Q}$-curves are interesting for many reasons; for instance they have an isogeny class that is fixed by $\mathrm{Gal}(\overline{\mathbb{Q}}/\mathbb{Q})$. Therefore $\mathbb{Q}$-curves are natural generalizations of elliptic curves over $\mathbb{Q}$, since it is classical to study elliptic curves up to isogeny or properties of elliptic curves that are invariant under isogeny.

Moreover, some 'twisted' forms of the *Fermat equation*, $x^p + y^p = z^p$ give rise to $\mathbb{Q}$-curves whose mod-p Galois representations satisfy similar ramification conditions as the elliptic curve arising from the classical Fermat equation. Recently by extending techniques that are used in the proof of Fermat's last theorem, much progress has been made in solving generalized versions of the Fermat equation, for details see [**BMS16**].

2. Twists of $X_0(N)$

The simplest twist of $X_0(N)$, denoted by $X_0^d(N)$, is determined by a quadratic field $\mathrm{K} := \mathbb{Q}(\sqrt{d})$ and the level N. On $X_0(N)$ the action of σ, the generator of $\mathrm{Gal}(\mathrm{K}/\mathbb{Q})$, is twisted. Namely, the action of σ becomes $\sigma \circ w_N$, where w_N is the $\mathbb{Q}$-rational involution on $X_0(N)$, called the *Atkin-Lehner involution*. It sends $(E, \ker(\phi))$ to $(E', \ker(\hat{\phi}))$ where $\phi : E \to E'$ is a degree-N cyclic isogeny and $\hat{\phi}$ is

The author was partially supported by Boğaziçi University Research Fund Grant Number 10842 and TUBITAK Bideb 2232 Fellowship-114C026.

the dual isogeny. The action of w_N can be extended to $X_0(N)$, i.e. w_N permutes the cusps.

The two curves are isomorphic over K, via an automorphism $\alpha : X_0^d(N)\times_{\mathbb{Q}}\mathrm{K} \to X_0(N)\times_{\mathbb{Q}}\mathrm{K}$ such that $\alpha^\sigma = w_N\circ\alpha$. The map α gives us a bijection between rational points of $X_0^d(N)$ and K-rational points of $X_0(N)$ that are fixed by $\sigma\circ w_N$.

EXAMPLE 2.1. The modular curve $X_0(23)$ is a hyperelliptic curve of genus 2 and is given by the following sextic by Gonzalez in [**Gon91**]:

$$y^2 = (x^3 - 2x^2 - 17x - 25)(x^3 + 6x^2 + 11x + 7).$$

In [**Ogg74**] Ogg proves that w_{23} acts as hyperelliptic involution; it follows that the twist $X_0^d(23)$ is given by:

$$dy^2 = (x^3 - 2x^2 - 17x - 25)(x^3 + 6x^2 + 11x + 7).$$

The quadratic twist, $X_0^d(N)$ is not the only twist of $X_0(N)$. In general, twists of an algebraic curve C over a number field L are classified by the first Galois cohomology group, $\mathrm{H}^1(\mathrm{Gal}(\overline{\mathrm{L}}/\mathrm{L}), \mathrm{Aut}(C))$ where $\mathrm{Aut}(C)$ denotes the automorphisms of C. Since N is squarefree by hypothesis, the automorphism group of the modular curve $X_0(N)$ is generated by the Atkin-Lehner involutions w_{p_i} for each prime divisor p_i of N. Hence, for any cocycle $\zeta : \mathrm{Gal}(\overline{\mathbb{Q}}/\mathbb{Q}) \to \mathrm{Aut}(X_0(N))$ there is a corresponding twist of $X_0(N)$, denoted by $X_0^\zeta(N)$. Each such cocycle factors through $\mathrm{Gal}(\overline{\mathbb{Q}}/\mathrm{K})$ for a number field K. In particular $\mathbb{Q}$-rational points of $X_0^\zeta(N)$ are K-rational points of $X_0(N)$ fixed by $\zeta(\sigma)\circ\sigma$ for each σ in $\mathrm{Gal}(\mathrm{K}/\mathbb{Q})$. Note that rational points of $X_0^\zeta(N)$ parametrize certain $\mathbb{Q}$-curves. It is a result of Elkies [**Elk04**] that every $\mathbb{Q}$-curve without complex multiplication is geometrically isogenous to a $\mathbb{Q}$-curve defined over a polyquadratic field i.e. a field that is generated by quadratic fields. Therefore understanding rational points of $X_0^\zeta(N)$ gives information about $\mathbb{Q}$-curves as well.

Even though the twist $X_0^\zeta(N)$ and $X_0(N)$ are the 'same' curve over K, they are quite different as curves over $\mathbb{Q}$. The classical modular curve $X_0(N)$ has natural rational points, the cusps. Since the cusps are freely permuted by the Atkin-Lehner operators, they are not rational points of $X_0^\zeta(N)$. Therefore the answer to the following question is the first step towards understainding $X_0^\zeta(N)(\mathbb{Q})$.

QUESTION 2.2 ([**Ell04b**]). Let $X_0^d(N)$ be a quadratic twist of $X_0(N)$. For which d and N does $X_0^d(N)$ have rational points over every completion of $\mathbb{Q}$?

A first answer was given by the author for all (N,d,p) when p is not ramified in $\mathbb{Q}(\sqrt{d})$ and $\mathbb{Q}(\sqrt{-N})$ simultaneously.

THEOREM 2.3 (Ozman, [**Ozm12**]). *Let p be a prime, N square-free integer, $K = \mathbb{Q}(\sqrt{d})$. Then*

(1) $X_0^d(N)(\mathbb{Q}_p) \neq \varnothing$ for all p that split in K and for $p = \infty$.
(2) $X_0^d(N)(\mathbb{Q}_p) \neq \varnothing$ if p is inert in K and does not divide N.
(3) For all odd p inert in K and dividing N, $X_0^d(N)(\mathbb{Q}_p) \neq \varnothing$ if and only if $N = p$ or $2p$.
(4) For all odd p that is ramified in K and not dividing N, $X_0^d(N)(\mathbb{Q}_p) \neq \varnothing$ if and only if there is a prime of the ring class field $\mathbb{Q}(j(\sqrt{-N}))$ lying over p with residue degree 1.

(5) If 2 is inert in K and divides N, $X_0^d(N)(\mathbb{Q}_2) \neq \varnothing$ if and only if $N = 2\Pi q_i$ where $q_i \equiv 1$ modulo 4 for all i.
(6) If 2 is ramified in K and $N \equiv 3$ modulo 4 then $X_0^d(N)(\mathbb{Q}_2) \neq \varnothing$ implies that 2 splits completely in $\mathrm{H}/\mathbb{Q}(\sqrt{-N})$ where H is the Hilbert class field of $\mathbb{Q}(\sqrt{-N})$.

Later Theorem 2.3 have been generalized to Shimura curves in [**Sta**] and to higher degree(bi-quadratic twists) in [**Ozm13**].

The statement of Theorem 2.3 have many cases depending on the relationships between p and K and between p and N. The proof of each of these cases require different techniques and the main work is in the case where p is inert or ramified in K.

Let us give an idea of the ingredients of the proof. Let R be the valuation ring of K_ν where ν is a prime lying over p and K_ν is the completion of K at ν. In the inert case, we have a good model for $X_0^d(N)$ over R by etale descent; existence of points is shown by working over a finite field and using a geometric version of Hensel's Lemma. This is the general idea, but since the minimal regular R-model of $X_0(N)$ is not smooth when $p \mid N$, we need different techniques for each of the two subcases: $p \mid N$ and $p \nmid N$. When $p|N$, we give necessary and sufficient conditions for existence of a smooth point on the special fiber of the regular model given by Mazur, Deligne and Rapoport ([**Maz77**], [**Del74**]). For the case $p \nmid N$, we construct a point on the special fiber using theory of complex multiplication. In each case, we lift the smooth point on the generic fiber to a point on the generic fiber by Hensel's Lemma.

However, when p is ramified in $\mathbb{Q}(\sqrt{d})$ and unramified in $\mathbb{Q}(\sqrt{-N})$, we do not have a regular model of $X_0^d(N)$ over R. Therefore, we must use different techniques. In this case, we actually construct a point on the generic fiber that is fixed by both w_N and σ, thus a point with complex multiplication. Since $R/\mathbb{Z}_p$ is ramified, σ induces a trivial map on the residue fields. Then using results of Deuring and some facts from algebraic geometry, we can deduce that a $w_N \circ \sigma$-fixed point of the generic fiber of $X_0(N)_{/R}$ corresponds to a w_N-fixed point of the special fiber of $X_0(N)_{/R}$. Thus, the problem is reduced to determining for which primes p there exists a w_N-fixed point of $X_0(N)$ over $\mathbb{F}_p$.

3. Violation of the Hasse Principle

The precise statement of Theorem 2.3 is explicit enough to check existence of local points of curves easily. Thus we can easily produce examples with local points everywhere.

EXAMPLE 3.1. For instance, the twist of $X_0(23)$ by $\mathbb{Q}(\sqrt{17})$, which is given by $17y^2 = (x^3 - 2x^2 - 17x - 25)(x^3 + 6x^2 + 11x + 7)$, has local points everywhere. This example is particularly striking because we can show that it possesses no rational point. See [**Ozm12**] for details.

Understanding local points is the first step towards understanding global points of any curve. The Hasse Principle is the statement that a global point exists if and only if it exists for each local completion of the global field. This statement holds only for genus zero curves. It is known that there are counter-examples to this principle for higher genus curves, however in general it is not known how often this happens in a given family. For instance, in [**Cla07**] Clark proves that the density of quadratic twists of $X_0(N)$ which violate the Hasse principle has positive density. We improve this result, in [**Ozm12**] and [**Ozm13**] by giving an explicit asymptotic formula for the number of quadratic and bi-quadratic twists of $X_0(N)$ which violates the Hasse principle. We include a simpler version of this result below:

THEOREM 3.2 (Ozman, [**Ozm12**]). *Given a prime number $N \equiv 1 \bmod 4$ and a positive integer X, let A be the set of positive squarefree integers $d \leq X$ such that $X_0^d(N)(\mathbb{Q}_p)$ is non-empty for all p when there is no prime p simultaneously ramified in $\mathbb{Q}(\sqrt{d})$ and $\mathbb{Q}(\sqrt{-N})$. Then*

$$|A| = \frac{1}{2}B(N)\frac{X}{\log^{1-\alpha_N} X} + O\left(\frac{X}{\log^{2-\alpha_N} X}\right)$$

where α_N and $B(N)$ are constants that only depend on N.

In order to obtain the density result of Theorem 3.2 we need the following result of Serre:

THEOREM 3.3 (Serre, Theorem 2.8 in [**Ser76**]). *Let $0 < \alpha < 1$ be Frobenius density of a set of primes S and $N_S(X)$ is the number of squarefree integers in $[1 \ldots X]$ all of whose prime factors lie in S. Then $N_S(X) = c_S \frac{X}{\log^{1-\alpha} X} + O\left(\frac{X}{\log^{2-\alpha} X}\right)$ for some positive constant c_S.*

Using this result we obtain a density result for the twists which have local points at every prime p. Combining this with Faltings's finiteness results as in the proof of Theorem 2 in [**Cla07**] we obtain an explicit asymptotic for the set of quadratic twists of $X_0(N)$ violating the Hasse principle.

In some cases, the lack of global points can be explained using the *Mordell Weil sieve*. This is equivalent to *Brauer Manin obstruction* by the work of Scharaschkin [**Sch99**]. One can not apply the Mordell Weil sieve if $\mathrm{Pic}^1(X_0^\varsigma(N))(\mathbb{Q})$ is empty. This amounts to computing period and index of $\mathrm{Pic}^1(X_0^\varsigma(N))$, which is a torsor for the abelian variety $\mathrm{Pic}^0(X_0^\varsigma(N))$. Since we begin with curves with local points everywhere, period and index of $\mathrm{Pic}^1(X_0^\varsigma(N))$ are the same and either two or one, see Chapter 4 Section 3 of [**Ozm10**] and the references therein. Once we know the existence of a rational degree one divisor, one can attempt to apply the Mordell-Weil sieve to prove that there are no rational points on the twisted curve. Understanding the Picard group is usually hard for a generic curve. Note that we even do not have equations for an arbitrary member of the twisted family. In [**Ozm13**], we give sufficient conditions for $\mathrm{Pic}^1(X_0^\varsigma(N))(\mathbb{Q})$ to be nonempty in the case of the quadratic and some biquadratic cases. We also find families of cases where $\mathrm{Pic}^1(X_0^\varsigma(N))(\mathbb{Q}_p) = \varnothing$ when p and N satisfy certain arithmetic conditions.

3.1. For genus > 1, Mordell Weil Sieve. Let C be a smooth, projective, geometrically integral curve over $\mathbb{Q}$ of genus greater than or equal to 2 with a rational degree one divisor D. Then we can embed C into its jacobian J via the map $P \mapsto [P] - D$.

Let S be a finite set of primes where C has good reduction at and assume that we know the generators of the Mordell Weil group, $J(\mathbb{Q})$. Then for every p in S we can compute the finite abelian group $J(\mathrm{F}_p)$ and the set $C(\mathrm{F}_p)$. Let inj_p denote the injection from $C(\mathrm{F}_p)$ to $J(\mathrm{F}_p)$ and red_p be the reduction map from $J(\mathbb{Q})$ to

$J(\mathrm{F}_p)$. Then we obtain the following diagram:

$$\begin{array}{ccc} C(\mathbb{Q}) & \stackrel{P\mapsto[P]-D}{\longrightarrow} & J(\mathbb{Q}) \\ \downarrow & & \downarrow red=\prod_{p\in S} red_p \\ \prod_{p\in S} C(\mathrm{F}_p) & \stackrel{inj=\prod_{p\in S} inj_p}{\longrightarrow} & \prod_{p\in S} J(\mathrm{F}_p) \end{array}$$

If there is a P in $C(\mathbb{Q})$ then $red_p([P]-D)$ is in $inj_p(C(\mathrm{F}_p))$ for any p in S. In particular if images of red and inj do not intersect then $C(\mathbb{Q}) = \emptyset$.

If we suspect that $C(\mathbb{Q})$ is empty then we can try to show the non-existence of global points using the Mordell-Weil sieve. This explicit method can be very hard to apply in practice since one needs to know equation of C and generators of the Mordell Weil group of jacobian of C. For our family of twists, finding equations can be very tricky since we are dealing with modular curves which tend to have equations with big coefficients. Finding generators $J(\mathbb{Q})$, even when the coefficients of the equation of the curve are small can be very hard as shown in [**Fly04**] and [**BS10**]. Moreover, in order to start applying this method, we need to make sure that $\mathrm{Pic}^1(C)(\mathbb{Q}) \neq \emptyset$ which again is not easy in general even for a single curve given with an explicit equation. Moreover, we need to apply this to a family of curves most of which don't have explicit equation(at least with small coefficients). Therefore, we study $\mathrm{Pic}^1(C)(\mathbb{Q})$ in [**Ozm13**] and get the following results:

THEOREM 3.4 (Ozman, [**Ozm13**]). *Let $N > 3$ be a prime that is inert in $\mathrm{K} = \mathbb{Q}(\sqrt{d})$ and $X_0^d(N)$ be the quadratic twist of $X_0(N)$ by* K *and w_N. Then $Pic^1(X_0^d(N))(\mathbb{Q}_N)$ is empty if and only if $N \equiv 1 \mod 24$ or $N \equiv 17 \mod 24$.*

THEOREM 3.5 (Ozman, [**Ozm13**]). *Let* K *be a polyquadratic field and say p is inert in* K. *Let $N = p\prod q_i$ such that $p > 3$ and $q_i \equiv 3 \mod 4, q_j \equiv 2 \mod 3$ for some i, j. Then $Pic^1(X)(\mathbb{Q}_p) = \emptyset$, where X denotes the polyquadratic twist of $X_0(N)$.*

In order to obtain these results regarding the Picard group, we need the result of Ogg in [**Ogg73**] which gives the order of the cuspidal group in $\mathrm{Pic}(C)(\mathbb{Q})$. Then we use the description of the reduction of the jacobian given by [**Del74**] which in particular says that the group of connected components of the reduction of the jacobian is isomorphic to the cuspidal group. Finally, we get the result by studying the action of the relevant maps(i.e Galois and Atkin-Lehner automorphisms) on the cuspidal group.

3.2. For genus 1, Trace obstruction. Understanding rational points on twisted modular curves of genus one is much different than the study of higher genus curves, as it is usually the case for rationality questions about curves.

When the modular curve $X_0(N)$ has genus one, its quadratic twist by w_N and $\mathrm{K} = \mathbb{Q}(\sqrt{d})$ corresponds to a cocycle $\zeta_S \in \mathrm{H}^1(\mathrm{Gal}(\overline{\mathbb{Q}}/\mathbb{Q}), \mathrm{Aut}(\mathrm{E}))$ such that

$$\zeta_S(\sigma) = \begin{cases} (-1, S) & \text{if } \sigma(\sqrt{d}) = -\sqrt{d} \\ (1, O) & \text{otherwise.} \end{cases}$$

where S is in $X_0(N)(\mathrm{K})$ and $\mathrm{Gal}(\mathrm{K}/\mathbb{Q})$ is generated by σ. Therefore the twisted curve $X_0^{\zeta_S}(N)$ has a rational point if and only if S is in the image of the trace map: $X_0(N)(\mathrm{K}) \to X_0(N)(\mathbb{Q})$ sending P to $P + \sigma(P)$. A necessary condition for S to be global trace is being a local trace for all primes p. Therefore, the study of rational points of $X_0^{\zeta_S}(N)$ is a natural application of the local-global trace problem which is an interesting question itself for any given elliptic curve E and a rational point of E.

In [**ÇO15**] we answer the local-global trace obstruction question for elliptic curves over any number field hence as an application get the obstruction result for rational points on the twists of genus one $X_0(N)$ which corresponds to $N = 14, 17, 15$ and 21.

4. Quadratic Points on Modular Curves

Let ζ_d be the quadratic character associated to the field $\mathrm{K} = \mathbb{Q}(\sqrt{d})$. Since $X_0^{\zeta}(N)(\mathbb{Q}) \subset X_0(N)(\mathrm{K})$, if we can understand all K-rational points of a fixed $X_0(N)$ then we can determine $X_0^{\zeta}(N)(\mathbb{Q})$. The study of quadratic points of $X_0(N)$ is a widely open area. Recall that we have the results of Mazur about $X_0(N)(\mathbb{Q})$ and Merel about $X_1(N)(L)$ for any number field L mentioned in the introduction. However, not much is known for $X_0(N)(L)$ except the theorems below and a result of Bruin and Najman [**BN15**] which finds and parametrizes all quadratic points on all hyperelliptic $X_0(N)$ explicitly.

THEOREM 4.1 (Mazur, [**Maz78**]). *Let* K *be an imaginary quadratic number field and N be a big enough prime which is inert in K. Then $X_0(N)(\mathrm{K}) = \{cusps\}$.*

THEOREM 4.2 (Bars [**Bar99**],Harris-Silverman [**HS91**]). *Fix a positive integer N such that the genus of $X_0(N)$ is at least 2. The set $\Gamma_2(X_0(N), \mathbb{Q}) := \cup_{|\mathrm{K}:\mathbb{Q}|\leq 2} X_0(N)(\mathrm{K})$ is a finite set for all but 28 values of N.*

The methods to explicitly compute $X_0(N)(\mathrm{K})$ for a fixed modular curve $X_0(N)$ requires nice equations of $X_0(N)$, moreover $\mathrm{Jac}(X_0(N))(\mathbb{Q})$ has to be torsion and be explicitly computed. Say $C = X_0(N)$ satisfied both of these requirements and not hyperlliptic. The method starts with enumerating the elements of T in $\mathrm{Jac}(C)(\mathbb{Q})$. Then we consider the map ϕ from $C^{(2)} = (C \times C)/S_2$ the 2-fold symmetric product of C to $\mathrm{Jac}(C)(\mathbb{Q})$ given by $\phi(D) = [D - 2\infty_1]$. The points of $C^{(2)}$ is in one to one correspondence with efective degree two divisors of C. Note that since C is not hyperelliptic, ϕ is injective. Therefore any $D \in C^{(2)}(\mathbb{Q})$ is linearly equivalent to $D' + 2\infty_1$, for some $[D'] \in \mathrm{Jac}(\mathbb{Q})$. Finally, for each $[D'] \in \mathrm{Jac}(\mathbb{Q})$, we compute Riemann Roch space of $D' + 2\infty_1$ to see if it is linearly equivalent to an effective div of degree 2.

In order to have nice equations of $X_0(N)$ we needed to stick to modular curves with small level i.e. $N < 100$. However, the main difficuilty is the explicit computation of the rational points of the jacobian. It is a folklore conjecture that the rational torsion subgroup of $\mathrm{Jac}(X_0(N))$ is equal to the rational cuspidal group i.e. the rational subgroup of the group generated by the cusps of $X_0(N)$. However it has been only proved for prime levels. Using many different techniques such as quotients, sieving etc we get the following result:

THEOREM 4.3 (Ozman,Siksek [**OS**]). *For the values of N such that $X_0(N)$ is non-hyperelliptic, of genus g where $3 \leq g \leq 5$ and for which $J_0(N)(\mathbb{Q})$ is finite, $\mathrm{Jac}(X_0(N))(\mathbb{Q})_{tors}$ is equal to the rational cuspidal group and moreover, the finite list of quadratic fields* K *such that $X_0(N)$ has* K*-rational points is explicitly determined.*

A widely open question about the quadratic points of $X_0(N)$ is given below inspired by the similar results of Mazur:

QUESTION 4.4. Is there a bound B such that for all $|d| > D$, $X_0(N)(\mathbb{Q}(\sqrt{d}))$ consists of only cusps?

5. Further Questions

There are many other follow-up questions to this problem. The most natural one is dealing with the global points of $X_0^d(N)$ for various quadratic number fields. The aim is to prove that for sufficiently large N (compared to d) there are no rational points on $X_0^d(N)$. The most successful argument of this type is due to Mazur ([**Maz78**], [**Maz77**]). Mazur's idea is to bound rational points on a curve C using an morphism from C to a factor of its jacobian, say A, such that $|A(\mathbb{Q})|$ is finite. When each factor A of the jacobian is associated to a newform f of weight 2, level N with root number -1, $|A(\mathbb{Q})|$ is infinite assuming the conjecture of Birch-Swinnerton-Dyer. Hence there is no hope of applying Mazur's techniques. Using the formula of Weil given in [**Li75**], we can check the sign of the functional equation and see that this phenomenon of having negative root number occurs when N splits in an imaginary quadratic field or N is inert in a real quadratic field.

Therefore we restrict to the cases where K is real quadratic and N splits in K or K is imaginary quadratic and N is inert in K. The second case is covered by a result of Mazur([**Maz78**]) in which he says that $X_0(N)(\mathrm{K})$ consists of only cusps when N is sufficiently big compared to discriminant of K. Therefore we make the following conjecture:

CONJECTURE 5.1. *Let* $\mathrm{K} = \mathbb{Q}(\sqrt{d})$ *be a real quadratic number field in which* N *splits. Then, for* N *sufficiently big compared to discriminant of* K*,* $X_0^d(N)(\mathbb{Q})$ *is empty.*

By the results of Kolyvagin and Logachev, in order to show that there is a quotient of jacobian, say A_f with Mordell-Weil rank 0, it is enough to show that $L(f,s)$ at $s = 1$ in nonzero. This can be achieved by proving that the average special value is nonzero when N is big compared to the conductor of χ, as done in [**Ell04a**]. Another approach to find such a form f, is to use techniques of Ramakrishnan and Michel to compute a certain average of L-values. In [**MR12**] they give an exact formula for a suitably weighted sum of $L(f,1)L(f \otimes \chi, 1)$ where χ is a quadratic character and f varies over weight 2 cuspforms of level N. Hence an affirmative answer to the following question leads to results about global points on $X_0^d(N)$:

QUESTION 5.2. Let $\mathrm{K} = \mathbb{Q}(\sqrt{d})$ be a real quadratic number field in which N splits and χ be the quadratic Dirichlet character associated to $\mathrm{K}/\mathbb{Q}$. For N sufficiently large compared to the conductor of χ, is there a weight 2 newform of level N such that $w_N f = -f$ and $L(f \otimes \chi, 1) \neq 0$?

References

[Bar99] Francesc Bars, *Bielliptic modular curves*, J. Number Theory **76** (1999), no. 1, 154–165, DOI 10.1006/jnth.1998.2343. MR1688168

[BMS16] Michael Bennett, MihailescuPreda, and Siksek Samir, *The generalized fermat equation* (2016), 173–205.

[BN15] Peter Bruin and Filip Najman, *Hyperelliptic modular curves $X_0(n)$ and isogenies of elliptic curves over quadratic fields*, LMS J. Comput. Math. **18** (2015), no. 1, 578–602, DOI 10.1112/S1461157015000157. MR3389884

[BS10] Nils Bruin and Michael Stoll, *The Mordell-Weil sieve: proving non-existence of rational points on curves*, LMS J. Comput. Math. **13** (2010), 272–306, DOI 10.1112/S1461157009000187. MR2685127

[Cla07] Pete L. Clark, *Galois groups via Atkin-Lehner twists*, Proc. Amer. Math. Soc. **135** (2007), no. 3, 617–624, DOI 10.1090/S0002-9939-06-08493-0. MR2262856

[ÇO15] Mirela Çiperiani and Ekin Ozman, *Local to global trace questions and twists of genus one curves*, Proc. Amer. Math. Soc. **143** (2015), no. 9, 3815–3826, DOI 10.1090/proc/12560. MR3359573

[Del74] Pierre Deligne, *La conjecture de Weil. I* (French), Inst. Hautes Études Sci. Publ. Math. **43** (1974), 273–307. MR0340258

[Elk04] Noam D. Elkies, *On elliptic K-curves*, Modular curves and abelian varieties, Progr. Math., vol. 224, Birkhäuser, Basel, 2004, pp. 81–91. MR2058644

[Ell04a] Jordan S. Ellenberg, *Galois representations attached to $\mathbb{Q}$-curves and the generalized Fermat equation $A^4 + B^2 = C^p$*, Amer. J. Math. **126** (2004), no. 4, 763–787. MR2075481

[Ell04b] Jordan S. Ellenberg, *$\mathbb{Q}$-curves and Galois representations*, Modular curves and abelian varieties, Progr. Math., vol. 224, Birkhäuser, Basel, 2004, pp. 93–103. MR2058645

[Fly04] E. V. Flynn, *The Hasse principle and the Brauer-Manin obstruction for curves*, Manuscripta Math. **115** (2004), no. 4, 437–466. MR2103661 (2005j:11047)

[Gon91] Josep González, *Equations of hyperelliptic modular curves*, Ann. Inst. Fourier (Grenoble) **41** (1991), 779–795.

[HS91] Joe Harris and Joe Silverman, *Bielliptic curves and symmetric products*, Proc. Amer. Math. Soc. **112** (1991), no. 2, 347–356, DOI 10.2307/2048726. MR1055774

[Li75] Wen Ch'ing Winnie Li, *Newforms and functional equations*, Math. Ann. **212** (1975), 285–315, DOI 10.1007/BF01344466. MR0369263

[Maz77] B. Mazur, *Modular curves and the Eisenstein ideal*, Inst. Hautes Études Sci. Publ. Math. **47** (1977), 33–186 (1978). MR488287

[Maz78] B. Mazur, *Rational isogenies of prime degree (with an appendix by D. Goldfeld)*, Invent. Math. **44** (1978), no. 2, 129–162, DOI 10.1007/BF01390348. MR482230

[Mer96] Loïc Merel, *Bornes pour la torsion des courbes elliptiques sur les corps de nombres* (French), Invent. Math. **124** (1996), no. 1-3, 437–449, DOI 10.1007/s002220050059. MR1369424

[MR12] Philippe Michel and Dinakar Ramakrishnan, *Consequences of the Gross-Zagier formulae: stability of average L-values, subconvexity, and non-vanishing mod p*, Number theory, analysis and geometry, 2012, pp. 437–459. MR2867928

[Ogg73] A. P. Ogg, *Rational points on certain elliptic modular curves*, Analytic number theory (Proc. Sympos. Pure Math., Vol XXIV, St. Louis Univ., St. Louis, Mo., 1972), Amer. Math. Soc., Providence, R.I., 1973, pp. 221–231. MR0337974

[Ogg74] Andrew P. Ogg, *Hyperelliptic modular curves*, Bull. Soc. Math. France **102** (1974), 449–462. MR0364259

[OS] Ekin Ozman and Samir Siksek, *Quadratic points on modular curves*, to appear in Math. Comp.

[Ozm10] Ekin Ozman, *Points on quadratic twists of the classical modular curve*, ProQuest LLC, Ann Arbor, MI, 2010. Thesis (Ph.D.)–The University of Wisconsin - Madison. MR2814010

[Ozm12] Ekin Ozman, *Points on quadratic twists of $X_0(N)$*, Acta Arith. **152** (2012), no. 4, 323–348. MR2890545

[Ozm13] Ekin Ozman, *On polyquadratic twists of $X_0(N)$*, J. Number Theory **133** (2013), no. 10, 3325–3338, DOI 10.1016/j.jnt.2013.03.015. MR3071815

[Sch99] Victor Scharaschkin, *Local-global problems and the Brauer-Manin obstruction*, ProQuest LLC, Ann Arbor, MI, 1999. Thesis (Ph.D.)–University of Michigan. MR2700328

[Ser76] Jean-Pierre Serre, *Divisibilité de certaines fonctions arithmétiques*, Enseignement Math. (2) **22** (1976), no. 3-4, 227–260. MR0434996

[Sta] Jim Stankewicz, *Twists of shimura curves.* Thesis (Ph.D.)–University of Georgia-Athens.

Mathematics Department, Boğaziçi University, Bebek, 34342 Istanbul, Turkey

Contemporary Mathematics
Volume **732**, 2019
https://doi.org/10.1090/conm/732/14781

On the number of representations of certain quadratic forms in 8 variables

B. Ramakrishnan, Brundaban Sahu, and Anup Kumar Singh

ABSTRACT. In this paper, we find the number of representations of integers by certain quadratic forms in 8 variables by using the theory of modular forms. By expressing these formulas in terms of certain convolution sums of the divisor function and using our formulas, we deduce formulas for the convolution sums $W_{j,7}(n)$ for $j = 1, 2, 3, 4$.

1. Introduction

For positive integers a, b, n, define the convolution sum $W_{a,b}(n)$ by

$$W_{a,b}(n) := \sum_{\substack{l,m \in \mathbb{N} \\ al+bm=n}} \sigma(l)\sigma(m), \tag{1}$$

where $\sigma(n)$ is the divisor function. We note that $W_{a,1}(n) = W_{1,a}(n)$, which is denoted by $W_a(n)$. These type of sums were evaluated as early as the 19th century. For example, the sum $W_1(n)$ was evaluated by M. Besge, J. W. L. Glaisher and S. Ramanujan [**3, 6, 12**]. Some of the convlotution sums of the above type have been obtained by several authors (see for example [**7, 11, 13**] and also the works of K. S. Williams and his co-authors ([**16**] and the references therein)).

Let Q be the quadratic form in four variables defined by

$$Q : x_1^2 + x_1x_2 + 2x_2^2 + x_3^2 + x_3x_4 + 2x_4^2,$$

and

$$R_Q(n) = \text{card}\left\{(x_1, x_2, x_3, x_4) \in \mathbb{Z}^4 : Q(x_1, x_2, x_3, x_4) = n\right\}$$

be the number of representations of a positive integer n by the quadratic form Q. A formula for $R_Q(n)$ was obtained in [**15**] by using an elementary method, which is given by

$$R_Q(n) = 4\sigma(n) - 28\sigma\left(\frac{n}{7}\right), \tag{2}$$

2010 *Mathematics Subject Classification.* Primary 11E25, 11F11, 11F99; Secondary 11F20, 11F30.

Key words and phrases. Convolution sums of the divisor functions, representation numbers of quadratic forms, modular forms of one variable.

where we define $\sigma(x) = 0$, when x is not a positive integer. In this article we find a formula for $R_{Q\oplus jQ}(n)$ where

$$Q \oplus jQ := x_1^2 + x_1x_2 + 2x_2^2 + x_3^2 + x_3x_4 + 2x_4^2 + j(x_5^2 + x_5x_6 + 2x_6^2 + x_7^2 + x_7x_8 + 2x_8^2),$$

for $j = 1, 2, 3, 4$ using theory of modular forms. Next, by using the above formula for $R_Q(n)$, we express $R_{Q\oplus jQ}(n)$ in terms of the convolution sums $W_j(n)$, $W_{7j}(n)$ and $W_{j,7}(n)$. Since the convolutions $W_j(n)$, $1 \le j \le 4$ and $W_{7j}(n)$, $j = 1, 2, 4$ are already known, we use them along with $W_{21}(n)$ (which we prove using the theory of quasimodular forms), to give formulas for $W_{j,7}(n)$ for $1 \le j \le 4$. We note that $R_{Q\oplus Q}(n)$ has also been evaluated by K. S. Williams in [**15**], using the convolution sums method.

2. Preliminaries and Statement of Results

Let $M_k(N)$ be the space of modular forms of weight k for the congruence subgroup $\Gamma_0(N)$ and $S_k(N)$ be the subspace of cusp forms of weight k for the congruence subgroup $\Gamma_0(N)$.

Let $\mathscr{Q} : \mathbb{Z}^{2m} \to \mathbb{Z}$ be a positive definite quadratic form given by

$$\mathscr{Q}(\mathbf{x}) = \frac{1}{2}\mathbf{x}^t \mathcal{A}\mathbf{x},$$

where $\mathbf{x} \in \mathbb{Z}^{2m}$. Then the associated theta series, denoted by $\Theta_{\mathscr{Q}}(z)$, is defined as

$$\Theta_{\mathscr{Q}}(z) = \sum_{\mathbf{x}\in\mathbb{Z}^{2m}} e^{2\pi i \mathscr{Q}(\mathbf{x}) z}.$$

Let M be the smallest positive integer such that $M\mathcal{A}^{-1}$ is an even integral matrix (i.e., a matrix with diagonal entries as even integers and off-diagonal entries as integers) and $\chi_{m,M}$ be the quadratic character defined by $\left(\frac{(-1)^m M}{\cdot}\right)$. Then, it is known that $\Theta_{\mathscr{Q}}(z)$ is a modular form of weight m on $\Gamma_0(M)$ with character $\chi_{m,M}$. For more details on theta series associated to integral quadratic forms we refer to Chapter IX of [**14**] and [**4**, p.32].

For $k \ge 4$, let E_k denote the normalized Eisenstein series of weight k in $M_k(1)$ given by

$$E_k(z) = 1 - \frac{2k}{B_k}\sum_{n\ge1} \sigma_{k-1}(n)q^n \quad (q = e^{2\pi i z},\ \mathrm{Im}(z) > 0),$$

where B_k is the k-th Bernoulli number defined by $\displaystyle\frac{x}{e^x - 1} = \sum_{m=0}^{\infty} \frac{B_m}{m!}x^m$. The Eisenstein series $E_4(z) = 1+240\sum_{n\ge1}\sigma_3(n)q^n$ is used in our proof. For $k = 2$, the Eisenstein series $E_2(z)$ is a quasimodular form of weight 2, depth 1 on $SL_2(\mathbb{Z})$, which has the following Fourier expansion:

$$E_2(z) = 1 - 24\sum_{n\ge1}\sigma(n)q^n$$

and it is fundamental in the theory of quasimodular forms. In order to evaluate the convolution sum $W_{21}(n)$, we use the following structure theorem on $\widetilde{M}_k^{\le k/2}(N)$, the space of quasimodular forms of weight k, depth $\le k/2$ on $\Gamma_0(N)$ (see [**8, 10**]).

Theorem A (Kaneko-Zagier): *For an even integer k with $k \geq 2$, we have*

$$\tilde{M}_k^{\leq k/2}(N) = \bigoplus_{j=0}^{k/2-1} D^j M_{k-2j}(N) \oplus \mathbb{C}D^{k/2-1}E_2, \tag{3}$$

where the differential operator D is defined by $D := \frac{1}{2\pi i}\frac{d}{dz}$.

Using this theorem, one can express each quasimodular form of weight k and depth $\leq k/2$ as a linear combination of j-th derivatives of modular forms of weight $k-2j$ on $\Gamma_0(N)$, $0 \leq j \leq k/2-1$ and the $(k/2-1)$-th derivate of the quasimodular form E_2. For details on basic theory of modular forms and quasimodular forms, we refer the reader to [**4, 8, 10**].

Let d be the dimension of the space $S_k^{new}(N)$, which is spanned by the normalised Hecke eigenforms in the space $S_k(N)$. We denote these basis elements as $\{\Delta_{k,N;j}(z) : 1 \leq j \leq d\}$, where we denote their Fourier expansions by

$$\Delta_{k,N;j}(z) = \sum_{n\geq 1} \tau_{k,N;j}(n)q^n.$$

If $d = 1$, then we write the basis element as $\Delta_{k,N}$ and its Fourier coefficients are denoted as $\tau_{k,N}(n)$. For more details on the theory of newforms, we refer to [**2**]. We also need the following 9 cusp forms which are basis elements of the space $S_4(28)$. Here we use the same notation for their Fourier coefficients as in [**1**]. For $1 \leq j \leq 9$, the nine cusp forms $f_j(z) = \sum_{n\geq 1} c_j(n)q^n$ are defined by the following eta products/quotients.

$$\begin{aligned}
f_1(z) &= \frac{\eta^5(z)\eta^5(7z)}{\eta(2z)\eta(14z)}; & f_2(z) &= \eta^2(z)\eta^2(2z)\eta^2(7z)\eta^2(14z);\\
f_3(z) &= \frac{\eta^6(z)\eta^6(14z)}{\eta^2(2z)\eta^2(7z)}; & f_4(z) &= \frac{\eta^6(2z)\eta^6(7z)}{\eta^2(z)\eta^2(14z)};\\
f_5(z) &= \eta^2(4z)\eta^4(14z)\eta^2(28z); & f_6(z) &= \frac{\eta^6(2z)\eta^6(28z)}{\eta^2(4z)\eta^2(14z)};\\
f_7(z) &= \frac{\eta^4(2z)\eta^6(28z)}{\eta^2(4z)}; & f_8(z) &= \frac{\eta(z)\eta(2z)\eta(7z)\eta^8(28z)}{\eta^3(14z)};\\
f_9(z) &= \frac{\eta(2z)\eta(4z)\eta^9(28z)}{\eta^3(14z)},
\end{aligned}$$

where $\eta(z)$ is the Dedekind eta function given by $\eta(z) = q^{1/24}\prod_{n\geq 1}(1-q^n)$.

We need the following convolution sums $W_N(n)$, $N = 1, 2, 3, 4, 14, 28$. Among these, $W_{28}(n)$ is recently evaluated by A. Alaca, S. Alaca and E. Ntienjem [**1**] and the rest were evaluated by E. Royer [**13**].

Theorem 2.1. *(See [**1, 13**]) For a natural number n,*

$$W_1(n) = \frac{5}{12}\sigma_3(n) + \left(\frac{1}{12} - \frac{n}{2}\right)\sigma(n),$$

$$W_2(n) = \frac{1}{12}\sigma_3(n) + \frac{1}{3}\sigma_3\left(\frac{n}{2}\right) + \left(\frac{1}{24} - \frac{n}{8}\right)\sigma(n) + \left(\frac{1}{24} - \frac{n}{4}\right)\sigma\left(\frac{n}{2}\right),$$

$$W_3(n) = \frac{1}{24}\sigma_3(n) + \frac{3}{8}\sigma_3\left(\frac{n}{3}\right) + \left(\frac{1}{24} - \frac{n}{12}\right)\sigma(n) + \left(\frac{1}{24} - \frac{n}{4}\right)\sigma\left(\frac{n}{3}\right),$$

$$W_4(n) = \frac{1}{48}\sigma_3(n) + \frac{1}{16}\sigma_3\left(\frac{n}{2}\right) + \frac{1}{3}\sigma_3\left(\frac{n}{4}\right) + \left(\frac{1}{24} - \frac{n}{16}\right)\sigma(n) + \left(\frac{1}{24} - \frac{n}{4}\right)\sigma\left(\frac{n}{4}\right),$$

$$\begin{aligned}
W_{14}(n) = {} & \frac{1}{600}\sigma_3(n) + \frac{1}{150}\sigma_3\left(\frac{n}{2}\right) + \frac{49}{600}\sigma_3\left(\frac{n}{7}\right) + \frac{49}{150}\sigma_3\left(\frac{n}{14}\right) \\
& + \left(\frac{1}{24} - \frac{n}{56}\right)\sigma(n) + \left(\frac{1}{24} - \frac{n}{4}\right)\sigma\left(\frac{n}{14}\right) - \frac{3}{350}\tau_{4,7}(n) - \frac{6}{175}\tau_{4,7}\left(\frac{n}{2}\right) \\
& - \frac{1}{84}\tau_{4,14;1}(n) - \frac{1}{200}\tau_{4,14;2}(n), \\
W_{28}(n) = {} & \frac{1}{2400}\sigma_3(n) + \frac{1}{800}\sigma_3\left(\frac{n}{2}\right) + \frac{1}{150}\sigma_3\left(\frac{n}{4}\right) + \frac{49}{2400}\sigma_3\left(\frac{n}{7}\right) + \frac{49}{800}\sigma_3\left(\frac{n}{14}\right) \\
& + \frac{49}{150}\sigma_3\left(\frac{n}{28}\right) + \left(\frac{1}{24} - \frac{n}{112}\right)\sigma(n) + \left(\frac{1}{24} - \frac{n}{4}\right)\sigma\left(\frac{n}{28}\right) \\
& + \frac{1121}{67200}c_1(n) + \frac{2389}{22400}c_2(n) - \frac{1}{128}c_3(n) - \frac{3349}{67200}c_4(n) \\
& - \frac{101}{200}c_5(n) - \frac{17}{40}c_6(n) + \frac{13}{200}c_7(n) - \frac{433}{150}c_8(n) - \frac{254}{75}c_9(n).
\end{aligned}$$

Following the method of Royer [**13**], we evaluate the following convolution sum.

THEOREM 2.2. *For a natural number n,*

$$\begin{aligned}
W_{21}(n) = {} & \frac{1}{24}\sigma(n) + \frac{1}{24}\sigma\left(\frac{n}{21}\right) - \frac{1}{4}n\sigma\left(\frac{n}{21}\right) - \frac{1}{84}n\sigma(n) + \frac{1}{1200}\sigma_3(n) \\
& + \frac{3}{400}\sigma_3\left(\frac{n}{3}\right) + \frac{49}{1200}\sigma_3\left(\frac{n}{7}\right) + \frac{147}{400}\sigma_3\left(\frac{n}{21}\right) - \frac{1}{175}\tau_{4,7}(n) \\
& - \frac{9}{175}\tau_{4,7}\left(\frac{n}{3}\right) - \frac{1}{100}\tau_{4,21;2}(n) - \frac{5}{672}(\tau_{4,21;3}(n) + \tau_{4,21;4}(n)) \\
& - \frac{11}{12768}\sqrt{57}(\tau_{4,21;3}(n) - \tau_{4,21;4}(n)).
\end{aligned}$$

Note: The dimension of $S_4^{new}(21)$ is 4 and out of which two newforms have rational Fourier coefficients and the other two newforms do not have rational Fourier coefficients. In the latter case, the eigenvalues of the Hecke operators $T(p)$ for $p \neq 3, 7$ satisfy the quadratic polynomial $x^2 + 3x - 12$, and therefore the eigenvalues (and hence the Fourier coefficients) of these two newforms belong to the number field $\mathbb{Q}(\sqrt{57})$ (57 is the discriminant of the quadratic polynomial). More precisely, the Fourier coefficients $\tau_{4,21;3}(n)$ and $\tau_{4,21;4}(n)$ belong to the number field $\mathbb{Q}(\sqrt{57})$. However, we note that both $\tau_{4,21;3}(n) + \tau_{4,21;4}(n)$ and $\sqrt{57}(\tau_{4,21;3}(n) - \tau_{4,21;4}(n))$ are rational numbers.

The following is the main theorem of this section.

THEOREM 2.3. *For a natural number n,*

$$\begin{aligned}
R_{Q\oplus Q}(n) = {} & \frac{24}{5}\sigma_3(n) + \frac{1176}{5}\sigma_3\left(\frac{n}{7}\right) + \frac{16}{5}\tau_{4,7}(n), \\
R_{Q\oplus 2Q}(n) = {} & \frac{24}{25}\sigma_3(n) + \frac{96}{25}\sigma_3\left(\frac{n}{2}\right) + \frac{1176}{25}\sigma_3\left(\frac{n}{7}\right) + \frac{4704}{25}\sigma_3\left(\frac{n}{14}\right) \\
& + \frac{48}{25}\tau_{4,7}(n) + \frac{192}{25}\tau_{4,7}\left(\frac{n}{2}\right) + \frac{28}{25}\tau_{4,14;1}(n), \\
R_{Q\oplus 3Q}(n) = {} & \frac{12}{25}\sigma_3(n) + \frac{108}{25}\sigma_3\left(\frac{n}{3}\right) + \frac{588}{25}\sigma_3\left(\frac{n}{7}\right) + \frac{5292}{25}\sigma_3\left(\frac{n}{21}\right) \\
& + \frac{32}{25}\tau_{4,7}(n) + \frac{288}{25}\tau_{4,7}\left(\frac{n}{3}\right) + \frac{56}{25}\tau_{4,21;2}(n),
\end{aligned}$$

$$\begin{aligned}R_{Q\oplus 4Q}(n) &= \frac{6}{25}\sigma_3(n)+\frac{18}{25}\sigma_3\left(\frac{n}{2}\right)+\frac{96}{25}\sigma_3\left(\frac{n}{4}\right)+\frac{294}{25}\sigma_3\left(\frac{n}{7}\right)+\frac{882}{25}\sigma_3\left(\frac{n}{14}\right)\\ &+\frac{4704}{25}\sigma_3\left(\frac{n}{28}\right)-\frac{356}{175}c_1(n)-\frac{316}{25}c_2(n)+2c_3(n)+\frac{1014}{175}c_4(n)\\ &+\frac{1328}{25}c_5(n)+\frac{336}{5}c_6(n)-\frac{664}{25}c_7(n)+\frac{10432}{25}c_8(n)+\frac{10432}{25}c_9(n).\end{aligned}$$

REMARK 2.1. As mentioned in the introduction K. S. Williams [**15**] evaluated the formula for $R_{Q\oplus Q}(n)$ by using the convolution sum $W_7(n)$. In this paper we have computed this formula using the theory of modular forms. Therefore, as a consequence to our formula for $R_{Q\oplus Q}(n)$, we obtain the convolution sum $W_7(n)$ (see Corollary 2.4 below). The proof is demonstrated in §3.3.

REMARK 2.2. By the Atkin-Lehner theory of newforms [**2**], we see that the normalized newforms $\Delta_{4,7}(z)$ and $\Delta_{4,14;j}(z)$, $j=1,2$ are eigenforms under the Hecke operator for the prime 7 dividing the level with eigenvalues ± 7. Using this, we have for $a\geq 1$,

$$\tau_{4,7}(7^a) = (-1)^a 7^a;\ \tau_{4,14;1}(7^a) = (-1)^a 7^a;\ \tau_{4,14;2}(7^a) = 7^a.$$

Therefore, in Theorem 2.3, the formulas for $R_{Q\oplus jQ}(n)$ for $1\leq j\leq 3$ have elementary evaluations when $n=7^a$, $a\geq 1$.

In the following corollary, we use the formulas of Theorem 2.3 and the convolution sums $W_N(n)$, $N=1,2,3,4,14,21,28$ (given by Theorems 2.1 and 2.2), to obtain the convolution sums $W_7(n)$, $W_{2,7}(n)$, $W_{3,7}(n)$ and $W_{4,7}(n)$. We note that in [**5**], Chan and Cooper also used a different method to evaluate the convolution sums $W_p(n)$, $p=3,7,11,23$.

COROLLARY 2.4. *For a natural number n, we have*

$$\begin{aligned}W_7(n) &= \frac{1}{120}\sigma_3(n)+\frac{49}{120}\sigma_3\left(\frac{n}{7}\right)+\left(\frac{1}{24}-\frac{n}{28}\right)\sigma(n)+\left(\frac{1}{24}-\frac{n}{4}\right)\sigma\left(\frac{n}{7}\right)\\ &-\frac{1}{70}\tau_{4,7}(n),\\ W_{2,7}(n) &= \frac{1}{600}\sigma_3(n)+\frac{1}{150}\sigma_3\left(\frac{n}{2}\right)+\frac{49}{600}\sigma_3\left(\frac{n}{7}\right)+\frac{49}{150}\sigma_3\left(\frac{n}{14}\right)\\ &+\left(\frac{1}{24}-\frac{n}{28}\right)\sigma\left(\frac{n}{2}\right)+\left(\frac{1}{24}-\frac{n}{8}\right)\sigma\left(\frac{n}{7}\right)-\frac{3}{350}\tau_{4,7}(n)\\ &-\frac{6}{175}\tau_{4,7}\left(\frac{n}{2}\right)+\frac{1}{600}\tau_{4,14;1}(n)+\frac{1}{200}\tau_{4,14;2}(n),\\ W_{3,7}(n) &= \frac{1}{1200}\sigma_3(n)+\frac{3}{400}\sigma_3\left(\frac{n}{3}\right)+\frac{49}{1200}\sigma_3\left(\frac{n}{7}\right)+\frac{147}{400}\sigma_3\left(\frac{n}{21}\right)\\ &+\left(\frac{1}{24}-\frac{n}{28}\right)\sigma\left(\frac{n}{3}\right)+\left(\frac{1}{24}-\frac{n}{12}\right)\sigma\left(\frac{n}{7}\right)-\frac{1}{175}\tau_{4,7}(n)\\ &-\frac{9}{175}\tau_{4,7}\left(\frac{n}{3}\right)-\frac{1}{100}\tau_{4,21;2}(n)+\frac{5}{672}(\tau_{4,21;3}(n)+\tau_{4,21;4}(n))\\ &+\frac{11}{12768}\sqrt{57}(\tau_{4,21;3}(n)-\tau_{4,21;4}(n)),\end{aligned}$$

$$
\begin{aligned}
W_{4,7}(n) &= \frac{1}{2400}\sigma_3(n)+\frac{1}{800}\sigma_3\left(\frac{n}{2}\right)+\frac{1}{150}\sigma_3\left(\frac{n}{4}\right)+\frac{49}{2400}\sigma_3\left(\frac{n}{7}\right)+\frac{49}{800}\sigma_3\left(\frac{n}{14}\right)\\
&+\frac{49}{150}\sigma_3\left(\frac{n}{28}\right)+\left(\frac{1}{24}-\frac{n}{28}\right)\sigma\left(\frac{n}{4}\right)+\left(\frac{1}{24}-\frac{n}{16}\right)\sigma\left(\frac{n}{7}\right)\\
&+\frac{697}{470400}c_1(n)+\frac{139}{22400}c_2(n)-\frac{9}{896}c_3(n)-\frac{893}{470400}c_4(n)\\
&+\frac{43}{1400}c_5(n)-\frac{7}{40}c_6(n)+\frac{241}{1400}c_7(n)-\frac{881}{1050}c_8(n)-\frac{178}{525}c_9(n).
\end{aligned}
$$

3. Proofs

For the proofs of our theorems, we need the following newforms $\Delta_{k,N}(z)$, $(k,N)\in\{(2,14),(2,21),(4,7)\}$, $\Delta_{4,14;1}(z)$, $\Delta_{4,14;2}(z)$ and $\Delta_{4,21;j}(z)$, $1\le j\le 4$. Below we give their expressions in terms of Eisenstein series and eta products. We have used the L-functions and modular forms database [**9**] to get some of these expressions.

$$
\begin{aligned}
\Delta_{2,14}(z) &= \eta(z)\eta(2z)\eta(7z)\eta(14z),\\
\Delta_{2,21}(z) &= \frac{\eta(7z)}{2\eta^2(z)\eta(3z)\eta(9z)\eta(21z)}\left(3\eta^2(z)\eta^2(7z)\eta^4(9z)+3\eta^4(z)\eta^2(9z)\eta^2(63z)\right.\\
&\quad -\eta^5(3z)\eta(7z)\eta(9z)\eta(21z)+7\eta(z)\eta^2(3z)\eta(9z)\eta^4(21z)\\
&\quad \left.+3\eta^3(z)\eta(7z)\eta^3(9z)\eta(63z)-3\eta(z)\eta^5(3z)\eta(21z)\eta(63z)\right),\\
\Delta_{4,7}(z) &= \frac{(\eta^3(z)\eta^3(7z)+4\eta^3(2z)\eta^3(14z))\eta^2(z)\eta^2(7z)}{\eta(2z)\eta(14z)},\\
\Delta_{4,14;1}(z) &= -\frac{9}{4}\Delta_{4,7}(z)-9\Delta_{4,7}(2z)+\frac{\Delta_{2,14}(z)}{4}(24\Delta_{2,14}(z)+14E_2(14z)-E_2(z)),\\
\Delta_{4,14;2}(z) &= \Delta_{4,7}(z)-4\Delta_{4,7}(2z)-5\Delta_{2,14}^2(z),\\
\Delta_{4,21;1}(z) &= \Delta_{2,21}(z)\left(-\frac{1}{24}E_2(z)-\frac{1}{8}E_2(3z)+\frac{7}{24}E_2(7z)+\frac{7}{8}E_2(21z)\right),\\
\Delta_{4,21;2}(z) &= \Delta_{2,21}(z)\left(\frac{1}{12}E_2(z)-\frac{1}{4}E_2(3z)-\frac{7}{12}E_2(7z)+\frac{7}{4}E_2(21z)\right).
\end{aligned}
$$

We have not found expressions for the remaining two newforms $\Delta_{4,21;3}(z)$ and $\Delta_{4,21;4}(z)$ (of weight 4 and level 21), whose Fourier coefficients lie in the real quadratic field $\mathbb{Q}(\sqrt{57})$. (Refer to the Note before Theorem 2.3.) Below, we give their first few Fourier coefficients using the database [**9**].

$$
\begin{aligned}
\Delta_{4,21;3}(z) &= q+\frac{-3+\sqrt{57}}{2}q^2+3q^3+\frac{17-3\sqrt{57}}{2}q^4+(3-\sqrt{57})q^5\\
&\quad +\frac{-9+\sqrt{57}}{2}q^6+7q^7+\cdots,\\
\Delta_{4,21;4}(z) &= q+\frac{-3-\sqrt{57}}{2}q^2+3q^3+\frac{17+3\sqrt{57}}{2}q^4+(3+\sqrt{57})q^5\\
&\quad -\frac{9+\sqrt{57}}{2}q^6+7q^7+\cdots.
\end{aligned}
$$

We also need the following weight 2 Eisenstein series defined (for two positive integers $a, b, a < b$) by

$$\Phi_{a,b}(z) = \frac{1}{b-a}\left(bE_2(bz) - aE_2(az)\right).$$

3.1. Proof of Theorem 2.2. The vector space $M_2(21)$ is of dimension 4 with a basis

$$\{\Phi_{1,3}(z), \Phi_{1,7}(z), \Phi_{1,21}(z), \Delta_{2,21}(z)\},$$

and $M_4(21)$ is of dimension 10 with a basis

$$\{E_4(dz), d|21; \Delta_{4,7}(z), \Delta_{4,7}(3z), \Delta_{4,21;j}(z), 1 \le j \le 4\}.$$

We note that $E_2(z)E_2(21z) \in \tilde{M}_4^{\le 2}(21)$. Using Theorem A and the above basis, we have

$$\begin{aligned} E_2(z)E_2(21z) = {} & \frac{1}{500}E_4(z) + \frac{9}{500}E_4(3z) + \frac{49}{500}E_4(7z) + \frac{441}{500}E_4(21z) - \frac{576}{175}\Delta_{4,7}(z) \\ & -\frac{5184}{175}\Delta_{4,7}(3z) - \frac{144}{25}\Delta_{4,21;2}(z) + \left(-\frac{66}{133}\sqrt{57} - \frac{30}{7}\right)\Delta_{4,21;3}(z) \\ & + \left(\frac{66}{133}\sqrt{57} - \frac{30}{7}\right)\Delta_{4,21;4}(z) + \frac{40}{7}D\Phi_{1,21}(z) + \frac{4}{7}DE_2(z). \end{aligned}$$

By comparing the n-th Fourier coefficients of both the sides of the above identity, we get the required formula for $W_{21}(n)$.

REMARK 3.1. In this remark we derive an expression for the newform $\Delta_{4,14;1}(z)$ in terms of the Eisenstein series $E_2(z)$ and its derivative. We note that (see [**13**])

$$\begin{aligned} E_2(z)E_2(14z) &= \frac{1}{250}E_4(z) + \frac{2}{125}E_4(2z) + \frac{49}{250}E_4(7z) + \frac{98}{125}E_4(14z) \\ & \quad -\frac{864}{175}\Delta_{4,7}(z) - \frac{3456}{175}\Delta_{4,7}(2z) - \frac{48}{7}\Delta_{4,14;1}(z) - \frac{72}{25}\Delta_{4,14;2}(z) \\ & \quad +\frac{39}{7}D\Phi_{1,14}(z) + \frac{6}{7}DE_2(z). \end{aligned}$$

Now, consider the quasimodular form $E_2(2z)E_2(7z)$ in $\tilde{M}_4^{\le 2}(14)$, and using Theorem A, we have the following expression.

$$\begin{aligned} E_2(2z)E_2(7z) &= \frac{1}{250}E_4(z) + \frac{2}{125}E_4(2z) + \frac{49}{250}E_4(7z) + \frac{98}{125}E_4(14z) \\ & \quad -\frac{864}{175}\Delta_{4,7}(z) - \frac{3456}{175}\Delta_{4,7}(2z) + \frac{48}{7}\Delta_{4,14;1}(z) - \frac{72}{25}\Delta_{4,14;2}(z) \\ & \quad +\frac{18}{7}D\Phi_{1,7}(z) + \frac{39}{7}D\Phi_{1,14}(z) - \frac{36}{7}D\Phi_{2,14}(z) + \frac{6}{7}DE_2(z). \end{aligned}$$

Therefore, from the above two expressions, we get

$$E_2(2z)E_2(7z) - E_2(z)E_2(14z) = \frac{96}{7}\Delta_{4,14;1}(z) + \frac{18}{7}D\Phi_{1,7}(z) - \frac{36}{7}D\Phi_{2,14}(z).$$

This gives an expression for the newform $\Delta_{4,14;1}(z)$ in terms of Eisenstein series. More precisely, we have

$$\begin{aligned} \Delta_{4,14;1}(z) = {} & \frac{7}{96}E_2(2z)E_2(7z) - \frac{7}{96}E_2(z)E_2(14z) - \frac{7}{32}DE_2(7z) \\ & + \frac{1}{32}DE_2(z) + \frac{7}{16}DE_2(14z) - \frac{1}{16}DE_2(2z). \end{aligned} \tag{4}$$

3.2. Proof of Theorem 2.3. In this section, we shall obtain the formulas for $R_{Q\oplus jQ}(n)$ as given in Theorem 2.3. Let

$$\Theta_{Q\oplus jQ}(z) = \sum_{x_1,\cdots,x_8\in\mathbb{Z}} q^{(Q\oplus jQ)(x_1,\cdots,x_8)} \tag{5}$$

be the theta series associated with the quadratic forms $Q \oplus jQ$ for $j = 1,2,3,4$. Now using the description mentioned in the Preliminaries section, we see that the theta series $\Theta_{Q\oplus jQ}(z)$ belongs to the space $M_4(7j)$ and has the Fourier expansion

$$\Theta_{Q\oplus jQ}(z) = \sum_{n=0}^{\infty} R_{Q\oplus jQ}(n) q^n.$$

Therefore, it is sufficient to obtain explicit bases for the spaces of modular forms $M_4(7j)$. The required formulas will follow by expressing each theta series as a linear combination of the corresponding modular form basis and comparing the nth Fourier coefficients. We shall give below explicit bases for the spaces $M_4(7j)$ for $j = 1,2,3,4$ in tabular form.

Space	Dimension	Basis
$M_4(7)$	3	$\{E_4(az), a\vert 7; \Delta_{4,7}(z)\}$
$M_4(14)$	8	$\{E_4(az), a\vert 14; \Delta_{4,7}(bz), b\vert 2; \Delta_{4,14;i}(z), 1\le i\le 2\}$
$M_4(21)$	10	$\{E_4(az), a\vert 21; \Delta_{4,7}(bz), b\vert 3; \Delta_{4,21;i}(z), 1\le i\le 4\}$
$M_4(28)$	15	$\{E_4(az), a\vert 28; f_i(z), 1\le i\le 9\}$

(Note that the newforms $\Delta_{k,N}(z)$, $\Delta_{k,N;i}(z)$ appearing in the above table are defined in §3 and the forms $f_i(z)$ that apper for level 28 case are defined in §2.)

Using the above bases, we have the following:

$$\Theta_{Q\oplus Q}(z) = \frac{1}{50}E_4(z) + \frac{49}{50}E_4(7z) + \frac{16}{5}\Delta_{4,7}(z), \tag{6}$$

$$\begin{aligned}\Theta_{Q\oplus 2Q}(z) &= \frac{1}{250}E_4(z) + \frac{2}{125}E_4(2z) + \frac{49}{250}E_4(7z) + \frac{98}{125}E_4(14z)\\ &\quad + \frac{48}{25}\Delta_{4,7}(z) + \frac{192}{25}\Delta_{4,7}(2z) + \frac{28}{25}\Delta_{4,14;1}(z),\end{aligned} \tag{7}$$

$$\begin{aligned}\Theta_{Q\oplus 3Q}(z) &= \frac{1}{500}E_4(z) + \frac{9}{500}E_4(3z) + \frac{49}{500}E_4(7z) + \frac{441}{500}E_4(21z)\\ &\quad + \frac{32}{25}\Delta_{4,7}(z) + \frac{288}{25}\Delta_{4,7}(3z) + \frac{56}{25}\Delta_{4,21;2}(z),\end{aligned} \tag{8}$$

$$\begin{aligned}\Theta_{Q\oplus 4Q}(z) &= \frac{1}{1000}E_4(z) + \frac{3}{1000}E_4(2z) + \frac{2}{125}E_4(4z) + \frac{49}{1000}E_4(7z)\\ &\quad + \frac{147}{1000}E_4(14z) + \frac{98}{125}E_4(28z) - \frac{356}{175}f_1(z) - \frac{316}{25}f_2(z) + 2f_3(z)\\ &\quad + \frac{1014}{175}f_4(z) + \frac{1328}{25}f_5(z) + \frac{336}{5}f_6(z) - \frac{664}{25}f_7(z)\\ &\quad + \frac{10432}{25}f_8(z) + \frac{10432}{25}f_9(z).\end{aligned} \tag{9}$$

The theorem now follows by comparing the nth Fourier coefficients from the above expressions.

3.3. Proof of Corollary 2.4. In order to get the convolution sums $W_{a,b}(n)$ for $(a,b) = (1,7), (2,7), (3,7), (4,7)$ in the corollary, we first compute the formulas $R_{Q\oplus jQ}(n)$ $(1 \le j \le 4)$ using the convolution sums method.

Let $\mathbb{N}_0 = \mathbb{N} \cup \{0\}$. For $n \in N$ we know that (see [**15**])

$$R_Q(n) = 4\sigma(n) - 28\sigma\left(\frac{n}{7}\right). \tag{10}$$

Then for $1 \leq j \leq 4$, $R_{Q\oplus jQ}(n)$ is given by

$$\begin{aligned} R_{Q\oplus jQ}(n) &= \sum_{\substack{a,b\in\mathbb{N}_0 \\ a+jb=n}} \Big(\sum_{Q(x_1,\cdots,x_4)=a} 1\Big)\Big(\sum_{Q(x_5,\cdots,x_8)=b} 1\Big) \\ &= R_Q(n) + R_Q(n/j) + \sum_{\substack{a,b\in\mathbb{N} \\ a+jb=n}} R_Q(a)R_Q(b) \\ &= 4\sigma(n) - 28\sigma\left(\frac{n}{7}\right) + 4\sigma\left(\frac{n}{j}\right) - 28\sigma\left(\frac{n}{7j}\right) + \\ &\quad 16W_j(n) - 4\times 28W_{7j}(n) - 4\times 28W_{j,7}(n) + 28^2W_j\left(\frac{n}{7}\right). \end{aligned}$$

Comparing these formulas with the formulas given in Theorem 2.3 and using the convolution sums given in Theorem 2.1 and Theorem 2.2, we get the convolution sums $W_{j,7}(n)$, for $1 \leq j \leq 4$. This completes the proof.

Acknowledgments

We have used the open-source mathematics software SAGE (`www.sagemath.org`) to do our calculations. The second author is partially funded by SERB grant SR/FTP/MS-053/2012. He would like to thank HRI, Allahabad for the warm hospitality where this work has been carried out. The SAGE worksheets are available in the last author's website[1]. Finally, we would like to thank the referee for making helpful suggestions.

References

[1] A. Alaca, S. Alaca and E. Ntienjem, *The convolution sum* $\sum_{al+bm=n}\sigma(l)\sigma(m)$ *for* $(a,b) = (1,28),(4,7),(1,14),(2,7),(1,7)$, arXiv:1607.06039v1 [math.NT]

[2] A. O. L. Atkin and J. Lehner, *Hecke operators on* $\Gamma_0(m)$, Math. Ann. **185** (1970), 134–160, DOI 10.1007/BF01359701. MR0268123

[3] M. Besge, *Extrait d'une lettre de M. Besge á M. Liouville*, J. Math. Pures Appl. **7** (1862), 256.

[4] J. H. Bruinier, G. van der Geer, G. Harder, and D. Zagier, *The 1-2-3 of modular forms*, Universitext, Springer-Verlag, Berlin, 2008. Lectures from the Summer School on Modular Forms and their Applications held in Nordfjordeid, June 2004; Edited by Kristian Ranestad. MR2385372

[5] H. H. Chan and S. Cooper, *Powers of theta functions*, Pacific J. Math. **235** (2008), no. 1, 1–14, DOI 10.2140/pjm.2008.235.1. MR2379766

[6] J. W. L. Glaisher, *On the squares of the series in which the coefficients are the sums of the divisor of the exponents*, Mess. Math. **15** (1885), 1–20.

[7] J. G. Huard, Z. M. Ou, B. K. Spearman, and K. S. Williams, *Elementary evaluation of certain convolution sums involving divisor functions*, Number theory for the millennium, II (Urbana, IL, 2000), A K Peters, Natick, MA, 2002, pp. 229–274. MR1956253

[8] M. Kaneko and D. Zagier, *A generalized Jacobi theta function and quasimodular forms*, The moduli space of curves (Texel Island, 1994), Progr. Math., vol. 129, Birkhäuser Boston, Boston, MA, 1995, pp. 165–172, DOI 10.1007/978-1-4612-4264-2_6. MR1363056

[9] LMFDB, The database of L-functions, modular forms, and related objects, http://www.lmfdb.org/

[1] URL: https://sites.google.com/site/theanupkumarsingh/sage-worksheets

[10] F. Martin and E. Royer, *Formes modulaires et périodes* (French, with English and French summaries), Formes modulaires et transcendance, Sémin. Congr., vol. 12, Soc. Math. France, Paris, 2005, pp. 1–117. MR2186573

[11] B. Ramakrishnan and B. Sahu, *Evaluation of the convolution sums* $\sum_{l+15m=n}\sigma(l)\sigma(m)$ *and* $\sum_{3l+5m=n}\sigma(l)\sigma(m)$ *and an application*, Int. J. Number Theory **9** (2013), no. 3, 799–809, DOI 10.1142/S179304211250162X. MR3043615

[12] S. Ramanujan, *On certain arithmetical functions,* Trans. Cambridge Philos. Soc. 22 (1916) 159–184.

[13] E. Royer, *Evaluating convolution sums of the divisor function by quasimodular forms*, Int. J. Number Theory **3** (2007), no. 2, 231–261, DOI 10.1142/S1793042107000924. MR2333619

[14] B. Schoeneberg, *Elliptic modular functions: an introduction*, Springer-Verlag, New York-Heidelberg, 1974. Translated from the German by J. R. Smart and E. A. Schwandt; Die Grundlehren der mathematischen Wissenschaften, Band 203. MR0412107

[15] K. S. Williams, *On a double series of Chan and Ong*, Georgian Math. J. **13** (2006), no. 4, 793–805. MR2309261

[16] K. S. Williams, *Number theory in the spirit of Liouville*, London Mathematical Society Student Texts, vol. 76, Cambridge University Press, Cambridge, 2011. MR2731552

Harish-Chandra Research Institute, HBNI, Chhatnag Road, Jhunsi, Allahabad - 211 019, India

Email address: ramki@hri.res.in

School of Mathematical Sciences, National Institute of Science Education and Research, HBNI, Bhubaneswar PO: Jatni, Khurda, Odisha - 752 050, India

Email address: brundaban.sahu@niser.ac.in

Harish-Chandra Research Institute, HBNI, Chhatnag Road, Jhunsi, Allahabad - 211 019, India

Email address: anupsingh@hri.res.in

Contemporary Mathematics
Volume **732**, 2019
https://doi.org/10.1090/conm/732/14798

Level of Siegel modular forms constructed via sym^3 lifting

Manami Roy

ABSTRACT. Ramakrishnan-Shahidi proved a lifting from a non-CM elliptic curve E over $\mathbb{Q}$ to a degree 2 Siegel cusp form F of weight 3. We want to better understand the level of the Siegel cusp form F coming from their lifting. Moreover, we are interested in the level of F with respect to different congruence subgroups.

1. Introduction

Kim and Shahidi [**KS02**] proved the Langlands functoriality from $\mathrm{GL}(2,\mathbb{A})$ to $\mathrm{GL}(4,\mathbb{A})$ coming from the symmetric cube (sym^3) map from $\mathrm{GL}(2,\mathbb{C})$ to $\mathrm{GL}(4,\mathbb{C})$. Ramakrishnan and Shahidi [**RS07a**] proved the following theorem using the sym^3 lifting, which generalizes the lifting from a non-CM elliptic curve to a Siegel cusp form of degree 2 and weight 3.

THEOREM 1.1. *Let* $\pi \cong \bigotimes_p \pi_p$ *be a cuspidal automorphic representation of* $\mathrm{GL}(2,\mathbb{A})$ *defined by a holomorphic, non-CM newform* f *of even weight* $k \geq 2$, *level* N *with trivial central character. Then there exists a cuspidal automorphic representation* $\Pi \cong \bigotimes_p \Pi_p$ *of* $\mathrm{GSp}(4,\mathbb{A})$ *with trivial central character, which is unramified at any prime* p *not dividing* N, *such that*

(1) Π_∞ *is a holomorphic discrete series representation, with its parameter being* sym^3 *of the archimedean parameter of* π.
(2) $\mathrm{L}(\mathrm{s},\Pi) = \mathrm{L}(\mathrm{s},\pi,\mathrm{sym}^3)$.

In the above theorem, Ramakrishnan and Shahidi lift a cuspidal automorphic representation of $\mathrm{GL}(2,\mathbb{A})$ to a cuspidal automorphic representation of $\mathrm{GSp}(4,\mathbb{A})$. First, they use the functoriality from $\mathrm{GL}(2,\mathbb{A})$ to $\mathrm{GL}(4,\mathbb{A})$. Then via the descent method they obtain a cuspidal automorphic representation of $\mathrm{SO}(5,\mathbb{A})$ from $\mathrm{GL}(4,\mathbb{A})$. Then they construct a cuspidal automorphic representation of $\mathrm{Sp}(4,\mathbb{A})$ from the cuspidal automorphic representation of $\mathrm{SO}(5,\mathbb{A})$ and finally they get a cuspidal automorphic representation of $\mathrm{GSp}(4,\mathbb{A})$ from the representation of $\mathrm{Sp}(4,\mathbb{A})$.

By this construction they get a globally generic representation of $\mathrm{GSp}(4,\mathbb{A})$. To switch the archimedean component Π_∞ from generic to a holomorphic discrete series, they give a proof using ℓ-adic cohomology.

2010 *Mathematics Subject Classification.* Primary 11F46, 11F70.

2. Current work

2.1. Different proof of the theorem. Using results that were not available at the time, we can now give a streamlined proof of the Ramakrishnan-Shahidi theorem involving the following groups, dual groups and liftings.

$$
\begin{array}{ccc}
\mathrm{GL}(2,\mathbb{A}) & \xrightarrow{\text{functoriality}} & \mathrm{GSp}(4,\mathbb{A}) \\
\downarrow & & \uparrow \\
\mathrm{GL}(4,\mathbb{A}) & \underset{\text{functoriality}}{\overset{\text{descent}}{\rightleftarrows}} & \mathrm{SO}(5,\mathbb{A})
\end{array}
\qquad\qquad
\begin{array}{ccc}
\mathrm{GL}(2,\mathbb{C}) & \xrightarrow{\mathrm{sym}^3} & \mathrm{GSp}(4,\mathbb{C}) \\
\downarrow{\scriptstyle \mathrm{sym}^3} & & \uparrow \\
\mathrm{GL}(4,\mathbb{C}) & \hookleftarrow & \mathrm{Sp}(4,\mathbb{C})
\end{array}
$$

The image under the sym^3 map lies in $\mathrm{GSp}(4,\mathbb{C})$. We can see that the lifting from $\mathrm{GL}(2,\mathbb{A})$ to $\mathrm{GSp}(4,\mathbb{A})$ via sym^3 is functorial at each place, since all other liftings in the above diagram respect functoriality at each place and the local Langlands correspondence is true for GL(2) and GSp(4). We use Arthur's packet structure to see that the representation of $\mathrm{GSp}(4,\mathbb{A})$ is of "general" type, which helps us to prove the claims of the theorem.

2.2. Level under the paramodular group. The cuspidal automorphic representations of $\mathrm{GSp}(4,\mathbb{A})$ are connected to the theory of Siegel cusp forms. One can consider the Siegel cusp forms coming from the sym^3 lifting. We are interested in finding the level of these Siegel cusp forms under some suitable congruence subgroups. In the paper [**RS07a**], the level is measured in terms of principal congruence subgroups. However, for a correspondence between cuspidal automorphic representations of $\mathrm{GSp}(4,\mathbb{A})$ and Siegel cusp forms, it is more convenient to consider a different congruence subgroup known as the paramodular group:

$$
K(N) = \mathrm{Sp}(4,\mathbb{Q}) \cap \begin{bmatrix} \mathbb{Z} & N\mathbb{Z} & \mathbb{Z} & \mathbb{Z} \\ \mathbb{Z} & \mathbb{Z} & \mathbb{Z} & N^{-1}\mathbb{Z} \\ \mathbb{Z} & N\mathbb{Z} & \mathbb{Z} & \mathbb{Z} \\ N\mathbb{Z} & N\mathbb{Z} & N\mathbb{Z} & \mathbb{Z} \end{bmatrix}.
$$

There is a well understood connection between paramodular forms (Siegel modular forms with respect to the paramodular group) and cuspidal automorphic representations of $\mathrm{GSp}(4,\mathbb{A})$, and there is a nice newform theory for paramodular forms [**RS07b**]. We now focus on finding the level of the paramodular forms obtained by the sym^3 lifting. The following result, one of the results in [**Roy**], is about the paramodular forms coming from elliptic curves via the sym^3 lifting:

THEOREM 2.1. *Given a non-CM elliptic curve E over $\mathbb{Q}$ with the global minimal Weierstrass equation $y^2 + a_1xy + a_3y = x^3 + a_2x^2 + a_4x + a_6$ with coefficients a_1, a_2, a_3, a_4, a_6 in $\mathbb{Z}$, there exists a cuspidal paramodular newform F of weight 3 and level M such that*

$$
L(s,F) = L(s,E,\mathrm{sym}^3),
$$

and the level M can be determined in an explicit and elementary way from the given Weierstrass coefficients a_1, a_2, a_3, a_4, a_6 of E.

In the following table, we give some examples to illustrate Theorem 2.1.

$E/\mathbb{Q}$	Conductor of E	Level of F
$y^2 + y = x^3 - x^2$	11	$11^3 = 1331$
$y^2 + xy + y = x^3 - x - 2$	$50 = 2 \cdot 5^2$	$2^3 \cdot 5^2 = 200$
$y^2 + xy + y = x^3 + x^2 - 3x + 1$	$50 = 2 \cdot 5^2$	$2^3 \cdot 5^4 = 5000$
$y^2 + xy = x^3 - x^2 - 3x + 3$	$54 = 2 \cdot 3^3$	$2^3 \cdot 3^5 = 1944$
$y^2 + xy = x^3 + x^2 - 2x - 7$	$121 = 11^2$	$11^2 = 121$
$y^2 + y = x^3 - x^2 - 7x + 10$	$121 = 11^2$	$11^4 = 14641$
$y^2 + xy + y = x^3 - x^2 - 5x + 5$	$162 = 2 \cdot 3^4$	$2^3 \cdot 3^4 = 648$
$y^2 + xy = x^3 - x^2 + 3x - 1$	$162 = 2 \cdot 3^4$	$2^3 \cdot 3^6 = 5832$
$y^2 + xy = x^3 - x^2 + 3x + 5$	$486 = 2 \cdot 3^5$	$2^3 \cdot 3^7 = 17496$

Note that we may have different paramodular levels for elliptic curves with the same conductor. This indicates that the level of the paramodular form F does not only depend on the conductor E, but on some data involving the coefficients of the Weierstrass equation of E.

References

[KS02] Henry H. Kim and Freydoon Shahidi, *Functorial products for* $\mathrm{GL}_2 \times \mathrm{GL}_3$ *and the symmetric cube for* GL_2, Ann. of Math. (2) **155** (2002), no. 3, 837–893, DOI 10.2307/3062134. With an appendix by Colin J. Bushnell and Guy Henniart. MR1923967

[Roy] Manami Roy, *Elliptic curves and paramodular forms*, doctoral dissertation (in progress).

[RS07a] Dinakar Ramakrishnan and Freydoon Shahidi, *Siegel modular forms of genus 2 attached to elliptic curves*, Math. Res. Lett. **14** (2007), no. 2, 315–332, DOI 10.4310/MRL.2007.v14.n2.a13. MR2318628

[RS07b] Brooks Roberts and Ralf Schmidt, *Local newforms for GSp(4)*, Lecture Notes in Mathematics, vol. 1918, Springer, Berlin, 2007. MR2344630

Department of Mathematics, University of Oklahoma, 660 Parrington Oval, Norman, Oklahoma 73019

Email address: `m.roy@ou.edu`

Contemporary Mathematics
Volume **732**, 2019
https://doi.org/10.1090/conm/732/14780

Dimension formulas and kernel functions for Hilbert modular forms

Fredrik Strömberg

ABSTRACT. We will briefly review some classical results about dimension formulas for spaces of modular forms and give details about explicit computations. One of the methods we present, using the reproducing kernel, can be generalized to Hilbert modular forms and even to vector-valued Hilbert modular forms and we will present resent results on explicit dimension formulas for vector-valued Hilbert modular forms based on joint work with N.-P. Skoruppa.

1. Introduction

One of the most fascinating aspects about modular forms is that they appear in many different contexts ranging from classical number theory to theoretical physics. A typical example of a number-theoretical problem which can be studied using modular forms is to determine the number of ways a positive integer can be represented as a sum of squares. If $r_t(n)$ denotes the number of ways the positive integer n can be written as a sum of t squares then the generating series $\theta_t(q) = \sum r_t(n) q^n$ belongs to a certain space of modular forms and a detailed study of this space can be used to deduce both estimates and explicit formulas.

We will first recall some basic facts about modular forms, then discuss a few of the most important methods which are used to obtain dimension formulas. In the final section we will give the dimension formulas for Hilbert modular forms in a general vector-valued setting.

2. Classical modular forms

Let $\mathbb{H} = \{x + iy \mid y > 0\}$ be the complex upper half-plane and recall that the the group of orientation-preserving isometries of $\mathbb{H}$ equipped with the hyperbolic metric $ds^2 = y^{-2}\left(dx^2 + dy^2\right)$ is $\mathrm{PSL}_2(\mathbb{R}) \simeq \mathrm{SL}_2(\mathbb{R})/\{\pm 1\}$. We often do not distinguish (notationally) between matrices in $\mathrm{SL}_2(\mathbb{R})$ and the corresponding linear fractional transformations in $\mathrm{PSL}_2(\mathbb{R})$. In particular the action of $\gamma \in \mathrm{SL}_2(\mathbb{R})$ is

$$\gamma : z \mapsto \gamma z = \frac{a\tau + b}{c\tau + d} \quad \text{if} \quad \gamma = \left(\begin{smallmatrix} a & b \\ c & d \end{smallmatrix}\right) \quad \text{and} \quad z \in \mathbb{H},$$

Part of the work reported on has been carried out in connection with the EPSRC grant RA22F9, *Explicit methods for Jacobi forms over number fields*.

and this extends naturally to an action on the boundary $\partial\mathbb{H} = \mathbb{R} \cup \{i\infty\}$, where we use the symbol $i\infty$ to denote the "point at infinity" (the use of $i\infty$ instead of just ∞ is slightly misleading but a common practice in the theory of modular forms).

Elements of $\mathrm{SL}_2(\mathbb{R})$ can be classified according to their fixed points. If $c \neq 0$ the equation $\gamma\tau = \tau$ corresponds to a quadratic equation in τ , hence γ has at most two fixed points in $\mathbb{C}$ and we say that γ is *elliptic* if it has exactly one fixed point in $\mathbb{H}$ (and another in the lower half-plane), it is *hyperbolic* if it has two distinct fixed points, and *parabolic* if it has a unique fixed point in $\partial\mathbb{H}$. These properties can easily be translated into the property that the absolute value of the trace of γ (i.e. $|a+d|$) is smaller than, greater than, or equal to 2 or alternatively that γ is conjugate in $\mathrm{SL}_2(\mathbb{R})$ to an element of the form $\left(\begin{smallmatrix} \sin\theta & \cos\theta \\ -\cos\theta & \sin\theta \end{smallmatrix}\right)$, $\left(\begin{smallmatrix} \lambda & 0 \\ 0 & \lambda^{-1} \end{smallmatrix}\right)$ or $\left(\begin{smallmatrix} 1 & \mu \\ 0 & 1 \end{smallmatrix}\right)$.

If k is an even integer there is an induced action of $\mathrm{PSL}_2(\mathbb{R})$ on functions $f : \mathbb{H} \to \mathbb{C}$ given by by $f \mapsto f|_k\gamma(\tau) = j_\gamma(\tau)^{-k} f(\gamma\tau)$, where $j_\gamma(\tau) = (c\tau + d)$ is an example of an *automorphy factor*, in other words it satisfies a *cocycle* relation of the form $j_\gamma(\delta\tau) j_\delta(\tau) = j_{\gamma\delta}(\tau)$ for all $\gamma, \delta \in \mathrm{SL}_2(\mathbb{R})$ and $\tau \in \mathbb{H}$.

To be more concrete, from now on we consider the modular group $\Gamma = \mathrm{SL}_2(\mathbb{Z})$ and let $G \subseteq \Gamma$ be a finite-index subgroup. A modular form of even integer weight k for a subgroup G is a holomorphic function $f : \mathbb{H} \to \mathbb{C}$ which satisfy

$$f|_k\gamma(\tau) = f(\tau) \quad \text{for all} \quad \gamma \in G$$

and which extends to a holomorphic function on $\mathbb{H}^* = \mathbb{H} \cup \mathbb{Q} \cup \{i\infty\}$. If, furthermore, $f|_k\gamma(i\infty) := \lim_{y\to\infty} f|_k\gamma(x+iy) = 0$ for all $\gamma \in G$ then we say that f is a *cusp form*. We let $M_k(G)$ and $S_k(G)$ be the space of modular forms and cusp forms of weight k for G. If f and g are elements of $M_k(G)$ and at least one of them a cusp form then we define the *Petersson inner product* of f and g by

$$\langle f, g\rangle = \frac{1}{[\Gamma : G]} \int_{G\backslash\mathbb{H}} f(\tau)\overline{g(\tau)} y^k d\mu(z),$$

where the integral is taken over any (measurable) fundamental domain for the action of G on $\mathbb{H}$ and $d\mu(z) = y^{-2}dxdy$ is the $\mathrm{PSL}_2(\mathbb{R})$ invariant measure given by the hyperbolic metric ds^2 on $\mathbb{H}$.

It follows immediately from the definition that $M_k(G)$ has the structure of a complex vector space and it is an easy exercise to show that it is a Hilbert space together with the above inner product. The immediate question should now be whether the space $M_k(G)$ is finite or infinite-dimensional.

3. Dimension formulas

There are many different ways of obtaining dimension formulas for spaces of modular forms, exploiting the multitude of interpretations these space have. The two most common methods are Riemann-Roch-type theorems and trace formulas, in particular the So-called Eichler–Selberg trace formula. We will give a (very) brief overview of the first of these methods in the next section and we will then focus on the second for the remainder of our presentation.

The trace formulas which we will use are mainly obtained using analytic tools and some fundamental concepts from functional analysis. There is, however, also an algebro-geometric version of the trace formula by Shimura [**16**]. This can be interpreted as a generalization of the the Atiyah–Bott fixed point formula and is also similar to a generalized Riemann–Roch formula by Weil [**20**].

3.1. Riemann – Roch theorems. A first course in complex analysis should contain the beautiful theorem of Liouville: *If a holomorphic function is bounded in the entire complex plane then it is constant.* An alternative formulation of this theorem says that a holomorphic function on the *Riemann sphere* $\mathbb{C}^* = \mathbb{C} \cup \{\infty\}$ is constant, or in other words, the space of holomorphic functions on $\mathbb{C}^*$ has dimension one.

It is then a natural question to ask about the dimensions of other spaces of holomorphic functions, for instance: how many linearly independent meromorphic functions on $\mathbb{C}^*$ are there which have a simple pole at $z = \infty$ and a simple pole at $z = 0$? It is a simple exercise to answer this question (do it!) but when studying more complicated surfaces or imposing more general constraints it becomes more difficult to both formulate the question precisely and to obtain concrete answers.

The first steps taken to give an explicit answer to this type of more general questions was taken by Riemann and his student Roch and the type of theorems they obtained is now usually called Riemann–Roch theorems.

To give a taste of the theorems that can be obtained using these methods we let M be a compact Riemann surface and the *group of divisors* of M, $\mathrm{Div}(M)$, be the free abelian group on the points of M. The elements of $\mathrm{Div}\,(M)$, called *divisors* of M, can be represented as formal sums $D = \sum_{x\in M} \alpha_x[x]$ where $\alpha_x \in \mathbb{Z}$ and all but finitely many α_x are equal to 0. The *degree* of a divisor D is defined as $\deg(D) = \sum \alpha_x$, and the map $\deg : \mathrm{Div}\,(M) \to \mathbb{Z}$ is a group homomorphism. We write $D \geq 0$ if $\deg(D) \geq 0$ and $D \geq D'$ if $\deg(D' - D) \geq 0$. If $f : M \to \mathbb{C}$ is a meromorphic function then the divisor of f is defined as

$$\mathrm{Div}(f) = \sum_{x\in M} v_x(f)[x]$$

where $v_x(f)$ is the order of f at x, in other words the Laurent expansion of f with respect to any local coordinate z at the point x is $f(z) = c \cdot z^{v_x(f)} + O(z^{v_x(f)+1})$ for some $c \neq 0$. A divisor of the form $\mathrm{Div}(f)$ is called a *principal divisor.* Note that if f does not have a pole at x then $v_x(f) \geq 0$ and if f has a zero at x then $v_x(f) > 0$.

If M is compact then the analog of the *Residue theorem* for the surface M says that principal divisors have degree equal to zero, in other words, that the orders of all the zeros and poles of f sum to zero (when taken with multiplicities).

If D is a divisor on M we define the space $L(D)$ as the set of meromorphic functions on M with divisor $\mathrm{Div}(f) \geq -D$. So for instance, if $M = \mathbb{C}^*$ and $D = [0] + [\infty]$ then $L(D)$ consists of all meromorphic functions on $\mathbb{C}^*$ with at most a simple pole at 0 and ∞. If $\omega \in \Omega^1(M)$ is a meromorphic differential on M then the divisor $\mathrm{Div}\,(\omega)$ is said to be a *canonical divisor* on M. Note that in local coordinates such an ω can be written as $\omega\,(z) = f\,(z)\,dz$ where f transforms as a modular form of weight 2.

If we use $l(D)$ to denote the dimension of $L(D)$ then the Riemann–Roch theorem (see e.g. [**7**, Thm. 3.4.1]) states that

$$l(D) = \deg(D) - g + 1 + l\,(\mathrm{Div}\,(\omega) - D)$$

where $\mathrm{Div}(\omega)$ is a canonical divisor of M (the result is independent of choice of ω). A simple consequence ([**7**, Cor. 3.4.2]) gives the special cases $l\,(\mathrm{Div}(\omega)) = g$ and if $\deg(D) < 0$ then $l(D) = 0$, while if $\deg(D) > 2g - 2$ then $l(D) = \deg(D) - g + 1$.

There are two issues that appear when we try to use the Riemann–Roch theorem to calculate dimensions of modular forms on subgroups Γ of the modular group. The first is that the quotients $\Gamma\backslash\mathbb{H}$ are not compact and the second is that they are not smooth, i.e. they are technically speaking not Riemannian surfaces or manifolds according to standard modern terminology, they are *orbifolds*. The non-compactness is relatively easy to deal with by adding the "points" at infinity and work with the compact surface $\Gamma\backslash\mathbb{H}^*$ where $\mathbb{H}^* = \mathbb{H}\cup\mathbb{Q}\cup\{\infty\} = \mathbb{H}\cup\Gamma(i\infty)$. The second issue is a bit more tricky. The theory of divisors extends perfectly well to surfaces with marked points, i.e. points on $\Gamma\backslash\mathbb{H}$ with non-trivial stabilizers but the Riemann–Roch theorem requires divisors to have integral coefficients. It is shown in e.g. [**7**, Sec. 3.5] that the divisors of holomorphic modular forms can be approximated by integral divisors and this leads to a result of the following form:

THEOREM. *If $k\geq 2$ and $G\subseteq\Gamma$ is a finite index subgroup then*

$$\dim M_k(G) = (k-1)(g-1) + \left\lfloor\frac{k}{4}\right\rfloor v_2 + \left\lfloor\frac{k}{3}\right\rfloor v_3 + \frac{k}{2}v_\infty$$

where g is the genus, v_i the number of elliptic points of order i, and v_∞ the number of cusps of the orbifold $G\backslash\mathbb{H}$.

Of course, in specific cases one does not need the entire machinery of Riemann–Roch it is sufficient to use the much simpler argument principle:

3.2. Simple example. If f is a holomorphic modular form of weight k for the modular group and C is a closed contour in $\mathbb{H}$ such that f does not have a zero on C then the argument principle says that $\int_C f'(z)/f(z)dz = N$ where N is the number of zeros of f inside C. If $f\in M_k(\mathrm{SL}_2(\mathbb{Z}))$ we can choose a contour C which is close to the boundary of the standard fundamental domain for the modular group $\mathcal{F} := \left\{z = x+iy\in\mathbb{H} \mid |x|\leq\frac{1}{2},\ |z|\geq 1\right\}$ in the sense that we cut out ε-neighbourhoods of each zero on the contour (in a symmetrical way) and such that we enclose each zero inside the contour by a circle of radius ε and we truncate the contour at $Y=\varepsilon^{-1}$. A direct calculation leads to a formula of the form

$$v_{i\infty}(f) + \frac{1}{3}v_\rho(f) + \frac{1}{2}v_i(f) + \sum_{\tau\neq i,\rho,\tau\in\Gamma\backslash\mathbb{H}} v_\tau(f) = \frac{k}{12}.$$

For details see e.g. [**5**, Ch. 1]. Since the order of a holomorphic modular form at any point is non-negative it is easy to give upper bounds for the dimension of modular forms. From the existence of the holomorphic Eisenstein series E_k of weight k for $k=4,6,8,10$ and 12 together with the cusp form $\Delta = (E_4^3 - E_6^2)/1728$ we also obtain lower bounds. Combining these dimensions with the fact that $M_k(\Gamma) = S_k(\Gamma) + \mathbb{C}E_k$ and using the isomorphism between $M_{12-k}(\Gamma)$ and $S_k(\Gamma)$ defined by $f\mapsto\Delta\cdot f$ results in the following dimension formula

$$\dim(S_k(\Gamma)) = \begin{cases}\lfloor k/12\rfloor - 1 + \delta_{k,2}, & k\equiv 2 \pmod{12},\\ \lfloor k/12\rfloor, & k\not\equiv 2 \pmod{12},\end{cases}$$

where $\delta_{k,2}=1$ if $k=2$ and 0 otherwise.

3.3. Trace formulas. The following argument can be applied to more general settings but to avoid technical restrictions we simply consider the following prototypical case. Let F be a domain in the complex plane, let $d\tau$ be a non-negative measure on F and consider the associated Hermitian inner product

$$\langle f,g\rangle = \int_F f(z)\overline{g(z)}d\tau(z)$$

which is defined for all measurable functions $f, g : F \to \mathbb{C}$ such that the integral is convergent. Let V be a finite-dimensional subspace of the corresponding L^2-space of functions on F with norm $\langle \cdot, \cdot \rangle$ and let $\{f_i\}_{i=1}^d$ be an orthonormal basis of V.

Consider a two-variable function $k : F \times F \to \mathbb{C}$ with the property that for each fixed $w \in F$ the function $z \mapsto k(z, w)$ (which we will denote by $k(\cdot, w)$) is an element of V. The linear operator $K : V \to V$ which is defined by

$$K(f)(w) = \langle f, k(\cdot, w) \rangle = \int_F f(z)\, k(z, w) d\tau(z),$$

is said to be an *integral operator* and the function k its *kernel function.* Since we assumed that $k(\cdot, w)$ is an element of V we can express it in terms of the basis as

$$k(z, w) = \sum_{i=1}^{d} \overline{\langle f_i, k(\cdot, w) \rangle} f_i(z).$$

This expression can be used to obtain the following integral formula for the trace of the operator K:

$$\begin{aligned} \mathrm{Tr}K = \sum_{i=1}^{d} \langle f_i, K(f_i) \rangle &= \sum_{i=1}^{d} \int_F f_i(w)\, \overline{\langle f_i, k(\cdot, w) \rangle} d\tau(w) \\ &= \int_F k(z, z) d\tau(z). \end{aligned}$$

If we assume that k is chosen so that the *spectral theorem* holds for K then we have an alternative expression for the trace of K:

$$\mathrm{Tr}K = \sum_{i=1}^{d} \lambda_i,$$

where $\{\lambda_i\}$ is the set of eigenvalues of K. This leads to an identity of the form

$$\sum \lambda_i = \int_F k(z, z) d\tau(z)$$

which is an example of a *trace formula* for the operator K. The left-hand side is usually said to be *spectral,* and the right-hand side *geometric.* The subject of trace formulas in general and for automorphic forms and Fuchsian groups in particular is a vast subject and contains topics like the Selberg Trace Formula, The Eichler-Selberg Trace Formula, the Arthur Trace formula. For an historical overview of these types of formulas see for instance [**1**].

The problem of choosing kernel function, or equivalently, integral operator, is in general very intricate and any particular choice is usually tailored to a specific application. However, in the case when V is a *reproducing kernel* Hilbert space, in other words if the linear evaluation operator: $L_w : V \to \mathbb{C}$, defined by $L_w : f \mapsto f(w)$ for $w \in F$ is bounded and hence can be represented by an integral operator of the form above then L_w is in many aspects a distinguished choice. Since $(L_w f_i)(w) = f_i(w)$ for all functions it is clear that the eigenvalues of L_w are all equal to 1, in other words the trace formula above simply says that if the dimension of V is d then

$$d = \sum_{i=1}^{d} 1 = \int_D k(z, z) d\mu(z).$$

It is therefore possible to calculate the dimension of V by integrating the kernel function corresponding to L_w.

EXAMPLE. Consider the unit disk $D = \{z = x + iy \in \mathbb{C} \mid |z| \leq 1\}$ together with $\mathcal{O}(D)$, the space of holomorphic functions on D, and the inner product given by

$$\langle f, g \rangle = \int_D f(z)\overline{g(z)}dxdy.$$

The Bergman space $B^2(D)$ is defined as $B^2(D) = \{f \in \mathcal{O}(D) \mid \langle f, f \rangle < \infty\}$and it is not hard to see that this is a reproducing kernel Hilbert space with kernel function

$$k(z, w) = \frac{1}{\pi} \frac{1}{(1 - z\overline{w})^2}$$

(hint: look at the Taylor expansion of $\langle f, k(\cdot, w) \rangle$).

3.4. The trace formula for the modular group. Most details presented in this section can be found in slightly more generality in Oesterle [**12**] and Zagier [**11**, Appendix] (and the correction [**21**]).

Consider the Hilbert space $M_k(\Gamma)$ of modular forms for $\Gamma = \mathrm{SL}_2(\mathbb{Z})$ and let $k > 2$ be an even integer. We can obtain a reproducing kernel by averaging the function $k(z, w) = (z - \overline{w})^{-k}$ over Γ:

$$K(z, w) = \sum_{\alpha \in \Gamma} k(z, w)|_{z,k}\alpha = \sum_{\alpha \in \Gamma} j_\alpha(z)^{-k} (\alpha z - \overline{w})^{-k}.$$

It is not difficult to verify the following properties:

(1) $K(\cdot, w) \in S_k(\Gamma)$ for each fixed w, and
(2) $\langle f, c \cdot K(\cdot, w) \rangle = f(w)$ for all $f \in S_k(\Gamma)$, where $c = \frac{(k-1)(2i)^k}{4\pi}$.

Since $c \cdot K$ is a reproducing kernel we can also write it as

$$K(z, w) = c^{-1} \sum f_i(z)\overline{f}_i(w)$$

where the sum is taken over a fixed orthonormal basis $\{f_i\}$ of $M_k(\Gamma)$. It can also be shown that the Fourier expansion of $K(z, w)$ in the variable w is given by

$$K(z, w) = \frac{(-2\pi i)^k}{\Gamma(k)} \sum_{n \geq 1} n^{k-1} P_k^n(z) e^{-2\pi i n \overline{w}}$$

where P_k^n is the standard Poincaré series of index n with respect to ∞: $P_k^n(z) = \sum_{\gamma \in \Gamma_\infty \backslash \Gamma} e^{2\pi i n z}|_k\gamma$. For the proof of this the Lipchitz (a special case of the Poisson) summation formula:

$$\sum_{n \in \mathbb{Z}} (\tau + n)^{-k} = \frac{(-2\pi i)^k}{\Gamma(k)} \sum_{n \geq 1} n^{k-1} e^{2\pi i n \tau}$$

is very useful. The kernel function $K(z, w)$ appears in a very natural way when trying to construct modular forms and it can be found (although not used in the way we use it here) in original papers of both Poincaré and Petersson (see e.g. [**14**] and [**13**]) where the theory of Poincaré series and Petersson inner products were introduced.

As we saw previously we can compute the dimension, $d = \dim S_k(\Gamma)$, by

$$\dim S_k(\Gamma) = \sum_{i=1}^{d} \langle f_i, f_i \rangle = c \int_F K(z, z) d\mu(z),$$

where we let F denote a fundamental domain for the action of Γ on $\mathbb{H}$. The last integral is then evaluated by first partitioning the elements of Γ into conjugacy classes. The conjugacy class of a given non-identity element $\gamma_0 \in \Gamma$ is defined as

$$[\gamma_0] = \left\{ g\gamma g^{-1} \mid \gamma \in \Gamma_{\gamma_0} \setminus \{I\},\ g \in \Gamma/\Gamma_{\gamma_0} \right\}$$

where Γ_{γ_0} is the centralizer of γ_0 in Γ (i.e. the set of all elements which commute with γ_0). Since we only consider even weights it is sufficient to study to $\mathrm{PSL}_2(\mathbb{Z}) = \Gamma \cap \mathrm{PSL}_2(\mathbb{R})$ and in this case it is well-known that we can choose non-conjugate representatives of the following form:

- $I_2 = \left(\begin{smallmatrix} 1 & 0 \\ 0 & 1 \end{smallmatrix}\right)$ – the identity element.
- $S = \left(\begin{smallmatrix} 0 & -1 \\ 1 & 0 \end{smallmatrix}\right)$ – elliptic with fixed point i and Γ_S a finite group of order 2.
- $R = \left(\begin{smallmatrix} 0 & -1 \\ 1 & 1 \end{smallmatrix}\right)$ and $R^2 = \left(\begin{smallmatrix} -1 & -1 \\ 1 & 0 \end{smallmatrix}\right)$ – elliptic with fixed point $e^{2\pi i/3}$ and Γ_R and Γ_{R^2} both finite groups of order 3.
- $T = \left(\begin{smallmatrix} 1 & 1 \\ 0 & 1 \end{smallmatrix}\right)$ – parabolic with Γ_T an infinite cyclic group.
- $\gamma_0 \in \Gamma$ a primitive hyperbolic element, i.e. γ_0 can not be written as a power of another hyperbolic element, and Γ_{γ_0} is an infinite cyclic group.

We now obtain a corresponding decomposition of the function K as well as the integral in the dimension formula:

$$\dim S_k(\Gamma) = c \cdot I_{\text{id}} + c \cdot I_{\text{ell}} + c \cdot I_{\text{par}} + c \cdot I_{\text{hyp}},$$

where each of the terms can be explicitly evaluated using standard methods in calculus and complex analysis. We remark again that most of the calculations here can also be found in Oesterle and Zagier [loc. cit.] and in particular we refer to the details about convergence of the integrals involved here. The identity term:

$$c \cdot I_{Id} = \int_{\mathcal{F}} (z-\overline{z})^{-k} y^k d\mu = c \cdot (2i)^{-k} \int_{\mathcal{F}} d\mu = c \cdot (2i)^{-k} \mu(\mathcal{F}) = \frac{(k-1)}{4\pi} \mu(\mathcal{F}) = \frac{k-1}{12}.$$

Using the cocycle condition $j_A(Bz) j_B(z) = j_{AB}(z)$, it is easy to show that if we set

$$\beta_\alpha(z) = (\alpha z - \overline{z})^{-k} j_\alpha(z)^{-k} y^k$$

then $\beta_\alpha(gz) = \beta_\alpha(z)$ for all matrices $g \in \Gamma_\alpha$ and

$$\beta_{g\gamma g^{-1}}(z) = \left(g\gamma g^{-1} z - \overline{z}\right)^{-k} j_{g\gamma g^{-1}}(z)^{-k} y^k = \beta_\gamma\left(g^{-1} z\right).$$

It follows that if the integral and summation can be interchanged (in all except the parabolic case this can be justified) then, for any representative γ_0 of a conjugacy class and $\gamma \in \Gamma_{\gamma_0} \setminus \{I\}$ we get a contribution

$$\begin{aligned} I_\gamma = \int_{\Gamma\backslash\mathbb{H}} \sum_{g \in \Gamma/\Gamma_{\gamma_0}} \beta_{g\gamma g^{-1}}(z) d\mu(z) &= \sum_{g \in \Gamma/\Gamma_{\gamma_0}} \int_{\Gamma\backslash\mathbb{H}} \beta_\gamma\left(g^{-1} z\right) d\mu(z) \\ &= \int_{\Gamma_{\gamma_0}\backslash\mathbb{H}} \beta_\gamma(z)\, d\mu(z), \end{aligned}$$

where the the last integral can be taken over *any* fundamental domain for the action of Γ_{γ_0} on $\mathbb{H}$. When Γ_{γ_0} has finite order we can use the invariance of β_γ to find that

$$|\Gamma_{\gamma_0}| \int_{\Gamma_{\gamma_0}\backslash\mathbb{H}} \beta_{\gamma_0}(z)\, d\mu(z) = \sum_{g \in \Gamma_{\gamma_0}} \int_{\Gamma_{\gamma_0}\backslash\mathbb{H}} \beta_{\gamma_0}(gz)\, d\mu(gz) = \int_{\mathbb{H}} \beta_{\gamma_0}(z)\, d\mu(z).$$

If Γ_{γ_0} has infinite order we have to be more careful and we might need to compute the integral over the actual domains.

To illustrate the type of calculations involved we will demonstrate how to compute the integral for the case of the elliptic element S explicitly:

$$I_S = \int_{\Gamma_S\backslash\mathbb{H}} \beta_S(z)\,d\mu = \frac{1}{2}\int_{\mathbb{H}} \beta_S(z)\,d\mu = \frac{1}{2}\int_0^\infty\int_{-\infty}^\infty \left(1+x^2+y^2\right)^{-k} y^{k-2}dxdy$$
$$= \frac{1}{2}\int_0^\pi \cos^{k-2}\theta d\theta \int_0^\infty \left(1+r^2\right)^{-k} r^{k-1}dr.$$

The integral over r can be computed using standard "tricks" which we find useful to remind the reader of: first set

$$J_n(A) = \int_0^\infty \frac{r}{(A+r^2)^n}dr$$

and use differentiation with respect to A to obtain the value $J_{n+1}(A) = \frac{1}{2n}A^{-n}$. Then use repeated integration by parts to find that

$$\int_0^\infty \frac{r^{k-1}}{(1+r^2)^k}dr = \frac{\Gamma(k/2)\Gamma(k/2+1)}{\Gamma(k)}J_{k/2+1}(1) = \frac{\Gamma(k/2)^2}{2\Gamma(k)}.$$

The final computation is then simply $\int_0^\pi \cos^{k-2}\theta d\theta$ which can be computed by, for instance the simple recursion $L_n = \frac{n-1}{n}L_{n-2}$ together with the starting value $L_2 = \frac{\pi}{2}$ and evaluates to

$$\int_0^\pi \cos^{k-2}\theta d\theta = \frac{\pi\Gamma(k-2)}{2^{k-3}\Gamma\left(\frac{k}{2}-1\right)\Gamma\left(\frac{k}{2}\right)}$$

for any even $k>2$. The total contribution for S can therefore be simplified to

$$I_S = \frac{1}{2}\frac{\Gamma(k/2)^2}{2\Gamma(k)}\frac{\pi\Gamma(k-2)}{2^{k-3}\Gamma\left(\frac{k}{2}-1\right)\Gamma\left(\frac{k}{2}\right)} = \frac{\pi}{2^{k-1}}\frac{(k/2-1)}{(k-1)(k-2)} = \frac{\pi}{2^k(k-1)}$$

so that

$$c\cdot I_S = \frac{i^k}{4} = \frac{1}{2}\begin{cases}1, & k\equiv 0 \pmod 4,\\ -1, & k\equiv 2 \pmod 4.\end{cases}$$

In a similar manner we find that the contribution of the other elliptic elements is

$$c\cdot\left(I_R + I_{R^2}\right) = \frac{i}{3\sqrt{3}}\left\{e^{\pi i(k-1)/3} + e^{2\pi i(k-1)/3}\right\} = \frac{1}{3}\begin{cases}1, & k\equiv 0 \pmod 6,\\ -1, & k\equiv 2 \pmod 6,\\ 0, & k\equiv 4 \pmod 6.\end{cases}$$

It turns out that the hyperbolic elements do not contribute to the dimension formula. To see this explicitly, consider a fixed primitive element γ_0 with centralizer Γ_{γ_0} and assume that $A\in \mathrm{SL}_2(\mathbb{R})$ is such that $A\gamma_0A^{-1} = g = \left(\begin{smallmatrix}\lambda & 0\\ 0 & \lambda^{-1}\end{smallmatrix}\right)$ for some $\lambda>1$. By a change of variable $z\mapsto A^{-1}z$ in the integral corresponding to γ_0 so that we instead integrate over a fundamental domain for $\langle g\rangle$, for instance $C := \left\{z\in\mathbb{H} \mid 1\le |z|\le \lambda^2\right\}$ a short computation (using again the cocycle relation) gives

$$\begin{aligned}
I_{\gamma_0} &= \int_{\Gamma_\gamma\backslash\mathbb{H}} \beta_\gamma(z)\, d\mu(z) = \int_{A\Gamma_{\gamma_0}A^{-1}\backslash\mathbb{H}} \beta_\gamma\left(A^{-1}z\right) d\mu(A^{-1}z)\\
&= \int_C \left(\lambda z - \lambda^{-1}\overline{z}\right)^{-k} y^k d\mu(z)\\
&= \{\text{introduce polar coordinates}\}\\
&= \int_1^{\lambda^2}\int_0^{\pi} \left(\lambda e^{i\theta} - \lambda^{-1}e^{-i\theta}\right)^{-k} \sin^{k-2}(\theta)\, r^{-1} d\theta dr\\
&= \ln\left(\lambda^2\right)\int_0^{\pi} \left(\lambda e^{i\theta} - \lambda^{-1}e^{-i\theta}\right)^{-k} \sin^{k-2}(\theta)\, d\theta.
\end{aligned}$$

That the last integral vanish follows easily by using invariance of the integrand under the transformation $\theta \mapsto \theta + \pi$ together with the residue theorem with respect to the variable $\xi = e^{i\theta}$ for the completed contour integral over $-\pi \le \theta \le \pi$.

There is now only one integral left to compute, that of the parabolic conjugacy class corresponding to T. As mentioned above, the main problem in this case is that we are not allowed to immediately interchange summation and integration. We use a standard trick and introduce a factor y^{-s} where $s \in \mathbb{C}$ has real part greater than 0 and which makes all sums absolutely and uniformly convergent in the appropriate regions. The formula obtained is then extended to $s = 0$ by analytic continuation (see also Zagier and [**8**, Ch. II, SS3]). We thus obtain the contribution:

$$I_T = \int_{\Gamma\backslash\mathbb{H}} \sum_{\substack{\gamma\in\langle T\rangle\backslash\{I\}\\ g\in\Gamma/\Gamma_T}} \beta_\gamma\left(g^{-1}z\right) y^{-s} d\mu(z) = \sum_{n\neq 0}\int_{\Gamma_\infty\backslash\mathbb{H}} (n+2iy)^{-k}\, y^{k-2-s} dx dy$$

where the interchange of summation and integration is justified by uniform convergence. By a change of variable to $u = y/|n|$ we then obtain

$$\begin{aligned}
I_T &= \lim_{s\to 0^+}\sum_{n\geq 1}\frac{1}{n^{1+s}}\left\{\int_0^\infty \frac{u^{k-2}}{(1+2iu)^{k+s}} du + (-1)^{-s}\int_0^\infty \frac{u^{k-2}}{(-1+2iu)^{k+s}} du\right\}\\
&= i\frac{2^{2-k}}{k-1}\lim_{s\to 0^+}\sum_{n\geq 1}\frac{1}{n^{1+s}}\left(1-e^{-\pi i s}\right) = i\frac{2^{1-k}}{k-1}\pi i,
\end{aligned}$$

where we used the following standard "trick" for evaluating the integral: Because of uniform convergence we can insert an auxiliary parameter t and differentiate with respect to this parameter:

$$\begin{aligned}
\int_0^\infty \frac{u^{k-2}}{(\pm 1+2iu)^{k+s}} du &= \frac{\Gamma(2+s)}{\Gamma(s+k)}\frac{1}{i^{k-2}}\left(\frac{d}{dt}\right)^{k-2}\int_0^\infty \frac{1}{(\pm 1+itu)^{2+s}} du\,\big|_{t=2}\\
&= \pm\frac{\Gamma(2+s)}{\Gamma(s+k)} i^{2-k}\left(\frac{d}{dt}\right)^{k-2}\frac{1}{(1+s)it}\,\big|_{t=2}\\
&= \pm\frac{\Gamma(2+s)}{\Gamma(s+k)}\frac{(k-2)!}{(1+s)i2^{k-1}} \to \pm i\frac{2^{1-k}}{k-1} \quad \text{as } s\to 0^+.
\end{aligned}$$

In the final step we also used that $\zeta(s)$ has a simple pole at $s=1$ with residue 1. The parabolic contribution is therefore

$$c\cdot I_T = i\frac{2^{1-k}}{k-1}\pi i\cdot\frac{(k-1)(2i)^k}{4\pi} = -\frac{1}{2}$$

and we can summarize the result into a dimension formula for $S_k(\Gamma)$:

THEOREM. *If $k>2$ then the dimension of the space of cusp forms of weight k for the modular group is given by:*

$$\dim S_k(\Gamma) = \frac{k-1}{12} + \frac{i^k}{4} + \frac{i}{3\sqrt{3}}\left\{e^{\pi i(1-k)/3} + e^{2\pi i(1-k)/3}\right\} - \frac{1}{2}.$$

The beauty about this simple formula (which you can, and should, check always returns an integer!) is that it can easily be extended to other more general situations. Both to more general operators, in particular Hecke operators, and to dimension formulas for other types of modular forms. For instance, let $\rho:\Gamma\to GL(V)$ be a n-dimensional complex representation of the modular group whose kernel is a finite index subgroup. For simplicity also assume that ρ is unitary and that $\rho(T)$ is a diagonal matrix with elements $e^{2\pi i\alpha_i}$ (with $\alpha_i\in[0,1)$) . Define the space, $M_k(\rho)$, of *vector-valued modular forms* for ρ as the vector space of holomorphic functions $f:\mathbb{H}\to V$ which satisfy

$$f(\gamma\tau) = j_\gamma(\tau)^k\,\rho(\gamma)\,f(\tau)$$

for all $\gamma\in\Gamma$ and which extends to a holomorphic function on $\partial\mathbb{H}$. There is a natural definition of cusp forms and we can define a Petersson inner-product $\langle\cdot,\cdot\rangle$ so that the space of cusp forms $S_k(\rho)$ is again a reproducing kernel Hilbert space and similar calculations as above lead to the following formula:

$$\dim S_k(\rho) = \dim V\cdot\frac{k-1}{12} + \frac{i^k}{4}\chi(S) + \frac{1}{3i\sqrt{3}}\left\{e^{\pi i(1-k)/3}\chi(R) + e^{2\pi i(1-k)/3}\chi(R^2)\right\}$$
$$+\sum_{\alpha_j\neq 0}\left(\frac{1}{2}-\alpha_j\right) - \frac{1}{2}\delta,$$

where $\chi=\mathrm{Tr}(\rho)$ and $\delta=1$ if $\rho(T)$ is the identity matrix and otherwise 0. See for instance [**9**, Thm. 4.1, p. 482] where this is shown under the additional assumption that ρ is irreducible but allowing non-integral weights, or [**2**] where it is shown for the very important special case when ρ is a Weil representation associated with an even integral lattice. An immediate generalization of the methods illustrated above also applies for Hilbert modular groups as we will briefly sketch in the next section.

REMARK. To compare the above formula with the formula for Hilbert modular forms in the next section note that the elliptic terms can be written in the form:

$$I_E = \frac{1}{|\Gamma_E|}\frac{\rho^{1-k}}{\rho-\overline{\rho}}$$

where E is an elliptic element and ρ is the fixed point of E in the upper half-plane.

4. Dimension formula for Hilbert modular forms

In this section we will give a (very) brief introduction to Hilbert modular groups and forms and then present the general result in terms of a dimension formula. For all statements which we present without giving precise references we refer to the textbooks of Freitag [**8**] and van der Geer [**19**]. More concise introductions to Hilbert modular forms can also be found in [**6**, Sect. 14.4] or [**5**, Ch. 2].

Let K be a totally real number field of degree n over $\mathbb{Q}$, with ring of integers $\mathcal{O}_K$, group of units $\mathcal{O}_K^\times$. The *Hilbert modular group* over K is defined as

$$\Gamma_K := \mathrm{SL}_2(\mathcal{O}_K) = \left\{ \left(\begin{smallmatrix} \alpha & \beta \\ \gamma & \delta \end{smallmatrix}\right), \alpha, \beta, \gamma, \delta \in \mathcal{O}_K, \alpha\delta - \beta\gamma = 1 \right\}.$$

Let $\sigma_i : K \to \mathbb{R}$, $i = 1, \dots, n$ be the embeddings of K into $\mathbb{R}$. We often use $\alpha^{(i)} = \sigma_i(\alpha)$ for $\alpha \in K$ and define the norm and trace on $K/\mathbb{Q}$ by $\mathrm{N}(\alpha) := \mathrm{N}_{K/\mathbb{Q}}(\alpha) = \prod \sigma_i(\alpha)$ and $\mathrm{Tr}(\alpha) := \mathrm{Tr}_{K/\mathbb{Q}}(\alpha) = \sum \sigma_i(\alpha)$. To avoid cluttering the notation with embeddings we introduce the ring $\mathbb{C}_K = \mathbb{C} \otimes_{\mathbb{Q}} K$ and the subring $\mathbb{R}_K = \mathbb{R} \otimes_{\mathbb{Q}} K$. We view $\mathbb{C}_K$ as an algebra over both $\mathbb{C}$ and K with operations defined in the natural way by:

$$(z \otimes a)(z' \otimes a') = zz' \otimes aa', \quad z(z' \otimes a) = zz' \otimes a \quad \text{and} \quad (z \otimes a)a' = z \otimes aa'$$

for $z, z' \in \mathbb{C}$ and $a, a' \in K$. Then $\mathbb{R}_K$ is naturally a subalgebra of $\mathbb{C}_K$ and we also define an upper half-plane $\mathbb{H}_K$ consisting of all $z \in \mathbb{C}_K$ with $\Im z \gg 0$, where we set $\Im z = \left(\Im z^{(1)}, \dots, \Im z^{(n)}\right) \in \mathbb{R}_K$ for $z = \left(z^{(1)}, \dots, z^{(n)}\right) \in \mathbb{C}_K$ and we write $x \gg y$ if $x - y$ is totally positive, in other words, if $x^{(i)} - y^{(i)} > 0$ for $i = 1, \dots, n$.

It is easy to see that the group $\mathrm{GL}_2^+(K)$, consisting of 2-by-2 matrices with totally positive determinants, acts on $\mathbb{H}_K$ by fractional linear transformations

$$Az = (az + b)(cz + d)^{-1} \quad \text{for } z \in \mathbb{H}_K \quad \text{and} \quad A = \left(\begin{smallmatrix} a & b \\ c & d \end{smallmatrix}\right) \in \mathrm{GL}_2^+(K).$$

This action also extends naturally to the boundary $\partial\mathbb{H}_K = \mathbb{R}_K \cup \{\infty\}$ and in particular to the cusps $\Gamma_K\infty = \mathbb{P}^1(K) = K \cup \{\infty\}$. If $k \in \mathbb{Z} \otimes_{\mathbb{Q}} K$ we can define a weight-k action of Γ_K on a function $f : \mathbb{H}_K \to \mathbb{C}$ by

$$f|_k A(z) = J_A(z, -k) f(\gamma z)$$

where the automorphy factor is $J_A(z, -k) = \mathrm{N}((cz + d)^{-k})$. A *Hilbert modular form* of weight k for Γ_K s a holomorphic function $f : \mathbb{H}_{\mathbb{K}} \to \mathbb{C}$ which satisfy

$$f|_k A(z) = f(z) \quad \text{for all} \quad A \in \Gamma_K,$$

and it has a Fourier expansion with respect to the cusp ∞ of the form

$$f(z) = a(0) + \sum_{0 \ll v \in \mathfrak{d}^{-1}} a(v) e^{2\pi i \mathrm{Tr}_{K/\mathbb{Q}}(vz)}$$

where $\mathfrak{d}^{-1} = \left\{ v \in K \mid \mathrm{Tr}_{K/\mathbb{Q}}(vx) \in \mathbb{Z}, \forall x \in \mathcal{O}_K \right\}$ is the codifferent ideal. That $a(v) = 0$ if v is not 0 or $\gg 0$ for $K \neq \mathbb{Q}$ follows from the Koecher principle (cf. [**8**, p. 51]). If $a(0) = 0$ (and similar for other cusps) we say that f is a *cusp form* and we let $M_k(\Gamma_K)$ and $S_k(\Gamma_K)$ denote the vector spaces of Hilbert modular forms and cusp forms. If $f, g \in S_k(\Gamma_K)$ the Petersson inner-product of f and g is

$$\langle f, g \rangle = \int_{\Gamma_K \backslash \mathbb{H}_K} f(z) \overline{g(z)} \mathrm{N}\left(y^k\right) d\mu_K(z),$$

where $d\mu_K(z) = \mathrm{N}\left(y^{-2}\right) \mathrm{N}(dy) \mathrm{N}(dx)$. It turns out that $S_k(\Gamma_K)$ together with this inner product is again a reproducing kernel Hilbert space with kernel function as above. Let $k \in \mathbb{Z} \otimes_{\mathbb{Q}} K$ with $k \gg 2$ and for $z, w \in \mathbb{H}_K$ define

$$R_K(z, w) = \sum_{\alpha \in \Gamma_K} J_\alpha(z, -k) \mathrm{N}\left((\alpha z - \overline{w})^{-k}\right).$$

Then

(1) $R_K(\cdot, w) \in S_k(\Gamma_K)$ for each fixed $w \in \mathbb{H}_K$, and
(2) $\langle f, c_K \cdot R_K(\cdot, w)\rangle = f(w)$ for all $f \in S_k(\Gamma_K)$, where $c_K = \mathrm{N}(\frac{k-1}{4\pi}(2i)^k)$.

A completely analogous argument to the above (see also [**8**, Ch. 2] and [**15**] where this formula was first developed for Hilbert modular forms) gives that if $k \gg 2$ then

$$\dim S_k(\Gamma_K) = \int_{\Gamma_K\backslash\mathbb{H}_K} R_K(z, z)\,\mathrm{N}\left(y^k\right) d\mu_K(z)$$

and if we decompose the sum into conjugacy classes we end up with

$$\dim S_k(\Gamma_K) = c \cdot I_{\text{id}} + c \cdot I_{\text{ell}} + c \cdot I_{\text{par}}.$$

The first term simply corresponds to the identity element and I_{ell} correspond to conjugacy classes of *elliptic elements*, in other words, $\alpha \in \Gamma_K$ where all embeddings of α are elliptic in $\mathrm{SL}_2(\mathbb{R})$. The term I_{par} corresponds to conjugacy classes of parabolic elements, i.e. elements $\alpha \in \Gamma_K$ with exactly one fixed point in $\partial\mathbb{H}_K$. It can be shown (see [**8**, Ch. 2, SS3] or [**15**]) that the contributions of all other types of elements vanish. The individual terms in the dimension formula can be evaluated by using the same techniques as in the previous section. The main term is

$$c \cdot I_{id} = \mathrm{N}\left(\frac{k-1}{4\pi}\right) \cdot \mu_K(\Gamma_K\backslash\mathbb{H}_K) = \mathrm{N}(k-1) \cdot 2^{1-n} \cdot (-1)^n \cdot \zeta_K(-1)$$

where $\mu_K(\Gamma_K\backslash\mathbb{H}_K)$ is the volume of the fundamental domain of Γ_K acting on $\mathbb{H}_K$ and the relationship with ζ_K was obtained by Siegel [**18**]. If we let R denote a set of inequivalent conjugacy classes of elliptic elements in Γ_K then

$$c \cdot I_{\text{ell}} = \sum_{[\alpha]\in R} \frac{1}{|\Gamma_{K,\alpha}|} \sum_{\pm 1 \neq A \in [\alpha]} \mathrm{N}\left(\frac{\rho_A^{1-k}}{\rho_A - \overline{\rho}_A}\right)$$

where ρ_A is the fixed point of A in $\mathbb{H}_K$. One of the biggest differences from $\mathrm{SL}_2(\mathbb{Z})$ is that even the full Hilbert modular group can have more than one inequivalent cusp. There is in fact a correspondence between cusps and elements of the ideal class group $\mathrm{Cl}(K)$. In particular, the number of inequivalent cusps of Γ_K is equal to h_K, the ideal class number of K. For each cusp class we choose a representative $\lambda_i \in \mathbb{P}^1_K$ and ideal $\mathfrak{c}_i$ of the corresponding class as well as a cusp normalizing map $A_i \in \mathrm{SL}_2(K)$ with $A_i(\lambda_i) = \infty$. The parabolic contribution can now be written

$$c \cdot I_{\text{par}} = |D_K|^{\frac{1}{2}} \sum_{i=1}^{h_K} \lim_{s\to 1^+} L(s, \mathfrak{c}_i)$$

where D_K is the discriminant of K and $L(s, \mathfrak{a})$ is the Shimizu L-function:

$$L(s, \mathfrak{a}) = \frac{\mathrm{N}\left(\mathfrak{a}^{-2}\right)}{\mathrm{N}(-2\pi i)} \sum_{0\neq a \in \mathfrak{a}^{-2}/\mathcal{O}_K^{\times 2}} \frac{\mathrm{sign}\left(\mathrm{N}(a)\right)}{|\mathrm{N}(a)|^s}.$$

Similar calculations can also be performed for vector-valued Hilbert modular forms. Let $\rho : \Gamma_K \to GL(V)$ be a finite dimensional complex unitary representation of Γ_K such that the kernel of ρ is a finite index subgroup $\Gamma' \subseteq \Gamma_K$ and assume that $k \in \mathbb{Z} \otimes_{\mathbb{Q}} K$ satisfies the consistency condition $\rho(\alpha) = J_\alpha(z, -k) \cdot 1_V$ for $\alpha = -I$.

Let $M_k(\rho)$ be the space of modular forms of weight k and representation ρ for Γ_K, i.e. of holomorphic functions $f : \mathbb{H}_K \to V$ satisfying

$$f|_k A = \rho(A) f \quad \text{for all} \quad A \in \Gamma_K.$$

The subspace of cusp forms, $S_k(\rho)$, consist of those f whose components are all (scalar-valued) Hilbert cusp forms for Γ'. If $\chi = \mathrm{Tr}\rho$ then we have the following main result from joint work with H. Boylan and N.-P. Skoruppa [**4**]:

THEOREM. *With the assumptions above and also assuming that $k \gg 2$ we have*

$$\begin{aligned}\dim S_k(\rho) = \quad & \dim V \cdot \mathrm{N}(k-1) \cdot 2^{1-n} \cdot (-1)^n \cdot \zeta_K(-1) \\ & + \sum_{[\alpha]} \frac{1}{|\Gamma_{K,\alpha}|} \sum_{\pm 1 \neq A \in [\alpha]} \chi(A) \mathrm{N}\left(\frac{\rho_A^{1-k}}{\rho_A - \rho_A^{-1}} \right) \\ & + |D_K|^{\frac{1}{2}} \sum_{i=1}^{h} \lim_{s \to 1^+} L(s, \mathfrak{c}_i, \chi)\end{aligned}$$

where we define the twisted Shimizu L-function

$$L(s, \mathfrak{c}_i, \chi) = \frac{\mathrm{N}(\mathfrak{c}_i)}{\mathrm{N}(-2\pi i)} \sum_{0 \neq a \in \mathfrak{c}_i^{-2}/U^2} \chi\left(A_i \begin{pmatrix} 1 & a \\ 0 & 1 \end{pmatrix} A_i^{-1} \right) \frac{\mathrm{sign}(\mathrm{N}(a))}{|\mathrm{N}(a)|^s}.$$

Our main aim when developing the general dimension formula above (which we also prove for half-integral weights) was to have a completely *explicit*formula which we could implement on a computer in order to make computer experiments with Jacobi forms over number fields. For an introduction to this topic see e.g. [**3**]. General forms of the formula in the scalar case has been given previously by e.g. Freitag [**8**] and Shimizu [**15**].

Computational remarks. The main term is easy to evaluate numerically to any desired precision using known algorithms for the Dedekind zeta function ζ_K.

For the elliptic terms the main problem is to determine an explicit set of inequivalent elliptic elements. The first step is to determine the orders of elements and numbers of elements of given orders. There are only complete tables of these for real quadratic cf. e.g. [**10**] and [**19**, p. 267]. It is not hard to find the possible orders: a simple analysis of the possible traces and the imaginary quadratic fields generated by the fixed points shows that the order m of an elliptic element must satisfy $\phi(m) = 2d$ where ϕ is Euler's ϕ-function and d is a divisor of the degree of K. It can be shown that if Γ_K has only one cusp then the set of *all* elliptic fixed points of elements with trace t are given by

$$H(t) = \left\{ z = \frac{x + \mathrm{sign}(y)\sqrt{t^2 - 4}}{2y} : x, y \in \mathcal{O}_K,\ x^2 - t^2 - 4 \in 4y\mathcal{O}_K \right\}.$$

Our approach is to use a reduction algorithm to find a set of inequivalent fixed points inside this set. If there are more cusps there will be a union of similar sets.

Using known algorithms for computing class groups it is easy to find a set of ideal class and corresponding cusp representatives. It is, however, not straightforward to evaluate the Shimizu L-functions. In the calculations for $\mathrm{SL}_2(\mathbb{Z})$ we used the fact that the Riemann zeta function has a pole at $s = 1$ with residue 1 to find the parabolic contribution. In contrast $L(s, \mathfrak{c}_i, \chi)$ is not absolutely convergent at $s = 1$ but has an analytic continuation to an entire function in $\mathbb{C}$ and thus the method above can not be immediately adapted. However, $L(s, \mathfrak{c}_i)$ is an example of a so-called *ray class*, or *Hecke* L-function, corresponding to a certain Hecke character with infinity part sign $\circ$ N and the twisted function $L(s, \mathfrak{c}_i, \chi)$ can be expressed as a sum of such. It is possible to use an explicit method introduced by Siegel [**17**] to evaluate certain Ray class L-functions at $s = 1$. Unfortunately Siegel's method relies on Kronecker's limit formulas for quadratic fields and all known extensions to higher degree are much less explicit. In ongoing work joint with M. Neururer we aim to address this problem using special values of Shintani zeta functions.

Acknowledgments. The author would like to thank N.-P. Skoruppa for many discussions in connection with [**4**], as well as the anonymous referee for carefully reading through the manuscript and providing helpful comments.

References

[1] J. Arthur, *An introduction to the trace formula*, Harmonic analysis, the trace formula, and Shimura varieties, Clay Math. Proc., vol. 4, Amer. Math. Soc., Providence, RI, 2005, pp. 1–263. MR2192011

[2] R. E. Borcherds, *Reflection groups of Lorentzian lattices*, Duke Math. J. **104** (2000), no. 2, 319–366, DOI 10.1215/S0012-7094-00-10424-3. MR1773561

[3] H. Boylan, *Jacobi forms, finite quadratic modules and Weil representations over number fields*, Lecture Notes in Mathematics, vol. 2130, Springer, Cham, 2015. With a foreword by Nils-Peter Skoruppa. MR3309829

[4] H. Boylan, N.-P. Skoruppa and F. Strömberg. A dimension formula for vector valued Hilbert modular forms. In preparation.

[5] J. H. Bruinier, G. van der Geer, G. Harder, and D. Zagier, *The 1-2-3 of modular forms*, Universitext, Springer-Verlag, Berlin, 2008. Lectures from the Summer School on Modular Forms and their Applications held in Nordfjordeid, June 2004; Edited by Kristian Ranestad. MR2385372

[6] H. Cohen and F. Strömberg, *Modular forms: A classical approach*, Graduate Studies in Mathematics, vol. 179, American Mathematical Society, Providence, RI, 2017. MR3675870

[7] F. Diamond and J. Shurman, *A first course in modular forms*, Graduate Texts in Mathematics, vol. 228, Springer-Verlag, New York, 2005. MR2112196

[8] E. Freitag, *Hilbert modular forms*, Springer-Verlag, Berlin, 1990. MR1050763

[9] D. A. Hejhal, *The Selberg trace formula for* PSL(2, **R**). *Vol. 2*, Lecture Notes in Mathematics, vol. 1001, Springer-Verlag, Berlin, 1983. MR711197

[10] H. Helling, *Bestimmung der Kommensurabilitätsklasse der Hilbertschen Modulgruppe* (German), Math. Z. **92** (1966), 269–280, DOI 10.1007/BF01112194. MR0228437

[11] S. Lang, *Introduction to modular forms*, Springer-Verlag, Berlin-New York, 1976. Grundlehren der mathematischen Wissenschaften, No. 222. MR0429740

[12] J. Oesterle. *Sur la trace des opérateurs de Hecke.* PhD thesis, L'Université de Paris-Sud, 1977.

[13] H. Petersson, *Über eine Metrisierung der automorphen Formen und die Theorie der Poincaréschen Reihen* (German), Math. Ann. **117** (1940), 453–537, DOI 10.1007/BF01450028. MR0002624

[14] H. Poincare, *Mémoire sur les fonctions fuchsiennes* (French), Acta Math. **1** (1882), no. 1, 193–294, DOI 10.1007/BF02391845. MR1554584

[15] H. Shimizu, *On discontinuous groups operating on the product of the upper half planes*, Ann. of Math. (2) **77** (1963), 33–71, DOI 10.2307/1970201. MR0145106

[16] G. Shimura, *On the trace formula for Hecke operators*, Acta Math. **132** (1974), no. 3-4, 245–281, DOI 10.1007/BF02392117. MR0562508

[17] C. L. Siegel, *Advanced analytic number theory*, 2nd ed., Tata Institute of Fundamental Research Studies in Mathematics, vol. 9, Tata Institute of Fundamental Research, Bombay, 1980. MR659851

[18] C. L. Siegel, *The volume of the fundamental domain for some infinite groups*, Trans. Amer. Math. Soc. **39** (1936), no. 2, 209–218, DOI 10.2307/1989745. MR1501843

[19] G. van der Geer, *Hilbert modular surfaces*, Ergebnisse der Mathematik und ihrer Grenzgebiete (3) [Results in Mathematics and Related Areas (3)], vol. 16, Springer-Verlag, Berlin, 1988. MR930101

[20] A. Weil. Généralisation des fonctions abéliennes. *J. Math. Pures Appl. (9)*, 17:47–87, 1938.

[21] D. Zagier, *Correction to: "The Eichler-Selberg trace formula on* $\mathrm{SL}_2(\mathbf{Z})$*" (Introduction to modular forms, Appendix, pp. 44–54, Springer, Berlin, 1976) by S. Lang*, Modular functions of one variable, VI (Proc. Second Internat. Conf., Univ. Bonn, Bonn, 1976), Springer, Berlin, 1977, pp. 171–173. Lecture Notes in Math., Vol. 627. MR0480354

School of Mathematics, University of Nottingham, University Park, NG7 2RD, Nottingham, United Kingdom

Email address: fredrik.stromberg@nottingham.ac.uk

Contemporary Mathematics
Volume **732**, 2019
https://doi.org/10.1090/conm/732/14811

An explicit evaluation of the Hauptmoduli at elliptic points for certain arithmetic groups

Holger Then

ABSTRACT. Let N be a square-free integer such that the arithmetic group $\Gamma_0(N)^+$ has genus zero. Using the Schwarzian derivative, we have proven that the Hauptmodul j_N at any elliptic point e of the surface associated to $\Gamma_0(N)^+$ is an algebraic integer. In this note we summarize our previous findings and extend the results from Jorgenson, Smajlović et al. by evaluating $j_N(z)(e)$ explicitly in terms of radicals.

1. Introduction

1.1. Klein's j-invariant. In a poll conducted by *Mathematics Intelligencer* in 1990, readers voted Euler's identity, which is the relation that $e^{\pi i}+1=0$, as the "most beautiful theorem in mathematics"; see page 2 of [**14**]. Going beyond Euler's identity, it is well-known at this time that the transcendental function $f(z)=q$ where $q=e^{2\pi iz}$ is an algebraic integer whenever z is a rational number.

Of inherent interest to mathematics is to determine additional transcendental functions which have the property of yielding algebraic values for prescribed values of their arguments. One of the most important such examples of such a transcendental function is Klein's j-invariant, which is an appropriately scaled generator of the function field associated to the genus zero Riemann surface $\Gamma\backslash\mathbb{H}$ where $\mathbb{H}$ is the hyperbolic upper half plane and $\Gamma=\mathrm{PSL}(2,\mathbb{Z})$. We refer to [**10**] and references therein for a review of some of the fascinating properties of Klein's j-invariant. For this article, we are most interested in the *singular moduli* of the j-invariant, which by definition are the values of j at imaginary quadratic arguments in $\mathbb{H}$. The value of the j-invariant at singular moduli is an algebraic integer. Furthermore, in [**16**] it is proven that if $\tau\in\mathbb{H}$ is an algebraic number which is not an imaginary quadratic rational, then $j(\tau)$ is transcendental.

1.2. Schwarzian derivative. Let f be a holomorphic function of one complex variable. The *Schwarzian derivative* $S(f)$ of f is a classically defined function given by

$$S(f)(z)=\left(\frac{f''(z)}{f'(z)}\right)'-\frac{1}{2}\left(\frac{f''(z)}{f'(z)}\right)^2,$$

where, as is the convention, the prime $'$ denotes differentiation with respect to the holomorphic parameter z. If f and g are holomorphic functions, then the

2010 *Mathematics Subject Classification.* Primary 11F12, 11R06.

Schwarzian of the composition $f \circ g$ satisfies the relation

$$S(f \circ g) = (S(f) \circ g)(g')^2 + S(g).$$

In addition, one can show that $S(g) = 0$ if and only if g is a fractional linear transformation. Therefore, if the function f is a holomorphic automorphic form with respect to some discrete group $\Gamma \subseteq \mathrm{PSL}(2, \mathbb{R})$, then the Schwarzian $S(f)$ is a meromorphic automorphic form of weight four with respect to Γ.

In [**13**] it is proven that Klein's invariant $j(z)$ satisfies the differential equation

$$S(j)(z) + R(j(z))(j'(z))^2 = 0, \tag{1.1}$$

where

$$R(y) = \frac{y^2 - 480y + 1743552}{2(y + 744)^2(y - 984)^2}.$$

1.3. Certain other genus zero groups. In order to study generalizations of Klein's j-invariant, one can look at appropriately scaled function field generators of genus zero Riemann surfaces obtained as $\Gamma\backslash\mathbb{H}$ for various discrete groups Γ. In this regard, for any positive integer N, let

$$\Gamma_0(N)^+ = \left\{ e^{-1/2} \begin{pmatrix} a & b \\ c & d \end{pmatrix} \in \mathrm{SL}(2, \mathbb{R}) : \quad a, b, c, d, e \in \mathbb{Z}, \quad e \mid (N, a, d), \quad N \mid c \right\}$$

and let $\overline{\Gamma_0(N)^+} = \Gamma_0(N)^+/\{\pm \mathrm{Id}\}$, where Id denotes the identity matrix. There are 44 square-free integers $N \geq 1$ such that the quotient space $X_N := \overline{\Gamma_0(N)^+}\backslash\mathbb{H}$ has genus zero (see [**4**]). For each such N, the surface X_N has one cusp which we take to be at $i\infty$ and to have width one. The analogue of Klein's j-invariant is defined to be the (uniquely defined) generator j_N of the function field of X_N with a first order pole at $i\infty$ whose residue is equal to one and the constant term in the q-expansion of j_N vanishes identically. We refer to j_N as the Hauptmodul of $\Gamma_0(N)^+$.

For $N > 1$ such that $\Gamma_0(N)^+$ has genus zero, the singular moduli were studied by I. Chen and N. Yui [**1**]. Analogous to Schneider's result, the authors in [**1**] have shown that if τ is a complex multiplication point, or CM point for short, which satisfies $az^2 + bz + c = 0$ with $(a, N) = 1$, then the singular moduli $j_N(\tau)$ is an algebraic integer. The results from [**1**] were further generalized in [**3**] and [**2**]. The results from [**1**] do not apply to study $j_N(\tau_\gamma)$ for *elliptic fixed points* τ_γ of $\Gamma_0(N)^+$.

2. Main results

Let us write the Schwarzian derivative of j_N as

$$S(j_N)(z) + \frac{P_N(j_N(z))}{Q_N(j_N(z))}(j_N'(z))^2 = 0. \tag{1.1a}$$

The following result was proven in [**10**]; see also [**12**] for earlier results in this direction.

THEOREM 2.1. *With notation as above, the polynomials P_N and Q_N have the following properties.*

(a) *The polynomial P_N is a monic polynomial with* $\deg(P_N) = \deg(Q_N) - 2$. *Furthermore,*

$$Q_N(j_N(z)) = 2 \prod_{e \in \mathcal{E}_N} (j_N(z) - j_N(e))^2$$

where $\mathcal{E}_N$ is the set of inequivalent elliptic points on $\mathbb{H}$ with respect to the action by $\Gamma_0(N)^+$.

(b) *The degree of Q_N is bounded by*

$$n_N \leq 2 \cdot \#\{e \in \mathcal{E}_N\}.$$

(c) *The coefficients of P_N and Q_N are integers.*

(d) *If we write $Q_N = 2(h_N)^2$, then h_N is a monic polynomial with integer coefficients.*

Therefore, for any $e \in \mathcal{E}_N$, $j_N(e)$ is an algebraic integer.

In order to evaluate explicitly values of $j_N(e)$ for $e \in \mathcal{E}_N$ one needs to perform the following computations.

First, by taking results from [**8**, **9**] for the Hauptmoduli, we used symbolic algebra of PARI/GP [**15**] to perform most of the algorithm for computation of coefficients of polynomials P_N and Q_N. We refer to [**10**] for a detailed explanation of this algorithm, which essentially reduces to solving the equation

$$\frac{1}{2}Q_N(j_N)\frac{2S(j_N)(j_N')^2}{(2\pi)^4} + P_N(j)\frac{(j_N')^4}{(2\pi)^4} = 0 \tag{1.1b}$$

for the coefficients of P_N and Q_N. To do so, one simply equates each coefficient in the q-expansion on the left hand side of (1.1b) to be equal to zero. Since we had it readily available, we used our own C-code linked against the GMP Bignum Library [**6**] to solve (1.1a) in rational arithmetic for the polynomials P_N and Q_N, see §3 for illustration of the algorithm when $N = 29$.

Then, we factored Q_N and h_N, respectively, into irreducible polynomials using a combination of an approximate root finding algorithm of PARI/GP [**15**] and our own C-code that converts the approximate roots into irreducible polynomials. Thereby, all subsets of approximate roots are used to compute approximate polynomials which are then rounded to polynomials with integer coefficients and tested in integer arithmetic whether they indeed divide h_N. Furthermore, our C-code inspects the set of divisors and selects the irreducible polynomials.

The factorization of the polynomials $h_N(y)$ into irreducible polynomials over $\mathbb{Z}$ is listed in Table 1, which we quote from [**10**]. For each irreducible polynomial which divides h_N, we used RadiRoot [**5**] based on GAP [**7**] to find exact values for the roots of h_N expressed in terms of radicals.

On the other hand, using the q-expansions of j_N, which were obtained in [**8**], we have published a list of approximate values of $j_N(e)$ for $e \in \mathcal{E}_N$ [**10**, Appendix C]. Finally, pairing the values of the roots of h_N with the approximate values of $j_N(e)$ results in exact expressions for the values of $j_N(e)$ (see §5).

In all instances, we found that $j_N(e)$ can be expressed in radicals.

3. An example: $N = 29$

Let us consider the case when $N = 29$. In [**8**] the Hauptmodul $j_{29}(z)$ was computed in terms of the Eisenstein series $E_k^{(29)}$ of weight k and the Kronecker limit function Δ_{29}^6. From that expression, we substituted the q-expansions of the Eisenstein series and of the Kronecker limit function to obtain the q-expansion of the Hauptmodul $j_{29}(z)$. From [**4**] we have that the surface X_{29} has seven elliptic points; therefore, we need to consider the q-expansion of j_{29} out to $O(q^{26})$. Using the results obtained in [**8**], the desired q-expansion for j_{29} is

$j_{29}(z) = q^{-1} + 3q + 4q^2 + 7q^3 + 10q^4 + 17q^5 + 22q^6 + 32q^7 + 44q^8 + 62q^9 + 80q^{10} + 112q^{11} + 144q^{12} + 193q^{13} + 248q^{14} + 323q^{15} + 410q^{16} + 530q^{17} + 664q^{18} + 845q^{19} + 1054q^{20} + 1324q^{21} + 1634q^{22} + 2037q^{23} + 2498q^{24} + 3082q^{25} + O(q^{26})$.

By computing the q-expansion of equation (1.1b) and comparing coefficients for $q^{-15}, \ldots, q^{10}$, one gets a set of 26 linear equations for the 26 unknown coefficients of P_{29} and Q_{29}. As it turns out, these equations are linearly independent and, after solving them, we arrive at the expressions that

$P_{29}(y) = y^{12} + 8y^{11} + 38y^{10} + 296y^9 + 2091y^8 + 9000y^7 + 24526y^6 + 45520y^5 + 59625y^4 + 53152y^3 + 28472y^2 + 7248y + 624$

and

$Q_{29}(y) = 2y^{14} + 16y^{13} - 20y^{12} - 608y^{11} - 2122y^{10} + 968y^9 + 27740y^8 + 95176y^7 + 175058y^6 + 197896y^5 + 140088y^4 + 60736y^3 + 15584y^2 + 2176y + 128.$

Note that P_{29} and $\frac{1}{2}Q_{29}$ are monic polynomials with integer coefficients.

For each square-free integer N such that the arithmetic group $\Gamma_0(N)^+$ has genus zero, we have evaluated the polynomials P_N and Q_N as described above. The polynomials P_N and Q_N are listed in [**10**, Appendix A]. We immediately observe that all coefficients are integers. Moreover, P_N and $\frac{1}{2}Q_N$ are monic polynomials with integer coefficients, and we conclude that their roots are algebraic integers. Therefore, all values of Hauptmoduli j_N at elliptic points e are algebraic integers.

4. Hauptmodul values at elliptic points

Having proven that the values of j_N at elliptic points are algebraic integers [**10**], we computed these algebraic integers explicitly. As above, let us write $h_N(y) := \left(\frac{1}{2}Q_N(y)\right)^{1/2}$. From Theorem 2.1(a), the function h_N is a monic polynomial and has the same roots as the polynomial Q_N.

We begin by factoring h_N into irreducible polynomials and use computer algebra to find explicit expressions for the roots in terms of radicals. From the expressions for Q_N, we obtain the following list for h_N, see Table 1.

Table 1: The list of monic polynomials h_N factored into irreducible polynomials.

$h_1(y) = (y + 744)(y - 984)$
$h_2(y) = (y + 104)(y - 152)$
$h_3(y) = (y + 42)(y - 66)$
$h_5(y) = (y + 16)(y^2 - 12y - 464)$
$h_6(y) = (y + 14)(y + 10)(y - 22)$
$h_7(y) = (y + 10)(y + 9)(y - 18)$
$h_{10}(y) = (y + 8)(y + 4)(y - 12)$
$h_{11}(y) = (y + 6)(y^3 - 2y^2 - 76y - 212)$
$h_{13}(y) = (y + 4)(y + 3)(y^2 - 4y - 48)$
$h_{14}(y) = (y + 6)(y + 2)(y^2 - 6y - 23)$
$h_{15}(y) = (y + 4)(y - 8)(y^2 + 6y + 13)$
$h_{17}(y) = (y + 2)(y^4 + 2y^3 - 39y^2 - 176y - 212)$
$h_{19}(y) = (y + 4)(y + 3)(y^3 - 4y^2 - 16y - 12)$
$h_{21}(y) = (y + 4)(y - 0)(y^2 - 2y - 27)$
$h_{22}(y) = (y + 2)(y - 6)(y^3 + 6y^2 + 8y + 4)$
$h_{23}(y) = (y^3 + 6y^2 + 11y + 7)(y^3 - 2y^2 - 17y - 25)$

$h_{26}(y) = (y+4)(y-0)(y^3-2y^2-15y-16)$
$h_{29}(y) = (y+2)(y^6+2y^5-17y^4-66y^3-83y^2-32y-4)$
$h_{30}(y) = (y+4)(y+3)(y-0)(y-1)(y-5)$
$h_{31}(y) = (y-0)(y^3+4y^2+3y+1)(y^3-17y-27)$
$h_{33}(y) = (y-0)(y^2-2y-11)(y^3+4y^2+8y+4)$
$h_{34}(y) = (y+2)(y+1)(y^2+3y-2)(y^2-5y+2)$
$h_{35}(y) = (y+2)(y^3-2y^2-4y-20)(y^2+2y+5)$
$h_{38}(y) = (y-0)(y^3+4y^2+4y+4)(y^3-2y^2-7y-8)$
$h_{39}(y) = (y+3)(y-1)(y^2+3y-1)(y^2-5y+3)$
$h_{41}(y) = (y-0)(y^8+4y^7-8y^6-66y^5-120y^4-56y^3+53y^2+36y-16)$
$h_{42}(y) = (y+3)(y-0)(y-1)(y-4)(y^2+3y+4)$
$h_{46}(y) = (y^2-2y-7)(y^3+2y^2+y+1)(y^3+2y^2-3y+1)$
$h_{47}(y) = (y^5+4y^4+7y^3+8y^2+4y+1)(y^5-5y^3-20y^2-24y-19)$
$h_{51}(y) = (y+2)(y^3-2y^2-4y-4)(y^4+2y^3+3y^2-2y+1)$
$h_{55}(y) = (y+1)(y^2+3y+1)(y^2-5y+5)(y^3+3y^2-y-7)$
$h_{59}(y) = (y^3+2y^2+1)(y^9+2y^8-4y^7-21y^6-44y^5-60y^4-61y^3-46y^2-24y-11)$
$h_{62}(y) = (y^3+4y^2+5y+3)(y^3+y-1)(y^4-2y^3-3y^2-4y+4)$
$h_{66}(y) = (y+3)(y-0)(y-1)(y^2-y-8)(y^3-4y+4)$
$h_{69}(y) = (y^3+4y^2+7y+5)(y^3-y+1)(y^4-2y^3-5y^2+6y-3)$
$h_{70}(y) = (y+2)(y+1)(y-3)(y^2-y+2)(y^3+2y^2+4)$
$h_{71}(y) = (y^7+4y^6+5y^5+y^4-3y^3-2y^2+1)(y^7-7y^5-11y^4+5y^3+18y^2+4y-11)$
$h_{78}(y) = (y+1)(y-3)(y^2+y+1)(y^2+y-3)(y^3+y^2-4)$
$h_{87}(y) = (y^3+2y^2+3y+3)(y^3-2y^2-y-1)(y^6+2y^5+7y^4+6y^3+13y^2+4y+8)$
$h_{94}(y) = (y^4-2y^3-3y^2+4y-4)(y^5+4y^4+3y^3-2y^2+2y+5)(y^5-y^3+2y^2-2y+1)$
$h_{95}(y) = (y-1)(y^3+y^2-y+3)(y^4+y^3-2y^2+2y-1)(y^4+y^3-6y^2-10y-5)$
$h_{105}(y) = (y-1)(y^2+3y+3)(y^2-y-1)(y^2-y-5)(y^3+y^2-y-5)$
$h_{110}(y) = (y-1)(y^2+y+3)(y^2+y-1)(y^3+y^2+3y-1)(y^3-y^2-8)$
$h_{119}(y) = (y^4+2y^3+3y^2+6y+5)(y^5+2y^4+3y^3+6y^2+4y+1)(y^5-2y^4+3y^3-6y^2-7)$

Roots for the linear and quadratic factors in h_N are easily obtained using, at most, the quadratic formula. In [**10**, Appendix B], for each level N, we state each irreducible factor of h_N of degree 3 and higher and then write the roots in terms of radicals. In all instances, $\zeta_n = \exp(2\pi i/n)$ denotes the primitive n-th root of unity.

For each level N, one then needs to identify different roots of $h_N(y)$ and match the roots with approximate values of the Hauptmoduli $j_N(z)$. In the following example we explain how to match the numbers presented in [**10**, Appendices B and C].

5. An example: $N = 29$

The polynomial h_{29} has seven roots which coincides with the number of elliptic points on X_{29}. One root is $y = -2$, the other roots result from the irreducible polynomial $g(y) := y^6 + 2y^5 - 17y^4 - 66y^3 - 83y^2 - 32y - 4$. Set

$$\omega_1 = \exp\big(\frac{1}{2}[\log(29) + 2\pi i\alpha]\big) \text{ for } \alpha \in \{0, 1\};$$

$$\omega_2 = \exp\big(\frac{1}{2}[\log(-470 + 42\omega_1) + 2\pi i\beta]\big) \text{ for } \beta \in \{0, 1\};$$

and

$$\omega_3 = \exp\big(\frac{1}{3}[\log((-\frac{233}{9}\zeta_3 - \frac{214}{9}\zeta_3^2) + (-\frac{164}{27}\zeta_3 - \frac{65}{27}\zeta_3^2)\omega_1 + (-\frac{427}{618}\zeta_3 - \frac{673}{618}\zeta_3^2)\omega_2 + (-\frac{61}{1854}\zeta_3 - \frac{155}{1854}\zeta_3^2)\omega_1\omega_2) + 2\pi i\gamma]\big)$$

for $\gamma \in \{0, 1, 2\}$. Then the roots of $g(y)$ are

$$\begin{aligned}
y_{\alpha,\beta,\gamma} = -\frac{1}{3} + \frac{1}{3}\omega_1 &+ \left(-\frac{4675663}{69945938}\zeta_3 + \frac{9170235}{34972969}\zeta_3^2\right)\omega_3 \\
&+ \left(\frac{230955}{69945938}\zeta_3 + \frac{1143895}{34972969}\zeta_3^2\right)\omega_1\omega_3 \\
&+ \left(\frac{14913792}{3602215807}\zeta_3 + \frac{7468392}{3602215807}\zeta_3^2\right)\omega_2\omega_3 \\
&+ \left(-\frac{949738}{3602215807}\zeta_3 - \frac{2272628}{3602215807}\zeta_3^2\right)\omega_1\omega_2\omega_3 \\
&+ \left(\frac{81809110004592}{1223108560674961}\zeta_3 - \frac{84662907394671}{1223108560674961}\zeta_3^2\right)\omega_3^2 \\
&+ \left(\frac{15630127276428}{1223108560674961}\zeta_3 + \frac{20828508011880}{1223108560674961}\zeta_3^2\right)\omega_1\omega_3^2 \\
&+ \left(\frac{291290361453255}{125980181749520983}\zeta_3 + \frac{146530487024733}{251960363499041966}\zeta_3^2\right)\omega_2\omega_3^2 \\
&+ \left(-\frac{109445587541901}{125980181749520983}\zeta_3 - \frac{134076636577647}{251960363499041966}\zeta_3^2\right)\omega_1\omega_2\omega_3^2.
\end{aligned}$$

We can choose two signs for ω_1 and ω_2 each, and three branches for ω_3. In total, there are 12 possible combinations of signs and branches. These 12 combinations result in the 6 different roots of the polynomial under consideration, with each root repeated twice; see Table 2.

Table 2: Numerical approximations of the algebraic expression for the roots of the irreducible polynomial $y^6 + 2y^5 - 17y^4 - 66y^3 - 83y^2 - 32y - 4$ for all 12 different combinations of signs and branches.

α	β	γ	numerical approximation of $y_{\alpha,\beta,\gamma}$
0	0	0	$-0.26966805687175861246116334 + i0.07411556647545578849796605 8$
0	0	1	$4.92450092087802125617303717$
0	0	2	$-0.26966805687175861246116334 - i0.074115566475455788497966058$
0	1	0	$-0.26966805687175861246116334 - i0.074115566475455788497966058$
0	1	1	$4.92450092087802125617303717$
0	1	2	$-0.26966805687175861246116334 + i0.074115566475455788497966058$
1	0	0	$-1.66789684020963080052582196 + i0.789806563261420529068981098$
1	0	1	$-1.66789684020963080052582196 - i0.789806563261420529068981098$
1	0	2	$-3.04937112671524243019906657$
1	1	0	$-1.66789684020963080052582196 - i0.789806563261420529068981098$
1	1	1	$-1.66789684020963080052582196 + i0.789806563261420529068981098$
1	1	2	$-3.04937112671524243019906657$

The surface X_{29} has seven elliptic points $e \in \mathcal{E}_{29}$, all of order two. If we identify the numerical values of the Hauptmodul at the elliptic points with the values of the roots of h_{29}, we determine which value of $j_{29}(e)$ corresponds to which root. Table 3 summarizes the end results.

Table 3: Identification of the values of $j_{29}(e)$ with the exact roots of the polynomial h_{29}.

$j_{29}(e)$	numerical approximation	exact value
$j_{29}(i\frac{\sqrt{29}}{29})$	4.92450092088	$y_{0,0,1}$
$j_{29}(\frac{1}{5}+i\frac{\sqrt{29}}{145})$	$-0.26966805687 - i0.07411556648$	$y_{0,0,2}$
$j_{29}(\frac{1}{3}+i\frac{\sqrt{29}}{87})$	$-1.66789684021 - i0.78980656326$	$y_{1,0,1}$
$j_{29}(\frac{12}{29}+i\frac{1}{29})$	-2.00000000000	-2
$j_{29}(\frac{1}{2}+i\frac{\sqrt{29}}{58})$	-3.04937112672	$y_{1,0,2}$
$j_{29}(\frac{2}{3}+i\frac{\sqrt{29}}{87})$	$-1.66789684021 + i0.78980656326$	$y_{1,0,0}$
$j_{29}(\frac{4}{5}+i\frac{\sqrt{29}}{145})$	$-0.26966805687 + i0.07411556648$	$y_{0,0,0}$

6. Further evaluations

In the case when the group $\Gamma_0(N)^+$ is not a genus zero group, two functions, say $x_N(z)$ and $y_N(z)$, are needed to generate the function field of all meromorphic $\Gamma_0(N)^+$-invariant functions, see [**8**] and [**9**] for an explicit evaluation of the q-expansions of $x_N(z)$ and $y_N(z)$ when genus of the X_N is equal to 1, 2 and 3. In this setting it is possible to adjust the above methods and explicitly evaluate $x_N(e)$ at all elliptic points e using the equation satisfied by the Schwarzian derivative of x_N.

For illustration, assume that genus is one. Then x_N' vanishes at all elliptic points of X_N and possibly at at most three other points, say a_i, $i = 1,2,3$ of order $m(a_i)$. Let $m_N(e)$ denote the order of vanishing of $x_N'(z)$ at $e \in \mathcal{E}_N$. Now, $S(x_N)(z)(x_N'(z))^2$ is a weight eight holomorphic form on $\mathbb{H}$ whose only pole is at $i\infty$ and of order four. Furthermore, its q-expansion begins with $(2\pi)^4 \cdot \frac{-1}{2q^4}$. Moreover, the set of poles of the weight zero form $S(x_N)(z)/(x_N'(z))^2$ is exactly $\mathcal{E}_N \cup \{a_1, a_2, a_3\}$ with orders $2(m_N(e)+1)$, for $e \in \mathcal{E}_N$ and $2(m(a_i)+1)$, $i = 1,2,3$.

Let

$$Q_N(y) = 2\prod_{e\in\mathcal{E}_N}(y - x_N(e))^2 \prod_{i=1}^{3}(y - x_N(a_i))^2.$$

Then the polynomial $Q_N(x_N(z))$ vanishes exactly at points of the set $\mathcal{E}_N \cup \{a_1, a_2, a_3\}$ with the order of vanishing equal to $2(m_N(e)+1)$, for $e \in \mathcal{E}_N$ and $2(m(a_i)+1)$, $i = 1,2,3$. Proceeding analogously as above, we get that

$$\text{(6.1)} \qquad \frac{S(x_N)(z)(x_N'(z))^2}{(x_N'(z))^4} = -\frac{P_{N,0}(x_N(z)) + P_{N,1}(x_N(z))y_N(z)}{Q_N(x_N(z))},$$

where $\deg(P_{N,0}(y)) = \deg(Q_N(y)) - 2$, while the degree of $P_{N,1}$ is strictly less than the degree of $P_{N,0}$. Moreover, the lead coefficient in $P_{N,0}$ is one.

Using the q-expansions of $x_N(z)$ and $y_N(z)$ from [**8**], one may solve the equation

$$\text{(6.1a)} \qquad \frac{1}{2}Q_N(x_N)\frac{2S(x_N)(x_N')^2}{(2\pi)^4} + (P_{N,0}(x_N) + P_{N,1}(x_N)y_N)\frac{(x_N')^4}{(2\pi)^4} = 0$$

for the coefficients of $P_{N,0}(x)$, $P_{N,1}(x)$ and $Q_N(x)$. As above the coefficients of $P_{N,0}(x)$, $P_{N,1}(x)$ and Q_N are obtained by setting each coefficient in the q-expansion of (6.1a) to zero. The evaluation of $x_N(e)$ for all $e \in \mathcal{E}_N$ is conducted in the way explained in §5.

We believe that this method of evaluation of $x_N(e)$ yields generating polynomials with small coefficients of corresponding class fields of imaginary quadratic orders over $\mathbb{Q}[\sqrt{-N}]$, i.e. gives rise to "higher genus" explicit class field theory. Such questions are addressed in [**11**].

Acknowledgments

All the presented material grew out of discussions and collaborations with Jay Jorgenson and Lejla Smajlović. It is my honour to acknowledge Prof. Jay Jorgenson and Prof. Lejla Smajlović for their inspiring ideas.

References

[1] I. Chen and N. Yui, *Singular values of Thompson series*, Groups, difference sets, and the Monster (Columbus, OH, 1993), Ohio State Univ. Math. Res. Inst. Publ., vol. 4, de Gruyter, Berlin, 1996, pp. 255–326. MR1400423

[2] S. Y. Choi and J. K. Koo, *Class fields from the fundamental Thompson series of level $N = o(g)$*, J. Korean Math. Soc. **42** (2005), no. 2, 203–222, DOI 10.4134/JKMS.2005.42.2.203. MR2121496

[3] D. Cox, J. McKay, and P. Stevenhagen, *Principal moduli and class fields*, Bull. London Math. Soc. **36** (2004), no. 1, 3–12, DOI 10.1112/S0024609303002583. MR2011972

[4] C. J. Cummins, *Congruence subgroups of groups commensurable with* $\mathrm{PSL}(2,\mathbb{Z})$ *of genus 0 and 1*, Experiment. Math. **13** (2004), no. 3, 361–382. MR2103333

[5] A. Distler, *Ein Algorithmus zum Lösen einer Polynomgleichung durch Radikale*, Diploma Thesis, TU Braunschweig, 2005.

[6] T. Granlund, *GMP – The GNU Multiple Precision Arithmetic Library, Version 5.0.5*; 2012, `https://gmplib.org/`

[7] The GAP Group, *GAP – Groups, Algorithms, and Programming, Version 4.7.9*; 2015, `http://www.gap-system.org`

[8] J. Jorgenson, L. Smajlović, and H. Then, *Kronecker's limit formula, holomorphic modular functions, and q-expansions on certain arithmetic groups*, Exp. Math. **25** (2016), no. 3, 295–319, DOI 10.1080/10586458.2015.1053161. MR3478329

[9] J. Jorgenson, L. Smajlović, and H. Then, data page `http://www.efsa.unsa.ba/~lejla.smajlovic/`

[10] J. Jorgenson, L. Smajlović, and H. Then, The Hauptmodul at elliptic points of certain arithmetic groups, arXiv:1602.07426 (2016).

[11] J. Jorgenson, L. Smajlović, and H. Then, On the evaluation of singular invariants for canonical generators of certain genus one arithmetic groups, to appear in Exp. Math. (2018), https://doi.org/10.1080/10586458.2017.1422161

[12] B. H. Lian and J. L. Wiczer, Genus zero modular functions, arXiv:math/0611291v1 (2006).

[13] D. Masser, *Heights, transcendence, and linear independence on commutative group varieties*, Diophantine approximation (Cetraro, 2000), Lecture Notes in Math., vol. 1819, Springer, Berlin, 2003, pp. 1–51, DOI 10.1007/3-540-44979-5_1. MR2009828

[14] P. J. Nahin, *Dr. Euler's fabulous formula: Cures many mathematical ills*, Princeton University Press, Princeton, NJ, 2011. Paperback reprint of the 2006 edition [MR2223305]. MR2799221

[15] The PARI Group, *PARI/GP version 2.5.1*, Bordeaux, 2014, `http://pari.math.u-bordeaux.fr/`

[16] T. Schneider, *Arithmetische Untersuchungen elliptischer Integrale* (German), Math. Ann **113** (1937), no. 1, 1–13, DOI 10.1007/BF01571618. MR1513075

Alemannenweg 1, 89537 Giengen, Germany

Contemporary Mathematics
Volume **732**, 2019
https://doi.org/10.1090/conm/732/14802

Torsion groups of elliptic curves over quadratic fields

Antonela Trbović

ABSTRACT. We describe methods for determining all possible torsion subgroups of an elliptic curve defined over a fixed quadratic field. We give a brief overview of known results and methods and state theorems on all torsion subgroups that can appear over four real quadratic fields.

Let K be a number field and let $E(K)$ be an elliptic curve defined over K. It is known, by the Mordell-Weil theorem, that $E(K)$ is a finitely generated abelian group. Now, by the structure theorem for finitely generated abelian groups, we know that $E(K)$ is isomorphic to $E(K)_{tors} \oplus \mathbb{Z}^r$, where r is a non-negative integer, the rank of $E(K)$, and $E(K)_{tors}$ is the torsion subgroup. There is no known algorithm guaranteed to determine the rank of an elliptic curve, there exist only algorithms for computing upper and lower bounds for the rank, but in comparison, the study of torsion is effective.

In the case of $K = \mathbb{Q}$, Mazur's theorem [**M**] gives us the list of all possible torsion subgroups of $E(\mathbb{Q})$:

$$\mathbb{Z}/n\mathbb{Z}, \quad n = 1, ..., 10, 12,$$

$$\mathbb{Z}/2\mathbb{Z} \oplus \mathbb{Z}/2n\mathbb{Z}, \quad n = 1, 2, 3, 4.$$

Similarly, we have a result by Kamienny, Kenku and Momose [**K, KM**] concerning possible torsion subgroups of elliptic curves defined over any quadratic field: the complete list consists of the following 26 groups:

$$\mathbb{Z}/n\mathbb{Z}, \quad n = 1, ..., 16, 18,$$

$$\mathbb{Z}/2\mathbb{Z} \oplus \mathbb{Z}/2n\mathbb{Z}, \quad n = 1, ..., 6,$$

$$\mathbb{Z}/3\mathbb{Z} \oplus \mathbb{Z}/3n\mathbb{Z}, \quad n = 1, 2,$$

$$\mathbb{Z}/4\mathbb{Z} \oplus \mathbb{Z}/4\mathbb{Z}.$$

We are interested in which torsion subgroups occur when we fix a certain quadratic field K. In order to see what happens over a fixed field, one would have to go through each of the 26 groups mentioned above and check whether that is a possible torsion.

First of all, if K is a fixed quadratic field, the groups mentioned in Mazur's theorem have to appear as a possible torsion group over K. Also, the properties of Weil

---

2010 *Mathematics Subject Classification.* Primary 11G05.

The author was supported by the QuantiXLie Centre of Excellence, a project co-financed by the Croatian Government and European Union through the European Regional Development Fund - the Competitiveness and Cohesion Operational Programme (Grant KK.01.1.1.01.0004).

pairing settle the question on what happens with the groups $\mathbb{Z}/3\mathbb{Z}\oplus\mathbb{Z}/3n\mathbb{Z}$, $n = 1, 2$ and $\mathbb{Z}/4\mathbb{Z} \oplus \mathbb{Z}/4\mathbb{Z}$ over K.

For the rest of the groups $\mathbb{Z}/m\mathbb{Z}\oplus\mathbb{Z}/n\mathbb{Z}$, we have to check if the modular curve $X_1(m,n)$ has $K-$rational points that are not cusps. If it does, we know that an elliptic curve with that torsion subgroup exists over K, since points on modular curves $X_1(m,n)$ parameterize isomorphism classes of elliptic curves with torsion $\mathbb{Z}/m\mathbb{Z} \oplus \mathbb{Z}/n\mathbb{Z}$.

If $X_1(m,n)$ is an elliptic curve we have to determine the rank over K. If the rank is positive, we have infinitely many elliptic curves with torsion $\mathbb{Z}/m\mathbb{Z}\oplus\mathbb{Z}/n\mathbb{Z}$ over K. If the rank is 0, we have to see if there exists a torsion point that is not a cusp.

If $X_1(m,n)$ is a hyperelliptic curve, we act similarly, but with the Jacobian of the curve. We can sometimes compute the rank of the Jacobian in MAGMA [**BCP**] by using 2-descent described in [**S**]. This will often work, even though it is not an algorithm.

A detailed description of these methods can be found in [**KN**]. One could also take a look at [**BN**] where different techniques are presented for ruling out torsion subgroups.

Now we state some results obtained in ways described above.

THEOREM 0.1. *Torsion subgroups appearing over* $\mathbb{Q}(\sqrt{6})$ *are the following:*

$$\mathbb{Z}/n\mathbb{Z}, \quad n = 1, ..., 11, 12, 14,$$
$$\mathbb{Z}/2\mathbb{Z} \oplus \mathbb{Z}/2n\mathbb{Z}, \quad n = 1, 2, 3, 4, 5, 6.$$

THEOREM 0.2. *Torsion subgroups appearing over* $\mathbb{Q}(\sqrt{7})$ *are the following:*

$$\mathbb{Z}/n\mathbb{Z}, \quad n = 1, ..., 11, 12, 14, 15,$$
$$\mathbb{Z}/2\mathbb{Z} \oplus \mathbb{Z}/2n\mathbb{Z}, \quad n = 1, 2, 3, 4, 6.$$

THEOREM 0.3. *Torsion subgroups appearing over* $\mathbb{Q}(\sqrt{11})$ *are the following:*

$$\mathbb{Z}/n\mathbb{Z}, \quad n = 1, ..., 11, 12, 15,$$
$$\mathbb{Z}/2\mathbb{Z} \oplus \mathbb{Z}/2n\mathbb{Z}, \quad n = 1, 2, 3, 4.$$

THEOREM 0.4. *Torsion subgroups appearing over* $\mathbb{Q}(\sqrt{13})$ *are the following:*

$$\mathbb{Z}/n\mathbb{Z}, \quad n = 1, ..., 11, 12, 15,$$
$$\mathbb{Z}/2\mathbb{Z} \oplus \mathbb{Z}/2n\mathbb{Z}, \quad n = 1, 2, 3, 4, 5, 6.$$

REMARK 0.5. This paper was the starting point, and can be consider an anouncement of the author's subsequent paper [**T**] in which more can be found about this topic. We omit the proofs of theorems 0.1-0.4 as the detailed descriptions of methods used in the proofs, and also MAGMA code used for computations in both papers can be found there.

References

[BCP] W. Bosma, J. Cannon, and C. Playoust, *The Magma algebra system. I. The user language*, J. Symbolic Comput. **24** (1997), no. 3-4, 235–265, DOI 10.1006/jsco.1996.0125. Computational algebra and number theory (London, 1993). MR1484478

[BN] P. Bruin and F. Najman, *A criterion to rule out torsion groups for elliptic curves over number fields*, Res. Number Theory **2** (2016), Art. 3, 13, DOI 10.1007/s40993-015-0031-5. MR3501016

[K] S. Kamienny, *Torsion points on elliptic curves and q-coefficients of modular forms*, Invent. Math. **109** (1992), no. 2, 221–229, DOI 10.1007/BF01232025. MR1172689

[KM] M. A. Kenku and F. Momose, *Torsion points on elliptic curves defined over quadratic fields*, Nagoya Math. J. **109** (1988), 125–149, DOI 10.1017/S0027763000002816. MR931956

[KN] S. Kamienny and F. Najman, *Torsion groups of elliptic curves over quadratic fields*, Acta Arith. **152** (2012), no. 3, 291–305, DOI 10.4064/aa152-3-5. MR2885789

[M] B. Mazur, *Rational isogenies of prime degree (with an appendix by D. Goldfeld)*, Invent. Math. **44** (1978), no. 2, 129–162, DOI 10.1007/BF01390348. MR482230

[S] M. Stoll, *Implementing 2-descent for Jacobians of hyperelliptic curves*, Acta Arith. **98** (2001), no. 3, 245–277, DOI 10.4064/aa98-3-4. MR1829626

[T] A. Trbović, *Torsion groups of elliptic curves over quadratic fields* $\mathbb{Q}(\sqrt{d}), 0 < d < 100$, preprint, https://arxiv.org/pdf/1806.05993.pdf

DEPARTMENT OF MATHEMATICS, UNIVERSITY OF ZAGREB, BIJENIČKA CESTA 30, 10000 ZAGREB, CROATIA

Email address: antonela.trbovic@math.hr

Contemporary Mathematics
Volume **732**, 2019
https://doi.org/10.1090/conm/732/14813

Maass space for lifting from SL(2,$\mathbb{R}$) to GL(2,B) over a division quaternion algebra

Siddhesh Wagh

ABSTRACT. Muto, Narita and Pitale constructed liftings from SL(2,$\mathbb{R}$) to GL(2,B) using relations between coefficients analogous to Saito-Kurokawa lifts. We want to characterize these lifts explicitly in terms of recurrence relations between their Fourier coefficients.

1. Introduction:

Saito and Kurokawa ([**1**]) found numerical counterexamples to the Generalized Ramanujan Conjecture (GRC). Maass, Adrianov and Zagier ([**1**]) showed that the lifts from $S_{2k-2}(SL_2(\mathbb{Z}))$ to $S_k^{(2)}(Sp_4(\mathbb{Z}))$ always violate GRC. Explicitly, if $f \in S_{2k-2}(SL_2(\mathbb{Z}))$ is mapped to $g(z) = \sum_n c(n)e^{2\pi inz}$ in the Kohnen plus space $S^+_{k-1/2}(\Gamma_0(4))$, the Saito-Kurokawa (S-K) lift is a Siegel modular form $F_f \in S_k^{(2)}(Sp_4(\mathbb{Z}))$ with Fourier coefficients $A(T)$ given by

$$A\left(T = \begin{bmatrix} n & r/2 \\ r/2 & m \end{bmatrix}\right) = \sum_{d|(n,r,m)} d^{k-1}c\left(\frac{\det(2T)}{d^2}\right). \tag{1.1}$$

THEOREM 1.1 (Maass). *F is a S-K lift $\Leftrightarrow$ Its Fourier coefficients satisfy the recurrence relation (Maass relations)*

$$A(n,r,m) = \sum_{d|(n,r,m)} d^{k-1}A\left(\frac{nm}{d^2}, \frac{n}{d}, 1\right) \qquad (\forall n \geq 0, m \geq 0, r \in \mathbb{Z}).$$

Muto, Narita and Pitale ([**2**]) constructed similar lifting from SL(2,$\mathbb{R}$) to GL(2,B). Let B be the Hamiltonian quaternions and let $\mathcal{O}$ be the Hurwitz order. Define $\varpi_2 := (1+i)$ and $S := \varpi_2\mathcal{O}$. It is easy to show that every $\beta \in S$ can be written uniquely as $\beta = \varpi_2^u d\beta_0$ where $\beta_0 \in S_{prim} := \{\beta = (a + ib + jc + ijd) \in S \mid g.c.d.(a,b,c,d) = 1\}$. Let $\epsilon \in \{\pm 1\}$ be the eigenvalue of the Atkin-Lehner involution.

Then, in terms of the Fourier coefficients $c(N)$ of $f \in S(\Gamma_0(2); -(\frac{1}{4}+\frac{r^2}{4}))$, we can define the Fourier coefficients $A(\beta)$ for $\beta \in S$ of the lift $F_f \in \mathcal{M}(GL_2(\mathcal{O}); r)$ by

$$A(\beta) := |\beta| \sum_{t=0}^{u} \sum_{n|d} (-\epsilon)^t c\Big(-\frac{|\beta|^2}{2^{t+1}n^2}\Big) \tag{1.2}$$

These relations (1.2) are similar to those in (1.1). However, these relations are obtained from the Maass converse theorem rather than through Jacobi expansion.

2. Current work:

We want to characterize the lifts in terms of recurrence relations between Fourier coefficients. To find the Maass relations in this case, we cannot use the techniques of Maass, since they rely on the intermediate spaces of Jacobi forms, an object unavailable to us.

The first observation we make is that $A(\beta)$ doesn't actually depend fully on β but only on $|\beta|, u$ and d as defined before. If $|\beta| = |\beta'|$ and they have the same u and d then $A(\beta) = A(\beta')$. Hence we can write the Fourier coefficients as $A(K, u, d)$ where $K = |\beta|^2$ and u and d as defined before.

Second, we notice that the condition which asserts that β is primitive change be written as $|\beta|^2 \equiv 2 \mod 4$ and $d = 1$. Now, using the fact that $A(\beta)$'s are given by (1.2) above, for $N = 4^a b$ the Fourier coefficients $c(N)$ of $f \in S(\Gamma_0(2); -(\frac{1}{4} + \frac{r^2}{4}))$ can be written as

$$c(-N) = \begin{cases} \frac{A(2N,2a,1)}{\sqrt{2N}} + \epsilon \frac{A(N,2a-1,1)}{\sqrt{N}}, & \text{if } b \equiv 1,3 \bmod 4 \\ \frac{A(2N,2a+1,1)}{\sqrt{2N}} + \epsilon \frac{A(N,2a,1)}{\sqrt{N}}, & \text{if } b \equiv 2 \bmod 4 \end{cases} \tag{2.1}$$

Using this relation, we showed in [**3**]

PROPOSITION 2.1. *Let $A(K, u, d)$ be the Fourier coefficients of the lift $F_f \in \mathcal{M}(GL_2(\mathcal{O}); r)$ a Hecke eigenform at $p = 2$ with eigenvalue $-3\sqrt{2}\epsilon$ for $\epsilon \in \{\pm 1\}$. Then they satisfy the recurrence relation*

$$A(K, u, d) = \sum_{n|d} nA(K/n^2, u, 1) \tag{2.2}$$

and

$$A(K, u, d) = \frac{-3\epsilon}{\sqrt{2}} A(\frac{K}{2}, u-1, d) - A(\frac{K}{4}, u-2, d) \tag{2.3}$$

Now that we have possible criterion for the Maass relations, we were able to show

THEOREM 2.2. *F satisfies relations* (2.2) *and* (2.3) $\Leftrightarrow$ *F is lift.*

The proof uses representation theory and Jacquet-Langlands correspondence ([**3**]).

References

[1] D. Lanphier, *A local criterion for the Saito-Kurokawa lifting of cuspforms with characters*, Math. Res. Lett. **16** (2009), no. 3, 421–438, DOI 10.4310/MRL.2009.v16.n3.a4. MR2511623

[2] M. Muto, H.-A. Narita, and A. Pitale, *Lifting to GL(2) over a division quaternion algebra, and an explicit construction of CAP representations*, Nagoya Math. J. **222** (2016), no. 1, 137–185, DOI 10.1017/nmj.2016.15. MR3509225

[3] S. Wagh - *Maass space for lifting from SL(2) to GL(2) over a division quaternion algebra*, arXiv:1810.11004 [math.NT]

UNIVERSITY OF OKLAHOMA, NORMAN, OKLAHOMA 73019

Contemporary Mathematics
Volume **732**, 2019
https://doi.org/10.1090/conm/732/14809

On the occurrence of large positive Hecke eigenvalues for GL(2)

Nahid Walji

ABSTRACT. Let π be a self-dual cuspidal automorphic representation for $\mathrm{GL}(2)/\mathbb{Q}$. We show that there exists a positive upper Dirichlet density of primes at which the associated Hecke eigenvalues of π are larger than a specified positive constant.

1. Introduction

Let π be a cuspidal automorphic representation for $\mathrm{GL}(2)/\mathbb{Q}$ that is self-dual. To each prime p at which π is not ramified is associated a Hecke eigenvalue, denoted by $a_p = a_p(\pi)$. The values taken by sequences $(a_p(\pi))_p$ of Hecke eigenvalues have been long-studied from various points of view. In 1994, J.-P. Serre asked (see appendix of [**8**]) whether it is possible to find positive constants c, c' such that, for all $\epsilon > 0$, there exist infinitely many a_p greater than $c - \epsilon$ and infinitely many a_p less than $-c' + \epsilon$. He then proved such results for the case of (certain) modular forms, and asked if similar results can be shown to hold in the case of Maass forms.

In [**10**] we proved, for any self-dual cuspidal automorphic representation π for $\mathrm{GL}(2)/\mathbb{Q}$, that for any positive ϵ there exist infinitely many primes p such that

$$a_p > 0.905... - \epsilon$$

and if π is non-dihedral, then there exist infinitely many primes p such that

$$a_p < -1.164... + \epsilon$$

(precise expressions for the constants are available in [**10**]). Note that a cuspidal automorphic representation π for GL(2) is said to be *dihedral* if it is associated to a 2-dimensional irreducible Artin representation ρ that is of dihedral type, meaning that the image of ρ in $\mathrm{PGL}_2(\mathbb{C})$ is isomorphic to a dihedral group. Furthermore, π is said to be of *solvable polyhedral type* if it is associated to a 2-dimensional irreducible Artin representation ρ that is of dihedral, tetrahedral, or octahedral type (which means that the projective image of ρ in $\mathrm{PGL}_2(\mathbb{C})$ is isomorphic to a dihedral group, A_4, or S_4, respectively).

A related question is to ask whether it is possible to obtain similar statements for not just an infinitude of primes but a positive upper Dirichlet density of primes.

2010 *Mathematics Subject Classification.* Primary 11F30.

Recall that the *upper Dirichlet density* of a set S of primes is defined to be

$$\overline{\delta}(S) := \limsup_{s\to 1^+} \frac{\sum_{p\in S} p^{-s}}{\log(1/(s-1))}.$$

The results of [**10**] relied on determining lower bounds on the asymptotic growth of certain Dirichlet series, but, because of the lack of knowledge of the Ramanujan conjecture (or indeed the non-existence of any known uniform bound on the Satake parameters), these will not directly lead to a positive density result.

In this paper, we outline a method to circumvent this issue and obtain positive upper Dirichlet density results. By way of example, we have

THEOREM 1.1. *Let π be a self-dual cuspidal automorphic representation for* GL(2) *over $\mathbb{Q}$ that is not of solvable polyhedral type. Then for any $\epsilon > 0$, the set*

$$\{p \mid a_p(\pi) > 0.778... - \epsilon\}.$$

has an upper Dirichlet density of at least 1/100.

The exact value of the constant in the set condition is $0.36729^{1/4}$, which is determined in Section 3. The method we use would also allow a change in this constant to a smaller value so as to obtain a mild increase in the lower bound of the density. The proof relies in part on the deep work of Gelbart–Jacquet [**2**], Kim–Shahidi [**5, 6**], and Kim [**4**] on the automorphy of symmetric power lifts. In the next section, we will outline the ingredients used in the proof, and in Section 3 we will prove the theorem.

2. Background

The proof will rely on the study of the asymptotic behaviour of various Dirichlet series, which we will briefly outline here and refer the reader to [**10**] for a more detailed explanation.

Given a cuspidal automorphic representation π for GL(2)/$\mathbb{Q}$, let T be the (finite) set consisting of the archimedean place and the finite places at which π is ramified. Then the incomplete L-function (with respect to T) that is associated to π can be defined in a right-half plane via an Euler product:

$$L^T(s,\pi) = \prod_{p\notin T} \det\left(I_2 - A_p(\pi)p^{-s}\right)^{-1},$$

where I_2 is the 2×2 identity matrix and $A_p(\pi) = \mathrm{diag}(\alpha_p(\pi), \beta_p(\pi)) \in \mathrm{GL}_2(\mathbb{C})$ is the matrix of Satake parameters associated to π at p (see Section 3.5 of [**1**] for further background on Satake parameters). Note that the Hecke eigenvalue $a_p(\pi)$ is equal to the sum of the Satake parameters $\alpha_p(\pi)$ and $\beta_p(\pi)$.

For any two cuspidal automorphic representations π_1 for GL(n)/$\mathbb{Q}$ and π_2 for GL(m)/$\mathbb{Q}$, one can define (again in a suitable right-half plane) their incomplete Rankin–Selberg L-function:

$$L^T(s,\pi_1\times\pi_2) = \prod_{p\notin T} \det\left(I_{nm} - A_p(\pi_1)\otimes A_p(\pi_2)p^{-s}\right)^{-1}.$$

This L-function converges absolutely for $\mathrm{Re}(s) > 1$. At $s = 1$ it has a simple pole iff π_1 is dual to π_2 [**3**], otherwise the L-function is invertible at that point [**7**].

In general one can define, for any positive integer $k \leq 8$, an incomplete kth product L-function as follows:

$$L^T(s,\pi^{\times k}) = \prod_{p\notin T} \det\left(I_{2^k} - A_p(\pi)^{\otimes k} p^{-s}\right)^{-1}.$$

One can also define the following incomplete symmetric power L-functions:

$$L^T(s,\mathrm{Sym}^2\pi) = \prod_{p\notin T} \det\left(I_3 - \begin{pmatrix} \alpha_p^2 & & \\ & \alpha_p\beta_p & \\ & & \beta_p^2 \end{pmatrix} p^{-s}\right)^{-1},$$

$$L^T(s,\mathrm{Sym}^3\pi) = \prod_{p\notin T} \det\left(I_4 - \begin{pmatrix} \alpha_p^3 & & & \\ & \alpha_p^2\beta_p & & \\ & & \alpha_p\beta_p^2 & \\ & & & \beta_p^3 \end{pmatrix} p^{-s}\right)^{-1},$$

$$L^T(s,\mathrm{Sym}^4\pi) = \prod_{p\notin T} \det\left(I_5 - \begin{pmatrix} \alpha_p^4 & & & & \\ & \alpha_p^3\beta_p & & & \\ & & \alpha_p^2\beta_p^2 & & \\ & & & \alpha_p\beta_p^3 & \\ & & & & \beta_p^4 \end{pmatrix} p^{-s}\right)^{-1}.$$

For the kth product L-functions, where $k = 3, 4, 6$, and 8, we have the following L-function identities, using Clebsch–Gordon decompositions (for details of these decompositions for $L^T(s,\pi^{\times 3})$ and $L^T(s,\pi^{\times 4})$ when π has trivial central character, see p74–75 and p70–71 (respectively) of [**9**] and note that the other cases follow in the same way):

$$\begin{aligned}
L^T(s,\pi^{\times 3}) =& L^T(s,\mathrm{Sym}^3\pi)L^T(s,\pi\otimes\omega)^2,\\
L^T(s,\pi^{\times 4}) =& L^T(s,\mathrm{Sym}^4\pi)L^T(s,\mathrm{Sym}^2\pi\otimes\omega)^3 L^T(s,\omega^2)^2,\\
L^T(s,\pi^{\times 6}) =& L^T(s,\mathrm{Sym}^3\pi\times\mathrm{Sym}^3\pi)L^T(s,\mathrm{Sym}^3\pi\times\pi\otimes\omega)^4 L^T(s,\pi\times\pi\otimes\omega^2)^4,\\
L^T(s,\pi^{\times 8}) =& L^T(s,\mathrm{Sym}^4\pi\times\mathrm{Sym}^4\pi)L^T(s,\mathrm{Sym}^4\pi\times\mathrm{Sym}^2\pi\otimes\omega)^6\\
&\cdot L^T(s,\mathrm{Sym}^2\pi\otimes\omega\times\mathrm{Sym}^2\pi\otimes\omega)^9 L^T(s,\mathrm{Sym}^4\pi\otimes\omega^2)^4\\
&\cdot L^T(s,\mathrm{Sym}^2\pi\otimes\omega^3)^{12} L^T(s,\omega^4)^4,
\end{aligned}$$

where ω is the central character of π.

From here on, we assume that π is self-dual and that it is not of solvable polyhedral type. We use the equations above in conjunction with the results of Gelbart–Jacquet [**2**], Kim–Shahidi [**5, 6**], and Kim [**4**] on the automorphy of the symmetric second, third, and fourth power lifts to obtain the following:

For $k = 3, 4, 6$, and 8, the incomplete L-function $L^T(s,\pi^{\times k})$ has an absolutely convergent Euler product for $s > 1$. If k is even, then the incomplete L-function has a pole of order $m(k)$ at $s = 1$, where $m(k) = 2, 5, 14$ for $k = 4, 6, 8$, respectively. If $k = 3$, then the L-function is invertible at $s = 1$.

Using the bounds towards the Ramanujan conjecture obtained by Kim–Sarnak (Appendix 2 of [**4**]), which imply that $|a_p| \leq 2p^{7/64}$ for all primes p, we obtain the following results:

As $s \to 1^+$, for $k = 3$ or 4, we have

$$\sum_{p \notin T} \frac{a_p^k}{p^s} = \log L^T(s, \pi^{\times k}) + O(1)$$

and for $k = 6$ or 8, using the positivity of the coefficients of p^{-ns} in the expansion of $\log L^T(s, \pi^{\times k})$, we have

$$\sum_{p \notin T} \frac{a_p^k}{p^s} \leq \log L^T(s, \pi^{\times k}).$$

Given the orders $m(k)$ of the poles of the (incomplete) product L-functions as determined above, we conclude:

$$\sum_{p \notin T} \frac{a_p^k}{p^s} = \begin{cases} O(1) & \text{for} \quad \mathrm{k} = 3 \\ 2\log(1/(s-1)) + O(1) & \text{for} \quad \mathrm{k} = 4 \end{cases}$$

and

$$\sum_{p \notin T} \frac{a_p^k}{p^s} \leq \begin{cases} 5\log(1/(s-1)) + O(1) & \text{for} \quad \mathrm{k} = 6 \\ 14\log(1/(s-1)) + O(1) & \text{for} \quad \mathrm{k} = 8 \end{cases}$$

as $s \to 1^+$.

We also make use of the following identities:

Let $f(x), g(x)$ be real-valued functions, and fix some point $u \in \mathbb{R}$. Then,

$$\lim_{x \to u^+} \sup (f(x) + g(x)) \leq \lim_{x \to u^+} \sup f(x) + \lim_{x \to u^+} \sup g(x)$$
$$\lim_{x \to u^+} \sup(-f(x)) = - \lim_{x \to u^+} \inf f(x)$$

and furthermore if g, f are non-negative functions, then

$$\lim_{x \to u^+} \inf (f(x) \cdot g(x)) \leq \lim_{x \to u^+} \inf f(x) \cdot \lim_{x \to u^+} \sup g(x)$$
$$\lim_{x \to u^+} \sup (f(x) \cdot g(x)) \leq \lim_{x \to u^+} \sup f(x) \cdot \lim_{x \to u^+} \sup g(x).$$

3. Proof

In this subsection we will prove Theorem 1.1.

First define the sets $A := \{p \text{ prime } \mid a_p > 0\}$ and $B := \{p \text{ prime } \mid a_p \leq 0\}$. From the previous section, we know that

$$\lim_{s \to 1^+} \sup \frac{\sum_p \frac{|a_p|^4}{p^s}}{\log\left(\frac{1}{s-1}\right)} = 2,$$

which implies

$$\lim_{s \to 1^+} \sup \frac{\sum_{p \in A} \frac{|a_p|^4}{p^s}}{\log\left(\frac{1}{s-1}\right)} + \lim_{s \to 1^+} \sup \frac{\sum_{p \in B} \frac{|a_p|^4}{p^s}}{\log\left(\frac{1}{s-1}\right)} \geq 2.$$

Let us define

$$d := \lim_{s \to 1^+} \sup \frac{\sum_{p \in B} \frac{|a_p|^4}{p^s}}{\log\left(\frac{1}{s-1}\right)},$$

and thus we can write

$$\lim_{s\to 1^+}\sup \frac{\sum_{p\in A}\frac{|a_p|^4}{p^s}}{\log\left(\frac{1}{s-1}\right)} \geq 2-d.$$

Now define $S_\beta \subset A$ to be exactly the set of primes p such that $|a_p|^4 \geq (2-d)\beta$, where $0<\beta<1$ is a constant to be fixed later. We will assume that S_β has an upper Dirichlet density smaller than $1/100$.

We have the bound

$$\limsup \left(\frac{\sum_{p\in A-S_\beta}\frac{|a_p|^4}{p^s}}{\log\left(\frac{1}{s-1}\right)}\right) \leq (2-d)\beta.$$

Since

$$\limsup \left(\frac{\sum_{p\in A-S_\beta}\frac{|a_p|^4}{p^s}}{\log\left(\frac{1}{s-1}\right)}\right) + \limsup \left(\frac{\sum_{p\in S_\beta}\frac{|a_p|^4}{p^s}}{\log\left(\frac{1}{s-1}\right)}\right) \geq (2-d),$$

we have

$$\limsup \left(\frac{\sum_{p\in S_\beta}\frac{|a_p|^4}{p^s}}{\log\left(\frac{1}{s-1}\right)}\right) \geq (2-d)\left(1-\beta\right).$$

Using Cauchy–Schwarz

$$\left(\limsup \frac{\sum_{p\in S_\beta}\frac{|a_p|^4}{p^s}}{\log\left(\frac{1}{s-1}\right)}\right)^2 \leq \left(\limsup \frac{\sum_{p\in S_\beta}\frac{|a_p|^8}{p^s}}{\log\left(\frac{1}{s-1}\right)}\right)\left(\limsup \frac{\sum_{p\in S_\beta}\frac{|a_p|^0}{p^s}}{\log\left(\frac{1}{s-1}\right)}\right),$$

where the third limit supremum can be bounded above by $1/100$ and we obtain

$$(2-d)^2\left(1-\beta\right)^2 \leq \frac{14}{100}. \tag{3.1}$$

We appeal to a result from [**10**], which we include here as a lemma.

LEMMA 3.1. *For A, B, and d defined as above, we have*

$$\limsup \frac{\sum_{p\in A}\frac{|a_p|^3}{p^s}}{\log\left(\frac{1}{s-1}\right)} \geq \frac{d^{5/4}}{(14-(2-d)^2)^{1/4}}.$$

PROOF. A proof of this Lemma essentially arises in [**10**]. We include a proof below for the convenience of the reader.

Using Holder's inequality and taking the limit supremum as $s\to 1^+$,

$$\begin{aligned}\lim_{s\to 1^+}\sup \frac{\sum_{p\in A}\frac{|a_p|^3}{p^s}}{\log\left(\frac{1}{s-1}\right)} &\leq \left(\lim_{s\to 1^+}\sup \frac{\sum_{p\in A}\frac{|a_p|^4}{p^s}}{\log\left(\frac{1}{s-1}\right)}\right)^{3/4}\left(\lim_{s\to 1^+}\sup \frac{\sum_{p\in A}\frac{1}{p^s}}{\log\left(\frac{1}{s-1}\right)}\right)^{1/4}\\ &\leq (2-d)^{3/4}\cdot 1^{1/4}.\end{aligned}$$

Similarly,

$$\lim_{s\to 1^+}\sup\frac{\sum_{p\in A}\frac{|a_p|^4}{p^s}}{\log\left(\frac{1}{s-1}\right)}\le\left(\lim_{s\to 1^+}\sup\frac{\sum_{p\in A}\frac{|a_p|^8}{p^s}}{\log\left(\frac{1}{s-1}\right)}\right)^{1/5}\left(\lim_{s\to 1^+}\sup\frac{\sum_{p\in A}\frac{|a_p|^3}{p^s}}{\log\left(\frac{1}{s-1}\right)}\right)^{4/5}$$

$$\Rightarrow\qquad 2-d\le\left(\lim_{s\to 1^+}\sup\frac{\sum_{p\in A}\frac{|a_p|^8}{p^s}}{\log\left(\frac{1}{s-1}\right)}\right)^{1/5}(2-d)^{3/5}$$

$$\Rightarrow\qquad (2-d)^2\le\lim_{s\to 1^+}\sup\frac{\sum_{p\in A}\frac{|a_p|^8}{p^s}}{\log\left(\frac{1}{s-1}\right)},$$

From the results in the previous section, we have

$$\lim_{s\to 1^+}\sup\frac{\sum_{p\in A}\frac{|a_p|^8}{p^s}}{\log\left(\frac{1}{s-1}\right)}+\lim_{s\to 1^+}\inf\frac{\sum_{p\in B}\frac{|a_p|^8}{p^s}}{\log\left(\frac{1}{s-1}\right)}\le 14,$$

and so

$$\lim_{s\to 1^+}\inf\frac{\sum_{p\in B}\frac{|a_p|^8}{p^s}}{\log\left(\frac{1}{s-1}\right)}\le 14-(2-d)^2. \tag{3.2}$$

We also have

$$\sum_{p\in B}\frac{|a_p|^{8/5}|a_p|^{12/5}}{p^s}\le\left(\sum_{p\in B}\frac{|a_p|^8}{p^s}\right)^{1/5}\left(\sum_{p\in B}\frac{|a_p|^3}{p^s}\right)^{4/5}.$$

We divide the equation above by $\log(1/(s-1))$ and take the limit infimum as $s\to 1^+$,

$$\lim_{s\to 1^+}\inf\frac{\sum_{p\in B}\frac{|a_p|^4}{p^s}}{\log\left(\frac{1}{s-1}\right)}\le\left(\lim_{s\to 1^+}\inf\frac{\sum_{p\in B}\frac{|a_p|^8}{p^s}}{\log\left(\frac{1}{s-1}\right)}\right)^{1/5}\left(\lim_{s\to 1^+}\sup\frac{\sum_{p\in B}\frac{|a_p|^3}{p^s}}{\log\left(\frac{1}{s-1}\right)}\right)^{4/5}.$$

We apply equation 3.2 to obtain

$$d\le(14-(2-d)^2)^{1/5}\left(\limsup\frac{\sum_{p\in B}\frac{|a_p|^3}{p^s}}{\log\left(\frac{1}{s-1}\right)}\right)^{4/5}$$

$$\Rightarrow\qquad\frac{d^{5/4}}{(14-(2-d)^2)^{1/4}}\le\limsup\frac{\sum_{p\in B}\frac{|a_p|^3}{p^s}}{\log\left(\frac{1}{s-1}\right)}.$$

For $s>1$, we have

$$\sum_{p\in A}\frac{|a_p|^3}{p^s}+\left(-\sum_p\frac{a_p^3}{p^s}\right)=\sum_{p\in B}\frac{|a_p|^3}{p^s},$$

and so

$$\lim_{s\to 1^+} \sup \frac{\sum_{p\in A} \frac{|a_p|^3}{p^s}}{\log\left(\frac{1}{s-1}\right)} \geq \lim_{s\to 1^+} \sup \frac{\sum_{p\in B} \frac{|a_p|^3}{p^s}}{\log\left(\frac{1}{s-1}\right)}.$$

Therefore

$$\lim_{s\to 1^+} \sup \frac{\sum_{p\in A} \frac{|a_p|^3}{p^s}}{\log\left(\frac{1}{s-1}\right)} \geq \frac{d^{5/4}}{(14-(2-d)^2)^{1/4}}.$$

□

Now we define $T_\alpha \subset A$ to be the set of primes p such that

$$|a_p|^3 \geq \left(\frac{d^{5/4}}{(14-(2-d)^2)^{1/4}}\right)\alpha,$$

where $0 < \alpha < 1$ is a constant to be fixed later.

Let us assume that the upper Dirichlet density of T_α is less than $1/100$. We have

$$\limsup \left(\frac{\sum_{p\in A-T_\alpha} \frac{|a_p|^3}{p^s}}{\log\left(\frac{1}{s-1}\right)}\right) \leq \left(\frac{d^{5/4}}{(14-(2-d)^2)^{1/4}}\right)\alpha.$$

Lemma 3.1 implies that

$$\limsup \left(\frac{\sum_{p\in A-T_\alpha} \frac{|a_p|^3}{p^s}}{\log\left(\frac{1}{s-1}\right)}\right) + \limsup \left(\frac{\sum_{p\in T_\alpha} \frac{|a_p|^3}{p^s}}{\log\left(\frac{1}{s-1}\right)}\right) \geq \frac{d^{5/4}}{(14-(2-d)^2)^{1/4}}$$

and therefore

$$\limsup \left(\frac{\sum_{p\in T_\alpha} \frac{|a_p|^3}{p^s}}{\log\left(\frac{1}{s-1}\right)}\right) \geq \frac{d^{5/4}}{(14-(2-d)^2)^{1/4}}(1-\alpha).$$

We have

$$\left(\limsup \frac{\sum_{p\in T_\alpha} \frac{|a_p|^3}{p^s}}{\log\left(\frac{1}{s-1}\right)}\right)^2 \leq \left(\limsup \frac{\sum_{p\in T_\alpha} \frac{|a_p|^6}{p^s}}{\log\left(\frac{1}{s-1}\right)}\right)\left(\limsup \frac{\sum_{p\in T_\alpha} \frac{|a_p|^0}{p^s}}{\log\left(\frac{1}{s-1}\right)}\right).$$

The second limit supremum can be bounded from above by 5 and the third limit supremum can be bounded from above by $1/100$, so

$$\left(\frac{d^{5/4}}{(14-(2-d)^2)^{1/4}}\right)^2 (1-\alpha)^2 \leq \frac{5}{100}. \tag{3.3}$$

Given some value for the constant β, we want to fix α such that

$$((2-d)\beta)^{1/4} = \left(\frac{d^{5/4}}{(14-(2-d)^2)^{1/4}}\alpha\right)^{1/3}. \tag{3.4}$$

Given equation 3.4, if we set $\beta = 0.495$, we have that if $d \leq 1.258$, then equation 3.1 is false, contradicting the assumption that S_β has an upper Dirichlet density smaller than $1/100$, and if $d > 1.258$, then equation 3.3 is false, and so T_α would have an upper Dirichlet density of at least $1/100$. Either way, since for

$\beta = 0.495$ and $d = 1.258$ the value of equation 3.4 is $0.36729^{1/4} = 0.778...$, this implies that the set of primes

$$\{p \mid a_p(\pi) > 0.778... - \epsilon\}.$$

has an upper Dirichlet density at least $1/100$.

REMARK 1. We determined our choice of β by solving the following simultaneous equations

$$(2-d)(1-\beta) = \frac{\sqrt{14}}{10}$$
$$\frac{d^{5/4}}{(14-(2-d)^2)^{1/4}}(1-\alpha) = \frac{\sqrt{5}}{10},$$

along with equation 3.4, and we obtained $\beta = 0.4957...$ and $d = 1.2581...$.

Acknowledgments. This work began at the University of Zürich, where the author was supported by Forschungskredit grant K-71116-01-01 of the University of Zürich and partially supported by grant SNF PP00P2-138906 of the Swiss National Foundation.

References

[1] D. Bump, *Automorphic forms and representations*, Cambridge Studies in Advanced Mathematics, vol. 55, Cambridge University Press, Cambridge, 1997. MR1431508

[2] S. Gelbart and H. Jacquet, *A relation between automorphic representations of* GL(2) *and* GL(3), Ann. Sci. École Norm. Sup. (4) **11** (1978), no. 4, 471–542. MR533066

[3] H. Jacquet and J. A. Shalika, *On Euler products and the classification of automorphic forms. II*, Amer. J. Math. **103** (1981), no. 4, 777–815, DOI 10.2307/2374050. MR623137

[4] H. H. Kim, *Functoriality for the exterior square of* GL_4 *and the symmetric fourth of* GL_2, J. Amer. Math. Soc. **16** (2003), no. 1, 139–183, DOI 10.1090/S0894-0347-02-00410-1. With appendix 1 by Dinakar Ramakrishnan and appendix 2 by Kim and Peter Sarnak. MR1937203

[5] H. H. Kim and F. Shahidi, *Functorial products for* $\mathrm{GL}_2 \times \mathrm{GL}_3$ *and functorial symmetric cube for* GL_2 (English, with English and French summaries), C. R. Acad. Sci. Paris Sér. I Math. **331** (2000), no. 8, 599–604, DOI 10.1016/S0764-4442(00)01677-3. MR1799096

[6] H. H. Kim and F. Shahidi, *Cuspidality of symmetric powers with applications*, Duke Math. J. **112** (2002), no. 1, 177–197, DOI 10.1215/S0012-9074-02-11215-0. MR1890650

[7] F. Shahidi, *On certain L-functions*, Amer. J. Math. **103** (1981), no. 2, 297–355, DOI 10.2307/2374219. MR610479

[8] F. Shahidi, *Symmetric power L-functions for* GL(2), Elliptic curves and related topics, CRM Proc. Lecture Notes, vol. 4, Amer. Math. Soc., Providence, RI, 1994, pp. 159–182. MR1260961

[9] N. Walji, *Supersingular distribution, congruence class bias, and a refinement of strong multiplicity one*, ProQuest LLC, Ann Arbor, MI, 2011. Thesis (Ph.D.)–California Institute of Technology. MR3153250

[10] N. Walji, *On the distribution of Hecke eigenvalues for cuspidal automorphic representations for* GL(2), Int. Math. Res. Not. IMRN **10** (2018), 3155–3172, DOI 10.1093/imrn/rnw331. MR3805199

THE AMERICAN UNIVERSITY OF PARIS, 102 RUE SAINT-DOMINIQUE, 75007 PARIS, FRANCE
Email address: nwalji@aup.edu

Contemporary Mathematics
Volume **732**, 2019
https://doi.org/10.1090/conm/732/14810

Representations by quadratic forms and the Eichler Commutation Relation

Lynne H. Walling

The Eichler Commutation Relation gives a relation between theta series whose Fourier coefficients are representation numbers of quadratic forms. We review a proof of this, and sketch an extension to Siegel theta series, deducing relations on "average" representation numbers. We take Q to be a positive definite quadratic form on a lattice $L = \mathbb{Z}x_1 \oplus \cdots \oplus \mathbb{Z}x_m$, and the question we address is the following. *Given an n-dimensional quadratic form T, on how many sublattices Λ of L does Q restrict to T?*

We approach this question by combining theory from quadratic forms (see [**2**], [**3**], [**5**]), Siegel modular forms (see [**1**]), and Hecke operators on Siegel modular forms (see [**4**]). Full details of the results on Siegel theta series can be found in [**6**], [**7**]. We use standard quadratic forms notation and terminology, found in [**2**], [**3**], [**5**]. For convenience, we assume that L is even integral, meaning that $Q(L) \subseteq 2\mathbb{Z}$, and we take B to be the symmetric bilinear form associated to Q so that $B(x, x) = Q(x)$. First we define theta series attached to L.

Set $A = (B(x_i, x_j))$. For $n \in \mathbb{Z}_+$ and $\tau = X + iY$ with X, Y symmetric, real $n \times n$ matrices where $Y > 0$ (meaning that Y represents a positive definite quadratic form), we set

$$\theta^{(n)}(L; \tau) = \sum_{C \in \mathbb{Z}^{m,n}} \mathrm{e}\{\,{}^tCAC\tau\}$$

where $\mathrm{e}\{*\} = \exp(\pi i Tr(*))$. Since $A, Y > 0$, this series is absolutely convergent and in fact is an analytic function (in each variable of the matrix τ). The Fourier coefficients of $\theta(L; \tau)$ are "representation numbers":

$$\theta^{(n)}(L; \tau) = \sum_T r(L, T)\,\mathrm{e}\{T\tau\}$$

where T varies over even integral, $n \times n$ symmetric matrices that represent positive semi-definite quadratic forms, and

$$r(L, T) = \#\{C \in \mathbb{Z}^{m,n} : \ {}^tCAC = T\ \}.$$

To relate this to the question posed above, take $\Lambda = \mathbb{Z}y_1 + \cdots + \mathbb{Z}y_n \subseteq L$ and $C \in \mathbb{Z}^{m,n}$ so that

$$(y_1 \ \cdots \ y_n) = (x_1 \ \cdots \ x_m)C;$$

2010 *Mathematics Subject Classification.* Primary 11F41.
Key words and phrases. Quadratic forms, Siegel theta series, Hecke operators.

then we have ${}^tCAC = (B(y_i, y_j))$. Set $d = \operatorname{rank}\Lambda$. When $d = n$, we have $\Lambda = \mathbb{Z}y_1 \oplus \cdots \oplus \mathbb{Z}y_n$ and we set $T_\Lambda = {}^tCAC$. When $d < n$, there is some $E \in SL_n(\mathbb{Z})$ so that

$$(y_1 \ \cdots \ y_n)E = (y_1' \ \cdots \ y_d' \ 0 \ \cdots \ 0),$$

and hence $\Lambda = \mathbb{Z}y_1' \oplus \cdots \oplus \mathbb{Z}y_d'$; with $T_\Lambda = {}^tE(B(y_i, y_j))E$, we have

$$T_\Lambda = \begin{pmatrix} T_\Lambda' & 0 \\ 0 & 0 \end{pmatrix}$$

where $T_\Lambda' = (B(y_i', y_j'))$ is $d \times d$. We define

$$\mathrm{e}\{\Lambda\tau\} = \sum_G \mathrm{e}\{{}^tGT_\Lambda G\tau\} \text{ where } G \text{ varies over } \left\{\begin{pmatrix} I_d & 0 \\ * & * \end{pmatrix}\right\} \backslash GL_n(\mathbb{Z}).$$

(Note that $\mathrm{e}\{\Lambda\tau\}$ is independent of the choice of basis $(y_1' \ \ldots \ y_d')$.) Then we have

$$\theta^{(n)}(L;\tau) = \sum_{\substack{\mathrm{cls}\,\Lambda \\ \mathrm{rank}\,\Lambda \le n}} r(L,\Lambda)o(\Lambda)\,\mathrm{e}\{\Lambda\tau\}$$

where $o(\Lambda)$ is the order of the orthogonal group of Λ.

The reason it is useful to encode these representation numbers $r(L,\Lambda)$ as Fourier coefficients is that $\theta^{(n)}(L;\tau)$ is a Siegel modular form of degree n, weight $m/2$, some level N and quadratic character χ modulo N. This means that $\theta^{(n)}(L;\tau)$ is analytic (in all variables of τ), and

$$\theta^{(n)}(L;(A\tau+B)(C\tau+D)^{-1}) = \chi(\det D)\det(C\tau+D)^{m/2}\theta^{(n)}(L;\tau)$$

for all $(2n \times 2n)$ $\begin{pmatrix} A & B \\ C & D \end{pmatrix} \in Sp_n(\mathbb{Z})$ with $N|C$.

To avoid discussing the technical issues involved with taking square roots of complex numbers, from now on we assume that $m = 2k$ where k is an integer. (The case that m is odd is treated in [**8**].)

We know that for a prime p, $p \nmid N$ if and only if $\mathbb{Z}_pL$ is unimodular, and for $p \nmid N$, $\chi(p) = 1$ if and only if $\mathbb{Z}_pL$ is hyperbolic. (When $p \nmid N$, there are only two possible local structures for L, even when $p = 2$ as we have assumed that L is even integral.)

The Fourier coefficients of Hecke eigenforms satisfy arithmetic relations, so we want to explore how these operators act on our theta series. Ideally, we would like to describe the image of our theta series under a Hecke operator as a linear combination of theta series so that all these theta series lie in the same space of modular forms.

We first consider the case that $n = 1$. So $\tau \in \mathbb{C}$ with $\Im\tau > 0$ and

$$\theta(L;\tau) := \theta^{(1)}(L;\tau) = \sum_{x\in L} \mathrm{e}\{Q(x)\tau\}.$$

Take a prime p. Then with $T(p)$ the Hecke operator associated to p, we have

$$\theta(L;\tau)|T(p) = \sum_{\substack{x\in L \\ 2p|Q(x)}} \mathrm{e}\{Q(x)\tau/p\} + \chi(p)p^{k-1}\sum_{x\in L} \mathrm{e}\{Q(x)p\tau\}.$$

In the second sum, we replace $x \in L$ by $x \in pL$ to get

$$\theta(L;\tau)|T(p) = \sum_{\substack{x\in L\\ 2p|Q(x)}} \mathrm{e}\{Q(x)\tau/p\} + \chi(p)p^{k-1}\sum_{x\in pL}\mathrm{e}\{Q(x)\tau/p\}.$$

If $\chi(p) = 0$, we really can't hope to get a nice description of $\theta(L;\tau)|T(p)$ in terms of other theta series, so we assume that $\chi(p) \neq 0$, and in fact we consider the case that $\chi(p) = 1$. (We later comment on the case $\chi(p) = -1$.) We proceed to sort the vectors $x \in L$ being counted in $\theta(L;\tau)|T(p)$ into sublattices of L corresponding to maximal totally isotropic subspaces of L/pL (here the bilinear form on L/pL is induced by B, and the quadratic form on L/pL is induced by Q when p is odd, and by $\frac{1}{2}Q$ when $p = 2$). To build these subspaces of L/pL, we begin by choosing an isotropic vector $v_1 \in L/pL$; we have $(p^k - 1)(p^{k-1} + 1)$ choices (see [**2**], [**3**]). Now we choose $u_1 \in L/pL$ so that $B(v_1, u_1) \neq 0$ (possible since L/pL is regular, in the sense that it has no radical). Thus $\langle v_1, u_1\rangle$ (the subspace spanned by v_1 and u_1) is a hyperbolic plane, and hence splits L/pL. Next, we choose an isotropic vector $v_2 \in \langle v_1\rangle^\perp \smallsetminus \langle v_1\rangle$; we have $p(p^{k-1} - 1)(p^{k-2} + 1)$ choices. Then we choose $u_2 \in \langle v_1, u_1\rangle^\perp$ so that $B(v_2, u_2) \neq 0$. So $\langle v_2, u_2\rangle$ is a hyperbolic plane. Continuing in this fashion, we build

$$[(p^k - 1)(p^{k-1} + 1)][p(p^{k-1} - 1)(p^{k-2} + 1)]\cdots[p^{k-1}(p^1 - 1)(p^0 + 1)]$$

bases for maximal totally isotropic subspaces of L/pL. (It is easy to check that this process gives us all the maximal totally isotropic subspaces of L/pL.) As each dimension k space over $\mathbb{Z}/p\mathbb{Z}$ has $(p^k - 1)(p^k - p)\cdots(p^k - p^{k-1})$ bases, we build $(p^{k-1} + 1)\beta$ maximal totally isotropic subspaces of L/pL, where

$$\beta = (p^{k-2} + 1)\cdots(p^0 + 1).$$

To count how many of these subspaces contain a given isotropic vector $v_1 \in L/pL$, we simply fix v_1 as the first basis vector in the above process, showing that there are β maximal totally isotropic subspaces of L/pL that contain v_1.

Now let K vary over the preimages in L of these maximal totally isotropic subspaces of L/pL (so $pL \subseteq K$ for all such K). Thus

$$\begin{aligned}\sum_K \theta(K;\tau/p) &= \beta \sum_{\substack{x\in L\smallsetminus pL\\ 2p|Q(x)}} \mathrm{e}\{Q(x)\tau/p\} + (p^{k-1}+1)\beta\sum_{x\in pL}\mathrm{e}\{Q(x)\tau/p\}\\ &= \beta\sum_{\substack{x\in L\\ 2p|Q(x)}}\mathrm{e}\{Q(x)\tau/p\} + p^{k-1}\beta\sum_{x\in pL}\mathrm{e}\{Q(x)\tau/p\}\\ &= \beta\,\theta(L;\tau)|T(p),\end{aligned}$$

which is the Eichler Commutation Relation.

Note that $\theta(K;\tau/p) = \theta(K^{1/p};\tau)$ (where $K^{1/p}$ means that we have scaled the quadratic form on K by $1/p$). What can we say about the local structures of the lattices $K^{1/p}$? A basis $(v_1 \cdots v_k\, u_1 \cdots u_k)$ for L/pL that we constructed as above pulls back to a basis for L, and adjusting the u_i, we can ensure that

$$L \simeq \begin{pmatrix} 0 & I_k \\ I_k & * \end{pmatrix} \pmod{p}.$$

Then we get

$$K \simeq \begin{pmatrix} p* & pI_k \\ pI_k & 0 \end{pmatrix} \pmod{p^2},$$

and consequently we have that $\mathbb{Z}_p K^{1/p}$ is hyperbolic. Thus $\mathbb{Z}_p K^{1/p} \simeq \mathbb{Z}_p L$. For any prime $q \neq p$, we have $\mathbb{Z}_q K = \mathbb{Z}_q L$. However, we do not have $\mathbb{Z}_q K^{1/p} \simeq \mathbb{Z}_q L$ for $q \neq p$ if $\mathbb{Z}_q L$ has an odd rank Jordan component and p is not a square in $\mathbb{Z}_q$. So we may not have $K^{1/p} \in \operatorname{gen} L$. Still, these lattices K all lie in the same genus, which we call $\operatorname{gen} M$.

Now we sort these lattices K into isometry classes. We get

$$\beta\, \theta(L;\tau)|T(p) = \sum_{\operatorname{cls} M'} \frac{\#\{\text{isometry } \sigma :\ pL \subseteq \sigma M' \subseteq L\ \}}{o(M')}\, \theta(M';\tau/p)$$

where the sum is over $\operatorname{cls} M' \in \operatorname{gen} M$, and $o(M')$ is the order of the orthogonal group of M'. Next, we average over $\operatorname{gen} L$, to get

$$\begin{aligned} &\sum_{\substack{\operatorname{cls} L' \in \operatorname{gen} L \\ \operatorname{cls} M' \in \operatorname{gen} M}} \frac{\#\{\text{isometry } \sigma :\ pL' \subseteq \sigma M' \subseteq L'\ \}}{o(L')o(M')}\, \theta(M';\tau/p) \\ &= \sum_{\operatorname{cls} M' \in \operatorname{gen} M} \left(\sum_{\operatorname{cls} L' \in \operatorname{gen} L} \frac{\#\{\sigma :\ pM' \subseteq \sigma pL' \subseteq M'\}}{o(L')} \right) \frac{\theta(M';\tau/p)}{o(M')}. \end{aligned}$$

In this last expression, the inner sum over $\operatorname{cls} L' \in \operatorname{gen} L$ is counting the number of maximal totally isotropic subspaces of M'/pM' where the quadratic form on this space has been scaled by $1/p$. Thus with

$$\theta(\operatorname{gen} L;\tau) = \sum_{\operatorname{cls} L' \in \operatorname{gen} L} \frac{1}{o(L')}\theta(L';\tau),$$

we get

$$\theta(\operatorname{gen} L;\tau)|T(p) = (p^{k-1}+1)\theta(\operatorname{gen} M^{1/p};\tau).$$

So when $\operatorname{gen} L = \operatorname{gen} M^{1/p}$, $\theta(\operatorname{gen} L)$ is a $T(p)$-eigenform, and then we have the following relation on average representation numbers:

$$(p^{k-1}+1)\, r(\operatorname{gen} L, t) = r(\operatorname{gen} L, tp) + p^{k-1}\, r(\operatorname{gen} L, t/p).$$

Further, by the theory of modular forms, $\theta(\operatorname{gen} L)$ is known to lie in the space of Eisenstein series, and it is known that $\theta(\operatorname{gen} L) = \operatorname{mass} L \cdot \theta(L) + \text{cusp form}$. So

$$r(\operatorname{gen} L, 2t) \sim \operatorname{mass} L \cdot r(L, 2t) \text{ as } t \to \infty.$$

(Note: When $\chi(p) = 1$ and $\operatorname{gen} L \neq \operatorname{gen} M^{1/p}$, it's not hard to see that $\theta(\operatorname{gen} L)|T(p)^2 = (p^{k-1}+1)^2\, \theta(\operatorname{gen} L)$. Also, when $\chi(p) = -1$, we can use a two-step construction of lattices sort of like K to get $\theta(\operatorname{gen} L)|T(p)^2 = (p^{k-1}-1)^2\, \theta(\operatorname{gen} L)$.)

We now consider the case $1 < n \le k$, and we still suppose that $\chi(p) = 1$. Then using [**4**] we have

$$\theta^{(n)}(L;\tau)|T(p) = \sum_{\substack{\Lambda\subseteq L\\ \operatorname{rank}\Lambda\le n}} \left(\sum_{\substack{p\Lambda\subseteq\Omega\subseteq\Lambda\\ \Omega^{1/p}\text{ integral}}} p^{E(\Lambda,\Omega)}\,\mathrm{e}\{\Omega\tau/p\} \right)$$
$$= \sum_{\substack{\Omega\subseteq L\\ \Omega^{1/p}\text{ integral}}} \left(\sum_{p\Omega\subseteq p\Lambda\subseteq\Omega} p^{E(\Lambda,\Omega)} \right) \mathrm{e}\{\Omega\tau/p\}$$

where $\mathbb{E}(\Lambda,\Omega)$ is given by an explicit expression, depending on the index of Ω in Λ. To compute this inner sum, we note that those Λ with $[\Omega : p\Lambda] = p^d$ correspond to $n-d$-dimensional subspaces of $\Omega/p\Omega$. So we can easily count all these Λ. Contrastingly, we compare $\theta^{(n)}(L;\tau)|T(p)$ to $\theta^{(n)}(K;\tau/p)$, where K varies as in our discussion when n was 1. We count how often any given Ω lies in some K, and we find that

$$\theta^{(n)}(L;\tau)|T(p) = \gamma \sum_K \theta^{(n)}(K;\tau/p)$$

where $\gamma = (p^{k-n-1}+1)\cdots(p^0+1)$. Then averaging over the genus of L, the exact same argument as before gives us that

$$\theta^{(n)}(\operatorname{gen} L;\tau)|T(p) = (p^{k-1}+1)\cdots(p^{k-n}+1)\,\theta^{(n)}(\operatorname{gen} M^{1/p};\tau)$$

where $\operatorname{gen} M$ is as before (meaning that all these K lie in $\operatorname{gen} M$). So when $\operatorname{gen} L = \operatorname{gen} M^{1/p}$, we get

$$(p^{k-1}+1)\cdots(p^{k-n}+1)\,r(\operatorname{gen} L,\Lambda) = \sum_{p\Lambda\subseteq\Omega\subseteq\Lambda} p^{E(\Lambda,\Omega)}\,r(\operatorname{gen} L,\Omega^{1/p}).$$

But wait – when $n > 1$ there are more (algebraically independent) Hecke operators attached to p! Now we allow $1 < n \le 2k$. For $1 \le j \le n$, with $j \le k$ if $\chi(p) = 1$ and $j < k$ if $\chi(p) = -1$, we compare $\theta^{(n)}(L;\tau)|T_j(p^2)$ to $\sum_{K_j} \theta^{(n)}(K_j;\tau)$ where $K_j \in \operatorname{gen} L$ and the invariant factors of K_j in L consist of j terms $1/p$, j terms p, and $2(k-j)$ terms 1. Directly applying to $\theta^{(n)}(L;\tau)$ the matrices from [**4**] giving the action of $T_j(p^2)$, we get

$$\theta^{(n)}(L;\tau)|T_j(p^2) = \sum_{\Lambda\subseteq L} \left(\sum_{\substack{p\Lambda\subseteq\Omega\subseteq\frac{1}{p}\Lambda\\ \Omega\text{ integral}}} \chi(p^{e_j(\Lambda,\Omega)})p^{E_j(\Lambda,\Omega)}\alpha'_j(\Lambda,\Omega)\,\mathrm{e}\{\Omega\tau\} \right).$$

Here $e_j(\Lambda,\Omega), E_j(\Lambda,\Omega)$ are given in terms of the invariant factors of Ω in Λ, and $\alpha'_j(\Lambda,\Omega)$ comes from an incomplete character sum. To complete the character sum, we replace $T_j(p^2)$ by $\widetilde{T}_j(p^2)$, which is a linear combination of $T_\ell(p^2)$ with $0 \le \ell \le j$

($T_0(p^2)$ being the identity map). Then we get

$$\theta^{(n)}(L;\tau)|\widetilde{T}_j(p^2) = \sum_{\Lambda\subseteq L}\left(\sum_{\substack{p\Lambda\subseteq\Omega\subseteq\frac{1}{p}\Lambda\\ \Omega\text{ integral}}}\chi(p^{e_j(\Lambda,\Omega)})p^{E_j(\Lambda,\Omega)}\alpha_j(\Lambda,\Omega)\,\mathrm{e}\{\Omega\tau\}\right)$$

$$= \sum_{\substack{\Omega\subseteq\frac{1}{p}L\\ \Omega\text{ integral}}}\left(\sum_{p\Omega\subseteq\Lambda\subseteq\Delta}\chi(p^{e_j(\Lambda,\Omega)})p^{E_j(\Lambda,\Omega)}\alpha_j(\Lambda,\Omega)\right)\mathrm{e}\{\Omega\tau\}$$

where $\Delta = \frac{1}{p}\Omega\cap L$, and with $\Lambda = \Lambda_0\oplus\Lambda_1\oplus\Lambda_2$, $\Omega = \frac{1}{p}\Lambda_0\oplus\Lambda_1\oplus p\Lambda_2$, $\alpha_j(\Lambda,\Omega)$ is the number of codimension $n-j$ totally isotropic subspaces of $\Lambda_1/p\Lambda_1$. To compare this to $\sum_{K_j}\theta^{(n)}(K_j;\tau)$, we count how often Ω is in some K_j. This depends on the structure of Ω. We write $\Omega = \frac{1}{p}\Omega_0\oplus\Omega'$ where Ω_0 is primitive in L modulo p (meaning that the rank of Ω_0 is the rank of the image of Ω_0 in L/pL). When $\operatorname{rank}\Omega_0 = i$, the coefficients of $\mathrm{e}\{\Omega\tau\}$ in $\theta^{(n)}(L;\tau)|T_j(p^2)$ and in $\sum_{K_j}\theta^{(n)}(K_j;\tau)$ differ by a constant independent of the quadratic structure of Ω, but it's not the same constant for different i. So we adjust, eventually yielding the relation

$$p^{j(k-n)+j(j-1)/2}\beta_p(n,j)(p^{k-1}+\chi(p))\cdots(p^{k-j}+\chi(p))r(\operatorname{gen}L,T_\Lambda)$$
$$=\sum_{\Omega}\left(\sum_{i=0}^{j}(-1)^{j-i}\chi(p^{e_i(\Lambda,\Omega)})p^{E_{j,i}(\Lambda,\Omega)}\beta_p(n-i,j-i)\alpha_i(\Lambda,\Omega)\right)r(\operatorname{gen}L,T_\Omega)$$

where Ω varies over all lattices so that

$$p\Lambda\subseteq\Omega\subseteq\frac{1}{p}\Lambda,$$

$E_{j,i}(\Lambda,\Omega) = p^{j-i)(j-i-1)/2}E_i(\Lambda,\Omega)$, and $\beta_p(s,r)$ is the number of r-dimensional subspaces of an s-dimensional space over $\mathbb{Z}/p\mathbb{Z}$. When $\chi(p)=1$ and $k<j\le n$, or when $\chi(p)=-1$ and $k\le j\le n$, we get

$$\theta^{(n)}(L;\tau)|T_j'(p^2) = 0,$$

which means that the above sum on Ω is 0.

(As $T(p)^2$ is a linear combination of $1, T_1(p^2),\dots,T_n(p^2)$, for $k<n\le 2k$ this also gives us

$$\theta^{(n)}(\operatorname{gen}L;\tau)|T(p)^2 = ((p^{k-1}+\chi(p))\cdots(p^{k-n}+\chi(p))^2\theta^{(n)}(\operatorname{gen}L;\tau),$$

but this will give us no new information at this point.)

Thus we see that the (generalized) Eichler Commutation Relation gives us relations among the average representation numbers of the genus of L, although they are complicated. It would be very interesting to explore what these relations might reveal.

References

[1] A. N. Andrianov, *Quadratic forms and Hecke operators*, Grundlehren der Mathematischen Wissenschaften [Fundamental Principles of Mathematical Sciences], vol. 286, Springer-Verlag, Berlin, 1987. MR884891

[2] E. Artin, *Geometric algebra*, Wiley Classics Library, John Wiley & Sons, Inc., New York, 1988. Reprint of the 1957 original; A Wiley-Interscience Publication. MR1009557

[3] L. J. Gerstein, *Basic quadratic forms*, Graduate Studies in Mathematics, vol. 90, American Mathematical Society, Providence, RI, 2008. MR2396246

[4] J. L. Hafner and L. H. Walling, *Explicit action of Hecke operators on Siegel modular forms*, J. Number Theory **93** (2002), no. 1, 34–57, DOI 10.1006/jnth.2001.2711. MR1892929

[5] O. T. O'Meara, *Introduction to quadratic forms*, Springer-Verlag, New York-Heidelberg, 1971. Second printing, corrected; Die Grundlehren der mathematischen Wissenschaften, Band 117. MR0347768

[6] L. H. Walling, *Action of Hecke operators on Siegel theta series. I*, Int. J. Number Theory **2** (2006), no. 2, 169–186, DOI 10.1142/S1793042106000516. MR2240221

[7] L. H. Walling, *Action of Hecke operators on Siegel theta series. II*, Int. J. Number Theory **4** (2008), no. 6, 981–1008, DOI 10.1142/S1793042108001845. MR2483307

[8] L. H. Walling, *A formula for the action of Hecke operators on half-integral weight Siegel modular forms and applications*, J. Number Theory **133** (2013), no. 5, 1608–1644, DOI 10.1016/j.jnt.2012.10.007. MR3007125

School of Mathematics, University of Bristol, Bristol BS8 1TH, United Kingdom
Email address: `l.walling@bristol.ac.uk`

Contemporary Mathematics
Volume **732**, 2019
https://doi.org/10.1090/conm/732/14799

Degenerate principal series and Langlands classification

Shunsuke Yamana

ABSTRACT. We determine the Langlands data of a certain irreducible subquotient of the degenerate principal series representation of a metaplectic or a quasisplit even unitary group. The subquotient occurs as a local component of an automorphic representation generated by a certain holomorphic cusp form.

1. Introduction

In his long-awaited book [**1**], Arthur established a classification of irreducible admissible representations of symplectic and quasisplit special orthogonal groups over local fields of characteristic 0 as well as a description of the automorphic discrete spectrum of these groups over number fields. Its metaplectic analog formulated by Wee Teck Gan [**4**] describes the automorphic discrete spectrum $\mathscr{A}_{\mathrm{disc}}(\mathrm{Mp}_n)$ of the metaplectic group of a symplectic vector space of dimension $2n$ over a number field F with adèle ring $\mathbb{A}$ (cf. [**5**]). For a fixed additive nontrivial character ψ of $F\backslash\mathbb{A}$ one expects that $\mathscr{A}_{\mathrm{disc}}(\mathrm{Mp}_n)=\bigoplus_\Psi L^2_{\Psi,\psi}$, where $\Psi=\bigoplus_i \varPi_i\boxtimes\mathrm{sym}^{r_i-1}$ is a global discrete A-parameter for the odd orthogonal group SO_{2n+1}. Here $\varPi_i$ is an irreducible cuspidal automorphic representation of $\mathrm{GL}_{n_i}(\mathbb{A})$ such that if r_i is odd, then its exterior square L-function has a pole at $s=1$ while if r_i is even, then its symmetric square L-function has a pole at $s=1$. Moreover, $\sum_i r_in_i=2n$, sym^{r_i-1} is the irreducible r_i-dimensional representation of $\mathrm{SL}_2(\mathbb{C})$ and the summands $\varPi_i\boxtimes\mathrm{sym}^{r_i-1}$ are mutually distinct.

We specifically consider the case in which $\Psi=\pi\boxtimes\mathrm{sym}^{n-1}$, n is odd and π is an irreducible cuspidal automorphic representation of $\mathrm{PGL}_2(\mathbb{A})$. For each v one has a local A-packet $\Pi_{\Psi_v,\psi_v}=\Pi^+_{\Psi_v,\psi_v}\sqcup\Pi^-_{\Psi_v,\psi_v}$ of equivalence classes of unitary admissible representations of the local metaplectic group associated to the local A-parameter $\Psi_v=\pi_v\boxtimes\mathrm{sym}^{n-1}$ and ψ_v. The subset $\Pi^+_{\Psi_v,\psi_v}$ contains the Langlands quotient $A^+_{n,\psi_v}(\pi_v)$ of the standard module

$$\mathrm{Ind}^{\mathrm{Mp}_n}_{\tilde{\mathscr{P}}_n}(\pi_v^{\psi_v}\otimes\boldsymbol{\alpha}_{F_v}^{(n-1)/2})\boxtimes(\pi_v^{\psi_v}\otimes\boldsymbol{\alpha}_{F_v}^{(n-3)/2})\boxtimes\cdots\boxtimes(\pi_v^{\psi_v}\otimes\boldsymbol{\alpha}_{F_v})\boxtimes\theta^{\psi_v}(\pi_v).$$

Here $\boldsymbol{\alpha}_{F_v}$ denotes the normalized absolute value of F_v. We refer to the body of this paper for unexplained notations. It is conjectured that

$$L^2_{\Psi,\psi}\simeq\bigoplus_{\substack{\sigma_v\in\Pi^{\epsilon_v}_{\Psi_v,\psi_v}\\ \prod_v\epsilon_v=\varepsilon(1/2,\pi)}}\otimes'_v\sigma_v. \tag{1.1}$$

If F is totally real of degree d and π is generated by a Hilbert cusp form of weight 2κ, where $\kappa=(\kappa_v)\in\mathbb{Z}^d$ and $\kappa_v>\frac{n}{2}$, then the holomorphic discrete series representation with lowest weight $\kappa_v+\frac{n}{2}$ is contained in $\Pi^{(-1)^{(n-1)/2}}_{\Psi_v,\psi_v}$. If $\varepsilon\big(\frac{1}{2},\pi\big)=(-1)^{d(n-1)/2}$, then there should be a Hilbert-Siegel cusp form of weight $\kappa+\frac{n}{2}$ whose $\mathfrak{p}$-component is the small nontempered representation $A^+_{n,\psi_\mathfrak{p}}(\pi_\mathfrak{p})$. When $F=\mathbb{Q}$ and π is everywhere unramified, such a Siegel cusp form is constructed in [**7**, **8**].

In this paper we will study the representation $A^+_{n,\psi_\mathfrak{p}}(\pi_\mathfrak{p})$ for finite primes $\mathfrak{p}$. If $\pi_\mathfrak{p}\simeq I(\mu_\mathfrak{p},\mu_\mathfrak{p}^{-1})$, then $A^+_{n,\psi_\mathfrak{p}}(\pi_\mathfrak{p})$ is equivalent to a degenerate principal series representation $I_n^{\psi_\mathfrak{p}}(\mu_\mathfrak{p})$. When $\pi_\mathfrak{p}$ is a twist of the Steinberg representation by a nontrivial quadratic character $\nu_\mathfrak{p}$ of $F_\mathfrak{p}^\times$, we shall show that $A^+_{n,\psi_\mathfrak{p}}(\pi_\mathfrak{p})$ is equivalent to the unique irreducible subrepresentation of $I_n^{\psi_\mathfrak{p}}(\nu_\mathfrak{p}\boldsymbol{\alpha}_{F_\mathfrak{p}}^{1/2})$. If $\pi_\mathfrak{p}$ is the Steinberg representation, then $A^-_{n,\psi_\mathfrak{p}}(\pi_\mathfrak{p})\in\Pi^-_{\Psi_\mathfrak{p},\psi_\mathfrak{p}}$ is equivalent to the unique irreducible subrepresentation of $I_n^{\psi_\mathfrak{p}}(\boldsymbol{\alpha}_{F_\mathfrak{p}}^{1/2})$. Ikeda and the author proved a large portion of (1.1) in [**9**] a few years ago. The result of the present article will be an essential ingredient to generalize the work.

Acknowledgment. The author would like to thank the reviewer for a careful reading of the paper and for several useful suggestions. The author is partially supported by JSPS Grant-in-Aid for Scientific Research (C) 18K03210 and received funding from the European Research Council under the European Community's Seventh Framework Programme (FP7/2007-2013)/ERC Grant agreement no. 290766 (AAMOT).

Notation

The ground field F is a finite extension of $\mathbb{Q}_p$. Let $\Omega(F^\times)$ be the group of all continuous homomorphisms from $F^\times$ to $\mathbb{C}^\times$. The symbol $\boldsymbol{\alpha}_F\in\Omega(F^\times)$ denotes the normalized absolute value and $\mathbb{1}_{\mathrm{GL}_m}$ denotes the trivial representation of $\mathrm{GL}_m(F)$. For $j\leq m$ we let $Q^m_j=L^m_jU^m_j$ denote a standard parabolic subgroup of GL_m with Levi subgroup $L^m_j\simeq\mathrm{GL}_j\times\mathrm{GL}_{m-j}$. We associate to a pair of characters $\mu_1,\mu_2\in\Omega(F^\times)$ the representation $I^m_j(\mu_1,\mu_2)=\mathrm{Ind}^{\mathrm{GL}_m(F)}_{Q^m_j}\mu_1\boxtimes\mu_2$ of $\mathrm{GL}_m(F)$ on the space of all smooth functions f on $\mathrm{GL}_m(F)$ satisfying

$$f\left(\begin{pmatrix}a_1 & b\\ 0 & a_2\end{pmatrix}g\right)=\mu_1(\det a_1)\mu_2(\det a_2)\boldsymbol{\alpha}_F(\det a_1)^{(m-j)/2}\boldsymbol{\alpha}_F(\det a_2)^{-j/2}f(g)$$

for all $a_1\in\mathrm{GL}_j(F)$, $a_2\in\mathrm{GL}_{m-j}(F)$, $b\in\mathrm{M}_{j,m-j}(F)$ and $g\in\mathrm{GL}_m(F)$. We use the abbreviation $I_n(\mu_1,\mu_2)=I^{2n}_n(\mu_1,\mu_2)$ and $I(\mu_1,\mu_2)=I_1(\mu_1,\mu_2)$. Note that $I(\mu_1,\mu_2)$ is irreducible unless $\mu_1\mu_2^{-1}\in\{\boldsymbol{\alpha}_F,\boldsymbol{\alpha}_F^{-1}\}$. The symbol St stands for the Steinberg representation of $\mathrm{GL}_2(F)$. Recall the nonsplit short exact sequence:

$$0\to\mathrm{St}\to I(\boldsymbol{\alpha}_F^{1/2},\boldsymbol{\alpha}_F^{-1/2})\to\mathbb{1}_{\mathrm{GL}_2}\to 0. \tag{1.2}$$

Fix a nontrivial additive character ψ on F. For $\eta\in F^\times$ we define an additive character ψ^η of F by $\psi^\eta(x)=\psi(\eta x)$ and have an 8th root of unity $\gamma(\psi^\eta)$ such that for all Schwartz functions ϕ on F

$$\int_F\phi(x)\psi(\eta x^2)\,\mathrm{d}x=\gamma(\psi^\eta)\boldsymbol{\alpha}_F(2\eta)^{-1/2}\int_F\mathcal{F}\phi(x)\psi\left(-\frac{x^2}{4\eta}\right)\mathrm{d}x,$$

where $\mathrm{d}x$ is the self-dual Haar measure on F with respect to the Fourier transform

$$\mathcal{F}\phi(y) = \int_F \phi(x)\psi(xy)\,\mathrm{d}x.$$

2. Degenerate principal series representations and Weil representations

We treat metaplectic and unitary groups, which we refer to as Cases M and U. Let $x \mapsto x^\tau$ be the trivial automorphism of $E = F$ in Case M. Let E be a quadratic extension of F and $x \mapsto x^\tau$ the nontrivial automorphism of E over F in Case U. The character $\epsilon_{E/F} \in \Omega(F^\times)$ corresponds to E/F via class field theory. When Π is a representation of $\mathrm{GL}_k(E)$, let ${}^\tau\Pi$ denote the representation of $\mathrm{GL}_k(E)$ defined by ${}^\tau\Pi(A) = \Pi(A^\tau)$ for $A \in \mathrm{GL}_k(E)$. When π is an irreducible admissible representation of $\mathrm{GL}_k(F)$, we denote its base change to $\mathrm{GL}_k(E)$ by π^E.

We express various things by matrices. Put

$$G_n = \left\{ g \in \mathrm{GL}_{2n}(E) \,\middle|\, g\begin{pmatrix} 0 & -\mathbf{1}_n \\ \mathbf{1}_n & 0 \end{pmatrix} {}^t g^\tau = \begin{pmatrix} 0 & -\mathbf{1}_n \\ \mathbf{1}_n & 0 \end{pmatrix} \right\}.$$

The F-vector space $\mathrm{Her}_n = \{z \in \mathrm{M}_n(E) \mid {}^t z^\tau = z\}$ has dimension $\frac{n[E:F]}{2}\varrho_n$, where

(Case M) $\qquad \varrho_n = n+1,$

(Case U) $\qquad \varrho_n = n.$

For $k \leq n$ the group G_n has a maximal parabolic subgroup $P_k^n = M_k^n N_k^n$ given by

$$M_k^n = \left\{ \iota_k^n\left(A, \begin{pmatrix} \alpha & \beta \\ \gamma & \delta \end{pmatrix}\right) = \left[\begin{array}{cc|cc} A & & & \\ & \alpha & & \beta \\ \hline & & {}^t(A^{-1})^\tau & \\ & \gamma & & \delta \end{array}\right] \,\middle|\, \begin{array}{c} A \in \mathrm{GL}_k(E) \\ \begin{pmatrix} \alpha & \beta \\ \gamma & \delta \end{pmatrix} \in G_{n-k} \end{array} \right\},$$

$$N_k^n = \left\{ \mathbf{v}_k^n(x;y;z) = \left[\begin{array}{cc|cc} \mathbf{1}_k & x & z - y\,{}^t x^\tau & y \\ 0 & \mathbf{1}_{n-k} & {}^t y^\tau & \mathbf{0}_{n-k} \\ \hline \multicolumn{2}{c|}{\multirow{2}{*}{$\mathbf{0}_n$}} & \mathbf{1}_k & 0 \\ & & -{}^t x^\tau & \mathbf{1}_{n-k} \end{array}\right] \,\middle|\, \begin{array}{c} x, y \in \mathrm{M}_{k,n-k}(E) \\ z \in \mathrm{Her}_k \end{array} \right\}.$$

In Case M we denote the two-fold metaplectic cover of G_n by $\tilde{G}_n$ and identify it as a set with $G_n \times \{\pm 1\}$ in the standard way. When S is a subgroup of G_n, we write $\tilde{S}$ for the pull-back of S in $\tilde{G}_n$. Note that

$$\tilde{M}_k^n \simeq \bar{H}_k \times \tilde{G}_{n-k}/\{(\zeta, \zeta^{-1}) \mid \zeta \in \{\pm 1\}\},$$

where we consider the set $\bar{H}_k \simeq \mathrm{GL}_k(F) \times \{\pm 1\}$ with multiplication

$$(A_1, \zeta_1) \cdot (A_2, \zeta_2) = (A_1 A_2, \zeta_1 \zeta_2 (\det A_1, \det A_2)_F).$$

Here $(\ ,\ )_F : F^\times \times F^\times \to \{\pm 1\}$ denotes the Hilbert symbol. We can define a genuine character γ^ψ of $\bar{H}_k$ by $\gamma^\psi((A,\zeta)) = \zeta\gamma(\psi)/\gamma(\psi^{\det A})$ for $A \in \mathrm{GL}_k(F)$ and $\zeta \in \{\pm 1\}$. Given a representation σ of $\mathrm{GL}_k(E)$, we can obtain a genuine representation σ^ψ of $\bar{H}_k$ by $\sigma^\psi((A,\zeta)) = \gamma^\psi((A,\zeta))\sigma(A)$. To make our exposition uniform, we write

$$\tilde{G}_n = G_n, \qquad \tilde{S} = S, \qquad \bar{H}_k = \mathrm{GL}_k(E), \qquad \sigma^\psi = \sigma$$

in Case U. More generally, when $P = MN$ is a parabolic subgroup of G_n with Levi subgroup $M \simeq \mathrm{GL}_{n_1}(E) \times \cdots \times \mathrm{GL}_{n_r}(E) \times G_{n_0}$, σ_0 is a (genuine) representation of $\tilde{G}_{n_0}$ and σ_i is a representation of $\mathrm{GL}_{n_i}(E)$ for $1 \le i \le r$, we can consider the normalized induced representation

$$\mathrm{Ind}_{\tilde{M}N}^{\tilde{G}_n} \sigma_1^\psi \boxtimes \cdots \boxtimes \sigma_r^\psi \boxtimes \sigma_0.$$

We frequently regard $\chi \in \Omega(E^\times)$ as defining a character of $\mathrm{GL}_n(E)$ via $\chi(A) = \chi(\det A)$ and define the degenerate principal series representation $I_n^\psi(\chi) = \mathrm{Ind}_{\tilde{P}_n^n}^{\tilde{G}_n} \chi^\psi$.

Let $V = E^m$ be the space of column vectors with a nondegenerate Hermitian form $(\ ,\ )$. Letting $\{v_i\}_{i=1}^m$ be any E-basis for V, we will write disc$V$ for the determinant of the matrix $((v_i, v_j)) \in \mathrm{Her}_m$. In Case M we define $\chi_V \in \Omega(F^\times)$ by $\chi_V(a) = (a, (-1)^{n(n-1)/2}\mathrm{disc}V)_F$. In Case U we set $\chi_V = \chi_m$, where χ_m is a fixed character of $E^\times$ which restricts on $F^\times$ to $\epsilon_{E/F}^m$. The Weil representation ω_V^ψ of $\tilde{G}_n$ is realized in the space $\mathcal{S}(V^n)$ of Schwartz functions on V^n. Recall that $\iota_n^n(A) = \begin{pmatrix} A & 0 \\ 0 & {}^t(A^{-1})^\tau \end{pmatrix}$ and $\mathbf{v}_n^n(z) = \begin{pmatrix} \mathbf{1}_n & z \\ 0 & \mathbf{1}_n \end{pmatrix}$ for $A \in \mathrm{GL}_n(F)$ and $z \in \mathrm{Her}_n$. Assume m to be odd in Case M. The action of $\tilde{P}_n^n$ is given by the following formulas:

$$[\omega_V^\psi((\iota_n^n(A), \zeta))\phi](u) = \chi_V^\psi((A, \zeta))\boldsymbol{\alpha}_E(\det A)^{m/2}\phi(uA), \qquad A \in \mathrm{GL}_n(E),$$
$$[\omega_V^\psi(\mathbf{v}_n^n(z))\phi](u) = \psi(\mathrm{tr}(z(u, u)))\phi(u), \qquad z \in \mathrm{Her}_n,$$

where $(u, u) = ((u_i, u_j)) \in \mathrm{Her}_n$ for $u = (u_1, \dots, u_n) \in V^n$. Let $R_n^\psi(V, \chi_V)$ be the image of $\mathcal{S}(V^n)$ under the Rallis coinvariant map

$$\mathcal{S}(V^n) \to I_n^\psi(\chi_V \boldsymbol{\alpha}_E^{(m-\varrho_n)/2}), \qquad \phi \mapsto f_\phi(\tilde{g}) = [\omega_V^\psi(\tilde{g})\phi](0).$$

By the invariant distribution theorem of Rallis (see [**15, 16**]) $R_n^\psi(V, \chi_V)$ coincides with the maximal quotient of $\mathcal{S}(V^n)$ on which $\mathrm{O}(V)$ acts trivially.

Let $R_n^\psi(m, \nu)$ be the subrepresentation of $I_n^\psi(\nu \boldsymbol{\alpha}_E^{(m-\varrho_n)/2})$ generated by the submodules $R_n^\psi(V, \nu)$ with $\dim V = m$ and $\chi_V = \nu$. In Case U there are precisely two equivalence classes $V_m^\pm$ of nondegenerate Hermitian spaces of dimension m, having opposite signs $\epsilon_{E/F}((-1)^{m(m-1)/2}\mathrm{disc}V_m^\pm) = \pm 1$. Let $\xi \in \mathrm{Her}_n \cap \mathrm{GL}_n(E)$. We say that a representation σ of $\tilde{G}_n$ is (ψ, ξ)-generic if there is a nonzero $\mathbb{C}$-linear functional $\lambda : \sigma \to \mathbb{C}$ such that

$$\lambda(\sigma(\mathbf{v}_n^n(z))v) = \psi(\mathrm{tr}(\xi z))\lambda(v) \qquad (z \in \mathrm{Her}_n,\ v \in \sigma).$$

After Kudla and Rallis [**12**] gave a complete description of all points of reducibility and the composition series of the degenerate principal series representations for symplectic groups, analogous studies were done by Kudla and Sweet [**13**] for unitary groups and by Sweet [**17**] for metaplectic groups. These works include the following statements.

LEMMA 2.1 ([**13, 17**]). *We assume n to be odd in Case M. Let $\nu \in \Omega(E^\times)$.*

(1) *Assume that $E \ne F$ and $\nu|_{F^\times} = \epsilon_{E/F}^n$. Then $I_n^\psi(\nu)$ is the direct sum of the two irreducible representations $R_n^\psi(V_n^\pm, \nu)$. Moreover, $R_n^\psi(V_n^\pm, \nu)$ is (ψ, ξ)-generic if and only if $\epsilon_{E/F}((-1)^{n(n-1)/2}\det \xi) = \pm 1$.*

(2) *When $E = F$, we assume that $\nu^2 = 1$. When $E \ne F$, we assume that the restriction of ν to $F^\times$ is $\epsilon_{E/F}^{n-1}$. Then $I_n^\psi(\nu \boldsymbol{\alpha}_E^{1/2})$ has a unique irreducible*

subrepresentation, which we denote by $\mathrm{St}^\psi_n(\nu)$. *Moreover,*

$$\mathrm{St}^\psi_n(\nu) \simeq I^\psi_n(\nu\boldsymbol{\alpha}_E^{-1/2})/R^\psi_n(\varrho_n - 1, \nu) \tag{2.1}$$

Furthermore, $\mathrm{St}^\psi_n(\nu)$ *cannot appear as a quotient of* $I^\psi_n(\nu\boldsymbol{\alpha}_E^{1/2})$.

3. Main theorems

Let π be an irreducible infinite dimensional unitary representation of $\mathrm{GL}_2(F)$ whose central character is denoted by $\hat{\omega}$. When π is a discrete series representation, the Jacquet-Langlands correspondence associates to π an irreducible representation π^- of the multiplicative group of the quaternion division algebra over F. To unify notation, we set $\pi^+ = \pi$. We extend $\hat{\omega}$ to $\hat{\chi} \in \Omega(E^\times)$. We require $\hat{\omega} = \hat{\chi}$ to be the trivial character and view π^+ as a representation of the split special orthogonal group $\mathrm{PGL}_2(F) \simeq \mathrm{SO}(1,2)$ and π^- as a representation of the anisotropic special orthogonal group $\mathrm{SO}(3)$ in Case M.

Now we define irreducible representations $A^{\pm,\hat{\chi}}_{1,\psi}(\pi)$. In Case M we let $A^{\pm,\hat{\chi}}_{1,\psi}(\pi) = A^{\pm}_{1,\psi}(\pi) := \theta^\psi(\hat{\pi}^\pm)$ be the theta lift of $\hat{\pi}^\pm$ to $\tilde{G}_1$ with respect to ψ, where $\hat{\pi}^\pm$ denotes the unique extension of $\pi^\pm$ to a representation of the split or anisotropic orthogonal group having a nonzero theta lift to $\tilde{G}_1$. Here we set $A^-_{1,\psi}(\pi) = 0$ if π is not a discrete series representation. In Case U the representation $\hat{\chi}^{-1} \boxtimes \pi$ defines a representation of G_1 in view of the accidental isomorphism

$$G_1 \simeq \{(a,g) \in E^\times \times \mathrm{GL}_2(F) \mid \mathrm{N}^E_F(a) = \det g\}/\{(\eta,\eta) \mid \eta \in F^\times\}.$$

Since it is irreducible or the direct sum of two irreducible summands, we can write $\hat{\chi}^{-1} \boxtimes \pi \simeq A^{+,\hat{\chi}}_{1,\psi}(\pi) \oplus A^{-,\hat{\chi}}_{1,\psi}(\pi)$, where $A^{+,\hat{\chi}}_{1,\psi}(\pi)$ is the $(\psi,1)$-generic summand. Note that $A^-_{1,\psi}(\pi) \neq 0$ if and only if $\pi \otimes \epsilon_{E/F} \simeq \pi$. We take n to be odd. Then there is a parabolic subgroup $\mathscr{P}_n$ of G_n with Levi subgroup $\mathrm{GL}_2(E) \times \cdots \times \mathrm{GL}_2(E) \times G_1$. We define $A^{\pm,\hat{\chi}}_{n,\psi}(\pi)$ as the Langlands quotient of the standard module

$$\mathrm{Ind}^{\tilde{G}_n}_{\tilde{\mathscr{P}}_n}(\boldsymbol{\alpha}_E^{(n-1)/2}{}^\tau\hat{\chi}^{-1}\otimes\pi^E)^\psi \boxtimes (\boldsymbol{\alpha}_E^{(n-3)/2}{}^\tau\hat{\chi}^{-1}\otimes\pi^E)^\psi \boxtimes \cdots \boxtimes (\boldsymbol{\alpha}_E{}^\tau\hat{\chi}^{-1}\otimes\pi^E)^\psi \boxtimes A^{\pm,\hat{\chi}}_{1,\psi}(\pi).$$

When π is not supercuspidal, the following theorem describes $A^{\pm,\hat{\chi}}_{n,\psi}(\pi)$ in terms of degenerate principal series representations.

THEOREM 3.1. *Let n be odd and π an irreducible infinite dimensional unitary representation of* $\mathrm{GL}_2(F)$ *whose central character is trivial if* $E = F$.

(1) *If* $\pi \simeq I(\mu, \hat{\omega}\mu^{-1})$, *then*

$$A^{+,\hat{\chi}}_{n,\psi}(\pi) \simeq I^\psi_n({}^\tau\hat{\chi}^{-1}\mu^E), \qquad A^{-,\hat{\chi}}_{n,\psi}(\pi) = 0$$

unless $E \neq F$ *and* $\hat{\omega}^{-1}\mu^2 = \epsilon_{E/F}$, *in which case*

$$A^{\pm,\hat{\chi}}_{n,\psi}(\pi) \simeq R^\psi_n(V^\pm_n, {}^\tau\hat{\chi}^{-1}\mu^E).$$

(2) *If* $E = F$, $\pi \simeq \mathrm{St} \otimes \nu$, $\nu^2 = 1$ *and* $\nu \neq 1$, *then* $A^+_{n,\psi}(\pi) \simeq \mathrm{St}^\psi_n(\nu)$.

(3) *If* $E = F$ *and* $\pi \simeq \mathrm{St}$, *then* $A^-_{n,\psi}(\pi) \simeq \mathrm{St}^\psi_n(1)$.

(4) *If* $E \neq F$ *and* $\pi \simeq \mathrm{St} \otimes \nu$, *then* $A^{+,\hat{\chi}}_{n,\psi}(\pi) \simeq \mathrm{St}^\psi_n({}^\tau\hat{\chi}^{-1}\nu^E)$ *and* $A^{-,\hat{\chi}}_{n,\psi}(\pi) = 0$.

REMARK 3.2. One can derive analogous results for symplectic and split even orthogonal groups of even rank from [**2, 10, 12**] (cf. Theorem 8.2 of [**19**]).

We conclude this section by sketching the proof of Theorem 3.1(1). The proof uses some inductive structure of degenerate principal series. It is important to note that quadratic or Hermitian spaces of odd rank n has Witt index greater than or equal to $\frac{n-3}{2}$. As given in Proposition 1 of [**14**] we obtain an intertwining map

$$\Psi_n^{\hat{\chi}}(\mu) : I_n^{\psi}({}^{\tau}\hat{\chi}^{-1}\mu^E) \to \mathrm{Ind}_{\tilde{P}_2^n}^{\tilde{G}_n}(I({}^{\tau}\hat{\chi}^{-1}\mu^E, \hat{\chi}(\mu^E)^{-1})^{\psi} \otimes \boldsymbol{\alpha}_E^{(1-n)/2}) \boxtimes I_{n-2}^{\psi}({}^{\tau}\hat{\chi}^{-1}\mu^E)$$

(cf. [**20**, Proposition 4.1, Lemma 5.1]). For each $\xi \in \mathrm{Her}_n \cap \mathrm{GL}_n(E)$ one can construct a nonzero degenerate Whittaker functional on $I_n^{\psi}({}^{\tau}\hat{\chi}^{-1}\mu^E)$ as a Jacquet integral. When ξ is of the form $\begin{pmatrix} 0 & 1 \\ 1 & 0 \end{pmatrix} \oplus \xi'$ with $\xi' \in \mathrm{Her}_{n-2}$, Lemma 8 of [**14**] asserts that it factors through the map $\Psi_n^{\hat{\chi}}(\mu)$. In particular, $\Psi_n^{\hat{\chi}}(\mu)$ is nonzero. Actually, $\Psi_n^{\hat{\chi}}(\mu)$ is nonzero on each of the irreducible summands of $I_n^{\psi}({}^{\tau}\hat{\chi}^{-1}\mu^E)$ in light of Lemma 2.1(1). Note that $I_n^{\psi}({}^{\tau}\hat{\chi}^{-1}\mu^E)$ is irreducible unless $E \neq F$ and $\hat{\omega}^{-1}\mu^2 = \epsilon_{E/F}$ by [**13, 17**]. We therefore infer by induction that $I_n^{\psi}({}^{\tau}\hat{\chi}^{-1}\mu^E)$ is equivalent to a subrepresentation of

$$\mathrm{Ind}_{\tilde{\mathscr{P}}_n}^{\tilde{G}_n}(\boldsymbol{\alpha}_E^{(1-n)/2}{}^{\tau}\hat{\chi}^{-1} \otimes \pi^E)^{\psi} \boxtimes \cdots \boxtimes (\boldsymbol{\alpha}_E^{-1}{}^{\tau}\hat{\chi}^{-1} \otimes \pi^E)^{\psi} \boxtimes I_1^{\psi}({}^{\tau}\hat{\chi}^{-1}\mu^E),$$

which is unique in light of the Langlands classification. Theorem 3.1(1) can now be deduced from the case $n = 1$, which is easily proved (see Proposition 4(3) of [**18**]). Sections 4 and 5 are devoted to the proof of Theorem 3.1(2), (3) and (4).

4. Jacquet modules of degenerate principal series representations

When Π is a smooth representation of $\tilde{G}_n$, let $\Pi_{N_k^n}$ be the normalized Jacquet module of Π with respect to $\tilde{N}_k^n$, viewed as a representation of $\tilde{M}_k^n$. Our proof relies on extensive computations of the Jacquet module of degenerate principal series.

LEMMA 4.1. *The Jacquet module $I_n^{\psi}(\chi)_{N_k^n}$ has an $\tilde{M}_k^n$-stable filtration*

$$I_n^{\psi}(\chi)_{N_k^n} = I^k \supset I^{k-1} \supset \cdots \supset I^1 \supset I^0 \supset I^{-1} = 0$$

with successive quotients

$$I^r/I^{r-1} \simeq I_r^k(\chi\boldsymbol{\alpha}_E^{(r-n)/2}, {}^{\tau}\chi^{-1}\boldsymbol{\alpha}_E^{(k-r-n)/2})^{\psi} \boxtimes I_{n-k}^{\psi}(\chi\boldsymbol{\alpha}_E^{(2r-k)/2}).$$

PROOF. When $E \neq F$ and $k = 1$ or n, this result is proved in [**13**, Lemma 2.1, Proposition 5.1] (cf. [**12**, Lemma 2.1, Proposition 4.1]). Note that the unnormalized Jacquet module is considered in [**12, 13**]. The proof can be done by the same technique. Chapter 4 of [**6**] determines the orbit structure of $P\backslash G_n$ for the action of Q for any pair P, Q of maximal parabolic subgroups of G_n. Letting $j = \tilde{m}_k = n$, $\ell = k$ and $m' = 2n$, we see that G_n is the disjoint union of the double cosets $\Omega_r = P_n^n \epsilon_r P_k^n$ $(0 \leq r \leq k)$. The matrix $\epsilon_r = \epsilon_{r,s}$ is defined in (4.14) and (4.15) of [**6**]. Note that s is necessarily equal to $n + r - k$ by (4.12) of [**6**]. Put

$$S_{k,r}^n = \epsilon_r^{-1} P_n^n \epsilon_r \cap P_k^n.$$

By (4.19) of [**6**] we have $S_{k,r}^n = T_{k,r}^n U_{k,r}^n$, where

$$T_{k,r}^n = \left\{ \iota_k^n \left(\begin{pmatrix} \beta & * \\ 0 & \delta \end{pmatrix}, \begin{pmatrix} A & * \\ 0 & {}^t(A^{-1})^{\tau} \end{pmatrix} \right) \,\middle|\, \begin{matrix} \beta \in \mathrm{GL}_r(E),\ \delta \in \mathrm{GL}_{k-r}(E) \\ A \in \mathrm{GL}_{n-k}(E) \end{matrix} \right\},$$

$$U_{k,r}^n = \left\{ \mathbf{v}_k^n \left(\begin{pmatrix} x \\ 0 \end{pmatrix}; y; \begin{pmatrix} z_1 & z_2 \\ {}^t z_2^{\tau} & 0 \end{pmatrix} \right) \,\middle|\, \begin{matrix} x \in \mathrm{M}_{r,n-k}(E),\ y \in \mathrm{M}_{k,n-k}(E) \\ z_1 \in \mathrm{Her}_r,\ z_2 \in \mathrm{M}_{r,k-r}(E) \end{matrix} \right\}.$$

The closure of an orbit Ω_r is given by $\bigsqcup_{i\geq r}\Omega_i$. The stratification of G_n by Ω_r gives rise to an $\tilde{M}^n_k$-stable filtration

$$I^\psi_n(\chi)=J^k\supset J^{k-1}\supset\cdots\supset J^1\supset J^0\supset J^{-1}=0,$$

where J^r is the set of functions in $I^\psi_n(\chi)$ whose restriction to the closure of Ω_{r+1} vanishes. The successive quotients are given by

$$J^r/J^{r+1}\simeq \text{c-ind}^{\tilde{P}^n_k}_{\tilde{S}^n_{k,r}}\chi^\psi_r,$$

where

$$\chi_r\left(\iota^n_k\left(\begin{pmatrix}\beta & *\\ 0 & \delta\end{pmatrix},\begin{pmatrix}A & *\\ 0 & {}^t(A^{-1})^\tau\end{pmatrix}\right)\right)=(\chi\boldsymbol{\alpha}_E^{\varrho_n/2})((\det\beta)(\det\delta^\tau)^{-1}(\det A))$$

by (4.20) of [**6**]. Proposition 6.3.3 of [**3**] shows that

$$(\text{c-ind}^{\tilde{P}^n_k}_{\tilde{S}^n_{k,r}}\chi^\psi_r)_{N^n_k}\simeq(\text{ind}^{\tilde{M}^n_k}_{\tilde{T}^n_{k,r}}\chi^\psi_r\gamma_r)\otimes\delta_{P^n_k}^{-1/2},$$

where $\delta_{P^n_k}$ is the modulus character of P^n_k and γ_r is the modulus character of $T^n_{k,r}$ acting on $N^n_k/U^n_{k,r}$. Observe that

$$\delta_{P^n_k}(\iota^n_k(A,g))=\boldsymbol{\alpha}_E(\det A)^{2(n-k)+\varrho_k}, \tag{4.1}$$

$$\gamma_r\left(\iota^n_k\left(\begin{pmatrix}\beta & *\\ 0 & \delta\end{pmatrix},\begin{pmatrix}A & *\\ 0 & {}^t(A^{-1})^\tau\end{pmatrix}\right)\right)=\boldsymbol{\alpha}_E(\det\delta)^{n-k}\boldsymbol{\alpha}_E(\det A)^{r-k}\boldsymbol{\alpha}_E(\det\delta)^{\varrho_{k-r}}.$$

These complete our proof. □

When V is an isotropic Hermitian space of dimension m, let V' be the Hermitian space of dimension $m-2$ obtained by deleting a hyperbolic plane from V.

LEMMA 4.2. *If $m=\dim V<\varrho_n$ and $\chi=\chi_V$, then*

$$R^\psi_n(V,\chi)_{N^n_1}\simeq(\chi^\psi\boldsymbol{\alpha}_E^{(\rho_1-m)/2}\boxtimes R^\psi_{n-1}(V',\chi))\oplus(\chi^\psi\boldsymbol{\alpha}_E^{(m-\rho_1)/2+1-n}\boxtimes R^\psi_{n-1}(V,\chi)),$$

where $R^\psi_{n-1}(V',\chi)$ is taken to be zero if V is anisotropic.

PROOF. The exact sequence given in Proposition 5.3 of [**13**] is split by assumption, which gives the stated decomposition in Case U. The case of metaplectic groups can be verified in the parallel way. □

5. Proof of the main theorem

LEMMA 5.1. *Let $\nu\in\Omega(E^\times)$. Assume that $\nu^2=1$ in Case M. Assume that the restriction of ν to $F^\times$ is $\epsilon^{n-1}_{E/F}$ in Case U. Let $n\geq 3$. Put*

$$H^\psi_n(\nu)=\text{Ind}^{\tilde{G}_n}_{\tilde{P}^n_2}(\text{St}^E\otimes\nu\boldsymbol{\alpha}_E^{(1-n)/2})^\psi\boxtimes\text{St}^\psi_{n-2}(\nu),$$

$$J^\psi_n(\nu)=\text{Ind}^{\tilde{G}_n}_{\tilde{P}^n_2}(I(\boldsymbol{\alpha}_E^{1/2},\boldsymbol{\alpha}_E^{-1/2})\otimes\nu\boldsymbol{\alpha}_E^{(1-n)/2})^\psi\boxtimes\text{St}^\psi_{n-2}(\nu).$$

Then $\text{Hom}_{\tilde{G}_n}(I^\psi_n(\nu\boldsymbol{\alpha}_E^{-1/2}),J^\psi_n(\nu))=\text{Hom}_{\tilde{G}_n}(I^\psi_n(\nu\boldsymbol{\alpha}_E^{-1/2}),H^\psi_n(\nu))$.

PROOF. Lemma 4.1 applied to $I^\psi_n(\nu\boldsymbol{\alpha}_E^{-1/2})$ with $k=2$ shows that

$$I^\psi_n(\nu\boldsymbol{\alpha}_E^{-1/2})_{N^n_2}=I^2\supset I^1\supset I^0\supset I^{-1}=0, \tag{5.1}$$

where

$$\begin{aligned} I^0 &\simeq \nu^\psi \boldsymbol{\alpha}_E^{(3-n)/2} \boxtimes I^\psi_{n-2}(\nu\boldsymbol{\alpha}_E^{-3/2}), \\ I^1/I^0 &\simeq I(\boldsymbol{\alpha}_E^{-1/2}, \boldsymbol{\alpha}_E^{1/2}) \otimes \nu^\psi \boldsymbol{\alpha}_E^{(1-n)/2} \boxtimes I^\psi_{n-2}(\nu\boldsymbol{\alpha}_E^{-1/2}), \\ I^2/I^1 &\simeq \nu^\psi \boldsymbol{\alpha}_E^{(1-n)/2} \boxtimes I^\psi_{n-2}(\nu\boldsymbol{\alpha}_E^{1/2}). \end{aligned}$$

Note that ${}^\tau\nu^{-1} = \nu$ by assumption. By looking at the GL_2-factor, we see that

$$\mathrm{Hom}_{\tilde{M}_2^n}(I^r/I^{r-1}, \nu^\psi \boldsymbol{\alpha}_E^{(1-n)/2} \boxtimes \mathrm{St}^\psi_{n-2}(\nu)) = 0$$

for $r = 0, 1$. This is true for $r = 2$ by the last statement of Lemma 2.1(2). Put

$$K^\psi_n(\nu) = \mathrm{Ind}^{\tilde{G}_n}_{\tilde{P}_2^n} \nu^\psi \boldsymbol{\alpha}_E^{(1-n)/2} \boxtimes \mathrm{St}^\psi_{n-2}(\nu).$$

Then $\mathrm{Hom}_{\tilde{G}_n}(I^\psi_n(\nu\boldsymbol{\alpha}_E^{-1/2}), K^\psi_n(\nu)) = 0$ by the Frobenius reciprocity. Since the functor $\mathrm{Ind}^{\tilde{G}_n}_{\tilde{P}_2^n}$ is exact, we have an exact sequence

$$0 \to H^\psi_n(\nu) \to J^\psi_n(\nu) \to K^\psi_n(\nu) \to 0 \tag{5.2}$$

by (1.2), from which our statement follows. □

LEMMA 5.2. *Notations and assumptions being as in Lemma* 5.1, *we have*

$$\mathrm{Hom}_{\tilde{G}_n}(R^\psi_n(\varrho_n - 1, \nu), H^\psi_n(\nu)) = 0.$$

PROOF. Let V be a Hermitian space of dimension $\varrho_n - 1$ and character ν. Lemma 4.2 gives a decomposition

$$R^\psi_n(V,\nu)_{N_1^n} \simeq (\nu^\psi \boldsymbol{\alpha}_E^{(2-n)/2} \boxtimes R^\psi_{n-1}(V',\nu)) \oplus (\nu^\psi \boldsymbol{\alpha}_E^{-n/2} \boxtimes R^\psi_{n-1}(V,\nu)).$$

We identify the space $\mathrm{Hom}_{\tilde{G}_n}(R^\psi_n(V,\nu), J^\psi_n(\nu))$, by the Frobenius reciprocity, with

$$\mathrm{Hom}_{\tilde{G}_{n-1}}(R^\psi_{n-1}(V',\nu), \mathrm{Ind}^{\tilde{G}_{n-1}}_{\tilde{P}_1^{n-1}} \nu^\psi \boldsymbol{\alpha}_E^{-n/2} \boxtimes \mathrm{St}^\psi_{n-2}(\nu)).$$

Lemma 4.2 again gives a decomposition

$$R^\psi_{n-1}(V',\nu)_{N_1^{n-1}} \simeq (\nu^\psi \boldsymbol{\alpha}_E^{(4-n)/2} \boxtimes R^\psi_{n-2}(V'',\nu)) \oplus (\nu^\psi \alpha_E^{-n/2} \boxtimes R^\psi_{n-2}(V',\nu)).$$

Since $\dim V' = \varrho_{n-2} - 1$, we apply the last statement of Lemma 2.1(2) to $\mathrm{St}^\psi_{n-2}(\nu)$ and $R^\psi_{n-2}(V',\nu)$ to show that $\mathrm{Hom}_{\tilde{G}_n}(R^\psi_n(V,\nu), J^\psi_n(\nu)) = 0$, which implies our statement in view of (5.2). □

LEMMA 5.3. *Notations and assumptions being as in Lemma* 5.1, *we have*

$$\dim \mathrm{Hom}_{\tilde{G}_n}(\mathrm{St}^\psi_n(\nu), H^\psi_n(\nu)) = 1.$$

PROOF. Taking (2.1) and Lemmas 5.1, 5.2 into account, we see that

$$\begin{aligned} \mathrm{Hom}_{\tilde{G}_n}(\mathrm{St}^\psi_n(\nu), H^\psi_n(\nu)) &= \mathrm{Hom}_{\tilde{G}_n}(I^\psi_n(\nu\boldsymbol{\alpha}_E^{-1/2}), H^\psi_n(\nu)) \\ &= \mathrm{Hom}_{\tilde{G}_n}(I^\psi_n(\nu\boldsymbol{\alpha}_E^{-1/2}), J^\psi_n(\nu)). \end{aligned} \tag{5.3}$$

Lemma 4.1 applied to $I^\psi_n(\nu\boldsymbol{\alpha}_E^{-1/2})$ with $k = 1$ gives a decomposition

$$I^\psi_n(\nu\boldsymbol{\alpha}_E^{-1/2})_{N_1^n} \simeq (\nu^\psi \boldsymbol{\alpha}_E^{(2-n)/2} \boxtimes I^\psi_{n-1}(\nu\boldsymbol{\alpha}_E^{-1})) \oplus (\nu^\psi \boldsymbol{\alpha}_E^{-n/2} \boxtimes I^\psi_{n-1}(\nu)).$$

We apply the Frobenius reciprocity to identify $\mathrm{Hom}_{\tilde{G}_n}(I^\psi_n(\nu\boldsymbol{\alpha}_E^{-1/2}), J^\psi_n(\nu))$ with

$$\mathrm{Hom}_{\tilde{G}_{n-1}}(I^\psi_{n-1}(\nu\boldsymbol{\alpha}_E^{-1}), \mathrm{Ind}^{\tilde{G}_{n-1}}_{\tilde{P}_1^{n-1}} \nu^\psi \boldsymbol{\alpha}_E^{-n/2} \boxtimes \mathrm{St}^\psi_{n-2}(\nu)).$$

We see that

$$I^{\psi}_{n-1}(\nu\boldsymbol{\alpha}_E^{-1})_{N_1^{n-1}} \simeq (\nu^{\psi}\boldsymbol{\alpha}_E^{(4-n)/2} \boxtimes I^{\psi}_{n-2}(\nu\boldsymbol{\alpha}_E^{-3/2})) \oplus (\nu^{\psi}\boldsymbol{\alpha}_E^{-n/2} \boxtimes I^{\psi}_{n-2}(\nu\boldsymbol{\alpha}_E^{-1/2}))$$

by applying Lemma 4.1 to $I^{\psi}_{n-1}(\nu\boldsymbol{\alpha}_E^{-1})$ with $k=1$. We conclude that

$$\mathrm{Hom}_{\tilde{G}_n}(\mathrm{St}^{\psi}_n(\nu), H^{\psi}_n(\nu)) \simeq \mathrm{Hom}_{\tilde{G}_{n-2}}(I^{\psi}_{n-2}(\nu\boldsymbol{\alpha}_E^{-1/2}), \mathrm{St}^{\psi}_{n-2}(\nu)).$$

The last space is one-dimensional by Lemma 2.1(2). □

We will use the following result in a future study. Put

$$M^{\psi}_n(\nu) = \mathrm{Ind}^{\tilde{G}_n}_{\tilde{P}^n_2}(I(\boldsymbol{\alpha}_E^{1/2}, \boldsymbol{\alpha}_E^{-1/2}) \otimes \nu\boldsymbol{\alpha}_E^{(1-n)/2})^{\psi} \boxtimes I^{\psi}_{n-2}(\nu\boldsymbol{\alpha}_E^{1/2}).$$

COROLLARY 5.4. *Notations and assumptions being as in Lemma 5.1, we have*

$$\mathrm{Hom}_{\tilde{G}_n}(\mathrm{St}^{\psi}_n(\nu), H^{\psi}_n(\nu)) = \mathrm{Hom}_{\tilde{G}_n}(\mathrm{St}^{\psi}_n(\nu), M^{\psi}_n(\nu)).$$

PROOF. Put

$$L^{\psi}_n(\nu) = \mathrm{Ind}^{\tilde{G}_n}_{\tilde{P}^n_2}(I(\boldsymbol{\alpha}_E^{1/2}, \boldsymbol{\alpha}_E^{-1/2}) \otimes \nu\boldsymbol{\alpha}_E^{(1-n)/2})^{\psi} \boxtimes R^{\psi}_{n-2}(\varrho_{n-2}-1, \nu).$$

Since $R^{\psi}_{n-2}(\varrho_{n-2}-1,\nu)$ cannot appear as a quotient of $I^{\psi}_{n-2}(\nu\boldsymbol{\alpha}_E^{-1/2})$, we get

$$\mathrm{Hom}_{\tilde{G}_n}(I^{\psi}_n(\nu\boldsymbol{\alpha}_E^{-1/2}), L^{\psi}_n(\nu)) = 0$$

in view of (5.1). Since $J^{\psi}_n(\nu)$ sits in a short exact sequence

$$0 \to J^{\psi}_n(\nu) \to M^{\psi}_n(\nu) \to L^{\psi}_n(\nu) \to 0,$$

we obtain

$$\mathrm{Hom}_{\tilde{G}_n}(I^{\psi}_n(\nu\boldsymbol{\alpha}_E^{-1/2}), J^{\psi}_n(\nu)) = \mathrm{Hom}_{\tilde{G}_n}(I^{\psi}_n(\nu\boldsymbol{\alpha}_E^{-1/2}), M^{\psi}_n(\nu)).$$

By (5.3) we conclude that

$$\begin{aligned}\mathrm{Hom}_{\tilde{G}_n}(\mathrm{St}^{\psi}_n(\nu), H^{\psi}_n(\nu)) &= \mathrm{Hom}_{\tilde{G}_n}(I^{\psi}_n(\nu\boldsymbol{\alpha}_E^{-1/2}), J^{\psi}_n(\nu)) \\ &= \mathrm{Hom}_{\tilde{G}_n}(I^{\psi}_n(\nu\boldsymbol{\alpha}_E^{-1/2}), M^{\psi}_n(\nu)) \\ &= \mathrm{Hom}_{\tilde{G}_n}(\mathrm{St}^{\psi}_n(\nu), M^{\psi}_n(\nu)).\end{aligned}$$

In the last line we have used the fact that $\mathrm{Hom}_{\tilde{G}_n}(R^{\psi}_n(V,\nu), J^{\psi}_n(\nu)) = 0$, which was proved in the proof of Lemma 5.2. □

REMARK 5.5. By Corollary 5.4 the restriction of $\Psi^{\hat{\chi}}_n(\nu\boldsymbol{\alpha}_F^{1/2})$ gives an intertwining map $\mathrm{St}^{\psi}_n({}^{\tau}\hat{\chi}^{-1}\nu^E) \to H^{\psi}_n({}^{\tau}\hat{\chi}^{-1}\nu^E)$. One can show that this map is nonzero by employing Lemma 8 of [**14**].

Now we complete the proof of Theorem 3.1(2), (3), (4). Let $\pi \simeq \mathrm{St} \otimes \nu$. When $E=F$, we let ν be a nontrivial quadratic character to get $\mathrm{St}^{\psi}_1(\nu) \simeq A^{+}_{1,\psi}(\pi)$ by Proposition 4 of [**18**]. If $E \neq F$, then

$$A^{+,\hat{\chi}}_{1,\psi}(\pi) \simeq \hat{\chi}^{-1} \boxtimes \mathrm{St} \otimes \nu \simeq \mathrm{St}^{\psi}_1({}^{\tau}\hat{\chi}^{-1}\nu^E).$$

We may suppose by induction that $\mathrm{St}^{\psi}_{n-2}({}^{\tau}\hat{\chi}^{-1}\nu^E) \simeq A^{+,\hat{\chi}}_{n-2,\psi}(\pi)$. Then $A^{+,\hat{\chi}}_{n,\psi}(\pi)$ is equivalent to the unique irreducible subrepresentation of $H^{\psi}_n({}^{\tau}\hat{\chi}^{-1}\nu^E)$, which is equivalent to $\mathrm{St}^{\psi}_n({}^{\tau}\hat{\chi}^{-1}\nu^E)$ by Lemma 5.3. When $E=F$ and $\pi \simeq \mathrm{St}$, since $\mathrm{St}^{\psi}_1(1) \simeq A^{-}_{1,\psi}(\mathrm{St})$ by Proposition 8 of [**18**], one can obtain $A^{-}_{n,\psi}(\mathrm{St}) \simeq \mathrm{St}^{\psi}_n(1)$ by the parallel argument.

6. Degenerate principal series for general linear groups

The following result is a special case of Proposition 4.6 of [**21**].

LEMMA 6.1 ([**21**]). *The representation $I_n(\boldsymbol{\alpha}_F^{1/2}, \boldsymbol{\alpha}_F^{-1/2})$ has a unique irreducible submodule, denoted by* St_n. *Moreover,* $I_n(\boldsymbol{\alpha}_F^{1/2}, \boldsymbol{\alpha}_F^{-1/2})/\mathrm{St}_n \simeq I^{2n}_{n+1}(1,1) \not\simeq \mathrm{St}_n$.

Let $\mathscr{Q}_n$ be a parabolic subgroup of GL_{2n} whose Levi subgroup is isomorphic to $(\mathrm{GL}_2)^n$. When π is an irreducible infinite dimensional unitary representation of $\mathrm{GL}_2(F)$, we define $A_n(\pi)$ as the Langlands quotient of the standard module

$$I_n(\pi) = \mathrm{Ind}_{\mathscr{Q}_n}^{\mathrm{GL}_{2n}(F)}(\boldsymbol{\alpha}_F^{(n-1)/2} \otimes \pi) \boxtimes (\boldsymbol{\alpha}_F^{(n-3)/2} \otimes \pi) \boxtimes \cdots \boxtimes (\boldsymbol{\alpha}_F^{(1-n)/2} \otimes \pi).$$

THEOREM 6.2. (1) *If $\pi \simeq I(\mu_1, \mu_2)$, then $A_n(\pi) \simeq I_n(\mu_1, \mu_2)$.*
(2) *If $\pi \simeq \mathrm{St} \otimes \nu$, then $A_n(\pi) \simeq \mathrm{St}_n \otimes \nu$.*

Proposition 1 and Lemma 8 of [**14**] gives a nonzero intertwining map

$$I_n(\mu_1, \mu_2) \to \mathrm{Ind}_{Q_2^{2n}(F)}^{\mathrm{GL}_{2n}(F)}(I(\mu_1, \mu_2) \otimes \boldsymbol{\alpha}_F^{(1-n)/2}) \boxtimes (I_{n-1}(\mu_1, \mu_2) \otimes \boldsymbol{\alpha}_F^{1/2}),$$

by which we can prove Theorem 6.2(1) as in the proof of Theorem 3.1(1).

PROPOSITION 6.3. *The module $I^m_j(\mu_1, \mu_2)_{U^m_\ell}$ has an L^m_ℓ-stable filtration*

$$I^m_j(\mu_1, \mu_2)_{U^m_\ell} = \mathscr{I}^k \supset \mathscr{I}^{k-1} \supset \cdots \supset \mathscr{I}^1 \supset \mathscr{I}^0 \supset \mathscr{I}^{-1} = 0, \qquad k = \min\{j, \ell\}$$

with successive quotients

$$\mathscr{I}^r/\mathscr{I}^{r-1} \simeq I^\ell_r(\mu_1\boldsymbol{\alpha}_F^{(r-j)/2}, \mu_2\boldsymbol{\alpha}_F^{(j+\ell-m-r)/2}) \boxtimes I^{m-\ell}_{j-r}(\mu_1\boldsymbol{\alpha}_F^{r/2}, \mu_2\boldsymbol{\alpha}_F^{(\ell-r)/2}).$$

PROOF. The proof is similar to that of Lemma 4.1. Letting $N = m$, $\ell_1 = \ell$, $\ell_2 = m - \ell$ and $\ell_3 = 0$ in Proposition 4.2 of [**6**], we see that $\mathrm{GL}_m(F)$ is the disjoint union of the double cosets $\Psi_r = Q^m_j w_r Q^m_\ell$ $(0 \le r \le k)$. The matrix $w_r = w_{r,j}$ is defined in (4.17) of [**6**]. Put $S^{m,j}_{\ell,r} = w_r^{-1}Q^m_j w_r \cap Q^m_\ell$. Then $S^{m,j}_{\ell,r} = T^{m,j}_{\ell,r}U^{m,j}_{\ell,r}$ by (4.24) of [**6**], where $T^{m,j}_{\ell,r}$ and $U^{m,j}_{\ell,r}$ consist respectively of matrices

$$\left[\begin{array}{cc|cc} \beta & * & & \\ 0 & \delta & & \\ \hline & & A & * \\ & & 0 & D \end{array}\right], \qquad \left[\begin{array}{cc|cc} \mathbf{1}_r & 0 & z_1 & z_2 \\ 0 & \mathbf{1}_{\ell-r} & 0 & z_3 \\ \hline & & \mathbf{1}_{j-r} & 0 \\ & & 0 & \mathbf{1}_{m-\ell-j+r} \end{array}\right],$$

where $\beta \in \mathrm{GL}_r(F)$, $\delta \in \mathrm{GL}_{\ell-r}(F)$, $A \in \mathrm{GL}_{j-r}(F)$ and $D \in \mathrm{GL}_{m-\ell-j+r}(F)$. The stratification of $\mathrm{GL}_m(F)$ by Ψ_r gives rise to an L^m_ℓ-stable filtration

$$I^m_j(\mu_1, \mu_2) = \mathscr{J}^k \supset \mathscr{J}^{k-1} \supset \cdots \supset \mathscr{J}^1 \supset \mathscr{J}^0 \supset \mathscr{J}^{-1} = 0,$$

where $\mathscr{J}^r$ is the set of functions in $I^m_j(\mu_1, \mu_2)$ whose restriction to the closure of Ψ_{r+1} vanishes. The successive quotients are given by

$$\mathscr{J}^r/\mathscr{J}^{r+1} \simeq \text{c-ind}_{S^{m,j}_{\ell,r}}^{Q^m_\ell} \mu_r,$$

where

$$\mu_r\left(\mathrm{diag}\left[\begin{pmatrix} \beta & * \\ 0 & \delta \end{pmatrix}, \begin{pmatrix} A & * \\ 0 & D \end{pmatrix}\right]\right) = \frac{(\mu_1\boldsymbol{\alpha}_F^{(m-j)/2})((\det \beta)(\det A))}{(\mu_2^{-1}\boldsymbol{\alpha}_F^{j/2})((\det \delta)(\det D))}$$

by (4.25) of [**6**]. Proposition 6.3.3 of [**3**] gives

$$(\text{c-ind}_{S^{m,j}_{\ell,r}}^{Q^m_\ell} \mu_r)_{U^m_\ell} \simeq (\mathrm{ind}_{T^{m,j}_{\ell,r}}^{L^m_\ell} \mu_r\delta_r) \otimes \delta_{Q^m_\ell}^{-1/2},$$

where δ_r is the modulus character of $T^{m,j}_{\ell,r}$ acting on $U^m_\ell / U^{m,j}_{\ell,r}$. Observe that

$$\delta_r \left(\operatorname{diag} \left[\begin{pmatrix} \beta & * \\ 0 & \delta \end{pmatrix}, \begin{pmatrix} A & * \\ 0 & D \end{pmatrix} \right] \right) = \boldsymbol{\alpha}_F(\det \delta)^{j-r} \boldsymbol{\alpha}_F(\det A)^{r-\ell}.$$

The proof continues as in that of Lemma 4.1. □

One can derive the following results from Proposition 6.3 as in Section 5.

COROLLARY 6.4. *Let $n \geq 2$. Put*

$$H_n = \operatorname{Ind}^{\mathrm{GL}_{2n}(F)}_{Q^{2n}_2(F)} (\mathrm{St} \otimes \boldsymbol{\alpha}_F^{(1-n)/2}) \boxtimes (\mathrm{St}_{n-1} \otimes \boldsymbol{\alpha}_F^{1/2}),$$

$$J_n = \operatorname{Ind}^{\mathrm{GL}_{2n}(F)}_{Q^{2n}_2(F)} (I(\boldsymbol{\alpha}_F^{1/2}, \boldsymbol{\alpha}_F^{-1/2}) \otimes \boldsymbol{\alpha}_F^{(1-n)/2}) \boxtimes (\mathrm{St}_{n-1} \otimes \boldsymbol{\alpha}_F^{1/2}).$$

Then the space $\operatorname{Hom}_{\mathrm{GL}_{2n}(F)}(I_n(\boldsymbol{\alpha}_F^{-1/2}, \boldsymbol{\alpha}_F^{1/2}), J_n)$ *is one-dimensional. Moreover,*

$$\operatorname{Hom}_{\mathrm{GL}_{2n}(F)}(I_n(\boldsymbol{\alpha}_F^{-1/2}, \boldsymbol{\alpha}_F^{1/2}), J_n) = \operatorname{Hom}_{\mathrm{GL}_{2n}(F)}(\mathrm{St}_n, H_n).$$

Finally, we complete the proof of Theorem 6.2(2). We may assume by induction that $\mathrm{St}_{n-1} \simeq A_{n-1}(\mathrm{St})$. Then $H_n \otimes \nu$ is equivalent to a subrepresentation of $I_n(\mathrm{St})^\vee \otimes \nu$. Corollary 6.4 concludes that $\mathrm{St}_n \otimes \nu$ is equivalent to an irreducible subrepresentation of $I_n(\mathrm{St})^\vee \otimes \nu$, which is unique and equivalent to $A_n(\mathrm{St} \otimes \nu)$.

References

[1] J. Arthur, *The endoscopic classification of representations*, American Mathematical Society Colloquium Publications, vol. 61, American Mathematical Society, Providence, RI, 2013. Orthogonal and symplectic groups. MR3135650

[2] D. Ban and C. Jantzen, *Degenerate principal series for even-orthogonal groups*, Represent. Theory **7** (2003), 440–480, DOI 10.1090/S1088-4165-03-00166-3. MR2017065

[3] W. Casselman, Introduction to the theory of admissible representations of p-adic reductive groups, preprint.

[4] W. T. Gan, *Theta correspondence: recent progress and applications*, Proceedings of the International Congress of Mathematicians—Seoul 2014. Vol. II, Kyung Moon Sa, Seoul, 2014, pp. 343–366. MR3728618

[5] W. T. Gan and A. Ichino, *The Shimura-Waldspurger correspondence for* Mp_{2n}, Ann. of Math. (2) **188** (2018), no. 3, 965–1016, DOI 10.4007/annals.2018.188.3.5. MR3866889

[6] D. Ginzburg, S. Rallis, and D. Soudry, *The descent map from automorphic representations of* GL(n) *to classical groups*, World Scientific Publishing Co. Pte. Ltd., Hackensack, NJ, 2011. MR2848523

[7] S. Hayashida, *Fourier-Jacobi expansion and the Ikeda lift*, Abh. Math. Semin. Univ. Hambg. **81** (2011), no. 1, 1–17, DOI 10.1007/s12188-011-0053-4. MR2812029

[8] T. Ikeda, *On the lifting of elliptic cusp forms to Siegel cusp forms of degree* $2n$, Ann. of Math. (2) **154** (2001), no. 3, 641–681, DOI 10.2307/3062143. MR1884618

[9] T. Ikeda and S. Yamana, On the lifting of Hilbert-cusp forms to Hilbert-Siegel cusp forms, Ann. Sci. École Norm. Sup. (4), to appear.

[10] C. Jantzen, *Degenerate principal series for symplectic and odd-orthogonal groups*, Mem. Amer. Math. Soc. **124** (1996), no. 590, viii+100, DOI 10.1090/memo/0590. MR1346929

[11] S. S. Kudla, *Splitting metaplectic covers of dual reductive pairs*, Israel J. Math. **87** (1994), no. 1-3, 361–401, DOI 10.1007/BF02773003. MR1286835

[12] S. S. Kudla and S. Rallis, *Ramified degenerate principal series representations for* Sp(n), Israel J. Math. **78** (1992), no. 2-3, 209–256, DOI 10.1007/BF02808058. MR1194967

[13] S. S. Kudla and W. J. Sweet Jr., *Degenerate principal series representations for* U(n, n), Israel J. Math. **98** (1997), 253–306, DOI 10.1007/BF02937337. MR1459856

[14] E. M. Lapid and S. Rallis, *On the local factors of representations of classical groups*, Automorphic representations, L-functions and applications: progress and prospects, Ohio State Univ. Math. Res. Inst. Publ., vol. 11, de Gruyter, Berlin, 2005, pp. 309–359, DOI 10.1515/9783110892703.309. MR2192828

[15] C. Mœglin, M.-F. Vignéras, and J.-L. Waldspurger, *Correspondances de Howe sur un corps p-adique* (French), Lecture Notes in Mathematics, vol. 1291, Springer-Verlag, Berlin, 1987. MR1041060

[16] S. Rallis, *On the Howe duality conjecture*, Compositio Math. **51** (1984), no. 3, 333–399. MR743016

[17] W. J. Sweet, Jr., Functional equations of p-adic zeta integrals and representations of the metaplectic group, preprint (1995)

[18] J.-L. Waldspurger, *Correspondances de Shimura et quaternions* (French), Forum Math. **3** (1991), no. 3, 219–307, DOI 10.1515/form.1991.3.219. MR1103429

[19] S. Yamana, *Degenerate principal series representations for quaternionic unitary groups*, Israel J. Math. **185** (2011), 77–124, DOI 10.1007/s11856-011-0102-9. MR2837129

[20] S. Yamana, *L-functions and theta correspondence for classical groups*, Invent. Math. **196** (2014), no. 3, 651–732, DOI 10.1007/s00222-013-0476-x. MR3211043

[21] A. V. Zelevinsky, *Induced representations of reductive $\mathfrak{p}$-adic groups. II. On irreducible representations of* $\mathrm{GL}(n)$, Ann. Sci. École Norm. Sup. (4) **13** (1980), no. 2, 165–210. MR584084

Department of Mathematics, Kyoto University, Kitashirakawa Oiwake-cho, Sakyo-ku, Kyoto, 606-8502, Japan; and Hakubi Center, Yoshida-Ushinomiya-cho, Sakyo-ku, Kyoto, 606-8501, Japan

Email address: yamana07@math.kyoto-u.ac.jp

Selected Published Titles in This Series

732 **Samuele Anni, Jay Jorgenson, Lejla Smajlović, and Lynne Walling, Editors,** Automorphic Forms and Related Topics, 2019

728 **Nicolás Andruskiewitsch and Dmitri Nikshych, Editors,** Tensor Categories and Hopf Algebras, 2019

727 **André Leroy, Christian Lomp, Sergio López-Permouth, and Frédérique Oggier, Editors,** Rings, Modules and Codes, 2019

726 **Eugene Plotkin, Editor,** Groups, Algebras and Identities, 2019

725 **Shijun Zheng, Marius Beceanu, Jerry Bona, Geng Chen, Tuoc Van Phan, and Avy Soffer, Editors,** Nonlinear Dispersive Waves and Fluids, 2019

724 **Lubjana Beshaj and Tony Shaska, Editors,** Algebraic Curves and Their Applications, 2019

723 **Donatella Danielli, Arshak Petrosyan, and Camelia A. Pop, Editors,** New Developments in the Analysis of Nonlocal Operators, 2019

722 **Yves Aubry, Everett W. Howe, and Christophe Ritzenthaler, Editors,** Arithmetic Geometry: Computation and Applications, 2019

721 **Petr Vojtěchovský, Murray R. Bremner, J. Scott Carter, Anthony B. Evans, John Huerta, Michael K. Kinyon, G. Eric Moorhouse, and Jonathan D. H. Smith, Editors,** Nonassociative Mathematics and its Applications, 2019

720 **Alexandre Girouard, Editor,** Spectral Theory and Applications, 2018

719 **Florian Sobieczky, Editor,** Unimodularity in Randomly Generated Graphs, 2018

718 **David Ayala, Daniel S. Freed, and Ryan E. Grady, Editors,** Topology and Quantum Theory in Interaction, 2018

717 **Federico Bonetto, David Borthwick, Evans Harrell, and Michael Loss, Editors,** Mathematical Problems in Quantum Physics, 2018

716 **Alex Martsinkovsky, Kiyoshi Igusa, and Gordana Todorov, Editors,** Surveys in Representation Theory of Algebras, 2018

715 **Sergio R. López-Permouth, Jae Keol Park, S. Tariq Rizvi, and Cosmin S. Roman, Editors,** Advances in Rings and Modules, 2018

714 **Jens Gerlach Christensen, Susanna Dann, and Matthew Dawson, Editors,** Representation Theory and Harmonic Analysis on Symmetric Spaces, 2018

713 **Naihuan Jing and Kailash C. Misra, Editors,** Representations of Lie Algebras, Quantum Groups and Related Topics, 2018

712 **Nero Budur, Tommaso de Fernex, Roi Docampo, and Kevin Tucker, Editors,** Local and Global Methods in Algebraic Geometry, 2018

711 **Thomas Creutzig and Andrew R. Linshaw, Editors,** Vertex Algebras and Geometry, 2018

710 **Raphaël Danchin, Reinhard Farwig, Jiří Neustupa, and Patrick Penel, Editors,** Mathematical Analysis in Fluid Mechanics, 2018

709 **Fernando Galaz-García, Juan Carlos Pardo Millán, and Pedro Solórzano, Editors,** Contributions of Mexican Mathematicians Abroad in Pure and Applied Mathematics, 2018

708 **Christian Ausoni, Kathryn Hess, Brenda Johnson, Ieke Moerdijk, and Jérôme Scherer, Editors,** An Alpine Bouquet of Algebraic Topology, 2018

707 **Nitya Kitchloo, Mona Merling, Jack Morava, Emily Riehl, and W. Stephen Wilson, Editors,** New Directions in Homotopy Theory, 2018

706 **Yeonhyang Kim, Sivaram K. Narayan, Gabriel Picioroaga, and Eric S. Weber, Editors,** Frames and Harmonic Analysis, 2018

For a complete list of titles in this series, visit the
AMS Bookstore at **www.ams.org/bookstore/conmseries/**.